Color Image and Video Enhancement

M. Emre Celebi • Michela Lecca • Bogdan Smolka
Editors

Color Image and Video Enhancement

Springer

Editors
M. Emre Celebi
Louisiana State University
Shreveport
Louisiana
USA

Bogdan Smolka
Silesian University of Technology
Gliwice
Poland

Michela Lecca
Fondazione Bruno Kessler
Center for Information and Communication Technology
Trento
Italy

ISBN 978-3-319-09362-8 ISBN 978-3-319-09363-5 (eBook)
DOI 10.1007/978-3-319-09363-5

Library of Congress Control Number: 2015943686

Springer Cham Heidelberg New York Dordrecht London

Printed on acid-free paper

Springer International Publishing AG Switzerland is part of Springer Science+Business Media (www.springer.com)

Preface

Enhancement of digital images and video sequences is the process of increasing the quality of the visual information by improving its visibility and perceptibility. Enhancement is a necessary step in image/video processing applications when the conditions under which a scene is captured result in quality degradation, e.g., increased/decreased brightness and/or contrast, distortion of colors, and introduction of noise and other artifacts such as blotches and streaks. Unfortunately, most of the traditional enhancement methods are designed for monochromatic image/video data. The multivariate nature of color image/video data presents considerable challenges for researchers and practitioners as the numerous methods developed for single channel data are often not directly applicable to multichannel data.

The goal of this volume is to summarize the state-of-the-art in color image and video enhancement. The intended audience includes researchers and practitioners, who are increasingly using color images and videos.

The volume opens with two chapters related to image acquisition. In "Colorimetric Characterisation," Westland focuses on the problem of color reproduction in devices such as cameras, monitors, and printers. The author describes color spaces mainly used for representing colors by consumer technologies currently available, analyzes the device accuracy on the reproduction of real-world colors, and illustrates various color correction methods for matching the color gamuts of different devices. In "Image Demosaicing," Zhen and Stevenson present an overview of demosaicking methods. The authors introduce the fundamentals of interpolation and analyze the structure of various state-of-the-art approaches. In addition, they elaborate on the advantages and disadvantages of the examined techniques and evaluate their performance using popular image quality metrics. Finally, they discuss demosaicing combined with deblurring and super-resolution.

The volume continues with two chapters on noise removal. In "DCT-Based Color Image Denoising: Efficiency Analysis and Prediction," Lukin et al. discuss image denoising techniques based on the discrete cosine transform (DCT). The authors analyze noise models, discuss various image quality measures, describe various types of filters, and introduce the concept of image enhancement utilizing the DCT.

In "Impulsive Noise Filters for Colour Images," Morillas et al. give an overview of the impulsive noise reduction methods for color images. They analyze various models of impulsive noise contamination, introduce quality metrics used for the evaluation of filtering effectiveness, discuss various methods of vector ordering, and analyze the main types of noise reduction algorithms. The authors not only describe various approaches to impulsive noise reduction, but also evaluate their effectiveness and summarize their main properties.

The volume continues with seven chapters on color/contrast enhancement. In "Spatial and Frequency-Based Variational Methods for Perceptually Inspired Color and Contrast Enhancement of Digital Images," Provenzi considers perceptually inspired color correction algorithms that aim to reproduce the color sensation produced by the human vision system. These algorithms are based on the well-known Retinex model, introduced by Land and McCann about 45 years ago. The author shows that Retinex-like approaches can be embedded in a general variational framework, where these methods can be interpreted as a local, nonlinear modification of histogram equalization. In "The Color Logarithmic Image Processing (CoLIP) Antagonist Space," Gavet et al. present a survey of Color Logarithmic Image Processing, a perceptually-oriented mathematical framework for representing and processing color images. The authors also present various applications of this framework ranging from contrast enhancement to segmentation. In "Color Management and Virtual Restoration of Artworks," Maino and Monti present a survey of the use of color and contrast enhancement techniques in the virtual restoration of artworks such as paintings, mosaics, ancient archival documents, and manuscripts. Histogram equalization approaches, Retinex-like methods, and multi-spectral image processing algorithms are essential tools to analyse an artwork, to discover its history, to measure its conservation/degradation status, and to plan future physical restoration. The authors provide examples of applications of such digital techniques on several well-known Italian artworks. In "A GPU-Accelerated Adaptive Simultaneous Dynamic Range Compression and Local Contrast Enhancement Algorithm for Real-Time Color Image Enhancement," Tsai and Huang propose an adaptive dynamic range compression algorithm for color image enhancement. The authors demonstrate that a CUDA implementation of the proposed algorithm achieves up to 700% speed up when executed on an NVIDIA NVS 5200M GPU compared to a LUT-accelerated implementation executed on an Intel Core i7-3520M CPU. In "Color Equalization and Retinex," Wang et al. give an overview of several perceptually inspired color correction algorithms that attempt to simulate the human color constancy capability. The authors first describe two histogram equalization methods that modify the image colors by manipulating respectively the global and local color distributions. They then illustrate an automatic color equalization approach that enhances the color and contrast of an image by combining the Gray-World and White-Patch models. Finally, they describe the Retinex model and various implementations of it. In "Color Correction for Stereo and Multi-View Coding," Fezza and Larabi first present a survey of color correction methods for multi-view video. They then compare the quantitative/qualitative performance of some of the popular methods with respect to color consistency, coding performance, and rendering quality. Finally, in "Enhancement of Image Content for Observers with Colour Vision Deficiencies," Milić et al. present a survey of daltonization methods designed for enhancing the perceptual quality of color images for the benefit of observers with color vision deficiencies.

In "Computationally Efficient Data and Application Driven Color Transforms for the Compression and Enhancement of Images and Video," Minervini et al. deal with the problem of efficient coding and transmission of color images and videos. The RGB data recorded by camera sensors are typically redundant due to high correlation of the color channels. The authors describe two frameworks to obtain linear maps of the RGB data that minimize the loss of information due to compression. The first adapts to the image data and aims at reconstruction accuracy, representing an efficient approximation of the classic Karhunen-Loève transform. The second adapts to the application in which the images are used, for instance, an image classification task. A chapter entitled "Overview of Grayscale Image Colorization Techniques," by Popowicz and Smolka completes the volume. The authors first present a survey of semi-automatic grayscale image colorization methods. They then compare the performance of three semi-automatic and one fully-automatic method on a variety of images. Finally, they propose a methodology for evaluating colorization methods based on several well-known quality assessment measures.

As editors, we hope that this volume focused on color image and video enhancement will demonstrate the significant progress that has occurred in this field in recent years. We also hope that the developments reported in this volume will motivate further research in this exciting field.

M. Emre Celebi
Michcla Lccca
Bogdan Smolka

Contents

Contributors

Sergey Abramov National Aerospace University, Kharkov, Ukraine

Victoriya Abramova National Aerospace University, Kharkov, Ukraine

Jaakko Astola Tampere University of Technology, Tampere, Finland

Joan-Gerard Camarena Conselleria d'Educacio, Valencia, Spain

Kacem Chehdi University of Rennes 1, Lannion, France

Johan Debayle École Nationale Supérieure des Mines, Saint-Etienne, France

Karen Egiazarian Tampere University of Technology, Tampere, Finland

Sid Ahmed Fezza University of Oran 2, Oran, Algeria

Yann Gavet École Nationale Supérieure des Mines, Saint-Etienne, France

Valentín Gregori Instituto Universitario de Matemática Pura y Aplicada, Universitat Politècnica de València, Valencia, Spain

Chih-Hung Huang Department of Electrical Engineering, Tamkang University, Tamsui District, New Taipei City, Taiwan R.O.C.

Ruslan Kozhemiakin National Aerospace University, Kharkov, Ukraine

Mohamed-Chaker Larabi XLIM Institute, SIC Department, University of Poitiers, Poitiers, France

Vladimir Lukin National Aerospace University, Kharkov, Ukraine

Giuseppe Maino University of Bologna, Bologna, Italy

Neda Milić Faculty of Technical Sciences, University of Novi Sad, Novi Sad, Serbia

Branko Milosavljević Faculty of Technical Sciences, University of Novi Sad, Novi Sad, Serbia

Massimo Minervini IMT Institute for Advanced Studies, Lucca, Italy

Mariapaola Monti University of Bologna, Bologna, Italy

Samuel Morillas Instituto Universitario de Matemática Pura y Aplicada, Universitat Politècnica de València, Valencia, Spain

Dragoljub Novaković Faculty of Technical Sciences, University of Novi Sad, Novi Sad, Serbia

Jean-Charles Pinoli École Nationale Supérieure des Mines, Saint-Etienne, France

Nikolay Ponomarenko National Aerospace University, Kharkov, Ukraine

Adam Popowicz Silesian University of Technology, Gliwice, Poland

Edoardo Provenzi Laboratoire MAP5 (UMR CNRS 8145), Université Paris Descartes, Sorbonne Paris Cité, Paris, France

Bernardino Roig Instituto para la Gestión Integral de las zonas Costeras, Universitat Politècnica de València, Valencia, Spain

Alexey Rubel National Aerospace University, Kharkov, Ukraine

Cristian Rusu IMT Institute for Advanced Studies, Lucca, Italy

Almanzor Sapena Instituto Universitario de Matemática Pura y Aplicada, Universitat Politècnica de València, Valencia, Spain

Bogdan Smolka Silesian University of Technology, Gliwice, Poland

Robert L. Stevenson University of Notre Dame, Notre Dame, IN, USA

Sotirios A. Tsaftaris IMT Institute for Advanced Studies, Lucca, Italy

Chi-Yi Tsai Department of Electrical Engineering, Tamkang University, Tamsui District, New Taipei City, Taiwan R.O.C.

Mikhail Uss National Aerospace University, Kharkov, Ukraine

Benoit Vozel University of Rennes 1, Lannion, France

Liqian Wang School of Telecommunications and Information Engineering, Nanjing University of Posts and Telecommunications, Nanjing, China

School of Computer Science and Engineering, Nanjing University of Science and Technology, Nanjing, China

Zhihui Wei Key Lab of Intelligent Perception and Systems for High-Dimensional Information of Ministry of Education, Nanjing, China

Stephen Westland University of Leeds, Leeds, UK

Liang Xiao Key Lab of Intelligent Perception and Systems for High-Dimensional Information of Ministry of Education, Nanjing, China

Ruiwen Zhen University of Notre Dame, Notre Dame, IN, USA

Chapter 1
Colorimetric Characterization

Stephen Westland

1.1 Introduction

Human color vision is trichromatic. That is to say that most people's color vision is based upon the responses of three classes of light-sensitive receptors in the retina of the eye, each of which has broadband sensitivity but maximum sensitivity at different wavelengths. Consequently, the use of three color primaries, together in mixture, allows a wide range of colors to be reproduced. When colored lights are mixed together, a phenomenon known as additive color mixing and exemplified by digital display systems, the gamut of reproducible colors for a trichromatic system is limited and is always smaller than the gamut of all the colors possible in the world. However, the gamut is smaller or larger depending upon the choice of primaries. Pragmatically, the largest gamut is achieved when the additive primaries are red, green and blue (RGB).

Unfortunately there is no single set of RGB primaries that has achieved universal acceptance as RGB primaries have evolved over time in response to consumer demand and technological advancement. Color images are captured and displayed using a wide variety of devices and technologies. Although at each pixel, color is represented by the intensities of the red (R), green (G) and blue (B) primaries, different display technologies use different RGB primaries so that, unless adjustment is made to the RGB values to compensate for these differences, the same image displayed on different display devices is likely to have a very different color appearance. For image-capture devices such as cameras, color is represented by RGB values that represent the responses of three broadband color filters. However, different manufacturers use different filters and, again, the settings of the camera (exposure time, white balance etc.) will affect the RGB values that are captured for a particular scene. It is clear that, without some process for compensating for these differences in device primaries and states, significant color differences are likely to result between images of the same scene that are captured by various cameras and

S. Westland (✉)
University of Leeds, Leeds, UK
e-mail: S.Westland@leeds.ac.uk

M. E. Celebi et al. (eds.), *Color Image and Video Enhancement*,
DOI 10.1007/978-3-319-09363-5_1

displayed on various devices in various states. Color management can be thought of as a process of adjustment of color information to compensate for properties of each imaging device to ensure color fidelity in the image workflow.

This chapter explores issues around how the variability in RGB primaries between different displays (and, indeed, other imaging devices such as cameras) can be overcome to enable color fidelity.

1.2 CIE XYZ

Colorimetry is a branch of color science concerned with numerically specifying the color of physically defined stimuli such that two stimuli that look the same (under certain criteria) have identical specifications [37]. The Commission Internationale de lÉclairage (CIE) developed a system for the specification of color stimuli that was recommended for widespread use in 1931 [11]. The CIE system allows any color stimulus (usually expressed in terms of radiance at fixed wavelength intervals in the visible spectrum) to be represented by three so-called tristimulus values (XYZ); the XYZ values denote the amounts of the three CIE primaries that, on average, an observer would use in additive mixture to match the visual appearance of the stimulus. Of course, if two stimuli have the same XYZ values then they will match each other and this allows for a system for predicting stimulus equivalence, whereby two spectrally dissimilar stimuli are a visual match. The XYZ values can be calculated because of the amounts of the three primaries needed to match one unit of light at wavelength in the visible spectrum were experimentally determined and are known as the color-matching functions. A number of good texts exist that describe the development, principles and practical uses of the CIE system [4, 13, 22, 24].

The proportional amounts of the CIE tristimulus values are often used to create a chromaticity diagram. The x and y chromaticity coordinates are calculated as shown in Eqs. 1.1, 1.2, thus

$$x = X/(X + Y + Z) \tag{1.1}$$

$$y = Y/(X + Y + Z) \tag{1.2}$$

where XYZ are the CIE tristimulus values. Figure 1.1 shows the chromaticity diagram where the colors that are shown are illustrative only. The colors of individual wavelengths are on the horseshoe-shaped locus. The gamut of a typical RGB display is shown by the triangle. Of course, the chromaticity diagram is a 2-D projection of 3-D color space so that colors of identical chrominance but different luminance (e.g., white, black and grey) are all collapsed onto a single point in this diagram [25].

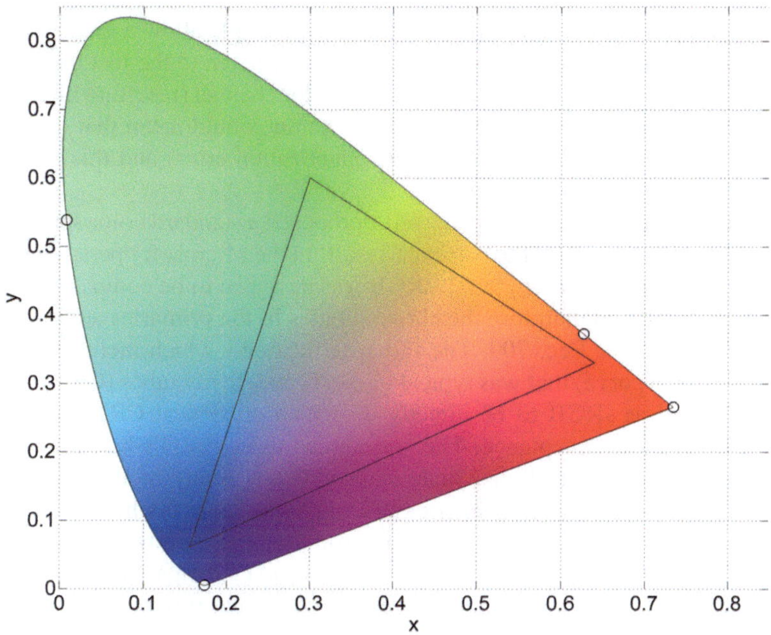

Fig. 1.1 The CIE chromaticity diagram. The gamut of reproducible colors for a typical RGB display is shown by the solid line where the vertices of the triangle are the chromaticity coordinates of each of the RGB primaries. The chromaticities of the single-wavelength stimuli at 400, 500, 600 and 700 nm are shown as open circles on the spectral locus

Table 1.1 The CIE *xy* chromaticities of the SMPTE-C, Rec. 601 and Rec. 709 primaries

	SMPTE-C			ITU-R BT.601			ITU-R BT.709–3		
	R	G	B	R	G	B	R	G	B
x	0.630	0.310	0.155	0.640	0.290	0.155	0.640	0.300	0.155
y	0.340	0.595	0.070	0.330	0.600	0.060	0.330	0.600	0.060

1.3 sRGB

No single RGB color space has achieved universal acceptance. Rather, many RGB standards and sets of RGB primaries have evolved over time in response to consumer demand, professional interests and technological advances. Table 1.1 shows the CIE chromaticity coordinates of three important RGB standards [35].

The SMPTE-C primaries have traditionally been used in the US, whereas the ITU-R BT.601 primaries have been used in Europe, Canada and Japan. In 1990, a new set of primaries was agreed upon for high-definition television (HDTV) known as ITU-R BT.709-3 or simply Rec. 709. It is currently possible to find displays that correspond to each of these standards and, indeed, to several others.

The use of more saturated primaries would in principle allow a greater gamut of reproducible colors; however, consideration must be given to the fact that in digital RGB systems it is normal to allocate 8 bits per color channel (resulting in 256 values for each of R, G and B). Using a wider RGB gamut would mean that digital steps would be more widely spaced (increasing quantization error) and this may not be desirable.

In 1996, Hewlett-Packard and Microsoft proposed a standard color space, sRGB, intended for widespread use but particularly within the Microsoft operating systems, HP products and the Internet [1,27]. sRGB was designed to be compatible with the Rec. 709 standard and therefore the chromaticities of the primaries are the same as those in Table 1.1 for Rec. 709. The full specification—which includes a transfer function (gamma curve) that was typical of most cathode ray tubes (CRTs)—allows images encoded as sRGB to be directly displayed on typical CRT monitors and this greatly aided its acceptance. The widespread use of sRGB has allowed some degree of convergence in digital imaging workflows. Many modern displays are based on the Rec. 709 (sRGB) primaries and this assists color fidelity to some extent. The color gamut of sRGB is too small for some purposes, however, and a wide-gamut RGB space known as Adobe RGB, designed to encompass most of the colors achievable on CMYK (cyan, magenta, yellow, key) color printers, has been in use since 1998 [30]. Most digital camera systems now allow captured images to be encoded in either sRGB or Adobe RGB color space.

1.4 Color Correction

Without color correction, it is unlikely that digital imaging systems would have been as successful as they have been for the past couple of decades. Although true color fidelity is still not achievable, color correction, ubiquitous but invisible to most casual users, enables a degree of color accuracy that is more than adequate for many purposes. Color correction, or color management, can be thought of as a process of adjustment of color information to compensate for properties of each imaging device to ensure color fidelity in the image workflow [36].

The role of the *International Color Consortium* (ICC) has been critical in terms increasing color fidelity across a wide range of imaging devices. The ICC is an industry consortium of more than 70 imaging and software vendors that was established in 1993 with the goal to "create, promote and encourage evolution of an open, vendor-neutral, cross-platform color management system architecture and components." The ICC process converts source-device color values to destination-device color values via a transform that must account for the color characteristics of both source and destination devices as illustrated in Fig. 1.2 [17].

For this system to be effective, each device should be associated with a device profile that describes the relationship between the device's color space and a device-independent color space (PCS). The ICC [23] develops and promotes standard color

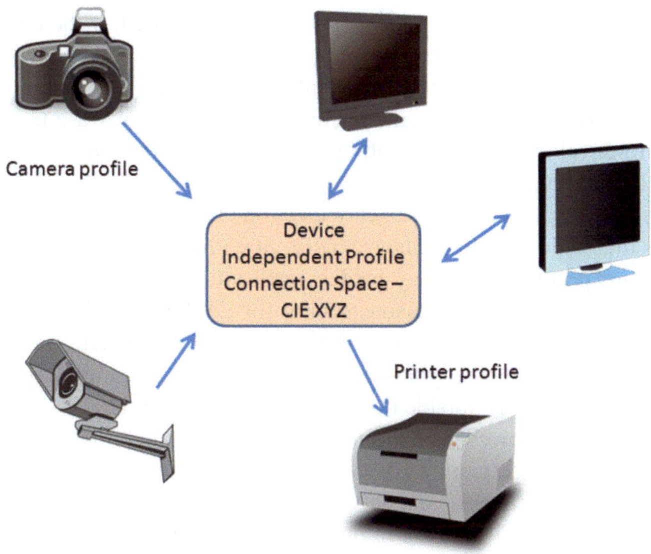

Fig. 1.2 The ICC color management workflow transforms color from a source device (such as a camera) to a destination-device (such as a printer) via a profile connection space (PCS) using ICC device profiles

profiles (ICC profile) whose specification is based on an earlier system called ColorSync (that was delivered on Apple computers). The system is efficient; for n devices only n transforms (or profiles) are required and adding a new device requires only one new color transform.

Consider for example, the workflow where a digital image is captured using a camera and transferred to a computer where it is displayed. The camera may specify the image in sRGB color space but the digital display color space may be, for example, based on Rec. 601 color space (see Table 1.1). If no adjustment (color management) is made then color fidelity will be lost since, for example, the RGB specification RGB = [255 0 0] refers to a different color in the camera and display color spaces. However, using the ICC system, the RGB values from the camera will be converted to the device-independent space XYZ and these will then be converted into RGB values for the display. The device profiles in this case are likely to be simply linear transforms that are computationally efficient. The device profile for the camera will be stored in the image-file header (for example, in the TIFF file) and the display profile will be stored on the host computer. The color management system in the computer's operating system software will perform the transforms.

Many image files do not have a profile associated with them; this most commonly occurs with image files that have been uploaded to the Internet. Most color management systems now assume that the default color space is sRGB if confronted with a digital color image that does not have a profile. For this reason, sRGB is often the color space of choice if creating images for display over the Internet on a variety of platforms. Note also that, in the example workflow just cited, if the camera and the display color spaces are both sRGB then no correction is needed.

Most imaging devices come with a default profile which is often installed as part of the device driver. However, methods (instruments and software tools) exist that allow users to develop their own profiles for a device which can take into account the settings of the device.

Although sRGB specification is used widely, the limited color gamut of the sRGB space means that wide-gamut displays consistent with, for example, the Adobe RGB color space, are preferred in many professional situations and color correction is then needed. Color correction is also essential for CMYK-based printers [2].

When the source and destination color gamuts do not match, part of the color correction must include a computational process known as gamut mapping. Gamut mapping is complex and is outside the scope of this chapter but readers are referred to the literature [25]. However, it should be noted that when the destination gamut is smaller than the source gamut, for example, then there is a choice between clipping the source gamut (in this case, colors outside the destination gamut would be desaturated but retain their hue) or compressing it (in this case, all of the colors in the source gamut are desaturated) so that it fits the destination gamut.

The following two sections consider the relationship between device RGB color and CIE XYZ color.

1.5 Theoretical Relationship between RGB and XYZ

If the luminance of the RGB channel values is linearly related to the channel input values, *RGB*, then there is a simple linear relationship between the display *RGB* values and the CIE *XYZ* values, thus

$$\boldsymbol{t} = \boldsymbol{M}\boldsymbol{d} \tag{1.3}$$

where $\boldsymbol{t}$ is a 3×1 column vector containing the XYZ values, $\boldsymbol{d}$ is a 3×1 column vector containing the RGB values, and $\boldsymbol{M}$ is a 3×3 matrix. If the values of $\boldsymbol{d}$ are in the range 0–1, then $\boldsymbol{M}$ is constructed as shown in Eq. 1.4,

$$\boldsymbol{M} = \begin{matrix} X_R & X_G & X_B \\ Y_R & Y_G & Y_B \\ Z_R & Z_G & Z_B \end{matrix} \tag{1.4}$$

where X_R, for example, is the CIE X value of the display when the R primary is at maximum intensity ($R = 1$) and the G and B primaries are at zero intensity (with similar definitions for the other terms). Therefore, if the CIE *XYZ* values of the primaries at maximum intensity are known for a display, then it is trivial to use Eq. 1.3 to predict the CIE color coordinates of any arbitrary 24-bit RGB stimulus on that display. It is therefore possible to take an image on one display and convert it to a second display without loss of color fidelity by using Eq. 1.3 (where $\boldsymbol{M}$ is derived for the first display) to convert the first display *RGB* values to *XYZ* values and then using the inverse of Eq. 1.3 (where $\boldsymbol{M}$ is derived for the second display) to convert the *XYZ* values into *RGB* values. Eq. 1.5 shows that it is possible to convert the *RGB* values for the first display, $\boldsymbol{d}_1$, into *RGB* values for the second display, $\boldsymbol{d}_2$, directly by calculating the scalar dot product of the inverse of $\boldsymbol{M}_2$ and $\boldsymbol{M}_1$ thus:

$$\boldsymbol{d}_2 = (\boldsymbol{M}_2^{-1}\boldsymbol{M}_1)\boldsymbol{d}_1 \tag{1.5}$$

However, the color correction illustrated by Eq. 1.5 is only valid if the *RGB* values of $\boldsymbol{d}_1$ and $\boldsymbol{d}_2$ are linearly related to the luminances of the displays. Normally, an additional computation stage is required to linearize the *RGB* values and this is often achieved using a gamma function [5]. The gamma model was effective for older CRT displays; for more modern LCD displays it is not clear whether the gamma model should be replaced by a more sigmoidal model [12].

For most trichromatic camera devices, Eq. 1.3 is usually not valid [19]. A camera is said to be colorimetric if it satisfies the Luther condition (also called the Maxwell-Ives criterion); that is, if the product of the spectral sensitivity of the three channels is a linear combination of the CIE color-matching functions. If this is the case, then there will be a linear transform between the camera *RGB* values and the CIE *XYZ* values assuming that any nonlinearity in the response is first accounted for. Spectral sensitivity here means the product of the spectral responsivity of the photoreceptor and the spectral transmittance of any filter or lens. Note, however, that it is normal that the RGB values recorded by a digital camera are subject to a nonlinear process which needs to be inverted before the linear transform to the CIE XYZ values can take place [7]. Various methods for estimating the nonlinearity in a camera system are described in the literature [3,31].

When a camera does satisfy the Luther condition, Eq. 1.3 can be used to convert the camera *RGB* values to CIE *XYZ* values. However, determination of the matrix $\boldsymbol{M}$ requires knowledge of the camera's spectral sensitivities. Unfortunately, it can be difficult to determine or to estimate the camera spectral sensitivities of the channels [31]. More seriously, it is often the case that digital cameras do not satisfy the Luther condition and are therefore not colorimetric [20]. In these situations, more empirical data-driven methods are often employed to convert camera *RGB* values into CIE *XYZ* values (and hence to enable color management) and these methods are discussed in the next section.

1.6 Data-Based Color-Correction Methods

Many digital cameras do not satisfy the Luther condition because of the practical production of filters, sensors and optical lenses and in such circumstances it is necessary to estimate the transform $\boldsymbol{M}$ in some optimal sense. Data-based correction methods use sets of imaged samples with known CIE values and corresponding captured RGB values to optimize the transform.

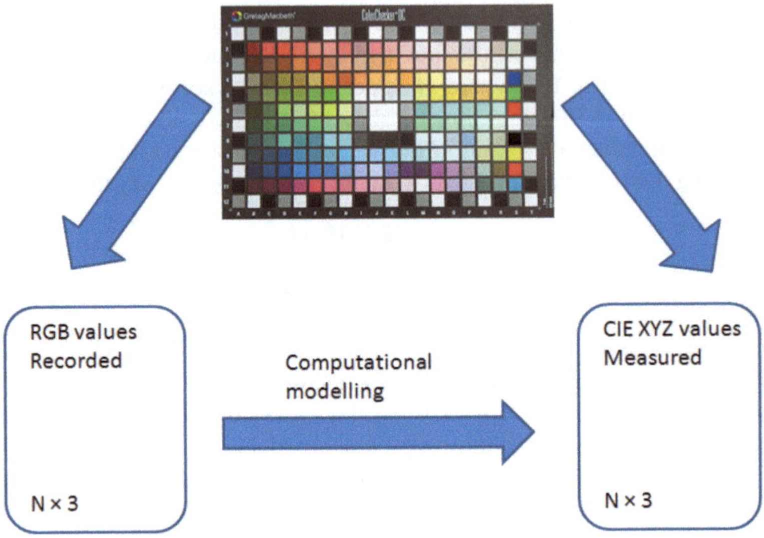

Fig. 1.3 Typical process for data-based camera color characterization. A chart containing a number of color patches is imaged to obtain the camera *RGB* values for each patch. The CIE *XYZ* values for each patch are obtained by direct colorimetric measurement. Computational models are derived to predict *XYZ* from *RGB*

Figure 1.3 shows a typical process for modeling the relationship between camera RGB and CIE XYZ. Standard test charts are available that contain a range of colored samples. These will be imaged using the camera system and also measured using a color-measurement device (such as a tristimulus colorimeter or a reflectance spectrophotometer) that is capable of reporting CIE XYZ values [4]. Computational models can then be constructed to predict XYZ from RGB. Two well-known methods for determining the transformation are the simple least-squares fit procedure and Vrhel's principal component method [14, 33]. However, various other methods have been used including pattern search optimization [6] and methods that exploit information about the illuminant estimation process [7, 8]. Some researchers have also considered whether it is better to minimize errors in a space that is more perceptually relevant than CIE XYZ space. It has been demonstrated that, for digital scanners, it is more accurate to minimize errors in CIE (1976) L*a*b space than in the tristimulus XYZ space [28].

Even when a camera does not satisfy the Luther condition, simple linear transforms are often preferred to more complex models because they can be computed quickly and can be easily inverted. However, higher order polynomials have also been used [15, 20] as well as artificial neural networks [10, 34, 38]. For more complex models, care must be taken to ensure that the model does not overfit the data. It is thus common to develop the model based on a training set of data and then to evaluate the model based on a test set of data. For example, in one study, [10] two different calibration charts were used (one for training containing 166 color patches and one for testing containing 50 color patches) and it was shown that good generalization of a third-order polynomial transform was possible with at least 100 training samples. It has also been demonstrated that minimizing the errors in CIE (1976) L*a*b color space is normally superior to XYZ space when using polynomials to predict RGB from CMYK space [29].

Simple linear transforms are preferred for ICC profiles but more complex models can give better performance and are often used in research laboratories where exceptional color fidelity is required. For printers, polynomial transforms, neural networks and look-up tables are commonly used [21, 26, 36].

Sometimes exceptional color fidelity is required, but only in a subspace of the full color space. In this case, the training and test data used for developing the models may consist of samples that are in that color subspace. Methods for optimizing the selection of color samples in a chart have been developed [9, 16, 18, 32] and may be useful for the design of such charts.

Several reviews of computational methods for determining the transformation between an imaging device and CIE colorimetric space are available in the literature [2, 18, 24].

1.7 Applications and Limitations

ICC color management is ubiquitous. Whenever, a consumer watches television, looks at their smart phone display, or goes to the cinema, for example, ICC color management is operating to provide good color fidelity. However, color fidelity is imperfect for a number of reasons but most noticeably:

(1) ICC profiles tend to use relatively simple transforms (linear transforms are common) which may not adequately describe the relationship between the device color space and CIE color space.
(2) ICC profiles are only valid under the conditions under which they were determined. If the device settings (e.g., brightness or contrast for a display, or exposure time for a camera) are changed then the profile needs to be reset. Many users rely upon default profiles which will likely be inaccurate. Professional users can build profiles that are consistent with a device's actual settings and this may improve performance.
(3) The mismatch in gamuts between the destination and source devices can mean that colorimetric image reproduction is not possible. Most color management systems provide so-called rendering-intent options to allow the user to choose between difficult options (e.g., gamut clipping and gamut compression).

The overwhelming majority of consumers probably find ICC color profiling to be acceptable. That is, they are reasonably content with the degree of color fidelity available in contemporary consumer imaging devices. However, there are many situations where color fidelity is critical. One such application is medical imaging where the correct reproduction of colors in a display is needed so that medical practitioners can make appropriate clinical assessments. Other applications include internet shopping and representation and conservation of artwork. For some products, very accurate representation on a consumer's display is essential if the shopping experience is to be practical. In such cases, state-of-the-art ICC color management is often inadequate.

1.8 Conclusions

Low-cost color-imaging devices for capture and display are ubiquitous and the accurate representation of color is important in the fields of entertainment, social media and medical imaging amongst others. Different image capture and display technologies use different RGB primaries so that, unless adjustment is made to the RGB values to compensate for these differences, the same image displayed on different display devices is likely to have a very different color appearance. Fortunately several open-source and cross-vendor standard procedures, such as the standard RGB color space known as sRGB and the color-management protocols known collectively as ICC, enable a level of color fidelity that is adequate for many purposes. This chapter introduced the theoretical framework for trichromatic color specification and the basis for the color management of images. In some fields, however, such as internet shopping, contemporary color fidelity is inadequate and further technological developments may be required. In addition, the public demand ever greater quality in the images that they consume which is likely to drive innovations in the next decade that will put greater strain on color-management systems of the future. The most likely of these will be more saturated color primaries and the introduction of systems with more than three primaries. Post-trichromatic color imaging, in particular, may generate more colorful images but poses significant challenges for color fidelity and for cross-media color reproduction.

References

1. Anderson, M., Motta, R., Chandraseka, S., Stokes, M.: Proposal for a standard default color space for the internet—sRGB. Proceedings of the IS&T 4th Color Imaging Conference, IS&T (Springfield, USA), 238–245 (1996)
2. Bala, R.: Device characterization. In: Sharma G (ed.) Digital color imaging handbook. CRC Press, Boca Raton (2003)
3. Barnard, K., Funt, B.: Camera characterization for color research. Color Res. Appl. **27**(3), 152–163 (2002)
4. Berns, R.S.: Billmeyer and Saltzman's principles of color technology. Wiley, New York (2000)
5. Berns, R.S., Motta, R.J., Gorzynski, M.E.: CRT colorimetry part 2. Metrology. Color Res. Appl. **18**, 315–325 (1993)

6. Bianco, S., Gasparini, F., Russo, A., Schettini, R.: A new method for RGB to XYZ transformation based on pattern search optimization. IEEE Trans. Consum. Electron. **53**(3), 1020–1028 (2007)
7. Bianco, S., Bruna, A.R., Naccari, F., Schettini, R.: Color space transformations for digital photography exploiting information about the illumination estimation process. J. Opt. Soc. Am. A. **29**(3), 374–384 (2012)
8. Bianco, S., Bruna, A.R., Naccari, F., Schettini, R.: Color correction pipeline optimization for digital cameras. J. Electron. Imaging. **22**(2), 023014–023014 (2013)
9. Cheung, V., Westland, S.: Methods for optimal color selection. J. Imaging Sci. Technol. **50**, 481–488 (2006)
10. Cheung, V., Westland, S., Connah, D., Ripamonti, C.: A comparative study of the characterization of colour cameras by means of neural networks and polynomial transforms. Color. Technol. **120**, 19–25 (2004)
11. CIE: Publication No. 15.2, Colorimetry. 2nd edition, Austria, Central Bureau of the CIE (1986)
12. Day, E.A., Taplin, L., Berns, R.S.: Colorimetric characterization of a computer-controlled liquid crystal display. Color Res. Appl. **29**, 365–373 (2004)
13. Fairchild, M.D.: Color appearance models, 3rd edn. Wiley, Hoboken (2013)
14. Finlayson, G.D., Drew, M.S.: Constrained least-squares regression in color spaces. J. Electron. Imaging. **6**(4), 484–493 (1997)
15. Finlayson, G.D., Mackiewicz, M., Hurlbert, A.: Root polynomial colour correction. Proceedings of the 19th IS & T Color and Imaging Conference, 114–119 (2011)
16. Gill, G.W.: Improved output device characterization test charts. Proceedings of the IS & T 12th Color Imaging Conference, IS & T (Springfield, USA) (2004)
17. Green, P., Holm, J., Li, W.: Recent developments in ICC color management. Color Res. Appl. **33**(6), 444–448 (2008)
18. Hardeberg, J.Y.: Acquisition and reproduction of colour images: colorimetric and multispectral approaches. Universal Publishers (2001)
19. Holm, G.: Capture color analysis gamuts. Proceedings of the IS & T 14th Color Imaging Conference, IS & T (Springfield, USA), 108–113 (2006)
20. Hong, G., Luo, M.R., Rhodes, P.A.: A study of digital camera colorimetric characterization based on polynomial modeling. Color Res. Appl. **26**, 76–84 (2000)
21. Hung, P.-C.: Colorimetric calibration in electronic imaging devices using a look-up table method and interpolations. J. Electron. Imaging. **2**(1), 53–61 (1993)
22. Hunt, R.W.G., Pointer, M.R.: Measuring colour, 4th edn. Wiley, Chichester (2011)
23. ICC: http://www.color.org (2014). Accessed 1 Sept 2014
24. Kang, H.R.: Computational Color Technology, The International Society for Optical Engineering, Bellingham, Washington (2006)
25. Morovič, J.: Color gamut mapping. Wiley, New York (2008)
26. Morovič, J., Arnabat, J., Richard, Y., Albarrán, Á.: Sampling optimization for printer characterization by greedy search. IEEE Trans. Image Process. **19**(10), 2705–2711 (2010)
27. Nielson, M., Stokes, M.: The creation of the sRGB ICC profile. Proceedings of the IS & T 6th Color Imaging Conference, IS & T (Springfield, USA), 253–257 (1998)
28. Shen, H.-L., Mou, T.-S., Xin, J.H.: Colorimetric characterization of scanner by measures of perceptual color error. J. Electron. Imaging. **15**(4), 041204–041204-5 (2006)
29. Sun, B., Liu, H., Zhou, S., Li, W.: Evaluating the performance of polynomial regression method with different parameters during color characterization. Mathematical Problems in Engineering, Article ID 418651. (2014). doi:10.1155/2014/418651
30. Süsstrunck, S., Buckley, R., Swen, S.: Standard RGB color spaces. Proceedings of the IS & T 7th Color Imaging Conference, IS & T (Springfield, USA), 127–134 (1999)
31. Thomson, M.G.A., Westland, S.: Color-imager calibration by parametric fitting of sensor responses. Color Res. Appl. **26**, 442–449 (2001)

32. Vanneschi, L., Castelli, M., Bianco, S., Schettini, R.: Genetic algorithms for training data and polynomial optimization in colorimetric characterization of scanners. In: Di Chio, C., Cagnoni, S., Cotta, C., Ebner, M., Ekart, A., Esparcia-Alcázar, A.I., Goh, C-K., Merelo, J.J., Neri, F., Preuss, M., Togelius, J., Yannakakis, G.N. (Eds.) Applications of evolutionary computation, pp. 282–291. Springer, Berlin (2010)
33. Vrhel, M.J., Trussell, H.J.: Color correction using principal components. Color Res. Appl. **17**(5), 328–338 (2007)
34. Watanabe, T.: High quality color correction method combining neural networks with genetic algorithms. Proceedings of the IEEE International Conference on Image Processing, 553–556 (2001)
35. Westland, S., Cheung, V.: RGB systems. In: Chen, J., Cranton, W., Fihn, M. (eds.) Handbook of visual display technology, pp. 147–154. Springer-Verlag, Berlin (2012)
36. Westland, S., Ripamonti, C., Cheung, V.: Computational colour science: Using MATLAB, 2nd edn. Wiley, New York. ISBN-10: 0470665696 (2012)
37. Wyszecki, G., Stiles, W.S.: Color science—concepts and methods, quantitative data and formulae, 2nd edn. Wiley, New York (1982)
38. Xu, X., Zhang, X., Cai, Y., Zhuo, L., Shen, L.: Supervised color correction based on QPSO-BP neural network algorithm. Proceedings of the IEEE 2nd International Congress on Image and Signal Processing, 1–5 (2009)

Chapter 2
Image Demosaicing

Ruiwen Zhen and Robert L. Stevenson

2.1 Introduction

Digital cameras are extremely popular and have replaced traditional film-based cameras in most applications. To produce a color image in a digital camera, there should be at least three color components at each pixel location. This can be achieved by three CCD (Charge-Coupled Devices) or CMOS (Complementary Metal-Oxide Semiconductor) sensors, each of which receives a specific primary color. However, the associated cost and space is prohibited in many situations. As a result, most digital cameras on the market use a single sensor covered by a color filter array (CFA) to reduce the cost and size.

The CFA consists of a set of spectrally selective filters that are arranged in an interleaving pattern so that each sensor pixel samples one of the three primary color components (Fig. 2.1a). These sparsely sampled color values are termed mosaic images or CFA images. To render a full-color image from the CFA samples, an image reconstruction process, commonly known as CFA demosaicing, is required to estimate the other two missing color values for each pixel. Among many possible CFA patterns, we focus on the widely used Bayer CFA pattern [8] shown in Fig. 2.1b. The Bayer pattern samples the green band using a quincunx grid, while red and blue are obtained by a rectangular grid. The green pixels are sampled at a higher rate since the green color approximates the brightness perceived by human eyes.

Before fully exploring the various demosaicing algorithms, we will introduce the basic knowledge about demosaicing in this section. We start from the formalism for demosaicing process and the simplest demosaicing method, bilinear interpolation, which allows us to introduce the demosaicing color artifacts, and major principles adopted by most demosaicing algorithms. After that, we show how to evaluate

R. Zhen (✉) · R. L. Stevenson
University of Notre Dame, 275 Fitzpatrick Hall, Notre Dame, IN 46556, USA
e-mail: rzhen@nd.edu

R. L. Stevenson
e-mail: rls@nd.edu

M. E. Celebi et al. (eds.), *Color Image and Video Enhancement*,
DOI 10.1007/978-3-319-09363-5_2

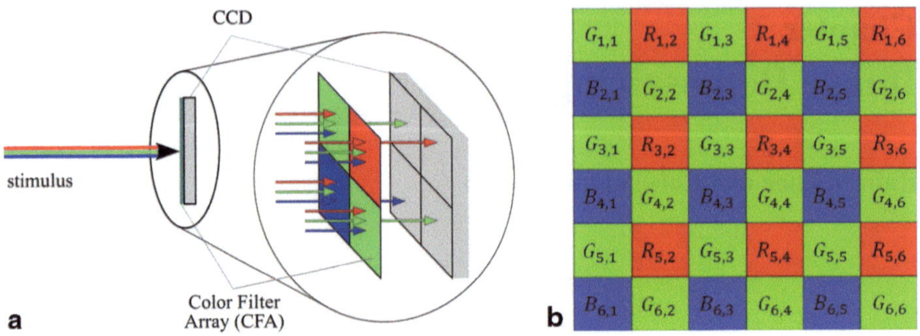

Fig. 2.1 **a** Single CCD sensor covered by a CFA [35], **b** Bayer CFA

demosaicing algorithms, including the test image database, objective measure, and subjective measure of quality.

2.1.1 Demosaicing

Let $I^{CFA}:\mathbb{Z}^2 \rightarrow \mathbb{Z}$ denote a $M \times N$ Bayer CFA image. Each pixel $I^{CFA}(i, j)$ with coordinates $i = 1, 2...M$, $j = 1, 2...N$ in the image I^{CFA} corresponds to a single color component. Assuming the sampling pattern is as Fig. 2.1b, then

$$I^{CFA}(i,j) = \begin{cases} R_{i,j} & \text{for } i \text{ odd and } j \text{ even} \\ B_{i,j} & \text{for } i \text{ even and } j \text{ odd} \\ G_{i,j} & \text{otherwise} \end{cases} \tag{2.1}$$

where $R_{i,j}, B_{i,j}, G_{i,j}$ values range from 0 to 255 if the image is quantized with 8-bit for each color channel. The demosaicing process is to estimate the missing two color values at each pixel location (i, j) for rendering a full color image $\hat{I}:\mathbb{Z}^2 \rightarrow \mathbb{Z}^3$:

$$\hat{I}(i,j) = \begin{cases} (R_{i,j}, \hat{G}_{i,j}, \hat{B}_{i,j}) & \text{for } i \text{ odd and } j \text{ even} \\ (\hat{R}_{i,j}, \hat{G}_{i,j}, B_{i,j}) & \text{for } i \text{ even and } j \text{ odd} \\ (\hat{R}_{i,j}, G_{i,j}, \hat{B}_{i,j}) & \text{otherwise} \end{cases} \tag{2.2}$$

Each triplet in Eq. (2.2) represents a color vector, in which $R_{i,j}, B_{i,j}, G_{i,j}$ are color components available in the CFA image I^{CFA} and $\hat{R}_{i,j}, \hat{B}_{i,j}, \hat{G}_{i,j}$ are estimated missing color components [35]. For use in later discussion, we also define the original full-color image as:

$$I(i,j) = (R_{i,j}, G_{i,j}, B_{i,j}) \quad \text{for } \forall i \text{ and } \forall j \tag{2.3}$$

Many algorithms have been proposed for CFA image demosaicing. The simplest demosaicing methods apply well-known interpolation techniques, such as nearest-neighbor replication, bilinear interpolation, and cubic spline interpolation, to each color plane separately. The remaining part of this subsection will introduce the bilinear interpolation demosaicing method. The bilinear approach is useful to understand since many advanced algorithms still adopt bilinear interpolation as an initial step; additionally, these algorithms usually use the results of bilinear interpolation for performance comparison.

The bilinear interpolation method fills the missing color values with weighted averages of their neighboring pixel values. Considering the CFA pattern in Fig. 2.1b, the missing blue and green values at pixel $R_{3,4}$ are estimated thanks to the following equations:

$$\hat{B}_{3,4} = \frac{1}{4}(B_{2,3} + B_{2,5} + B_{4,3} + B_{4,5}) \tag{2.4}$$

$$\hat{G}_{3,4} = \frac{1}{4}(G_{3,3} + G_{2,4} + G_{3,5} + G_{4,4}) \tag{2.5}$$

Similarly, the red and green components can be estimated at blue pixel locations. As for the green pixel location, for example $G_{3,3}$, the blue and red values are calculated as:

$$\hat{R}_{3,3} = \frac{1}{2}(R_{3,2} + R_{3,4}) \tag{2.6}$$

$$\hat{B}_{3,3} = \frac{1}{2}(B_{2,3} + B_{4,3}) \tag{2.7}$$

These interpolation operations can be easily implemented by convolution [6]. If we decompose the CFA image into three color planes, I_R^{CFA}, I_G^{CFA} , and I_B^{CFA}, as shown in Fig. 2.2, the convolution kernels for bilinear interpolating of each color plane are:

$$K_B = K_R = \frac{1}{4}\begin{bmatrix} 1 & 2 & 1 \\ 2 & 4 & 2 \\ 1 & 2 & 1 \end{bmatrix} \tag{2.8}$$

$$K_G = \frac{1}{4}\begin{bmatrix} 0 & 1 & 0 \\ 1 & 4 & 1 \\ 0 & 1 & 0 \end{bmatrix} \tag{2.9}$$

Figure 2.3 shows an example of bilinear interpolating of the image *lighthouse* with the above convolution kernels. Though the bilinear interpolation method is computationally efficient and easy to implement, we see that the demosaiced image in Fig. 2.3b suffers from severe visible artifacts, especially the image regions with high-frequency content.

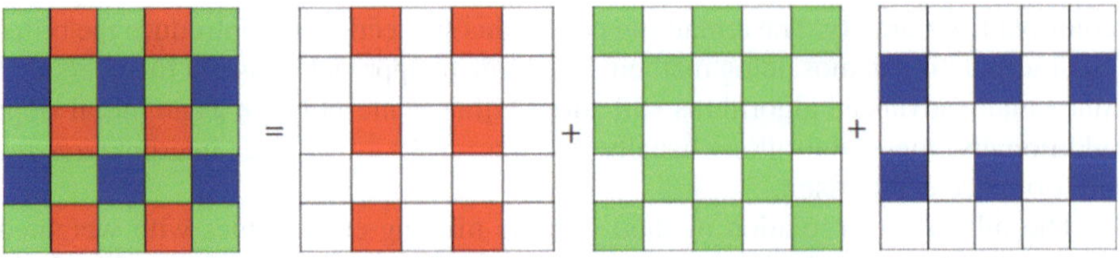

Fig. 2.2 CFA color plane decomposition

Fig. 2.3 **a** Original image, **b** Demosaiced image by bilinear interpolation

2.1.2 Demosaicing Artifacts

To analyze the demosaicing artifacts introduced by bilinear interpolation, Chang et al. [11] synthesized an image with a vertical edge (Fig. 2.4a) and obtained the corresponding bilinear interpolated result (Fig. 2.4c). The synthesized image has two homogeneous areas with different gray levels L and H ($L < H$), and the three color components in each gray area are equal. Figure 2.4b shows the Bayer CFA image yielded by sampling the synthesized image. The results of bilinear interpolating of each color plane are displayed in Figs. 2.4d–2.4f.

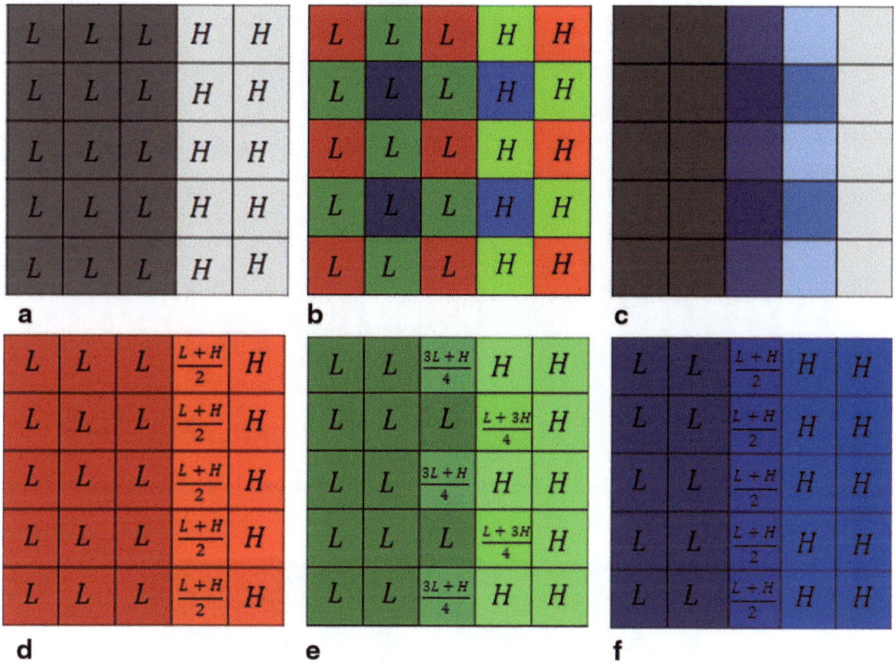

Fig. 2.4 **a** Synthesized gray image, **b** CFA samples of **a**, **c** Bilinear interpolation result, **d** Bilinear interpolated red plane, **e** Bilinear interpolated green plane, **f** Bilinear interpolated blue plane

We can see that the three interpolated color planes suffer from different errors due to their different sampling patterns. The green plane gives rise to the obvious grid error pattern while the red and blue planes produce an intermediate level between low and high intensity levels. Visually, two types of artifacts are generated in the demosaiced image: one is the pattern of alternating colors along the edge, called zipper effect, and the other is the noticeable color errors (the bluish tint in this example), called false color.

The zipper effect refers to the abrupt or unnatural changes of intensities over a number of neighboring pixels, manifesting as an "on–off" pattern in regions around edges [11]. Figure 2.5b shows that the fence bars in the bilinear interpolated *lighthouse* are corrupted by the zipper effects. They are primarily caused by improper averaging of neighboring color values across edges. Interpolation along an object boundary is always preferable to interpolation across it because the discontinuity of the signal at the boundary contains high-frequency components that are difficult to estimate. If an image is interpolated in the direction orthogonal to the orientation of the object boundary, the color that appears at the pixel of interest is unrelated to the physical objects represented in the image [27]. For this reason, many proposed demosaicing algorithms are edge-sensitive. Another reason that could influence zipper effects is the quincunx structure of the CFA green samples. According to Chang's experimental results, the zipper effects are more likely to occur around

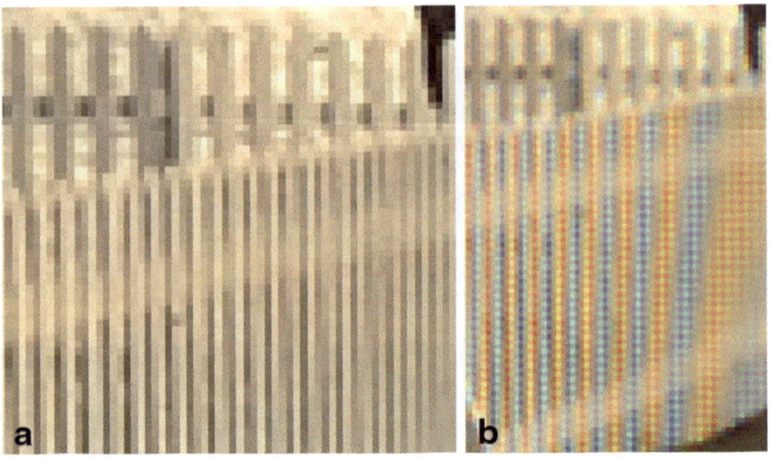

Fig. 2.5 Zipper effect. **a** Fence bars in the original image, **b** Fence bars in the bilinear interpolated image

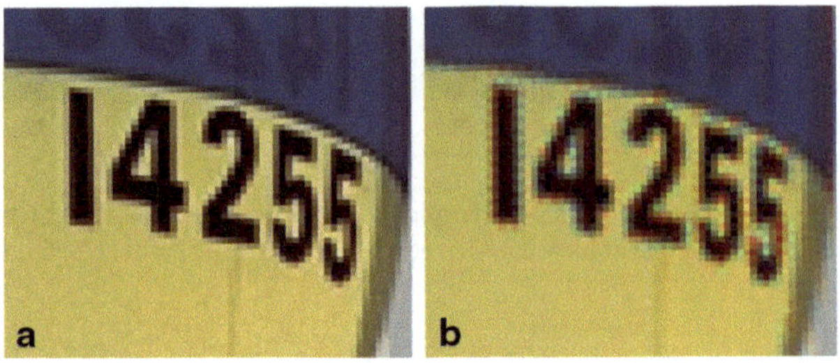

Fig. 2.6 False color. **a** Numbers in the original image, **b** Numbers in the bilinear interpolated image

edges not aligned in the diagonal direction along which the green values are fully sampled.

The false colors are spurious colors which are not present in the original image, as in Figs. 2.5b and 2.6b. They appear as sudden hue changes due to inconsistency among the three color planes. Such inconsistency usually results in the large intensity changes in the color difference planes [11]. Based on this observation, many algorithms attempt to utilize the spectral correlation between different planes and ensure that the hue or color difference plane is slowly varying.

Both the zipper effect and the false color are referred to as misguidance color artifacts, which are mainly caused by erroneous interpolation direction. These artifacts affect the regions with high-frequency content most. However, even with correct interpolation direction, the reconstructed image may still contain several errors called interpolation artifacts, and it is associated with limitations in the interpolation [27]. Normally, interpolation artifacts are far less noticeable than misguidance color artifacts.

2.1.3 Demosaicing Principles

The drawbacks brought by simple interpolation in separate planes motivated the appearance of more advanced algorithms specifically designed for the reconstruction of CFA images to improve the overall demosaicing performance. An excellent review of the demosaicing algorithms proposed in the past several decades can be found in [31,35,47]. In order to reduce the misguidance color artifacts, most of them are developed based on three principles: spectral correlation, spatial correlation, and green-plane-first rule.

The most popular principle in the demosiacing literature appears to be the green-plane-first rule, that is to interpolate the green plane first. The key motivation behind this principle is that the green component is less aliased than the other two. Thus, having a full-resolution green plane could facilitate the recovery of blue and red planes. In addition, human eyes are more sensitive to the change of the luminance component (green) than that of the chrominance components. The interpolation accuracy of the green plane is critical to the quality of the demosaiced image.

The spectral correlation of a color image dictates that there is a strong dependency among the pixel values of different color planes, especially in areas with high spatial frequencies [11]. This correlation is usually exploited by using the assumption that the differences (or ratios) between the pixel values in two color planes are likely to be constant within a local image region. In 1987, Cok [15] first proposed interpolation based on color hue constancy. Hue is understood as the ratio between chrominance and luminance, i.e., R/G and B/G. Following his work, several schemes [2, 29] were devised to estimate the missing color values with the aid of other color planes. The formal statement of the hue constancy is given below:

- The color ratios between green and red/blue channels satisfy:

$$\begin{aligned} \frac{R_{i,j}}{G_{i,j}} &= C_{i,j}^{rg} \text{ and} \\ \frac{B_{i,j}}{G_{i,j}} &= C_{i,j}^{bg} \end{aligned} \tag{2.10}$$

  where $C_{i,j}^{rg}$ and $C_{i,j}^{bg}$ are piecewise constant within the boundary of a given object.

However, later work asserted that the differences instead of the ratios between green and red/blue planes are slowly varying [3,21,22,49,59], i.e.,

- The color differences between green and red/blue channels satisfy:

$$\begin{aligned} R_{i,j} &= G_{i,j} + A_{i,j}^{rg} \text{ and} \\ B_{i,j} &= G_{i,j} + A_{i,j}^{bg} \end{aligned} \tag{2.11}$$

  where $A_{i,j}^{rg}$ and $A_{i,j}^{bg}$ are piecewise constant within the boundary of a given object.

Fig. 2.7 Compare ratio image and difference image. **a** Original image, **b** Green plane, **c** R/G ratio image, **d** $R - G$ difference image

This is because the inter-spectral correlation lies in the high-frequency spectrum and consequently, the difference image of two color planes contains low-frequency components only. Generally, the color difference presents some benefits in comparison to the color ratio. The latter is indeed error-prone when its denominator takes a low value. This happens, for instance, when saturated red/blue components lead to comparatively low values of green, making the ratio very sensitive to the small variations in the red/blue plane. Figure 2.7a shows a natural image which is highly saturated in red. The corresponding green plane G, ratio image R/G and difference image $R - G$ are given in Figs. 2.7b–2.7d respectively. It can be noticed that the ratio and difference images carry out less high-frequency information than the green plane. Moreover, in areas where red is saturated, the ratio image contains more high-frequency information than the difference image, which makes the interpolation result more artifact-prone [35].

The spatial correlation reflects the fact that within a homogeneous image region, neighboring pixels share similar color values [10]. One could use this principle to estimate the missing color components at any pixel location except the pixels near the edge since these pixels have neighbors which do not belong to the same homogeneous region. Therefore, the following assumption is proposed based on the spatial correlation [59]:

- The rate of change of neighboring pixel values along an edge direction is a constant. For example, the pixels along horizontal edges satisfy:

$$\begin{aligned} R_{i,j} - R_{i,j+1} &= R_{i,j+1} - R_{i,j+2} = dR \\ G_{i,j} - G_{i,j+1} &= G_{i,j+1} - G_{i,j+2} = dG \\ B_{i,j} - B_{i,j+1} &= B_{i,j+1} - B_{i,j+2} = dB \end{aligned} \tag{2.12}$$

  where dR, dG and dB are constants.

Following this assumption, many demosaicing methods first analyze the spatial structure of a local image neighborhood and then select a suitable direction for interpolation.

Depending on how the two correlations are exploited, existing demosaicing methods can be grouped into four classes [11]. The methods in the first class exploit neither correlation, applying the same interpolation scheme in each individual color plane, such as bilinear interpolation, nearest-neighbor replication, and cubic spline interpolation. The methods in the second class mainly exploit spatial correlation but little or no spectral correlation; they usually apply some adaptive interpolation scheme in each color plane separately. Examples of this class include Cok's pattern recognition interpolation (PRI) [14] and Adam's edge-sensing (ES) method [2]. Since this class does not fully utilize the spectral correlation, the methods in this class often result in excessive false colors. The methods in the third class mainly exploit image spectral correlation, including Cok's constant-hue interpolation [15], Freeman's median interpolation [20], Pei's effective color interpolation (ECI) [54], and Gunturk's alternating projections method (AP) [24]. Although capable of alleviating false color artifacts, these methods normally produce visible zipper effects around edges and details due to less usage of spatial correlation. The methods of the last class exploit both spatial and spectral correlations. Examples are Li's new edge-directed interpolation [32], Hamilton's adaptive color plane interpolation (ACPI) [5], Wu's primary-consistent soft-decision method (PCSD) [61], Hirakawa's adaptive homogeneity-directed demosaicing algorithm (AHD) [27], and so on.

In addition to the above classification, the demosaicing methods could also be divided into frequency-domain and spatial-domain [31], heuristic and nonheuristic [13], iterative and noniterative [57]. These classifications represent most demosaicing algorithms, but they are too general to capture each algorithm's main characteristics. Therefore, in the next section we will learn from Menon [47] and describe five representative methods to give readers a more comprehensive introduction of the existing demosaicing algorithms.

2.1.4 Evaluation Criteria

The common process for evaluating demosaicing algorithms consists of choosing color images that are captured using highly professional three-sensor cameras or

Fig. 2.8 Kodak image database. (These images are referred as Image 1 to Image 24 from *left* to *right* and *top* to *bottom*.)

color scanners, sampling them according to the Bayer CFA pattern to obtain mosaic images, interpolating the mosaic images back to full color images, and comparing the results with the original images [47]. This subsection will discuss the first and last step of the evaluation process.

Most work in the literature uses the Kodak image database [33] shown in Fig. 2.8 as a benchmark for performance comparison. The 24 images in this database are film captured and then digitized at the resolution of 512×768 with 8-bit depth per color component. The popularity of the Kodak image database is mainly due to the fact that the database contains natural real-life scenes and varies in complexity and color appearances. To increase the test difficulty, Li et al. [31] included a set of IMAX images with varying-hue and high-saturation edges and Lian et al. [34] added several classical images which are often used in other image processing fields. In addition to the real images, some synthetic images, such as starburst [36] and circular zone plate [39] shown in Figs. 2.9a and 2.9b respectively, were used as well to test the ability of the demosaicing algorithms in handling edges of various orientations and spatial resolutions.

In order to evaluate the demosaiced image, the Mean Square Error (MSE) is widely considered [35, 47, 57]. This criterion measures the mean quadratic error between the original image and the demosaiced image in each color plane. It is defined as:

$$MSE(k) = \frac{1}{MN} \sum_{i=1}^{M} \sum_{j=1}^{N} (\hat{I}_k(i, j) - I_k(i, j))^2 \qquad (2.13)$$

where $I_k(i, j)$ is a color component in the original image and $\hat{I}_k(i, j)$ is the corresponding color component in the demosaiced image, $k = R, G, B$. The MSE

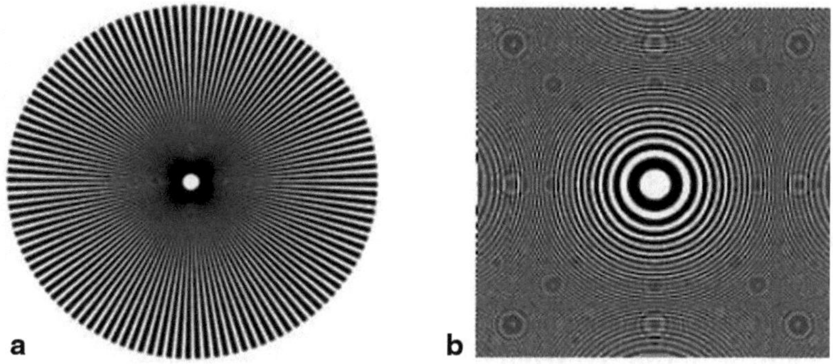

Fig. 2.9 Synthetic test images. **a** Starburst [36], **b** Circular zone plate [39]

criterion can also be used to measure the estimation error for the full color image,

$$CMSE(I,\hat{I}) = \frac{1}{3MN}\sum_{k=R,G,B}\sum_{i=1}^{M}\sum_{j=1}^{N}(\hat{I}_k(i,j) - I_k(i,j))^2 \tag{2.14}$$

Alternatively, the peak signal-to-noise ratio (PSNR) and the color peak signal-to-noise ratio (CPSNR) are often adopted to quantify the demosaicing performance. The PSNR is expressed in decibels as:

$$PSNR(k) = 10\log_{10}\left(\frac{255^2}{MSE(k)}\right) \tag{2.15}$$

The definition of CPSNR is similar to PSNR but replaces MSE with CMSE. Contrary to the MSE criterion, the higher the PSNR value is, the better is the demosaicing quality.

Though the preceding criteria could estimate the fidelity of the demosaiced image compared with the original image, they are not consistent with quality estimation provided by the Human Visual System (HVS). Therefore, the criteria operating in perceptual uniform color spaces CIELab and S-CIELab have been used [13,31,47, 57]. Let the color space conversion map from RGB to CIELab be π: $\begin{bmatrix}L, a, b\end{bmatrix} = \pi(\begin{bmatrix}R, G, B\end{bmatrix})$ where $\begin{bmatrix}L, a, b\end{bmatrix}$ is a CIELab value. $I(i,j)$ and $\hat{I}(i,j)$ represent the color vectors of the same pixel in the original image and the demosaiced image respectively. The distance between $I(i,j)$ and $\hat{I}(i,j)$ in CIELab color space is given below [43]:

$$D_{i,j} = \|\pi(I(i,j)) - \pi(\hat{I}(i,j))\| \tag{2.16}$$

where $\|\cdot\|$ indicates the ℓ_2 norm. The CIELab criterion is defined as the mean error processed with all image pixels [35]:

$$\Delta E^*_{Lab}(I,\hat{I}) = \frac{1}{MN}\sum_{i=1}^{M}\sum_{j=1}^{N}D_{i,j} \tag{2.17}$$

The lower ΔE^*_{Lab} is, the lower is the perceptual difference between the original image and the demosaiced image, and the higher is the demosaicing quality. The S-CIELab criterion is an extension of the CIELab color difference formula ΔE^*_{Lab} by using S-CIELab color space instead of CIELab color space. Compared with CIELab color space, S-CIELab adds a spatial pre-processing step which incorporates the pattern-color sensitivity measurements proposed by [63] in order to simulate the spatial blurring generated by the HVS.

2.2 Demosaicing Approaches

In this section, the main demosaicing approaches proposed in the literature are described. Similar to [47], we divide them into five categories, namely edge-sensitive methods, directional interpolation and decision methods, frequency-domain approaches, wavelet-based methods, and statistical reconstruction techniques. In each category, the related works are reviewed and some representative methods are analyzed in details. To show the advantages and drawbacks of the examined methods, we perform a comparison between different methods whose source codes are made available directly from the original authors.

2.2.1 Edge-Sensitive Methods

As mentioned in Sect. 2.1.3, the green plane is usually estimated before red and blue planes due to the double amount of green samples in a CFA image. After a reconstruction of the green plane, the red and blue planes are populated based on the spectral correlation, either color ratio constancy or color difference constancy. In this case, the green component estimation quality becomes critical in the overall demosaicing performance, since any error in the green plane estimation will be propagated in the following red and blue estimation steps. As a consequence, significant effort has been devoted to improve the accuracy of green plane interpolation. A general solution is to use the spatial correlation, i.e., interpolation along edges rather than across them, to reduce color artifacts, where to determine the edge direction from CFA samples becomes a key issue.

Hibbard [26] used the available green components in a local 3×3 window centered at a non-green pixel location to calculate horizontal and vertical gradients from which the edge direction is derived. If the horizontal gradient is greater than the vertical one, the missing green component is estimated using the green components along the vertical edge and vice versa. Laroche and Prescott [30] suggested to approximate the partial derivatives with the help of surrounding chrominance components in a 5×5 neighborhood. Adams et al. [5] took both approaches into account and proposed the adaptive color plane interpolation algorithm (ACPI) which uses

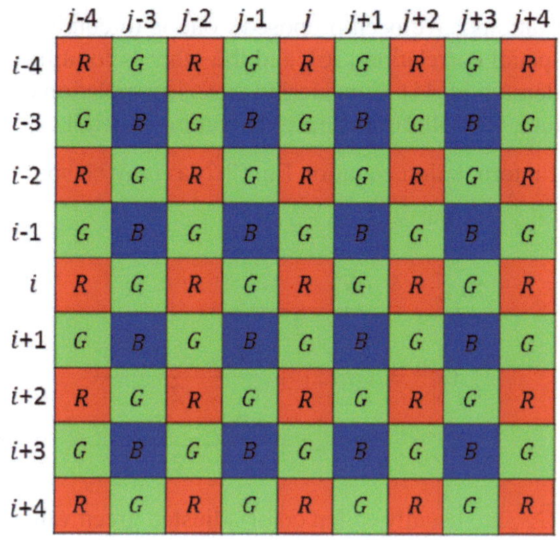

Fig. 2.10 A 9 × 9 CFA pattern centered at a red pixel

the mixed Laplacian operator to decide the edge direction. Without losing the generality, we consider the case in Fig. 2.10 to describe the green plane interpolation method in ACPI.

The horizontal gradient $\Delta H_{i,j}$ and the vertical gradient $\Delta V_{i,j}$ at position (i, j) are first estimated with the mixed Laplacian operator:

$$\Delta H_{i,j} = |G_{i,j-1} - G_{i,j+1}| + |2R_{i,j} - R_{i,j-2} - R_{i,j+2}| \tag{2.18}$$
$$\Delta V_{i,j} = |G_{i-1,j} - G_{i+1,j}| + |2R_{i,j} - R_{i-2,j} - R_{i+2,j}| \tag{2.19}$$

where $G_{m,n}$ and $R_{m,n}$ denote the available red and green samples at position (m, n) in the CFA image. Based on the values of $\Delta H_{i,j}$ and $\Delta V_{i,j}$, the missing green component $\hat{G}_{i,j}$ is interpolated as follows:

$$\hat{G}_{i,j} = \begin{cases} \dfrac{(G_{i,j-1} + G_{i,j+1})}{2} + \dfrac{(2R_{i,j} - R_{i,j-2} - R_{i,j+2})}{4} & \Delta H_{i,j} < \Delta V_{i,j} \quad (2.20a) \\ \dfrac{(G_{i-1,j} + G_{i+1,j})}{2} + \dfrac{(2R_{i,j} - R_{i-2,j} - R_{i+2,j})}{4} & \Delta H_{i,j} > \Delta V_{i,j} \quad (2.20b) \\ \dfrac{(G_{i,j-1} + G_{i,j+1} + G_{i-1,j} + G_{i+1,j})}{4} & \quad (2.20c) \\ + \dfrac{(4R_{i,j} - R_{i,j-2} - R_{i,j+2} - R_{i-2,j} - R_{i+2,j})}{8} & \Delta H_{i,j} = \Delta V_{i,j} \end{cases}$$

In Eq. (2.20), the direction with a smaller gradient is preferred for interpolation because a smaller gradient always implies smaller variations among the pixels on that direction. Obviously, the estimation using these highly correlated pixels will be more accurate. With regard to the interpolation equations, Eqs. (2.20a) and (2.20b), two different derivations have been proposed : one is to combine the spectral correlation and the spatial correlation [36], and the other is to utilize filter techniques in the frequency domain [4, 44]. Considering these interpolation equations are often used in the literature, we will discuss each derivation in detail.

1. The horizontal interpolation equation Eq. (2.20a) may be split into one left $\hat{G}^l_{i,j}$ and one right $\hat{G}^r_{i,j}$ side parts:

$$\begin{aligned} \hat{G}_{i,j} &= \frac{(\hat{G}^l_{i,j} + \hat{G}^r_{i,j})}{2} \\ \hat{G}^l_{i,j} &= G_{i,j-1} + \frac{(R_{i,j} - R_{i,j-2})}{2} \\ \hat{G}^r_{i,j} &= G_{i,j+1} + \frac{(R_{i,j} - R_{i,j+2})}{2} \end{aligned} \tag{2.21}$$

The $\hat{G}^l_{i,j}$ and $\hat{G}^r_{i,j}$ are usually called color-adjusted green values [36]. We derive the left side part $\hat{G}^l_{i,j}$, for example, and $\hat{G}^r_{i,j}$ is similar. The spectral correlation assumption suggests the following relationship:

$$R_{i,j} - \hat{G}_{i,j} = \hat{R}_{i,j-1} - G_{i,j-1} = R_{i,j-2} - \hat{G}_{i,j-2} \tag{2.22}$$

Since $\hat{R}_{i,j-1}$ is not available from the CFA samples, we can only use $R_{i,j}$ and $R_{i,j-2}$ to assist the estimation of $\hat{G}_{i,j}$. As a result, the above relationship is reorganized as:

$$\hat{G}_{i,j} - \hat{G}_{i,j-2} = R_{i,j} - R_{i,j-2} \tag{2.23}$$

The spatial correlation assumption gives another relationship along the left interpolation direction:

$$\hat{G}_{i,j} - G_{i,j-1} = G_{i,j-1} - \hat{G}_{i,j-2} \tag{2.24}$$

Combining Eqs.(2.23) and (2.24), we have

$$\begin{aligned} R_{i,j} - R_{i,j-2} &= \hat{G}_{i,j} - \hat{G}_{i,j-2} \\ &= (\hat{G}_{i,j} - G_{i,j-1}) + (G_{i,j-1} - \hat{G}_{i,j-2}) \\ &= 2(\hat{G}_{i,j} - G_{i,j-1}) \end{aligned} \tag{2.25}$$

which states that $\hat{G}_{i,j} = G_{i,j-1} + (R_{i,j} - R_{i,j-2})/2$. Therefore, $\hat{G}^l_{i,j}$ is in fact the estimate of $\hat{G}_{i,j}$ from the left interpolation direction.

2. A five-coefficient FIR filter designed by [4, 44] can also explain the above interpolation equations, for example, Eq. (2.20a). The green components in a CFA image is sampled with a factor of 2. In the frequency domain, this gives

$$G_s(\omega) = \frac{1}{2}G(\omega) + \frac{1}{2}G(\omega - \pi) \tag{2.26}$$

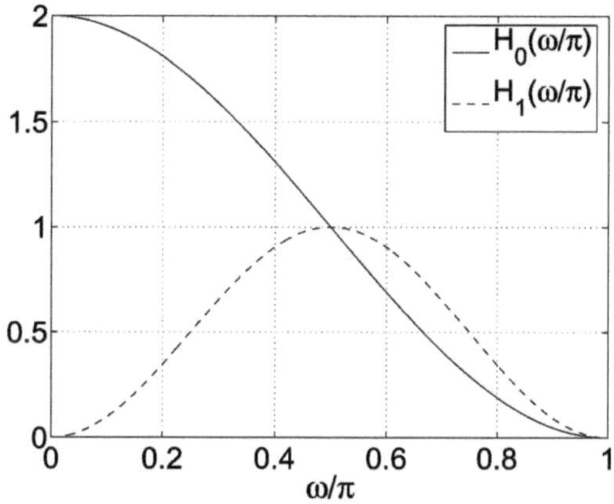

Fig. 2.11 Frequency response of the filters h_0 and h_1 [44]

where $G(\omega)$ and $G_s(\omega)$ denote the Fourier transform of the original green plane and of the subsampled green plane respectively. The low-pass filter $h_0 = \begin{bmatrix} 0.5, & 1.0, & 0.5 \end{bmatrix}$ used in bilinear interpolation is attempted to remove the aliasing component $1/2G(\omega - \pi)$, but as shown in Fig. 2.11 this filter is nonideal, so it cannot totally remove the aliasing. In fact, the result after filtering is:

$$G_s(\omega)H_0(\omega) = \frac{1}{2}G(\omega)H_0(\omega) + \frac{1}{2}G(\omega - \pi)H_0(\omega) \tag{2.27}$$

where $H_0(\omega)$ is the frequency response of h_0 and the second term denotes the aliasing component. To decrease the aliasing effect and improve the mid-frequency response, the information coming from the high-frequency bands of the red samples [1] is considered. In a green–red row, the red component is sampled with an offset of one sample with respect to the green component. Therefore, its Fourier transform is:

$$R_s(\omega) = \frac{1}{2}R(\omega) - \frac{1}{2}R(\omega - \pi) \tag{2.28}$$

where $R(\omega)$ and $R_s(\omega)$ are similarly defined as $G(\omega)$ and $G_s(\omega)$. If we interpolate the subsampled red plane with a filter h_1 and add the filtered result to Eq. (2.27),

[1] The information coming from the high-frequency bands of the blue samples can also be used to facilitate the recovery of the green plane. We use red samples just for illustrating the interpolation equation Eq. (2.20a).

the estimated green plane is:

$$\begin{aligned}\hat{G}(\omega) &= G_s(\omega)H_0(\omega) + R_s(\omega)H_1(\omega) \\ &= \frac{1}{2}G(\omega)H_0(\omega) + \frac{1}{2}G(\omega-\pi)H_0(\omega) \\ &+ \frac{1}{2}R(\omega)H_1(\omega) - \frac{1}{2}R(\omega-\pi)H_1(\omega)\end{aligned} \tag{2.29}$$

Considering the spectral correlation assumption that $R(\omega)-G(\omega)$ is slowly varying, if h_1 is designed such that $H_1(\omega)\approx 0$ at low frequencies and $H_0(\omega)\approx H_1(\omega)$ at high frequencies, we have,

$$\begin{aligned}R(\omega)H_1(\omega) &\approx G(\omega)H_1(\omega) \\ G(\omega-\pi)H_0(\omega) &\approx R(\omega-\pi)H_1(\omega)\end{aligned} \tag{2.30}$$

and Eq. (2.29) could be approximated by

$$\begin{aligned}\hat{G}(\omega) &\approx \frac{1}{2}G(\omega)H_0(\omega) + \frac{1}{2}R(\omega)H_1(\omega) \\ &\approx \frac{1}{2}G(\omega)(H_0(\omega)+H_1(\omega)) \\ &\approx G(\omega)\end{aligned} \tag{2.31}$$

A good choice for the filter h_1 that satisfies the constraints Eq. (2.30) is the five-coefficient FIR filter $\left[-0.25,\ 0,\ 0.5,\ 0,\ -0.25\right]$ shown in Fig. 2.11. Therefore, for each green–red row of the CFA image, the final filter coefficients are $\left[-0.25,\ 0.5,\ 0.5,\ 0.5,\ -0.25\right]$.

Though the interpolation equations Eqs. (2.20a) and (2.20b) are the approximated optimal CFA interpolators which have been proved by [27], the performance of ACPI is not good enough and there is great room for improvement. Chung et al. [13] proposed a more effective gradient test than ACPI using the variance of color differences (VCD). This algorithm first computes two parameters $L^H_{i,j}$ and $L^V_{i,j}$ that are similar to $\Delta H_{i,j}$ and $\Delta V_{i,j}$, but adds the interband information in a 5×5 window. Then the ratio of the two parameters is used to determine whether the window is a sharp edge block or not. If it is, the gradient test is performed between $L^H_{i,j}$ and $L^V_{i,j}$ and the corresponding interpolation scheme in ACPI is applied. If not, the VCD of the pixels along the horizontal direction $H^{\sigma^2_{i,j}}$, the vertical direction $V^{\sigma^2_{i,j}}$, and the diagonal direction $B^{\sigma^2_{i,j}}$ are evaluated in a 9×9 block. The one that provides the minimum variance among $H^{\sigma^2_{i,j}}$, $V^{\sigma^2_{i,j}}$, and $B^{\sigma^2_{i,j}}$ is the final interpolation direction.

The VCD algorithm is summarized below:

$$\hat{G}_{i,j} = \begin{cases} \text{if } e > T \begin{cases} \text{Eq.(2.20a)} & \text{if } L^H_{i,j} < L^V_{i,j} \\ \text{Eq.(2.20b)} & \text{if } L^H_{i,j} > L^V_{i,j} \end{cases} \\ \text{if } e <= T \begin{cases} \text{Eq.(2.20a)} & \text{if } H^{\sigma^2_{i,j}} = \min(H^{\sigma^2_{i,j}}, V^{\sigma^2_{i,j}}, B^{\sigma^2_{i,j}}) \\ \text{Eq.(2.20b)} & \text{if } V^{\sigma^2_{i,j}} = \min(H^{\sigma^2_{i,j}}, V^{\sigma^2_{i,j}}, B^{\sigma^2_{i,j}}) \\ \text{Eq.(2.20c)} & \text{if } B^{\sigma^2_{i,j}} = \min(H^{\sigma^2_{i,j}}, V^{\sigma^2_{i,j}}, B^{\sigma^2_{i,j}}) \end{cases} \end{cases} \quad (2.32)$$

$$\text{and } e = \max\left(\frac{L^V_{i,j}}{L^H_{i,j}}, \frac{L^H_{i,j}}{L^V_{i,j}}\right)$$

Chung et al. thought that the block which is not classified to be an edge block is usually in a flat or pattern region. Since it has been found that the color differences of pixels in such regions are more or less the same, $L^H_{i,j}$ and $L^V_{i,j}$ may no longer provide effective information. Therefore, the VCDs are considered as supplementary information. Similar work was also proposed by Tsai et al. [59], which improved the gradient test in ACPI through heterogeneity maps.

Another proposal based on ACPI comes from Su [57]. He adopted the same gradient test as ACPI, but interpolated the green plane as a weighted sum of values defined by Eqs. (2.20a) and (2.20b). Specifically the interpolation equation is:

$$\hat{G}_{i,j} = \begin{cases} w_1 * \text{Eq.(2.20a)} + w_2 * \text{Eq.(2.20b)} & \text{if } \Delta H_{i,j} < \Delta V_{i,j} \\ w_1 * \text{Eq.(2.20b)} + w_2 * \text{Eq.(2.20a)} & \text{if } \Delta H_{i,j} > \Delta V_{i,j} \\ \text{Eq.(2.20c)} \quad \text{otherwise} \end{cases} \quad (2.33)$$

where w_1 and w_2 are weighting factors, separately representing the weight of smooth transition term and the weight of sharp transition term. The optimal choices for the weighting factors are estimated by minimizing the average MSE of a large number of demosaiced images. In comparison with the interpolators of ACPI, incorporating the sharp transition term could allow to undertake high-frequency information in the green plane estimation [35].

The demosaicing methods introduced so far explicitly select one direction for interpolation. In this way, the information provided by other directions may be ignored. Hence, using a weighted sum of estimates from different directions instead was considered. Kimmel [29] first proposed an adaptive edge-weighted interpolation algorithm which evaluates each local estimate by a normalized weight. As a function of a directional gradient, the weights are defined in four directions, horizontal, vertical, left diagonal, and right diagonal and adapt to the local image pattern.

Lu et al. [36] modified the weight computation of Kimmel to boost algorithm speed. They use a Sobel filter to approximate the directional gradient and adopt the absolute value instead of the square value of gradients when calculating the weights. Such strategy is also implemented by Lukac et al. [38]. We still take the missing green component $G_{i,j}$ in Fig. 2.10 for example. In Lu's approach, $G_{i,j}$ is estimated as:

$$\hat{G}_{i,j} = \frac{\alpha_l \hat{G}^l_{i,j} + \alpha_r \hat{G}^r_{i,j} + \alpha_t \hat{G}^t_{i,j} + \alpha_b \hat{G}^b_{i,j}}{\alpha_l + \alpha_r + \alpha_t + \alpha_b} \tag{2.34}$$

where $\hat{G}^l_{i,j}$ and $\hat{G}^r_{i,j}$ have been defined in Eq. (2.21), and $\hat{G}^t_{i,j}$ and $\hat{G}^b_{i,j}$ denote the top and bottom side parts of Eq. (2.20b), i.e.,

$$\begin{aligned}\hat{G}^t_{i,j} &= G_{i-1,j} + \frac{(R_{i,j} - R_{i-2,j})}{2}\\ \hat{G}^b_{i,j} &= G_{i+1,j} + \frac{(R_{i,j} - R_{i+2,j})}{2}\end{aligned} \tag{2.35}$$

The weights α_l, α_r, α_t and α_b are computed with the following equations:

$$\begin{aligned}\alpha_l = 1/(1 &+ |G_{i,j+1} - G_{i,j-1}| + |G_{i,j-1} - G_{i,j-3}| + |R_{i,j} - R_{i,j-2}|\\ &+ |G_{i-1,j} - G_{i-1,j-2}|/2 + |G_{i+1,j} - G_{i+1,j-2}|/2)\end{aligned} \tag{2.36}$$

$$\begin{aligned}\alpha_r = 1/(1 &+ |G_{i,j-1} - G_{i,j+1}| + |G_{i,j+1} - G_{i,j+3}| + |R_{i,j} - R_{i,j+2}|\\ &+ |G_{i-1,j} - G_{i-1,j+2}|/2 + |G_{i+1,j} - G_{i+1,j+2}|/2)\end{aligned} \tag{2.37}$$

$$\begin{aligned}\alpha_t = 1/(1 &+ |G_{i+1,j} - G_{i-1,j}| + |G_{i-1,j} - G_{i-3,j}| + |R_{i,j} - R_{i-2,j}|\\ &+ |G_{i,j-1} - G_{i-2,j-1}|/2 + |G_{i,j+1} - G_{i-2,j+1}|/2)\end{aligned} \tag{2.38}$$

$$\begin{aligned}\alpha_b = 1/(1 &+ |G_{i-1,j} - G_{i+1,j}| + |G_{i+1,j} - G_{i+3,j}| + |R_{i,j} - R_{i+2,j}|\\ &+ |G_{i,j-1} - G_{i+2,j-1}|/2 + |G_{i,j+1} - G_{i+2,j+1}|/2)\end{aligned} \tag{2.39}$$

where 1 in the denominator is included to avoid division by zero and the other parts of the denominator are the gradient magnitude along the respective direction. Figure 2.12 illustrates how to calculate the gradient magnitude of the top interpolation direction using a Sobel filter.

2.2.2 *Directional Interpolation and Decision Methods*

The edge-sensitive demosaicing methods select the interpolation direction or estimate the weights using the CFA samples, which may lead to wrong results since the CFA image contains less information than the full color image. Therefore, a novel strategy was proposed. This strategy computes two estimation candidates for each missing component and the decision for the best one is made a posteriori.

The approach presented by Hirakawa et al. [27] is representative. They first reconstructed the green planes G^h and G^v along horizontal and vertical directions with the approximated optimal FIR. Then the missing red and blue components were

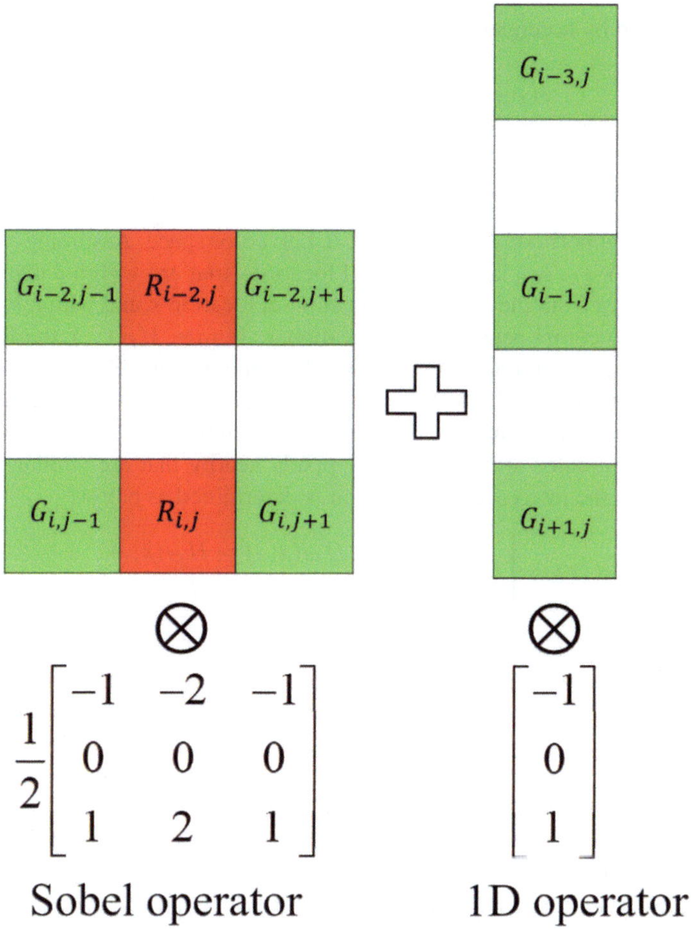

Fig. 2.12 Compute the gradient magnitude of the top interpolation direction

estimated in both directions with each of the green plane candidates. For example, the horizontally interpolated red plane R^h is calculated as:

$$R^h - G^h = L * (R_{CFA} - G^h_{CFA}) \tag{2.40}$$

where R_{CFA} denotes the known red samples in the CFA image and G^h_{CFA} is the sampled signal of G^h at the locations corresponding to the CFA red samples. Since the color difference plane $R^h - G^h$ is assumed to be piecewise constant, a 2-D low-pass filter L is applied on the sampled difference image to recover the full difference image. Now two full-color images, the horizontally interpolated image f^h and the vertically interpolated image f^v, have been generated. In order to combine f^h and f^v to obtain the final demosaiced image, Hirakawa proposed to select the interpolation result by comparing the level of color artifacts present in f^h and f^v, which can be evaluated by homogeneity values. The homogeneity H_f is defined as

$$H_f(x, \delta, \varepsilon_L, \varepsilon_C) = \frac{|U_f(x, \delta, \varepsilon_L, \varepsilon_C)|}{|B(x, \delta)|} \tag{2.41}$$

where

$$U_f(x, \delta, \varepsilon_L, \varepsilon_C) = B(x, \delta) \bigcap L_f(x, \varepsilon_L) \bigcap C_f(x, \varepsilon_C) \tag{2.42}$$

is the metric neighborhood measured in CIELab color space. Assuming x is a pixel location, $U_f(x, \delta, \varepsilon_L, \varepsilon_C)$ is a set of pixel locations that are within δ distance of x, within ε_L distance from the L component at pixel location x and within ε_C distance from the a and b components at pixel location x. If $x_0 \in U_f(x, \cdot)$, the RGB components of x_0 appear similar to those of x. The homogeneity H_f is a tool designed to analyze the behavior of U_f with respect to $B(x, \delta)$. Hirakawa found that regions with more color artifacts usually have smaller homogeneity values since a pixel marked by severe color artifacts has few pixels nearby that are similar. Based on this observation, the interpolation direction is determined as below:

$$f(x) = \begin{cases} f^h(x), & \text{if } H_{f^h}(x, \cdot) > H_{f^v}(x, \cdot) \\ f^v(x), & \text{if } H_{f^h}(x, \cdot) \leq H_{f^v}(x, \cdot) \end{cases} \tag{2.43}$$

where H_{f^h} and H_{f^v} are homogeneity maps calculated from f^h and f^v. Considering that frequently switching interpolation direction may introduce discontinuity in the output image, Hirakawa convolved the homogeneity maps with a spatial averaging kernel before using them for comparison. Figure 2.13 shows the homogeneity maps corresponding to the *lighthouse* image (brighter white means larger homogeneity value). Wu et al. [61] proposed a different posteriori selection criterion that performs a gradient test on the difference plane of the estimated full color images. In order to speed up the algorithm, Menon et al. [44] made the decision with the reconstructed green plane instead of a full color image and the criterion for selecting the interpolation direction takes into account the sum of gradients in a 5×5 neighborhood of the color difference plane.

Another approach attempts to fuse the horizontally and vertically interpolated images. Intuitively, this approach looks similar to the edge-weighted demosaicing algorithm, but there is still a difference between them. In the edge-weighted algorithm, the weights for each estimate candidate are computed from the CFA samples, while in this approach, the directionally reconstructed images are used for weights estimation. One work belonging to this category is proposed by Zhang et al. [62]. Regarding the poorly interpolated image as a noisy signal, they refined the reconstruction of the color difference planes in the horizontal and vertical directions with linear minimum mean square-error estimation (LMMSE) technique and fused two LMMSE estimates by minimizing the MSE of the final fused signal. Let $d(x)$, $\hat{d}_h(x)$ and $\hat{d}_v(x)$ represent the ground truth, the horizontally estimated, and the vertically estimated difference planes (for example, $G - R$) respectively and x is the position index of the pixels. The fused signal $\hat{d}_w(x)$ is defined as:

$$\hat{d}_w(x) = w_h(x) \cdot \hat{d}_h(x) + w_v(x) \cdot \hat{d}_v(x) \tag{2.44}$$

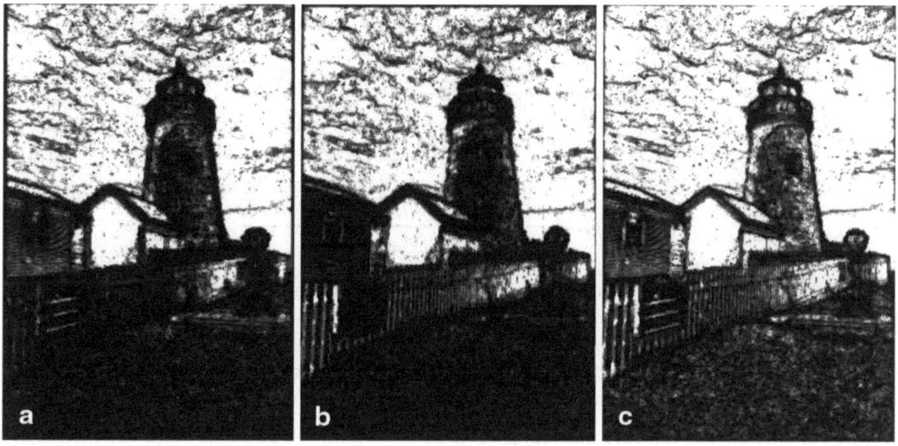

Fig. 2.13 Homogeneity maps of the reconstructed *lighthouse* image. **a** Horizontally interpolated image **b** Vertically interpolated image **c** Final output [27]

where $w_h(x)$ and $w_v(x)$ are weights and satisfy $w_h(x) + w_v(x) = 1$. In order to optimize the weights, the equation

$$\sigma^2_{\hat{d}_w} = E[(\hat{d}_w(x) - d(x))^2] \tag{2.45}$$

is minimized by using the signal statistical properties and setting the partial derivative to zero. Among other post-fusion methods, Menon et al. [45] proposed to fuse estimates by a suitable wavelet-based edge estimation on the luminance component and Paily et al. [52] adopted the local polynomial approximation (LPA) and the paradigm of the intersection of confidence intervals (ICI) to fuse two directional interpolations.

2.2.3 *Frequency-Domain Approaches*

Some recent demosaicing approaches rely on a frequency-domain analysis. The fundamental principle is to interpret the frequency representation of a Bayer CFA image as a combination of a luminance signal at base band and two chrominance signals modulated at high spatial frequencies. Appropriately designed filters are then applied on the CFA image to extract the luminance and the chrominances which are a linear transformation of the RGB components in the spatial domain.

As shown in Fig. 2.2, a Bayer CFA image I^{CFA} can be decomposed into three subsampled color planes $I_k^{CFA}, k \in \{R, G, B\}$. The sampling operation for each color plane can be expressed via multiplication by modulation function m_k:

$$
\begin{aligned}
m_R(i,j) &= \frac{(1+\cos(\pi i))(1+\cos(\pi j))}{4} \\
m_G(i,j) &= \frac{(1-\cos(\pi i)\cos(\pi j))}{2} \\
m_B(i,j) &= \frac{(1-\cos(\pi i))(1-\cos(\pi j))}{4}
\end{aligned}
\tag{2.46}
$$

Taking $m_R(i,j)$ for example, its Fourier transform denoted as $\hat{m}_R(u,v)$ is given by

$$
\begin{aligned}
\hat{m}_R(u,v) &= \frac{(\delta(u)+\frac{1}{2}(\delta(u+\frac{1}{2})+\delta(u-\frac{1}{2})))(\delta(v)+\frac{1}{2}(\delta(v+\frac{1}{2})+\delta(v-\frac{1}{2})))}{4} \\
&= \Delta(u)^T \begin{bmatrix} \frac{1}{16} & \frac{1}{8} & \frac{1}{16} \\ \frac{1}{8} & \frac{1}{4} & \frac{1}{8} \\ \frac{1}{16} & \frac{1}{8} & \frac{1}{16} \end{bmatrix} \Delta(v)
\end{aligned}
\tag{2.47}
$$

where $\Delta(u) = \left[\delta(u+\frac{1}{2}),\ \delta(u),\ \delta(u-\frac{1}{2})\right]^T$ and so is $\Delta(v)$. Similar expressions can be found for $m_G(i,j)$ and $m_B(i,j)$. Given the Fourier transform of the full-resolution color component $F_k(u,v), k \in R, G, B$, the Fourier transform of the subsampled color component $F_k^{CFA}(u,v)$ is obtained by convolving $\hat{m}_k(u,v)$ with $F_k(u,v)$, that is

$$
\begin{aligned}
F_R^{CFA}(u,v) &= \left(\Delta(u)^T \begin{bmatrix} \frac{1}{16} & \frac{1}{8} & \frac{1}{16} \\ \frac{1}{8} & \frac{1}{4} & \frac{1}{8} \\ \frac{1}{16} & \frac{1}{8} & \frac{1}{16} \end{bmatrix} \Delta(v) \right) * F_R(u,v) \\
F_G^{CFA}(u,v) &= \left(\Delta(u)^T \begin{bmatrix} -\frac{1}{8} & 0 & -\frac{1}{8} \\ 0 & \frac{1}{2} & 0 \\ -\frac{1}{8} & 0 & -\frac{1}{8} \end{bmatrix} \Delta(v) \right) * F_G(u,v) \\
F_B^{CFA}(u,v) &= \left(\Delta(u)^T \begin{bmatrix} \frac{1}{16} & -\frac{1}{8} & \frac{1}{16} \\ -\frac{1}{8} & \frac{1}{4} & -\frac{1}{8} \\ \frac{1}{16} & -\frac{1}{8} & \frac{1}{16} \end{bmatrix} \Delta(v) \right) * F_B(u,v)
\end{aligned}
\tag{2.48}
$$

These equations show that the Fourier transforms of the subsampled components are scaled, periodic replications of the Fourier transforms of the full-resolution components. For the green component, the replication takes place only along the diagonal directions; for the red and blue components, the replication is also in the horizontal and vertical directions, as shown in Fig. 2.14. This means the spectrum of the subsampled green component contains less aliasing than the other components. Glotzbach et al. [22] proposed to estimate the green component using a diamond

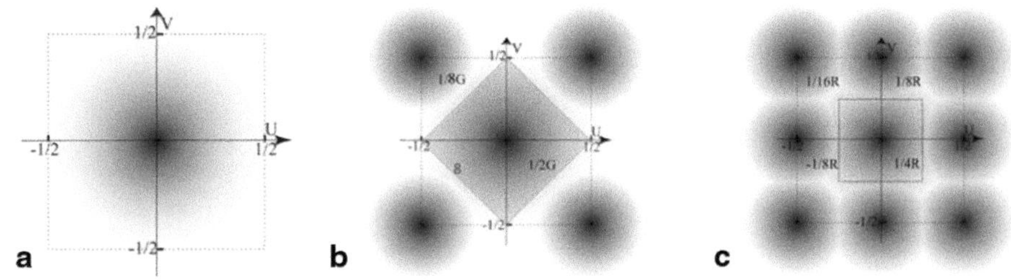

Fig. 2.14 Fourier transforms of **a** Full-resolution component, **b** Subsampled green component, **c** Subsampled red/blue component [34]

shape 2-D filter (Fig. 2.14b), while the red and blue components are computed using a rectangular filter (Fig. 2.14c). Since the high-frequency content in different color planes is strongly correlated, the high-frequency information of the green plane can be used to improve the reconstruction of the red and blue planes.

Consistent with the conclusion in the spatial domain, Alleysson et al. [7] and Lian et al. [34] observed that filtering a CFA image could better preserve high frequencies than filtering each color component separately. The Fourier transform of the CFA image $F^{CFA}(u, v)$ can be easily calculated using Eq. (2.48),

$$\begin{aligned} F^{CFA}(u, v) &= F_R^{CFA}(u, v) + F_G^{CFA}(u, v) + F_B^{CFA}(u, v) \\ &= L(u, v) + C_1(u, v) + C_2(u, v) \end{aligned} \tag{2.49}$$

where

$$\begin{aligned} L(u, v) &= \frac{1}{4}[2F_G + F_R + F_B](u, v) \\ C_1(u, v) &= \frac{1}{8}\sum_{s=-1,1}[F_R - F_B](u - \frac{s}{2}, v) + \frac{1}{8}\sum_{t=-1,1}[F_R - F_B](u, v - \frac{t}{2}) \\ C_2(u, v) &= \frac{1}{16}\sum_{s=-1,1}\sum_{t=-1,1}[F_R - 2F_G + F_B](u - \frac{s}{2}, v - \frac{t}{2}) \end{aligned} \tag{2.50}$$

$L(u, v)$ is the luminance component at the center of the spectrum, while $C_1(u, v)$ and $C_2(u, v)$ are chrominance components shifted to the sides and corners respectively. Figure 2.15 shows the energy distribution in the CFA spectrum. Since the color difference components C_1 and C_2 only contain low frequencies, they have a more compact spectrum that has less overlap with the luminance spectrum, as compared with the overlap between spectrum replicas in the subsampled components. Considering this, Alleysson et al. [7] proposed to estimate the full-resolution luminance component first, by low-pass filtering the CFA image. The design of this filter should balance between preserving the high-frequency information in the luminance component while reducing the aliased part as much as possible. Figure 2.16a is the filter designed by Alleysson, where the two stop bandwidths r_1

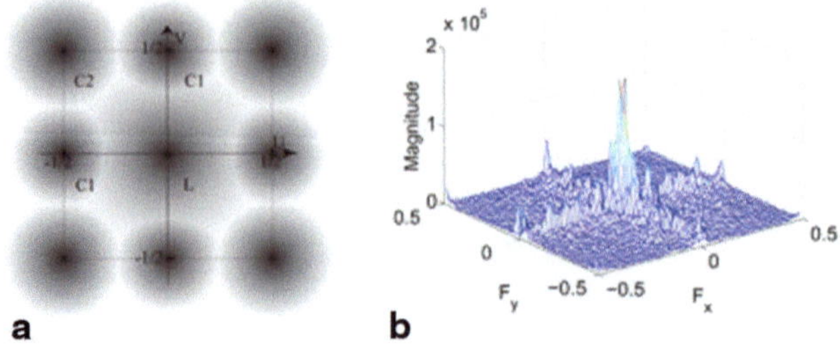

Fig. 2.15 **a** Fourier transform **b** Fourier spectrum of a CFA image [34]

and r_2 used to remove chrominance components C_1 and C_2 are estimated by optimizing CPSNR for four natural images. The chrominance component, obtained by subtracting the estimated luminance component from the CFA image, is the sum of three subsampled and modulated opponent color images $R-L$, $G-L$, and $B-L$. Alleysson demultiplexed it by multiplying the modulation functions $m_k(i,j)$ and then performed bilinear interpolation on each subsampled opponent signal. Finally the reconstructed opponent components are added to the luminance component to recover the RGB color planes.

However, Lian et al. [34] thought that the preceding filter for luminance estimation may not be optimal since this filter removes the high-frequency information along horizontal and vertical directions, which has a larger impact on the demosaiced image quality. Considering the phenomenon that the frequency components modulated at $(0, \pm\frac{1}{2})$ and $(\pm\frac{1}{2}, 0)$ of the subsampled green plane vanish, Lian et al. proposed to preserve this information through estimating the luminance component at green pixel locations. They designed a new low-pass filter shown in Fig. 2.16b to remove the corner components C_2 from the CFA image spectrum. The resultant image whose spectrum is denoted as $\hat{F}^{CFA}(u,v)$ is then subsampled at green pixel locations. As shown in Eq. (2.51), the side components in the horizontal and vertical directions will overlap and cancel out each other. In this case, the obtained luminance component at green pixel locations bears less aliasing while preserving the important high-frequency information along the horizontal and vertical axes.

$$\hat{F}^{CFA}(u,v)*\hat{m}_G(u,v) = \frac{1}{4}[2F_G+F_R+F_B](u,v)*\hat{m}_G(u,v) = L(u,v)*\hat{m}_G(u,v) \tag{2.51}$$

Nevertheless, the same method is not suitable for estimating the luminance at red/blue pixel locations. Lian et al. proposed to populate the estimated subsampled version of the luminance component by alternately interpolating the red/blue plane and the luminance plane with an adaptive edge-weighting strategy. With the fully populated luminance as a reference, the color components R, G, B can be estimated using bilinear interpolation.

Dubois [17] also noticed the significant spectral overlap of the luminance L and the modulated chrominance C_1 in Alleysson's work. By observing that the C_1 components modulated at $(0, \pm\frac{1}{2})$ and $(\pm\frac{1}{2}, 0)$ have different spectral overlap with the

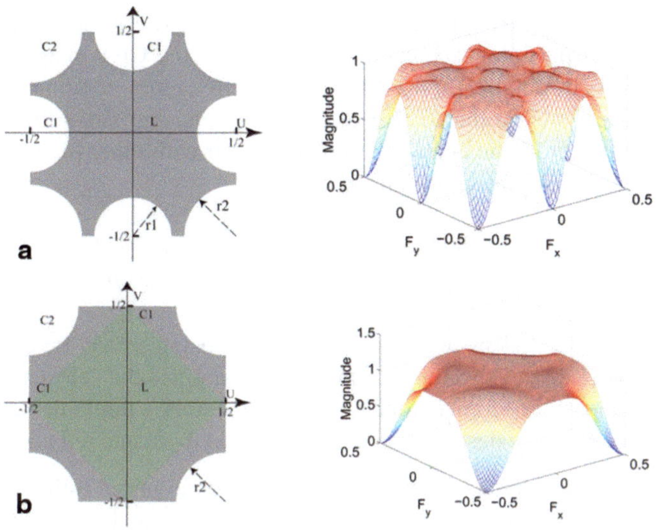

Fig. 2.16 Filters for luminance estimation **a** Alleysson et al. **b** Lian et al. [35]

luminance component, he proposed to use asymmetric filters to extract the vertical frequency component from C_1 at $(\pm\frac{1}{2}, 0)$ and the horizontal frequency component from C_1 at $(0, \pm\frac{1}{2})$ and combine them so as to recover the entire passband of C_1. The luminance component is finally obtained by subtracting C_1 from the CFA image spectrum in which the C_2 components have been removed.

2.2.4 *Wavelet-Based Methods*

The 2-D discrete wavelet transform (DWT) constructed from a low-pass filter and a high-pass filter are adopted by some demosaicing methods to decompose the image into four subbands: (LL) both rows and columns are low-pass filtered, (LH) rows are low-pass filtered, columns are high-pass filtered, (HL) rows are high-pass filtered, columns are low-pass filtered and (HH) both rows and columns are high-pass filtered. It has been proved that the high-frequency subbands (LH, HL, HH) in different color channels are strongly correlated, so the information provided by subbands in one color plane can be exploited to recover subbands in a different color plane. This is the principle behind the famous "projection-onto-convex-set" (POCS) algorithm proposed by Gunturk et al. [24]. The POCS algorithm refines the red and blue components iteratively via alternating the projection onto two convex constraint sets. The first constraint set is "observation" that enforces the interpolated image to be consistent with the observed data,

$$C_o = \{\hat{I}_k(i,j) : \hat{I}_k(i,j) = I_k^{CFA}(i,j), \forall (i,j) \in \Lambda_k, k = R, G, B\} \qquad (2.52)$$

where $\hat{I}_k(i, j)$ and $I_k^{CFA}(i, j)$ are defined as before, and Λ_k represents the set of pixel locations where color channel k is observed in the CFA image. The second constraint set called "detail" imposes that the high-frequency components of the red and blue channels are similar to that of the green channel since the spectrum of the subsampled green plane is less aliased compared with other planes' spectrums. Specifically, this constraint set is defined as:

$$C_d = \left\{ \begin{array}{c} \hat{I}_k(i, j) : |(W_m \hat{I}_k)(i, j) - (W_m \hat{I}_G)(i, j)| \leq T(i, j) \\ \forall (i, j), m = 2, 3, 4 \text{ and } k = R, B \end{array} \right\} \tag{2.53}$$

where $W_m, m = 1, 2, 3, 4$ are analysis filters decomposing each color plane into four subbands, and $T(i, j)$ denotes the threshold that quantifies the "closeness" between the detail subbands of different color channels. In Gunturk's implementation, $T(i, j)$ is set to be zero. Now we describe the technique used to update red and blue planes.

1. interpolate the red, green, and blue channels to obtain initial estimates
2. decompose all three channels into four subbands using analysis filters and update the high-frequency subbands of red and blue by the projection onto the "detail" constraint set.
3. reconstruct the red and blue channels with the synthesis filters $U_m, m = 1, 2, 3, 4$, which satisfy the following condition:

$$\hat{I}_k(i, j) = U_1(W_1 \hat{I}_k)(i, j) + U_2(W_2 \hat{I}_k)(i, j) + U_3(W_3 \hat{I}_k)(i, j) + U_4(W_4 \hat{I}_k)(i, j) \tag{2.54}$$

4. perform a projection onto the "observation" constraint set by inserting the observed data in their corresponding locations $\Lambda_k, k = R, B$
5. go to step 2, and repeat the procedure several times (eight suggested)

The POCS algorithm could achieve excellent results and is often considered as a reference in demosaicing benchmarks. However, its computation cost is rather high, and the performance depends on the quality of initial estimates.

At the beginning, Gunturk thought about using bilinear interpolation for the red and blue channels, and the edge-directed interpolation method [5] for the green channel to get the initial estimates. However, the result was not satisfying. In order to further improve the demosaicing quality, a method similar to the red and blue updating technique is adopted for updating the green plane. This method is based on an observation that there exists a strong correlation between the high-frequency subbands in different downsampled color planes, for instance, red and green [12]. The downsampled red plane is formed using the observed red samples in the CFA image while the downsampled green plane using the interpolated green samples at red pixel locations, respectively shown in Figs. 2.17a and 2.17b. As was projection onto the "detail" convex set, the high-frequency subbands of the downsampled green plane is replaced by those of the downsampled red plane to update the green samples at red pixel locations. The same procedure is repeated for the downsampled blue plane to update the green samples at blue pixel locations. Chen et al. [12] utilized this method to update all three channels initially estimated by bilinear interpolation.

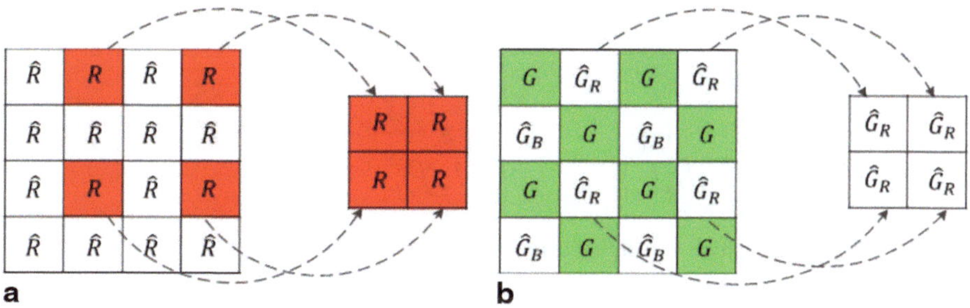

Fig. 2.17 **a** Extracting downsampled red plane using the observed red samples. **b** Extracting downsampled green plane using the interpolated green samples

With regard to the high demand of computation in the POCS algorithm, Lu et al. [37] provided a rigorous analysis of the convergence property of this algorithm and presented a method that can obtain the same results about eight times faster.

Another wavelet-based work is proposed by Driesen et al. [16]. They merged the detail coefficients of the wavelet-transformed luminance image into the wavelet-transformed color planes, in order to enhance the spatial resolution of the latter and at the same time preserve the color information. We also put the successive approximation (SA) method proposed by Li [33] in this category due to its similarity to the POCS algorithm, though his method doesn't use any wavelet transform or decomposition. The SA method updates the green plane and the red/ blue plane in color difference domain, so the constraint set based on the color difference rule is linear and the filtering operation is no longer needed, which largely reduces the computation time. In addition, a spatially adaptive stopping criterion is designed to suppress the artifacts caused by insufficiently or overly enforcing the color difference rule.

2.2.5 *Statistical Reconstruction Techniques*

The CFA image demosaicing is an image reconstruction problem, so the statistical technique widely used in the image reconstruction could be used to solve the demosaicing problem. Before we dive deep into the algorithms in this category, the model for the CFA image acquisition process is introduced first.

As shown in Fig. 2.18, the acquisition process in a digital camera is more complex than the ideal sampling. To begin with, each color component $I_k(i, j), k = R, G, B$ of the original scene projection $I(i, j), i = 1 \dots M, j = 1 \dots N$ is convolved with a low-pass filter $h_k(\cdot)$, which may represent the optical or the motion point spread function (PSF). Then it is sampled according to the CFA arrangements via multiplication by the modulation function $m_k(i, j)$. Finally, the CFA image $I^{CFA}(i, j)$ is recorded with additive noise $n(i, j)$. The relation between $I^{CFA}(i, j)$ and $I(i, j)$ is summarized in the following equation:

$$I^{CFA}(i, j) = \sum_{k=R,G,B} m_k(i, j)h_k * I_k(i, j) + n(i, j) \tag{2.55}$$

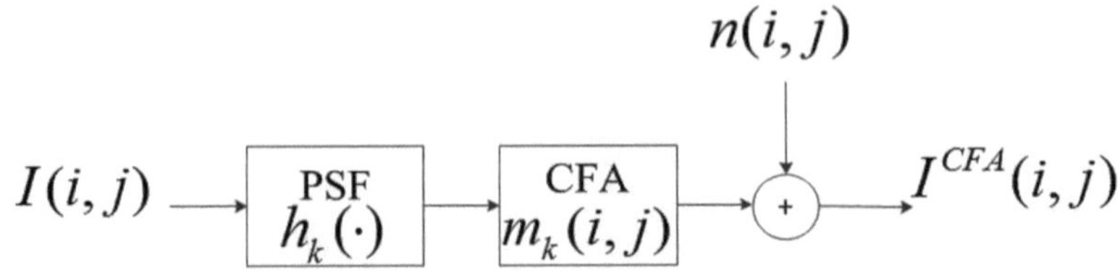

Fig. 2.18 Image acquisition in a digital camera

In the context of demosaicing, the impulse response of $h_k(\cdot)$ is usually assumed to be a Dirac function. The modulation functions $m_k(i, j)$ are periodic and satisfy $m_R(i, j) + m_G(i, j) + m_B(i, j) = 1$. If the CFA pattern is Bayer, $m_k(i, j)$ are orthogonal sampling functions taking 1 or 0 at each location (i, j), as defined in Sect. 2.2.3.

Generally, the above formulation is converted into a matrix-vector-multiplication form for easier computation. Let $\boldsymbol{r}, \boldsymbol{g}, \boldsymbol{b} \in \mathbb{R}^{MN\times 1}$ denote the lexicographically ordered vectors of the three full-resolution color components $I_k(i, j), k = R, G, B$, with the entire original image given by $\boldsymbol{i} = \left[\boldsymbol{r}^T, \boldsymbol{g}^T, \boldsymbol{b}^T\right]^T \in \mathbb{R}^{3MN\times 1}$. Similarly, $\boldsymbol{i}^{CFA} \in \mathbb{R}^{MN\times 1}$ and $\boldsymbol{n} \in \mathbb{R}^{MN\times 1}$ are defined as the stacking vectors of the CFA image $I^{CFA}(i, j)$ and the noise $n(i, j)$, respectively. Then we rewrite Eq. (2.55) as:

$$\boldsymbol{i}^{CFA} = \boldsymbol{C}\boldsymbol{i} + \boldsymbol{n} \tag{2.56}$$

where the matrix $\boldsymbol{C} \in \mathbb{R}^{MN\times 3MN}$ is given by

$$\boldsymbol{C} = \left[\boldsymbol{M}_R\boldsymbol{H}_R, \boldsymbol{M}_G\boldsymbol{H}_G, \boldsymbol{M}_B\boldsymbol{H}_B\right] \tag{2.57}$$

with the square matrices $\mathbf{H}_k \in \mathbb{R}^{MN\times MN}$ accounting for the filters $h_k(\cdot)$ and the square matrices $\mathbf{M}_k \in \mathbb{R}^{MN\times MN}$ consisting of the entries given by the modulation functions $m_k(i, j)$. In this form, demosaicing is a linear inverse problem, where the latent image $\boldsymbol{i}$ is estimated given the observed data $\boldsymbol{i}^{CFA}$ and the linear matrix $\mathbf{C}$.

One common solution to this inverse problem is the maximum a posteriori (MAP) estimation,

$$\hat{\boldsymbol{i}} = \underset{\boldsymbol{i}}{\text{argmax}}\; p(\boldsymbol{i}|\boldsymbol{i}^{CFA}) \tag{2.58}$$

where $p(\boldsymbol{i}|\boldsymbol{i}^{CFA})$ is the posteriori distribution specifying the conditional probability of $\boldsymbol{i}$ given $\boldsymbol{i}^{CFA}$. Using the Bayesian rule, the MAP estimate becomes:

$$\hat{\boldsymbol{i}} = \underset{\boldsymbol{i}}{\text{argmax}}\; p(\boldsymbol{i}^{CFA}|\boldsymbol{i})p(\boldsymbol{i}) \tag{2.59}$$

in which $p(\boldsymbol{i}^{CFA}|\boldsymbol{i})$ is the likelihood term and $p(\boldsymbol{i})$ represents the subjective priori knowledge about the original color image. Most of the time the noise $\boldsymbol{n}$ is assumed to be i.i.d. Gaussian with variance σ^2 and uncorrelated with $\boldsymbol{i}$, so the likelihood term is proportional to $e^{-\frac{1}{2\sigma^2}\|\boldsymbol{i}^{CFA}-\boldsymbol{C}\boldsymbol{i}\|^2}$. With regard to the priori information, Brainard [9] expressed the image $\boldsymbol{i}$ as a weighted sum of basis images that are chosen to be spatial sinusoids in each color band, and assumed that the weights follow a multivariate Gaussian distribution. Mukherjee et al. [48] adopted Markov random fields (MRF) for modeling the priori probability.

Since most probability distributions in Eq. (2.59) are natural exponential functions, we can apply the negative logarithm on it and the objective function becomes the form of a data-fidelity term $\Psi(\boldsymbol{i}, \boldsymbol{i}^{CFA})$ plus several regularization terms $J_s(\boldsymbol{i})$ weighted by λ_s

$$\hat{\boldsymbol{i}} = \underset{\boldsymbol{i}}{\operatorname{argmin}} \left\{ \Psi(\boldsymbol{i}, \boldsymbol{i}^{CFA}) + \sum_s \lambda_s J_s(\boldsymbol{i}) \right\} \tag{2.60}$$

Among different regularization approaches [28, 42, 50, 55], the one proposed by Menon et al. [46] is typical. They defined the data-fidelity term according to the least-squares approaches and considered the characteristics of the sensor noise $\boldsymbol{R_n}$,

$$\Psi(\boldsymbol{i}, \boldsymbol{i}^{CFA}) = \|\boldsymbol{i}^{CFA} - \boldsymbol{C}\boldsymbol{i}\|^2_{\boldsymbol{R}_n^{-1}} \tag{2.61}$$

where $\boldsymbol{R_n}$ denotes the autocorrelation matrix of the noise $\boldsymbol{n}$. The regularization terms are included in quadratic forms $J_s(\boldsymbol{i}) = \|\boldsymbol{G}_s\boldsymbol{i}\|^2$, which allows to find solutions with few computations. The first regularization constraint $J_1(\boldsymbol{i})$ is chosen in order to impose smoothness on each single color component, that is $\boldsymbol{G}_1 = \boldsymbol{I}_3 \otimes \boldsymbol{S}_1$, where $\boldsymbol{I}_3$ is the 3×3 identity matrix, $\otimes$ denotes the Kronecker operator and $\boldsymbol{S}_1$ represents a two-dimensional high-pass filter. The second constraint forces smoothness on the color difference components. This is consistent with the spectral correlation that requires a high correlation between the high frequencies of the three color bands, so $J_2(\boldsymbol{i})$ is defined as:

$$J_2(\boldsymbol{i}) = \|\boldsymbol{S}_2\boldsymbol{r} - \boldsymbol{S}_2\boldsymbol{g}\|^2 + \|\boldsymbol{S}_2\boldsymbol{g} - \boldsymbol{S}_2\boldsymbol{b}\|^2 + \|\boldsymbol{S}_2\boldsymbol{r} - \boldsymbol{S}_2\boldsymbol{b}\|^2 \tag{2.62}$$

with $\boldsymbol{S}_2$ representing a different high-pass filter. This constraint can also be expressed in the quadratic form $J_2(\boldsymbol{i}) = \|\boldsymbol{G}_2\boldsymbol{i}\|^2$, where

$$\boldsymbol{G}_2 = \begin{bmatrix} 1.547 & -0.577 & -0.577 \\ -0.577 & 1.547 & -0.577 \\ -0.577 & -0.577 & 1.547 \end{bmatrix} \otimes \boldsymbol{S}_2 \tag{2.63}$$

$J_1(\boldsymbol{i})$ in combination with $J_2(\boldsymbol{i})$ is enough to provide a solution to the demosaicing problem. However, only utilizing these global constraint terms may result in artifacts near the edges and details in a natural image, where adaptive strategies are usually found to give better results. As a result, Menon proposed an additional constraint term in a nonquadratic form to include the local adaptivity,

$$J_3(\boldsymbol{i}) = \|\boldsymbol{G}_3\boldsymbol{i}\|^2_{\boldsymbol{W}_i} \tag{2.64}$$

where $\boldsymbol{G}_3$ is set to be equivalent to $\boldsymbol{G}_1$, and $\boldsymbol{W_i}$ is a diagonal matrix estimated from the image $\boldsymbol{i}$ for adapting the penalty term to the local features of the image. If this regularization term is considered together with the previously defined quadratic penalties $J_1(\mathbf{i})$ and $J_2(\mathbf{i})$, the solution to Eq. (2.60) is found by solving:

$$(\boldsymbol{C}^T\boldsymbol{R}_n^{-1}\boldsymbol{C} + \lambda_1\boldsymbol{G}_1^T\boldsymbol{G}_1 + \lambda_2\boldsymbol{G}_2^T\boldsymbol{G}_2 + \lambda_3\boldsymbol{G}_3^T\boldsymbol{W}_i\boldsymbol{G}_3)\boldsymbol{i} - \boldsymbol{C}^T\boldsymbol{R}_n^{-1}\boldsymbol{i}^{CFA} = 0 \quad (2.65)$$

Since the matrix $\boldsymbol{W}_i$ depends on $\boldsymbol{i}$, which increases the difficulty to solve the above equation, an initial estimate $\boldsymbol{i}_0$ is used such that the value $\boldsymbol{G}_3^T\boldsymbol{W}_{i_0}\boldsymbol{G}_3\boldsymbol{i}_0$ approximates $\boldsymbol{G}_3^T\boldsymbol{W}_i\boldsymbol{G}_3\boldsymbol{i}$. Therefore, the final estimate is

$$\hat{\boldsymbol{i}} = (\boldsymbol{C}^T\boldsymbol{R}_n^{-1}\boldsymbol{C} + \lambda_1\boldsymbol{G}_1^T\boldsymbol{G}_1 + \lambda_2\boldsymbol{G}_2^T\boldsymbol{G}_2)^{-1}(\boldsymbol{C}^T\boldsymbol{R}_n^{-1}\boldsymbol{i}^{CFA} - \lambda_3\boldsymbol{G}_3^T\boldsymbol{W}_{i_0}\boldsymbol{G}_3\boldsymbol{i}_0) \quad (2.66)$$

In fact, Menon's work is not limited to the Bayer CFA image demosaicing. The proposed approach is applicable with any type of CFA arrangement and with sensors having non-ideal impulse response. Moreover, the regularization-based strategy allows to couple demosaicing with other frequent problems in image reconstruction and restoration.

2.2.6 Experimental Results

In this subsection, we evaluate the performance of 12 demosaicing algorithms by analyzing the CPSNR and the S-CIELab results over the Kodak benchmark data set. The selected demosaicing algorithms are: (1) heterogeneity-projection hard-decision adaptive interpolation (HPHD-AI) [59]; (2) variance of color differences (VCD) [13]; (3) adaptive homogeneity-directed (AHD) [27]; (4) directional linear minimum mean square-error estimation (DLMMSE) [62]; (5) directional filtering and a posteriori decision (DFAD) [44]; (6) local polynomial approximation (LPA) [52]; (7) alternating projections (AP or POCS) [24]; (8) Lu & Tan's method (LT) [36]; (9) successive approximation (SA) [33]; (10) Alleysson's method (FD) [7]; (11) complementary asymmetric filters with adaptive weights (AFAW) [17]; (12) regularization approach to demosaicing (RAD) [46]. The code or software of each algorithm is provided by the authors. Since some algorithms don't take care of the image boundary pixels, the outermost 16-pixel border is removed when we perform the quality assessment.

Table 2.1 reports the CPSNR values for each Kodak image, as well as the average CPSNR over the 24 images. It is easy to see that performance varies from one image to another for a given algorithm. This is consistent with the fact that the demosaiced quality depends on whether the image contains a lot of high-frequency areas. For example, the CPSNR value of image 23 is larger than those of other images since it contains many homogeneous regions. In addition, the highest CPSNR in each row is highlighted in bold. Obviously, the LPA algorithm [52] outperforms others by noticeable margin on 17 test images, achieving the best average performance. Good performance is also achieved by the VCD algorithm [13] and the DLMMSE algorithm [62]. In contrast, the average CPSNR values of the AHD [27] and FD [7] algorithms are around 2.5 dB lower than the best average value.

Table 2.1 CPSNR results for different demosaicing algorithms

	Method											
Image	[59]	[13]	[27]	[62]	[44]	[52]	[24]	[36]	[33]	[7]	[17]	[46]
1	36.09	38.55	35.05	39.67	36.83	**40.42**	37.75	36.37	38.37	36.38	38.07	38.23
2	40.25	40.38	39.08	**41.51**	40.38	41.30	39.60	40.39	39.97	38.25	39.97	39.80
3	42.10	42.60	41.00	42.35	42.00	**43.37**	41.50	42.15	41.26	39.85	41.55	42.00
4	39.81	40.53	38.54	40.65	39.82	**40.78**	40.07	40.33	39.59	38.99	40.55	40.61
5	36.54	37.89	35.29	37.68	37.28	37.48	37.46	37.33	36.52	35.90	37.79	**37.94**
6	38.16	40.01	37.43	40.51	39.12	**40.87**	38.51	37.12	39.14	37.67	39.96	39.76
7	41.61	42.22	40.40	41.89	41.70	**42.98**	41.70	42.63	41.53	39.95	42.08	42.39
8	34.01	36.36	33.66	36.76	35.27	**37.10**	35.23	34.36	35.90	32.53	35.16	35.97
9	42.04	43.04	40.90	42.77	42.46	**43.41**	41.75	41.77	42.01	39.86	41.97	42.29
10	41.60	42.49	40.42	42.55	42.06	**42.71**	42.03	41.87	41.79	41.03	42.22	42.61
11	38.31	39.88	37.34	40.25	39.07	**40.46**	39.16	38.51	38.98	37.89	39.67	39.63
12	42.45	43.48	41.49	43.81	42.91	**43.90**	42.55	42.07	42.44	41.06	42.99	43.11
13	32.31	34.87	31.24	35.49	33.12	**36.06**	34.31	32.50	34.94	34.88	35.02	34.79
14	36.36	36.92	35.06	36.46	36.46	36.82	35.67	**37.12**	34.63	33.49	35.72	35.97
15	39.32	39.79	37.86	39.99	39.20	**40.04**	39.33	39.12	38.99	38.62	39.63	39.85
16	41.86	43.66	41.25	**43.96**	42.82	43.91	41.72	40.34	41.67	40.46	43.59	43.17
17	40.33	41.36	39.06	**41.80**	40.63	41.68	41.34	40.59	41.05	41.01	41.47	41.55
18	36.32	37.32	34.79	**37.66**	36.61	37.60	37.46	36.62	37.18	36.92	37.54	37.64
19	39.54	41.00	38.21	41.14	39.91	**41.50**	39.71	39.36	39.99	37.46	40.31	40.14
20	39.97	41.15	38.99	41.08	40.21	**41.44**	40.64	40.67	40.54	39.60	40.35	40.97
21	37.20	39.16	36.41	39.54	37.76	**39.57**	38.85	37.74	38.98	37.75	38.66	39.16
22	37.67	38.11	36.16	38.18	37.48	**38.40**	37.73	38.18	37.74	36.76	38.00	38.20
23	42.62	42.97	41.46	42.85	42.44	**43.82**	41.80	42.74	41.85	39.72	42.06	42.29
24	34.59	35.20	33.05	35.56	34.39	35.44	34.72	34.26	34.82	35.04	35.43	**35.61**
Avg.	38.79	39.95	37.67	40.17	39.16	**40.46**	39.19	38.92	39.16	37.96	39.57	39.74

Similar conclusions can be obtained when comparing the performance of various algorithms in terms of S-CIELab metric, as shown in Table 2.2. The bold numbers indicate the lowest S-CIELab value for each image. However, we notice that the different criteria provide different performance rankings for some algorithms. For instance, the HPHD-AI algorithm [59] falls behind the SA algorithm [33] on average CPSNR values, while the former achieves a better S-CIELab performance than the latter.

A more effective evaluation is given by the visual inspection of the reconstructed images. Figure 2.19a shows the fence bars of test image 20, which contains many fine detail features, largely challenging the performance of demosaicing algorithms.

Table 2.2 S-CIELab results for different demosaicing algorithms

	Method											
Image	[59]	[13]	[27]	[62]	[44]	[52]	[24]	[36]	[33]	[7]	[17]	[46]
1	1.21	1.11	1.25	1.01	1.29	**0.94**	1.25	1.26	1.25	1.40	1.17	1.21
2	0.68	0.69	0.81	**0.57**	0.68	0.62	0.65	0.73	0.71	0.86	0.68	0.72
3	0.48	0.50	0.58	0.51	0.54	**0.46**	0.55	0.51	0.59	0.65	0.54	0.55
4	0.76	0.77	0.92	0.71	0.82	0.73	0.74	0.78	0.85	0.84	**0.70**	0.76
5	**1.04**	1.05	1.32	1.22	1.15	1.31	1.14	1.05	1.38	1.34	1.06	1.13
6	0.81	0.81	0.87	0.78	0.88	**0.77**	0.96	1.00	0.93	1.02	0.79	0.86
7	0.54	0.59	0.69	0.63	0.63	**0.52**	0.64	0.55	0.68	0.75	0.54	0.58
8	1.36	1.28	1.44	1.29	1.48	**1.23**	1.55	1.50	1.53	1.94	1.50	1.41
9	0.61	0.58	0.63	0.58	0.60	**0.55**	0.65	0.64	0.65	0.74	0.60	0.60
10	0.60	0.58	0.63	0.56	0.59	0.56	0.59	0.61	0.62	0.65	**0.55**	0.56
11	0.77	0.76	0.86	0.76	0.81	**0.73**	0.85	0.84	0.86	0.95	0.77	0.81
12	0.48	0.48	0.52	0.46	0.53	**0.45**	0.53	0.53	0.54	0.59	0.47	0.52
13	1.79	1.61	1.92	**1.52**	1.92	1.56	1.77	1.83	1.72	1.55	1.59	1.64
14	**0.96**	1.02	1.22	1.06	1.09	1.00	1.16	1.00	1.32	1.42	1.10	1.10
15	0.68	0.66	0.79	**0.63**	0.74	0.64	0.64	0.70	0.73	0.74	0.66	0.69
16	0.58	0.56	0.59	0.53	0.60	**0.53**	0.69	0.73	0.68	0.75	0.54	0.60
17	0.59	0.58	0.63	0.54	0.62	0.55	0.57	0.60	0.60	0.56	**0.53**	0.54
18	1.05	1.05	1.26	1.03	1.12	1.07	1.06	1.06	1.16	1.11	1.01	**1.00**
19	0.80	0.76	0.87	0.75	0.83	**0.71**	0.84	0.85	0.83	0.97	0.79	0.79
20	0.56	0.56	0.64	0.56	0.69	**0.53**	0.59	0.57	0.60	0.64	0.60	0.59
21	0.94	0.91	1.02	0.89	1.05	**0.89**	0.94	0.97	0.98	1.03	0.94	0.95
22	0.97	1.03	1.23	1.05	1.08	1.00	1.07	0.99	1.10	1.15	**0.95**	0.99
23	0.51	0.53	0.62	0.54	0.56	**0.47**	0.57	0.53	0.60	0.68	0.51	0.54
24	1.04	1.05	1.27	1.07	1.20	1.11	1.16	1.07	1.21	1.12	**1.00**	1.05
Avg.	0.82	0.81	0.94	0.80	0.90	**0.79**	0.88	0.87	0.92	0.98	0.82	0.84

Figures 2.19b–2.19h are respectively the demosaiced results obtained from algorithms [13, 17, 24, 27, 46, 52, 62]. It can be observed that the VCD, DLMMSE and LPA algorithms recover fine texture patterns and almost completely suppress the introduction of unpleasant false colors with respect to other methods. Another example is given in Fig. 2.20, which shows a cropped region of test image 10 and the corresponding demosaiced images. In this example, a different set of demosaicing algorithms is selected for comparison. We notice that most algorithms obtain a good perceptual result with sharper edges and fewer zipper artifacts except the AP, SA, and FD methods which produce severe zipper artifacts around the edge due to the use of isotropic interpolation.

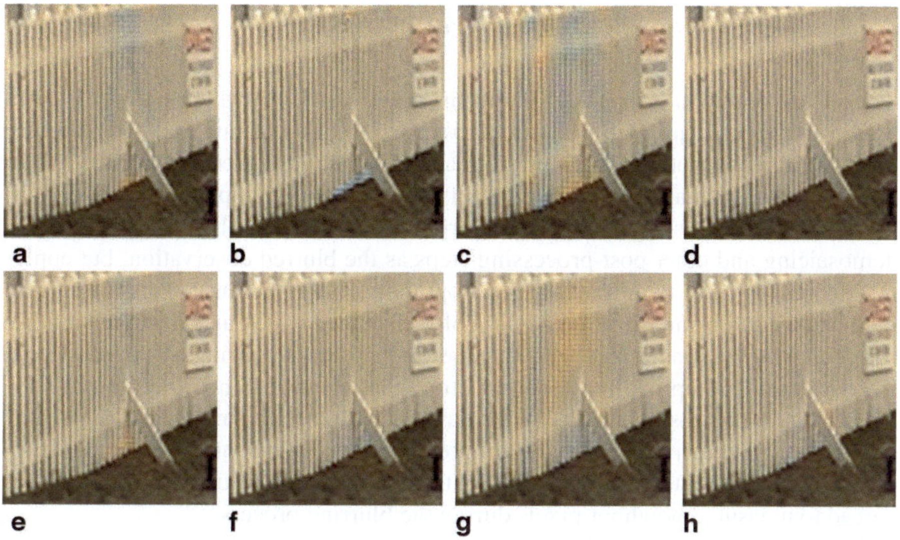

Fig. 2.19 Original and demosaiced results of a cropped region from test image 20. **a** original **b** VCD [13] **c** AHD [27] **d** DLMMSE [62] **e** LPA [52] **f** AP [24] **g** AFAW [17] **h** RAD [46]

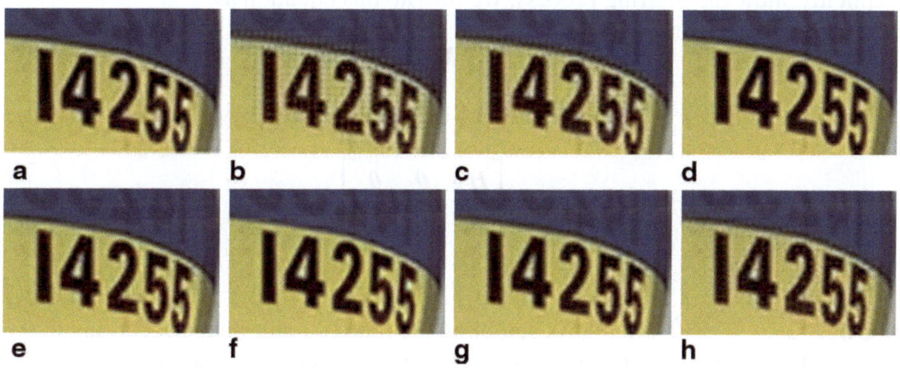

Fig. 2.20 Original and demosaiced results of a cropped region from test image 10. **a** original **b** HPHD-AI [59] **c** AHD [27] **d** DFAD [44] **e** LPA [52] **f** AP [24] **g** SA [33] **h** FD [7]

2.3 Advanced Topics

This section discusses two advanced topics which combine the image reconstruction problems, deblurring and super-resolution, with the demosaicing problem. This is still an active area of research with the soul of developing a unified framework so as to generate a better reconstruction result.

2.3.1 Joint Demosaicing and Deblurring

One of the common artifacts in digital images is the blur which may be caused by the non-ideal response of the camera sensor, the out-of-focus capturing, and the object or camera motion. The blurred image is usually modeled as the convolution of a sharp color image and a blur kernel (also called point spread function, PSF) plus noise. Most existing deblurring algorithms use the result that has gone through demosaicing and other post-processing steps as the blurred observation, but applying deblurring after demosaicing is suboptimal. This is because the demosaicing methods tend to blend the noise across color channels, which makes the input noise for deblurring no longer independent. Therefore, joint deblurring and demosaicing is considered to improve the reconstruction performance. Since the blurred CFA image has no processing artifacts, it can truly record the redistribution of light (i.e., blur kernel). In return, the blur kernel provides more information about the missing color components by considering that the information of unmeasured pixels is spread to measured neighbor pixels during the blurring process.

The matrix-vector-multiplication notation for the blurred CFA image acquisition process is already given in Eq. (2.56), which models the blur kernel with matrix $\boldsymbol{H}_k, k = R, G, B$, and the CFA subsampling operation with matrix $\boldsymbol{M}_k$. In order to explicitly show the blurring process, we rewrite this equation as:

$$\boldsymbol{i}^{CFA} = \boldsymbol{M}\boldsymbol{H}\boldsymbol{i} + \boldsymbol{n} \tag{2.67}$$

where $\boldsymbol{H} \in \mathbb{R}^{3MN \times 3MN}$ is a sparse block diagonal matrix:

$$\boldsymbol{H} = \begin{bmatrix} \boldsymbol{H}_R & \boldsymbol{0} & \boldsymbol{0} \\ \boldsymbol{0} & \boldsymbol{H}_G & \boldsymbol{0} \\ \boldsymbol{0} & \boldsymbol{0} & \boldsymbol{H}_B \end{bmatrix} \tag{2.68}$$

and $\boldsymbol{M} \in \mathbb{R}^{3MN \times 3MN}$ is a matrix sampling the blurred color image according to the Bayer CFA structure. In most literature, the blur kernel is assumed to be spatially-invariant and identical in all color channels, so $\boldsymbol{H}$ is a circulant-block-circulant matrix, usually denoted as a convolution kernel and implemented in the frequency domain.

As mentioned in Sect. 2.2.5, one possible solution to estimate $\boldsymbol{i}$ given $\boldsymbol{i}^{CFA}$ is the regularization method (or MAP) that exploits various regularizing constraints to impose image priori information, i.e.,

$$\hat{\boldsymbol{i}} = \underset{\boldsymbol{i}}{\operatorname{argmin}} \left\{ \|\boldsymbol{i}^{CFA} - \boldsymbol{M}\boldsymbol{H}\boldsymbol{i}\|^2 + \sum_s \lambda_s J_s(\boldsymbol{i}) \right\} \tag{2.69}$$

if the additive noise $\boldsymbol{n}$ is assumed to be i.i.d. Gaussian. The approaches proposed by [40, 56, 64] belong to this category. Similar to Menon et al. [46], Soulez et al. [56] considered a spatial regularization term and a spectral regularization term, which enforce smoothness on each single color component and on the color difference components respectively. However, to avoid oversmoothing sharp edges of the image, Soulez adopted the $\ell_2 - \ell_1$ norm for the spatial regularization term instead of the quadratic form. The mathematical expression of this term is:

$$J_1(\boldsymbol{i}) = \sum_{k=R,G,B} \sum_{x} \sum_{x' \in V_x} \varphi(\frac{I_k(x) - I_k(x')}{\ell(x, x')}, \lambda_k) \tag{2.70}$$

where $I_k(x)$ is the pixel value at location x in color channel k of the sharp color image $\boldsymbol{i}$, $x' \in V_x$ is a pixel location in the neighborhood of pixel x and $\ell(x, x')$ is the distance between the two pixels. The penalty function $\varphi(\cdot, \lambda_k)$ is defined as:

$$\varphi(\cdot, \lambda_k) = 2\lambda_k^2 \left[\frac{|\cdot|}{\lambda_k} - \log(1 + \frac{|\cdot|}{\lambda_k}) \right] \tag{2.71}$$

The parameter λ_k is the threshold beyond which the difference between neighboring pixels is most certainly due to an edge and must not be too smoothed. Luong et al. [40] presented a primal-dual algorithm that exploits the sparsity of both discrete gradient and shearlet coefficients as priori knowledge. In order to deal with the sparsity across color channels, they decorrelated the signals via the principle component analysis (PCA) transform before sparsifying them spatially. The common problem of the above approaches is that they assume the blur is known or just defocus blur of a simple shape. This is because the blurred CFA image has incomplete data which makes image-based blur kernel estimation methods always fail. However, the blur estimation methods with the help of hardware cannot be influenced by the image data. Inspired by this idea, Zhen et al. [64] proposed a novel approach to perform joint demosaicing and deblurring in the case of a more complex blur kernel. They designed an imaging system to capture the blurred CFA image and the inertial measurements simultaneously and utilized the motion data to estimate a blur kernel.

Another solution to the inverse problem of Eq. (2.67) is presented in [25,41,53], which performs demosaicing and deblurring alternately several times to strengthen the correlation between them. In this way, the state-of-the-art demosaicing techniques and deblurring techniques could be utilized. For instance, the algorithm proposed by Paliy et al. [53] proceeds in the following steps:

1. obtain the downsampled versions of the red, green, and blue channels from the blurred CFA image as Fig. 2.17a
2. apply the regularized inverse (RI) linear filter to each downsampled color plane separately in order to remove the blur
3. upsample the blur-free downsampled color planes and combine them together
4. utilize the modified LPA-ICI algorithm for interpolation of the noisy Bayer data [51]
5. sample the estimated full-color image according to the Bayer CFA structure
6. repeat step 1–4 but use the regularized Wiener inverse filter instead of the RI filter

Note that the filters in this algorithm use the downscaled blur kernel which matches the downsampled color planes in size, and that the modified LPA-ICI algorithm assumes the additive noise follows Poisson distribution. Another work that directly deblurred the CFA image before demosaicing is presented by Trimeche et al. [58]. In contrast, the iterative approach proposed by Ma et al. [41] first reconstructed the CFA image with a state-of-the-art demosaicing algorithm and then removed the blur using a least-squares regularization method. The deblurring process is performed in the luminance/chrominance domain rather than the RGB domain to achieve better performance. The advantage of using the luminance component is also taken by Har-Noy et al. [25]. They enforced the solution close to the demosaiced color planes while still being consistent with the extracted luminance component that has been deblurred.

2.3.2 *Joint Demosaicing and Super-Resolution*

In addition to the blurring effects, an image taken with a conventional digital camera also suffers from the spatial resolution limit brought by lens and sensors. Surpassing this limit can be achieved by the super-resolution (SR) technique, which produces a high-resolution (HR) image from several low-resolution (LR) images of the same scene with slight motion among them. In the past few decades, a variety of super-resolution algorithms have been proposed. A good review of the state-of-the-art can be found in [1].

Typically, the SR problem and the demosaicing problem are studied and solved independently. However, treating them separately is suboptimal since the aliasing artifacts introduced in the demosaicing process cannot be removed or even be amplified by the SR algorithm. Also, the CFA sampling leads to severe aliasing in the mosaiced image and the SR technique aims to fuse multiple aliased images to obtain a HR image. It is more reasonable to address the two reconstruction problems in a unified context. In this section, we present several methods for joint multiframe demosaicing and color super-resolution.

The complete model that takes SR process into account is given below:

$$\begin{aligned} \boldsymbol{i}^{CFA}(l) &= \boldsymbol{M}(l)\boldsymbol{D}(l)\boldsymbol{H}(l)\boldsymbol{F}(l)\boldsymbol{i}^{SR} + \boldsymbol{n}(l) \\ l &= 1, \ldots, L \end{aligned} \tag{2.72}$$

The vectors $\boldsymbol{i}^{SR} \in \mathbb{R}^{3r^2MN\times 1}$ and $\boldsymbol{i}^{CFA}(l) \in \mathbb{R}^{MN\times 1}$ represent the HR color image and the lth LR CFA image after lexicographic ordering, if we assume the image enhancement factor is r. The matrix $\boldsymbol{F}(l) \in \mathbb{R}^{3r^2MN\times 3r^2MN}$ is the geometric motion operator between the HR and LR images. The PSF, assumed to be lens blur only, is modeled by $\boldsymbol{H}(l) \in \mathbb{R}^{3r^2MN\times 3r^2MN}$. The matrices $\boldsymbol{D}(l) \in \mathbb{R}^{3MN\times 3r^2MN}$ and $\boldsymbol{M}(l) \in \mathbb{R}^{MN\times 3MN}$ denote the CCD downsampling operation and the CFA sampling operation respectively. The vector $\boldsymbol{n}(l) \in \mathbb{R}^{MN\times 1}$ is the additive noise and L is the number of available LR CFA images. Compared with the model in Eq. (2.67), the operators $\boldsymbol{F}(l)$ and $\boldsymbol{D}(l)$ are added to account for the SR process. The problem of interest is to estimate the HR image $\boldsymbol{i}^{SR}$ from the observed data $\boldsymbol{i}^{CFA}(l), l = 1, \ldots, L$.

Same as before, one of the solutions is the regularization method [18, 23] with $\Psi(\boldsymbol{i}^{SR})$ accounting for the data fidelity term and $J_s(\boldsymbol{i}^{SR})$ representing the regularization constraints,

$$\hat{\boldsymbol{i}}^{SR} = \underset{\boldsymbol{i}^{SR}}{\operatorname{argmin}} \left\{ \Psi(\boldsymbol{i}^{SR}) + \sum_s \lambda_s J_s(\boldsymbol{i}^{SR}) \right\} \tag{2.73}$$

In the approach proposed by Farsiu et al. [18], the data fidelity term is defined as the ℓ_1 norm of the residual vector instead of the most common ℓ_2 norm,

$$\Psi(\boldsymbol{i}^{SR}) = \sum_{l=1}^{L} \|\boldsymbol{M}(l)\boldsymbol{D}(l)\boldsymbol{H}(l)\boldsymbol{F}(l)\boldsymbol{i}^{SR} - \boldsymbol{i}^{CFA}(l)\|_1 \tag{2.74}$$

This definition can effectively remove outliers in the data and errors due to possibly inaccurate motion estimation. Farsiu et al. also came up with three regularization terms to reduce possible artifacts. Since the human eye is more sensitive to the details in the luminance component of an image than those in the chrominance components, the first regularization term is applying bilateral-total-variation regularization to the luminance component in order to preserve the edges in the luminance,

$$J_1(\boldsymbol{i}^{SR}) = \sum_{m=-P}^{P} \sum_{n=-P}^{P} \alpha^{|m|+|n|} \|\boldsymbol{i}_{Lu}^{SR} - S_x^m S_y^n \boldsymbol{i}_{Lu}^{SR}\|_1 \tag{2.75}$$

where S_x^m and S_y^n are the operators for shifting the luminance image $\mathbf{i}_{Lu}^{SR}$ by m pixels in horizontal direction and n pixels in vertical direction, respectively. The scalar α weights each term in the summation and the parameter P defines the size of the bilateral filter kernel. The expression $\boldsymbol{i}_{Lu}^{SR} - S_x^m S_y^n \boldsymbol{i}_{Lu}^{SR}$ actually computes derivatives of the luminance across multiple scales of resolution. Readers could refer to [19] for more information. Correspondingly, a regularization term should be applied on the chrominance components $\boldsymbol{i}_{C_1}^{SR}$ and $\boldsymbol{i}_{C_2}^{SR}$. This term imposes smoothness only, so the simple Tikhonov regularization is used,

$$J_2(\boldsymbol{i}^{SR}) = \|\Lambda \boldsymbol{i}_{C_1}^{SR}\|^2 + \|\Lambda \boldsymbol{i}_{C_2}^{SR}\|^2 \tag{2.76}$$

where Λ is a high-pass filter. The idea of the third regularization term comes from [28], which minimizes the vector product norm of any adjacent color pixels to make sure that different color bands have similar edge location and orientation. With some modifications to what was proposed in [28], the third term is defined as:

$$\begin{aligned} J_3(\boldsymbol{i}^{SR}) = \sum_{m=-1}^{1} \sum_{n=-1}^{1} [\|\boldsymbol{i}_G^{SR} \odot S_x^m S_y^n \boldsymbol{i}_B^{SR} - \boldsymbol{i}_B^{SR} \odot S_x^m S_y^n \boldsymbol{i}_G^{SR}\|^2 \\ + \|\boldsymbol{i}_B^{SR} \odot S_x^m S_y^n \boldsymbol{i}_R^{SR} - \boldsymbol{i}_R^{SR} \odot S_x^m S_y^n \boldsymbol{i}_B^{SR}\|^2 \\ + \|\boldsymbol{i}_R^{SR} \odot S_x^m S_y^n \boldsymbol{i}_G^{SR} - \boldsymbol{i}_G^{SR} \odot S_x^m S_y^n \boldsymbol{i}_R^{SR}\|^2] \end{aligned} \tag{2.77}$$

where $\boldsymbol{i}_k^{SR}, k = R, G, B$ represents each color plane in $\boldsymbol{i}^{SR}$ and $\odot$ is the element by element multiplication operator. In order to simplify the optimization problem, Farsiu et al. assumed spatially-invariant lens blur (also invariant among different LR images) and translational motion model. Therefore, the operators $\boldsymbol{H}$ and $\boldsymbol{F}(l)$ are commutative and Eq. (2.72) becomes

$$\begin{aligned} \boldsymbol{i}^{CFA}(l) &= \boldsymbol{M}(l)\boldsymbol{D}(l)\boldsymbol{F}(l)\boldsymbol{H}\boldsymbol{i}^{SR} + \boldsymbol{n}(l) \\ &= \boldsymbol{M}(l)\boldsymbol{D}(l)\boldsymbol{F}(l)\boldsymbol{z} + \boldsymbol{n}(l) \\ l &= 1, \ldots, L \end{aligned} \tag{2.78}$$

Now, the inverse problem is separated into two simpler subtasks:

1. fuse the available LR images into a blurred HR image $\boldsymbol{z}$ using shift-and-add operation
2. estimate the deblurred HR image $\hat{\boldsymbol{i}}^{SR}$ from $\boldsymbol{z}$ by optimizing Eq. (2.73) where the data fidelity term $\Psi(\boldsymbol{i}^{SR})$ is changed to measure the distance between $\boldsymbol{H}\hat{\boldsymbol{i}}^{SR}$ and $\boldsymbol{z}$.

In parallel to Farsiu's work, Gotoh et al. [23] proposed another MAP estimation method for solving the same joint demosaicing and SR problem. Their algorithm has much in common with Farsiu's algorithm, but Gotoh adopted the ℓ_2 norm in data fidelity term and used an unisotropic Tikhonov method in regularizing the luminance component. In addition, Gotoh examined direct registration of raw images based on an imaging model. Similarly, the joint algorithm presented by Vandewalle et al. [60] used the luminance information directly extracted from the LR CFA images for computing alignment parameters. After image registration, the HR luminance and chrominance images were interpolated using the data from the LR image set and combined afterwards to generate the final HR image.

2.4 Conclusion

An introduction to the demosaicing issue and its major solutions has been explored in this chapter. Focusing on the Bayer CFA pattern, we discussed the color artifacts generated by bilinear interpolation and described the widely used demosaicing principles and the common evaluation process. After that, five categories of demosaicing algorithms were analyzed in detail. In each category, we presented its characteristics and listed several algorithms as representations of the type. To show the advantages and drawbacks of the examined algorithms, we also reported experimental results of a performance comparison between selected demosaicing algorithms.

As an extension, the approaches that perform demosaicing jointly with deblurring and super-resolution were discussed. These approaches may allow us to find a more competitive solution with respect to regarding them as separate problems.

In this chapter, time analysis of each demosaicing algorithm is ignored due to lack of implementation details and different programming languages of the available codes. However, the time constraint is a very important problem in practice. Readers could refer to [34] for a computational complexity analysis.

References

1. EURASIP journal on applied signal processing, special issue on super-resolution. (2006)
2. Adams, J.E., Jr.: Interactions between color plane interpolation and other image processing functions in electronic photography. In: IS&T/SPIE's Symposium on Electronic Imaging: Science & Technology, pp. 144–151. International Society for Optics and Photonics (1995)
3. Adams, J.E., Jr.: Design of practical color filter array interpolation algorithms for digital cameras. In: Electronic Imaging'97, pp. 117–125. International Society for Optics and Photonics (1997)
4. Adams, J.E., Jr.: Design of practical color filter array interpolation algorithms for digital cameras .2. In: 1998 International Conference on Image Processing, 1998. Proceedings, vol 1, pp. 488–492. IEEE (1998)
5. Adams, J.E., Jr, Hamilton, J.F., Jr.: Adaptive color plan interpolation in single sensor color electronic camera. US Patent 5,629, 734, 13 May 1997
6. Alleysson D., Chaix De Lavarène B., Susstrunk S., Hérault, J.: Linear minimum mean square error demosaicking. In: Lukac, R. (ed.) Single-sensor imaging: methods and applications for digital cameras, pp. 213–237. CRC, Boca Raton (2008)
7. Alleysson D., Susstrunk S., Hérault, J.: Linear demosaicing inspired by the human visual system. IEEE Transac. Image Process. **14**(4), 439–449 (2005)
8. Bayer, B.E.: Color imaging array. US Patent 3,971, 065, 20 July 1976
9. Brainard, D.H. et al.: Bayesian method for reconstructing color images from trichromatic samples. In: Proceedings of the IS&T 47th Annual Meeting, pp. 375–380 (1994)
10. Chang, L., Tan, Y-P.: Effective use of spatial and spectral correlations for color filter array demosaicking. IEEE Transac. Consum. Elec. **50**(1), 355–365 (2004)
11. Chang, L., Tan, Y-P.: Hybrid color filter array demosaicking for effective artifact suppression. J. Elec. Imaging **15**(1), 013003–013003 (2006)
12. Chen, L., Yap, K-H., He, Y.: Color filter array demosaicking using wavelet-based subband synthesis. In: IEEE International Conference on Image Processing, 2005. ICIP 2005. vol 2, pp. II–1002. IEEE (2005)
13. Chung, K-H., Chan, Y-H.: Color demosaicing using variance of color differences. IEEE Transac. Image Process. **15**(10), 2944–2955 (2006)
14. Cok, D.R.: Signal processing method and apparatus for sampled image signals. US Patent 4,630, 307, 16 Dec 1986
15. Cok, D.R.: Signal processing method and apparatus for producing interpolated chrominance values in a sampled color image signal. US Patent 4,642, 678, 10 Feb 1987

16. Driesen, J., Scheunders, P.: Wavelet-based color filter array demosaicking. In: 2004 International Conference on Image Processing, 2004. ICIP'04. vol 5, pp. 3311–3314. IEEE (2004)
17. Dubois, E.: Frequency-domain methods for demosaicking of Bayer-sampled color images. IEEE Signal Process. Lett. **12**(12), 847–850 (2005)
18. Farsiu, S., Elad, M., Milanfar, P.: Multiframe demosaicing and super-resolution of color images. IEEE Transac. Image Process. **15**(1), 141–159 (2006)
19. Farsiu, S., Robinson, M.D., Elad, M., Milanfar, P.: Fast and robust multiframe super resolution. IEEE Transac. Image Process. **13**(10), 1327–1344 (2004)
20. Freeman, W.T.: Median filter for reconstructing missing color samples. US Patent 4,724, 395, 9 Feb 1988
21. Freeman, W.T.: Method and apparatus for reconstructing missing color samples. US Patent 4,774, 565, 27 Sept 1988
22. Glotzbach, J.W., Schafer, R.W., Illgner, K.: A method of color filter array interpolation with alias cancellation properties. In: 2001 International Conference on Image Processing, 2001. Proceedings. vol 1, pp. 141–144. IEEE (2001)
23. Gotoh, T., Okutomi, M.: Direct super-resolution and registration using raw CFA images. In: Proceedings of the 2004 IEEE Computer Society Conference on Computer Vision and Pattern Recognition, 2004. CVPR 2004. vol 2, pp. II–600. IEEE (2004)
24. Gunturk, B.K., Altunbasak, Y., Mersereau, R.M.: Color plane interpolation using alternating projections. IEEE Transac. Image Process. **11**(9), 997–1013 (2002)
25. Har-Noy, S., Chan, S.H., Nguyen, T.Q.: Demosaicking images with motion blur. In: 2010 IEEE International Conference on Acoustics Speech and Signal Processing (ICASSP). pp. 1006–1009. IEEE (2010)
26. Hibbard, R.H.: Apparatus and method for adaptively interpolating a full color image utilizing luminance gradients. US Patent 5,382, 976, 17 Jan 1995
27. Hirakawa, K., Parks, T.W.: Adaptive homogeneity-directed demosaicing algorithm. IEEE Transac. Image Process. **14**(3), 360–369 (2005)
28. Keren, D., Osadchy, M.: Restoring subsampled color images. Mach. Vis. Appl. **11**(4), 197–202 (1999)
29. Kimmel, R.: Demosaicing: image reconstruction from color CCD samples. IEEE Transac. Image Process. **8**(9), 1221–1228 (1999)
30. Laroche, C.A., Prescott, M.A.: Apparatus and method for adaptively interpolating a full color image utilizing chrominance gradients. US Patent 5,373, 322, 13 Dec 1994
31. Li, X., Gunturk, B., Zhang, L.: Image demosaicing: A systematic survey. In: Electronic Imaging 2008, pp. 68221J–68221J. International Society for Optics and Photonics (2008)
32. Li, X., Orchard, M.T.: New edge-directed interpolation. IEEE Transac. Image Process. **10**(10), 1521–1527 (2001)
33. Li, X.: Demosaicing by successive approximation. IEEE Transac. Image Process. **14**(3), 370–379 (2005) http://www.csee.wvu.edu/ xinl/demo/demosaic.html.
34. Lian, N-X., Chang, L., Tan, Y-P., Zagorodnov, V.: Adaptive filtering for color filter array demosaicking. IEEE Transac. Image Process. **16**(10), 2515–2525 (2007)
35. Losson, O., Macaire, L., Yang, Y.: Comparison of color demosaicing methods. Adv. Imaging Electron Phys. **162**, 173–265 (2010)
36. Lu, W., Tan, Y-P.: Color filter array demosaicking: new method and performance measures. IEEE Transac. Image Process. **12**(10), 1194–1210 (2003)
37. Lu, Y.M., Karzand, M., Vetterli, M.: Demosaicking by alternating projections: theory and fast one-step implementation. IEEE Transac. Image Process. **19**(8), 2085–2098 (2010)
38. Lukac, R., Plataniotis, K.N.: Universal demosaicking for imaging pipelines with an RGB color filter array. Pattern Recognit. **38**(11), 2208–2212 (2005)
39. Lukac, R., Plataniotis, K.N., Hatzinakos, D., Aleksic, M.: A novel cost effective demosaicing approach. IEEE Transac. Consumer Elec. **50**(1), 256–261 (2004)

40. Luong, H.Q., Goossens, B., Aelterman, J., Pizurica, A., Philips, W.: A primal-dual algorithm for joint demosaicking and deconvolution. In: 2012 19th IEEE International Conference on Image Processing (ICIP), pp. 2801–2804. IEEE (2012)
41. Ma, T., Reeves, S.J.: An iterative regularization approach for color filter array image restoration. In 2011 IEEE International Conference on Industrial Technology (ICIT), pp. 332–335. IEEE (2011)
42. Marino, B.E., Stevenson, R.L.: Improving the performance of single chip image capture devices. J. Elec. Imaging **12**(2), 209–218 (2003)
43. McLaren, K.: XIII - the development of the CIE 1976 (L* a* b*) uniform colour space and colour-difference formula. J. Soc. Dyers Colour. **92**(9), 338–341 (1976)
44. Menon, D., Andriani, S., Calvagno, G.: Demosaicing with directional filtering and a posteriori decision. IEEE Transac. Image Process. **16**(1), 132–141 (2007)
45. Menon, D., Calvagno, G.: Demosaicing based on wavelet analysis of the luminance component. In: IEEE International Conference on Image Processing, 2007. ICIP 2007. vol 2, pp. II–181. IEEE (2007)
46. Menon, D., Calvagno, G.: Regularization approaches to demosaicking. IEEE Transac. Image Process. **18**(10), 2209–2220 (2009)
47. Menon, D., Calvagno, G.: Color image demosaicking: an overview. Signal Process.: Image Commun. **26**(8), 518–533 (2011)
48. Mukherjee, J., Parthasarathi, R., Goyal, S.: Markov random field processing for color demosaicing. Pattern Recognit. Lett. **22**(3), 339–351 (2001)
49. Muresan, D.D., Parks, T.W.: Optimal recovery demosaicing. IASTED Signal and Image Processing (2002)
50. Omer, O.A., Tanaka, T.: Image demosaicking based on chrominance regularization with region-adaptive weights. In 2007 6th International Conference on Information, Communications & Signal Processing, pp. 1–5. IEEE (2007)
51. Paliy, D., Foi, A., Bilcu, R., Katkovnik, V.: Denoising and interpolation of noisy Bayer data with adaptive cross-color filters. In Electronic Imaging 2008, pp. 68221K–68221K. International Society for Optics and Photonics (2008)
52. Paliy, D., Katkovnik, V., Bilcu, R., Alenius, S., Egiazarian, K.: Spatially adaptive color filter array interpolation for noiseless and noisy data. Int. J. Imaging Syst. Tech. **17**(3), 105–122 (2007)
53. Paliy, D., Foi, A., Bilcu, R., Katkovnik, V., Egiazarian, K.: Joint deblurring and demosaicing of Poissonian Bayer data based on local adaptivity. In Proc. 16th EUropean SIgnal Process COnference (EUSIPCO). Citeseer (2008)
54. Pei, S-C., Tam, I-K.: Effective color interpolation in CCD color filter arrays using signal correlation. IEEE Transac. Circuits Syst. Video Tech. **13**(6), 503–513 (2003)
55. Saito, T., Komatsu, T.: Demosaicing approach based on extended color total-variation regularization. In 15th IEEE International Conference on Image Processing, 2008. ICIP 2008. pp. 885–888. IEEE (2008)
56. Soulez, F., Thiébaut, E.: Joint deconvolution and demosaicing. In 2009 16th IEEE International Conference on Image Processing (ICIP), pp. 145–148. IEEE (2009)
57. Su, C-Y. Highly effective iterative demosaicing using weighted-edge and color-difference interpolations. IEEE Transac. Consumer Elec. **52**(2), 639–645 (2006)
58. Trimeche, M., Paliy, D., Vehvilainen, M., Katkovnic, V.: Multichannel image deblurring of raw color components. In Electronic Imaging 2005, pp. 169–178. International Society for Optics and Photonics (2005)
59. Tsai, C-Y., Song, K-T.: Heterogeneity-projection hard-decision color interpolation using spectral-spatial correlation. IEEE Transac. Image Process. **16**(1), 78–91 (2007)
60. Vandewalle, P., Krichane, K., Alleysson, D., Süsstrunk, S.: Joint demosaicing and super-resolution imaging from a set of unregistered aliased images. In Electronic Imaging 2007, pp. 65020A–65020A. International Society for Optics and Photonics (2007)
61. Wu, X., Zhang, N.: Primary-consistent soft-decision color demosaicking for digital cameras (patent pending). IEEE Transac. Image Process. **13**(9), 1263–1274 (2004)

62. Zhang, L., Wu, X.: Color demosaicking via directional linear minimum mean square-error estimation. IEEE Transac. Image Process. **14**(12), 2167–2178 (2005)
63. Zhang, X., Wandell, B.A.: A spatial extension of CIELAB for digital color-image reproduction. J. Soc. Info. Display **5**(1), 61–63 (1997)
64. Zhen, R., Stevenson, R.L.: Joint deblurring and demosaicking of CFA image data with motion blur. In IS&T/SPIE Electronic Imaging, pp. 90290B–90290B. International Society for Optics and Photonics (2014)

Chapter 3
DCT-Based Color Image Denoising: Efficiency Analysis and Prediction

Vladimir Lukin, Sergey Abramov, Ruslan Kozhemiakin, Alexey Rubel, Mikhail Uss, Nikolay Ponomarenko, Victoriya Abramova, Benoit Vozel, Kacem Chehdi, Karen Egiazarian and Jaakko Astola

3.1 Introduction

Color images are widely met and used in everyday life [1]. Color photos are, probably, the most popular type [1,2], although color images are also exploited in medical diagnostics [3], remote sensing from airborne, and spaceborne platforms [4], etc. In practice, all acquired color images are inevitably noisy in larger or less extent due to basic principle of their forming (that deals with photon counting), circuitry noise, and other factors [2]. This noise can be of such low level that it is practically unnoticeable. Then, a color image might be perceived as having practically perfect (visual) quality if the influence of other degrading factors is negligible.

In other cases, that is, if the noise is visible, it often occurs to be a major (dominant) factor degrading the quality of images. Then, one needs to enhance a considered color image. Filtering (also called denoising) is a standard operation for this by suppressing the noise and preserving useful information in a processed image [1, 5]. Numerous denoising techniques have been designed so far for color image processing (see [1, 5–9]) and, more generally, for filtering multichannel images and videos [10–14]. Starting from intensive developments of nonlinear vector filters in 1980s–1990s [1, 15, 16] and their further modifications [16–20], many other noise removal filters have been designed. They include locally adaptive and transform-based filters (see e.g., [5, 6, 8, 10, 21–24]), fuzzy denoising methods [25–30], a family of nonlocal filtering approaches [7, 21, 31–36], etc. A brief analysis of existing methods and their shortcomings is presented in Sect. 3.3.

B. Vozel (✉) · K. Chehdi
University of Rennes 1, Lannion, France
e-mail: benoit.vozel@univ-rennes1.fr

V. Lukin · S. Abramov · R. Kozhemiakin · A. Rubel · M. Uss · N. Ponomarenko · V. Abramova
National Aerospace University, Kharkov, Ukraine

K. Egiazarian · J. Astola
Tampere University of Technology, Tampere, Finland

M. E. Celebi et al. (eds.), *Color Image and Video Enhancement*,
DOI 10.1007/978-3-319-09363-5_3

Such interest to design filters to process color images can be explained by the following reasons. First reason is that there are several requirements to design methods that are imposed simultaneously and that can be contradictory. Essentially, it is desirable to efficiently suppress the noise but it is also necessary not to distort the edges, textures, fine details, and color features [1, 7, 15, 16]. Meanwhile, it is often desired to carry out the processing quickly and in an automatic manner. Such a set of requirements is quite hard to fulfill [9, 23, 24].

Second reason is that most color images are subject to visual inspection and analysis. This shows that quality of color images has to be characterized using a metric adequate to human visual system (HVS). Here, it is worth stating the following: for many color image filtering methods, their design and performance analyses have been carried out using only standard quantitative criteria as output mean square error (MSE), signal-to-noise ratio (SNR) or peak signal-to-noise ratio (PSNR). Meanwhile, design and comparative analysis of visual quality metrics still continues [37, 38]. There is no HVS-metric jointly accepted by image processing community as the best one. Recently, researchers have started to exploit HVS-metrics for characterization of filtering efficiency for image denoising techniques [9, 14, 35, 39–42].

Third reason is that two main approaches to processing color images are possible, namely, component-wise filtering and vector (or 3D) denoising. Main advantages of the component-wise filtering approach are simplicity and better theoretical grounds [15, 42, 43]. A main advantage of the latter (vector) approach is that it potentially allows taking into account such an important feature of color images that color components (in conventional red-green-blue (RGB) representation) are highly correlated [1, 15]. Then, filtering can be more efficient [42, 44] but more complex.

A fourth reason just deals with other factors that complicate noise suppression. Noise often cannot be considered as pure additive and spatially uncorrelated (white) [23, 40, 45, 46]. Moreover, its intensity can be different for different color components [47]. Thus, a denoising method should be able to adapt to these obstacles and a priori information on properties of nonadditive and nonwhite noise has to be available. For many denoising techniques, especially 3D ones, such adaptation can be problematic.

Finally, before starting applying image denoising it is good to understand or to predict how efficient it can be and is it worth to perform such filtering method. Recently, some papers dealing with these questions have appeared [48–51] and experience obtained from them is to be summarized.

In this chapter, we mainly address two aforementioned aspects. A special attention is paid to exploiting inter-component correlation inherent for color images [15] considering both white and spatially correlated noise cases. Two groups of filters are mainly considered. First group includes discrete cosine transform (DCT)-based denoising techniques as an example of transform-based filters [24, 40–42, 46, 52, 53]. Second group includes state-of-the-art nonlocal filters, namely, BM3D [54] to be applied component-wise and its color extension is called C-BM3D [7]. These filters exploit both search of similar patches (blocks) and DCT denoising employed for the collected similar patches as aforementioned. The main goal of such an analysis is to show that basic problems of color image filtering are observed for both groups of filters.

One more peculiarity of studies performed in this chapter consists of the following: analysis is carried out using color images of the databases TID2008 [55] and a recently designed database TID2013 [38]. There are several reasons why we do this. First, these databases contain noise-free and already noisy images (available at www.ponomarenko.info). This allows to use them to check efficiency of denoising. Another peculiarity of this chapter is that the cases of i.i.d. and spatially correlated noise are studied where the latter case is quite rarely analyzed in literature. In these studies, special attention is paid to the analysis of visual quality. For this purpose, we apply recently introduced adequate metrics as PSNR-HA [56] or color version of FSIM [57] as well as some other metrics [58] which are among the best to characterize visual quality of noise-free (original) and processed (filtered) images.

3.2 Image and Noise Properties

3.2.1 Image Properties

Efficiency of image filtering considerably depends on adequacy of a used model of the noise and properties of images taken into account by a used filter. Thus, we first recall some basic properties of color images and the noise present in them. Note that we consider only one color system which is, probably, the most known and popular in color image processing, namely, RGB [1, 2]. Many researchers have stated that component images in R, G, and B possess high correlation (about 0.7 ... 0.8) [15, 16] and this property has been stressed in motivations of vector filter design. On the average, this statement is correct, although essential variations are possible. To give imagination about them, consider noise-free images from the database TID2008. This database contains 25 test color images, all of equal size 512 $\times$ 384 pixels.

Out of 25 images, 24 images are natural while, the last, 25th image is artificial (image index $n = 1, \ldots, 25$). For these images, we have calculated inter-component correlation factors $R_{kl} = \sum_{i=1}^{I}\sum_{j=1}^{J}(I_{kij}^{true} - \bar{I}_{k}^{true})(I_{lij}^{true} - \bar{I}_{l}^{true})/(IJ\sigma_{\mathrm{Im}\,k}^{2}\sigma_{\mathrm{Im}\,l}^{2})$ where I_{kij}^{true} denotes true (noise-free) value of ij-th pixel of k-th component, $k=$ 1,2,3; I and J define a processed image size, $\bar{I}_{k}^{true}$ and $\sigma_{\mathrm{Im}\,k}^{2}$ define mean and variance of the true image in k-th channel (e.g., $k=1$ and $l=2$ corresponds to correlation between R and G components). The obtained values are presented in Table 3.1 where n denotes a test image index in the database (indices are given below the images in Fig. 3.1). Interesting observations from Table 3.1 are the following: first, for many test images, correlation factors are high and even larger than 0.8. In particular, this happens for such highly textural (complex structure) images as ## 1, 5, 8, 13, and 19. Meanwhile, there are also images for which correlation factors are

Table 3.1 Inter-component correlation factors for images of the TID2008 database

n	1	2	3	4	5	6	7	8	9	10	11	12	13
RG	0.86	0.59	0.61	0.68	0.87	0.95	0.74	0.96	0.92	0.93	0.84	0.88	0.96
GB	0.96	0.90	0.27	0.92	0.88	0.97	0.88	0.95	0.80	0.95	0.95	0.95	0.95
BR	0.83	0.55	0.03	0.74	0.77	0.94	0.67	0.90	0.67	0.89	0.78	0.87	0.94
n	14	15	16	17	18	19	20	21	22	23	24	25	–
RG	0.79	0.86	0.94	0.97	0.88	0.95	0.99	0.87	0.86	0.62	0.92	0.77	–
GB	0.55	0.97	0.89	0.96	0.85	0.93	0.93	0.90	0.89	0.66	0.92	0.77	–
BR	0.25	0.84	0.80	0.94	0.75	0.87	0.93	0.67	0.77	0.51	0.87	0.86	–

Fig. 3.1 Noise-free test color images of TID2008

considerably smaller than 0.7 ... 0.8, for example, rather simple images ## 3 and 23. Second, cross-correlation factor between G and B components is usually the largest, while for R and B components this factor is often the smallest (although exceptions are possible). Factor values smaller than 0.3 sometimes happen, but rarely.

3.2.2 Noise Models

Now consider models of the noise used in color image processing. A simplest among them is additive white Gaussian noise (AWGN)

$$I_{kij}^{noisy} = I_{kij}^{true} + n_{kij},\ i = 1, \ldots, I,\ j = 1, \ldots, J \tag{3.1}$$

where I_{kij}^{noisy} is *ij*-th sample of noisy (original) *k*-th component of a color image in RGB representation ($k = 1,2,3$). Variance σ_{kij}^2 of the zero mean noise n_{kij} is assumed to be equal and noise is assumed to be independent in all three color components ($\sigma_{kij}^2 = \sigma_a^2$). Let us further consider this model as model 1. It is widely exploited in the scientific literature [1,4,7,15,31,42], although it possesses a limited adequacy.

In fact, model 1 can be accepted as a rough assumption (approximation). In practice, variance σ_{kij}^2 is usually not constant for all I_{kij}^{true} and it has smaller values for $I_{kij}^{true} < 25$ and $I_{kij}^{true} > 230$ for 8-bit images [59]. Moreover, noise variance in different color components can be nonidentical. Usually, it is slightly smaller in green component than in red and blue [47, 59]. However, these peculiarities of noise statistics can be, at the first stage, ignored since there is another, more important peculiarity of the noise in practice. This peculiarity deals with existing spatial correlation of the noise stemming from image interpolation from Bayer pattern to conventional RGB representation [2, 59, 60]. Then, an observed image can be still described by the expression (3.1), but the noise n_{kij} is no more white. For model 2, we assume that this noise has zero mean, fixed variance $\sigma_{kij}^2 = \sigma_a^2$ equal for all color components and is independent of color components. The main difference is that now the noise is also characterized by 2D autocorrelation function or spatial spectrum in a certain basis.

Conventional basis for correlation analysis is the Fourier transform [4]. However, here we exploit the DCT basis because all filters that are considered later perform denoising in DCT domain and are, in general, able to use information represented in the form of 8 × 8 block DCT spectrum. To avoid dependence of this spectrum on noise variance, consider the so-called normalized DCT spectrum $W_{norm}(p, q)$, $p = 0, \ldots, 7; q = 0, \ldots, 7$ [42]. In such spectrum, component $W_{norm}(0,0)$ relates to DC component and components with small *p* and *q* correspond to "low" spatial frequencies.

Table 3.2 Normalized amplitude 8 × 8 DCT spectrum of spatially correlated noise

2.69	2.37	2.04	1.51	0.93	0.51	0.59	0.81
2.39	2.09	1.80	1.35	0.85	0.45	0.52	0.75
2.04	1.80	1.53	1.14	0.72	0.38	0.43	0.62
1.52	1.35	1.15	0.85	0.53	0.29	0.33	0.48
0.93	0.85	0.70	0.53	0.33	0.18	0.20	0.29
0.51	0.45	0.38	0.29	0.18	0.10	0.11	0.16
0.58	0.52	0.44	0.33	0.21	0.11	0.12	0.18
0.84	0.75	0.63	0.48	0.29	0.16	0.18	0.26

Fig. 3.2 Test image #16 corrupted by AWGN (**a**) and spatially correlated noise (**b**)

To imagine how such DCT normalized spectrum looks like, consider correlated noise case for images in the database TID2008. Spatially correlated noise is distortion #3 in the databases TID2008 and TID2013 (there are totally 17 and 24 types of distortions in these databases, respectively). This noise is obtained in a very simple manner by applying 3 × 3 pixel mean filter to 2D zero mean AWGN and further providing a desired variance of the noise. Amplitude normalized spectrum $\sqrt{W_{norm}(p,q)}$, $p = 0,\ldots,7; q = 0,\ldots,7$ for this case is presented in Table 3.2 as 8 × 8 matrix below. As it is seen, there are a few low-frequency components with values greater than unity (where the noise main power is concentrated), while for most spatial frequencies the values are less than unity with a general tendency to decrease for larger frequencies. These are typical properties of DCT spectra for spatially correlated noise.

It is also worth demonstrating the difference in a visual appearance of images corrupted by AWGN and spatially correlated noise. Figure 3.2 presents two test images #16 from the database TID2013 where noise variance is the same and equals to 32.5 for all the color components (this corresponds to PSNR = 33 dB).

These examples simultaneously show several interesting phenomena. First, AWGN (Fig. 3.2a) with such variance is visible in homogeneous image regions as sky, while it is practically masked (not visible) in texture regions as water surface. Second, spatially correlated noise (Fig. 3.2b) is obviously more annoying and it is visible in both homogeneous and textural areas. Thus, under given variance of the noise, it is more important to suppress spatially correlated noise. Third, this example shows inadequacy of standard quality metrics—two images with the same PSNR in Fig. 3.2 obviously have different visual quality.

Here some results of the paper [42] are worth recalling. In case of AWGN, noise is practically not visible in original (noisy) images if its variance σ_a^2 does not exceed 15 (and even 20 ... 25 if the image is highly textural). Meanwhile, if noise is spatially correlated, then it might remain visible even if its variance is about 10. Only if the noise variance does not exceed 5, it can be practically guaranteed that the noise is invisible. These observations show when the filtering is not necessary for enhancing color images. If such a decision (to perform denoising or not) is to be undertaken automatically, one needs to have an accurate blind method for estimating noise variance and spatial spectrum. One technique for doing this can be found in the paper [61].

Noise can be also described by other, more complicated, models than model 1 or model 2. In particular, noise model can take into account dependence of noise variance σ_{kij}^2 on I_{kij}^{true}. Such dependence usually does not exist or is not obvious for RGB color images obtained at the output of image data processing chain [60–62], that is, after white balance, gamma correction, and other operations inside digital cameras or sensors. However, for raw color images, signal-dependent model (model 3) of the noise is adequate [63, 64]. Taking into account thermal noise component as well as photon-counting noise, noise statistics (local standard deviation) can be expressed as [63]

$$\sigma_{kij} = \sqrt{\sigma_{ak}^2 + \gamma_k I_{kij}^{true}} \tag{3.2}$$

where σ_{kij} denotes noise standard deviation in an *ij*-th pixel of a *k*-th component image, σ_{ak}^2 is signal-independent noise component variance for a *k*-th component, γ_k is signal-dependent component parameter. Expression (3.1) is still valid but a noise variance is described by (3.2). The signal-dependent noise still has zero mean and distribution which is Gaussian or close to Gaussian [63,64]. A specific feature is that if the case of raw color images is considered (nowadays such images can be retrieved and processed on computer for many types of modern digital cameras), the noise can be assumed as spatially uncorrelated and independent in color components.

We do not pay much attention to model 3 in our further studies, since it is necessary to model all stages (operations) of image data processing chain and to set parameters for operations employed in this chain. Besides, there is a limited number of denoising techniques adapted to model 3. Not all filtering methods considered below are adapted or can be easily modified to cope with images corrupted by model 3 noise. However, we believe that there is a great potential of image denoising applied just after obtaining raw image data. One reason is that it is usually simpler and more efficient to remove white noise than spatially correlated. In particular, such denoising can be performed by using variance stabilizing transforms with a properly set parameters [65].

3.3 Considered Filters and Quantitative Criteria

From the very beginning, we have to state that two main approaches to denoise color images are possible: component-wise and vector-based (3D) [15]. There are numerous filters that can be applied component-wise for noise of the model 1 [9,21, 31]. The best performance in such practical situations is provided by methods related to the group of nonlocal filters [21, 31, 32, 34, 35, 48, 54] and dictionary learning based methods (e.g., [36]). The BM3D filter [54] is usually declared as state of the art in AWGN suppression according to output MSE (or PSNR) [48]. Meanwhile, simulation results show that the filters from the studies [32, 34, 36] possess similar performance.

For the noise model 2, amount of designed filters is considerably smaller. Neglecting a fact that spatially correlated noise can lead to a sufficient drop in filtering efficiency [23, 42]. Therefore, a filtering algorithm has to be adapted to a spatial spectrum of the noise supposed to be known a priori or pre-estimated with an appropriate accuracy. This can be easily done for transform-based methods [23,42], details are given below.

For the noise model 3, component-wise processing can be done using either filters specially intended for removing signal-dependent noise [64–66] or using a variance stabilizing (homomorphic) transform (VST) with properly adjusted parameters [24, 67] based on estimated characteristics of the noise [68]. Then, after performing a direct VST, signal-dependent noise converts to an additive one turning a situation to model 1 (or model 2), allowing many filters to be applicable for this model as well.

Vector filters [15–20] and their modifications [1, 5, 25] have been, in the first order, intended on removing not AWGN but impulse or mixed noise (in the latter case, removal of outliers is more important than suppression of an additive component of noise). Here, we assume that impulse noise is either absent in acquired color images or it is removed at a special initial stage of image processing. In this sense, fuzzy set-based filters [25–30] are very efficient allowing to exploit information from one component to improve a processing in other one. However, a main intention in designing vector (3D) filters was, and still is, use of essential interchannel correlation [15] (see Table 3.1). In fact, availability of additional data in three channels (components) opens opportunities in data decorrelation and search for similar patches [4, 7, 11, 33]. Exploiting these opportunities is not a problem for model 1 and model 2 noises. For model 3, a use of the preliminary VST in each component solves the problem that arises if parameters of signal-dependent noise are different in components of multichannel images [24, 65, 66].

Let us now concentrate on the basic principles of 2D conventional DCT-based filtering, that is, component-wise processing of color images. In opposite to wavelet denoising, DCT-based filtering is performed in blocks [52, 53]. There are three main operations carried out with data in each block. First operation is a direct 2D DCT in *mn*-th block applied to image values $I_{ij}^{noise}, i = m, \ldots, m+7, j = n, \ldots, n+7$ (indices *mn* define the block left upper corner) and they are in the limits

$m = 1, \ldots, I-7, n = 1, \ldots, J-7$ for fixed block size of 8×8 pixels. Thus, DCT coefficients $D_{mn}(p,q), p = 0, \ldots, 7, q = 0, \ldots, 7$ are obtained, which then undergo a thresholding operation to obtain $D^{thr}_{mn}(p,q), p = 0, \ldots, 7, q = 0, \ldots, 7$. After this, inverse 2D DCT is applied to the thresholded coefficients. In this way, preliminary filtered values for all the image pixels belonging to a given block, that is, $I^{filt}_{ij}, i = m, \ldots, m+7, j = n, \ldots, n+7$ are obtained.

DCT-based denoising can be carried out in nonoverlapping, partly overlapping, or fully overlapping blocks. Filtering is more efficient (in terms of both standard and visual quality criteria) if fully overlapping blocks are used [53]. This means that each next block is shifted by only one pixel with respect to its neighbor. Then, for a given *ij*-th pixel, one has preliminary filtered values obtained from different positions of blocks that include this pixel ($m = i-7, \ldots, i, n = j-7, \ldots, j$). All these preliminary filtered values are averaged to obtain a final filtered value. In this chapter, we consider 8×8 blocks with a full overlap. The use of 8×8 pixel block allows accelerating the processing. Other block sizes (e.g., 16×16 or 9×9 pixels) might improve filtering performance but only slightly [69].

An important point in the DCT-based denoising is a thresholding. We have studied hard, soft [53], and combined [41] thresholding. Hard thresholding ($D^{thr}_{mn}(p,q) = D_{mn}(p,q),\ if\ |D_{mn}(p,q)| > T_{nm}(p,q)$ and zero otherwise, where $T_{mn}(p,q)$ denotes threshold that, in a general case, can vary locally and be frequency dependent) has demonstrated the best or equally good performance. It provides efficient noise suppression in homogeneous regions combined with rather good edge/detail/texture preservation and appropriate computational efficiency [41, 52, 53], even if processing with full overlapping of the blocks is used. Moreover, DCT-based filtering can be easily adapted to noise statistics and spatial correlation properties under the condition that necessary a priori information is available or retrieved from image under interest in interactive or blind manner with an appropriate accuracy [61, 64, 68].

If one deals with model 1 of the noise, the threshold is set fixed and frequency independent as $T_{mn}(p,q) = T = 2.6\sigma$ where σ denotes the noise standard deviation assumed a priori known. Following this recommendation allows producing output PSNR close to attainable optimum (maximum) for most considered test images and noise variance values [41]. If one wishes to provide optimal visual quality of filtered images, then it is recommended to apply $T \approx (2.2 \ldots 2.4)\sigma$ [41].

If the noise obeys model 2 and its normalized DCT spectrum $W_{norm}(p,q)$ is known or pre-estimated with desired accuracy, the frequency-dependent threshold has to be set: $T(p,q) = 2.6\sigma\sqrt{W_{norm}(p,q)}$. Recall that for all modifications of the DCT-based filters, the DC coefficients of DCT in blocks remain unchanged. To provide a slightly better visual quality of a processed image, the factor 2.6 in the expression above can be changed to smaller values, for example, 2.3 [69].

For the model 3, the threshold is set in a locally adaptive manner [52]. Assume that a dependence of local variance or standard deviation on the true value $\sigma_{kij} = f(I^{true}_{kij})$ is known or pre-estimated. Then, for a given block, T is calculated as $2.6\hat{\sigma}_{kmn} = 2.6f(\hat{I}^{mn}_{kmean})$ where $\hat{\sigma}_{kmn}$ is a local estimate of the nonstationary noise standard deviation obtained using a local mean $\hat{I}^{mn}_{kmean}$ estimate for an analyzed mn-th block. Again, an "optimal" visual quality is provided for a local threshold set as $T = 2.3f(\hat{I}_{kmean})$ [69].

3D version of the DCT-based filter for model 1 has been first proposed in [24]. It performs similarly to the 2D case in the sense of the sequence of basic operations. There are two main differences. First, $8 \times 8 \times 3$ block is used. Second, the first operation in each block is a "spectral" decorrelation (3-element DCT) of an RGB color image. After this, three 2D data arrays (layer images) are obtained which are processed separately as described above using fixed thresholds. After this, an inverse 3D DCT is performed.

This denoising approach can be easily extended to model 2 if all σ_k^2 are approximately equal to each other and the normalized spectra are almost the same for all color components (then averaged normalized spectra are exploited). A specific feature of 3D DCT-based filtering in this case [46] is that frequency-dependent thresholds are used as given above.

Finally, 3D DCT-based filtering can be also applied for model 3 noise. A distinctive feature of processing then consists in the usage of an appropriate variance stabilizing transform before exploiting the conventional 3D DCT-based denoising. Variance stabilization can be performed by logarithmic homomorphic transform [24], generalized Anscombe transform [65, 66], or other appropriate transforms depending on the properties of signal-dependent noise.

Although DCT-based filters are efficient enough [40,42,46], we have also studied two nonlocal mean filters; BM3D [54], oriented on denoising 2D (grayscale, component) images and its version C-BM3D [7], specially intended on filtering color images. The core of these two filters is a search of similar patches where C-BM3D carries out search for intensity image (Y component). Another peculiarity of these denoising techniques is that 3D filtering using DCT and Haar is performed for each obtained 3D cube built from obtained similar patches.

Two mechanisms of denoising produce, in aggregate, very efficient filtering. Conversion of RGB color images to another color space applied in the C-BM3D is, in fact, equivalent to decorrelation. Therefore, this filter can be considered as a 3D filter. Due to this, the C-BM3D filter occurs more efficient than the BM3D filter applied component-wise [7]. Note that BM3D was originally designed for model 1 noise. However, it can also be adapted to model 2 noise [45] and even to model 3 noise (after corresponding homomorpic transform [65]).

Any filtering should be characterized by a quality metric which is able to describe its efficiency adequately and quantitatively. First of all, it should be stressed that there is no commonly accepted standard among visual quality metrics; their design and testing continues [37, 38]. According to the latest results [38], the recently proposed color version of the metric FSIM (denote it as FSIM_c) [57] is currently the leader for all 24 types and 5 levels of distortions presented in the database TID2013. However, other visual quality metrics demonstrate efficiency (adequateness) close to FSIM_c [38], especially for particular types of distortions which are under interest in our case [38]. These metrics are PSNR-HMA and PSNR-HA [56] (available at http://www.ponomarenko.info/psnrhma.htm), PSNR-HVS-M [70], MSSIM [58], etc.

Since all existing visual quality metrics are not perfect, it is worth jointly considering several of them for providing substantial conclusions. Besides, it is desirable to have some background on behavior and properties of these metrics. The metrics $FSIM_c$ [57] and MSSIM [58] are based on the wavelets and have the values in the range from 0 (very bad visual quality) to 1 (perfect quality). The values of $FSIM_c$ are usually slightly greater than the corresponding values of MSSIM, where the values of the latter metric are equal to or greater than 0.99 relate to distortions that cannot be noticed by visual inspection [71]. If MSSIM values for two different images differ by 0.005 or more, then these images have different visual quality. Otherwise, their visual quality is of the same order [68]. Although the metric SSIM [72] is often used in the analysis of filtering efficiency [14, 35], we prefer MSSIM and $FSIM_c$ as they provide better performance [38].

The metrics PSNR-HMA, PSNR-HA [56] and PSNR-HVS-M [70] are all based on DCT in 8×8 pixel blocks. Their values are expressed in dB and the values exceeding 40 dB correspond to practical invisibility of distortions [71]. If the values of PSNR-HVS-M for two different images differ by 0.5 dB and more, then the difference in their visual quality is noticeable [68]. Otherwise, the images might appear as having the same visual quality.

It has also been established in recent papers [58] that the metrics that take into account different sensitivities of humans to distortions in intensity and color ($FSIM_c$, PSNR-HMA, and PSNR-HA) perform slightly better than the metrics calculated for each color component independently and then averaged (e.g., PSNR-HVS-M and MSSIM). Thus, keeping these properties in mind, we are equipped for analyzing the obtained denoising data.

3.4 Component-Wise Denoising and Prediction

As the simplest case, consider first component-wise denoising for images in the database TID2008 for the model 1 noise. Output PSNR values for AWGN with variance σ_a^2 equal to 65 are presented in plots in Fig. 3.3 (bar color corresponds to color component). Note that input PSNR ($PSNR_{inp}$) for all considered images equals to 30 dB. Thus, we can easily calculate how large the difference $PSNR_k(n) - PSNR_{inp}$ (improvement of PSNR—IPSNR) due to the filtering is.

Analysis of the represented plots shows the following. First, IPSNR is practically the same for all color components for a given image and noise variance. Second, IPSNR varies in rather wide limits—from about 1 dB by up to 7.5 dB for the considered plots. In fact, IPSNR can be smaller (for highly textural images and/or smaller noise variance [74]) and larger (for simple structure images and/or larger noise variance) [73]. Third, it is seen that an efficiency of image filtering (expressed, e.g., in terms of IPSNR) depends on the image "complexity," although this term is not either introduced verbally or exactly characterized quantitatively. IPSNR is the largest for

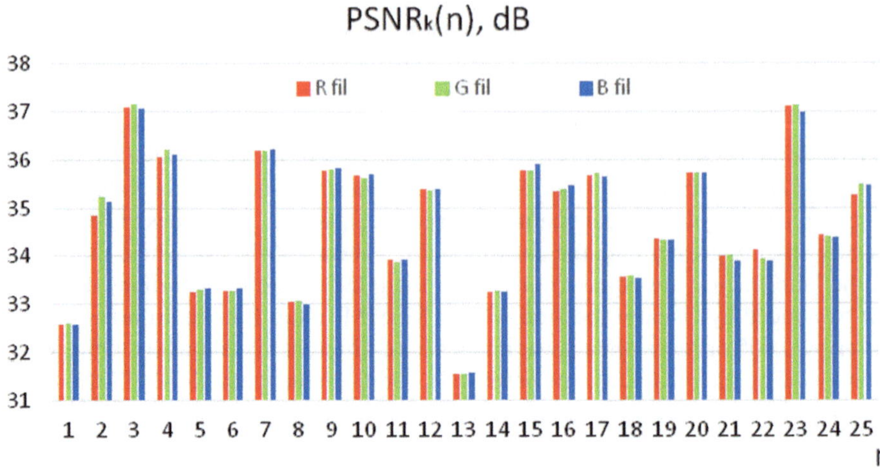

Fig. 3.3 Output *PSNR(n)* for the DCT-based filter applied component-wise for $\sigma_a^2 = 65$, AWGN

images of "simple structure" that contain rather large homogeneous regions as, for example, the test color images ## 3, 7, and 23. Meanwhile, IPSNR is comparatively small for images of "complex" structure that contain a lot of textural areas and/or many small-sized objects and edges. Examples of such images are the test images ## 1, 5, 6, 8, 13, 14 (see Fig. 3.1).

One might think that the problem of efficient noise removal in the latter case (for textural images) deals only with the considered filter and its principle of operation. Let us show that this is not true. Recently P. Chatterjee and P. Milanfar [48] have proposed a method for determining lower bound output MSE for nonlocal-based denoising for AWGN case under condition of available noise-free image and known variance (or standard deviation) of the noise. Using this method and software freely available from the authors of [48], we have obtained the values of lower bound MSE (denoted as MSE_{lb}) for many components of our color test images and noise variances. Some of them are presented in Table 3.3 taken from the paper [40] for R color component (more data are given in the papers [42] and [69]). Along with MSE_{lb}, we give output MSE for the considered DCT-based filter (denoted as MSE_{DCT}).

As it is seen, for the most textural image # 13, the ratio MSE_{lb}/ σ^2 is large. This means that, even potentially, filtering cannot have high efficiency. Meanwhile, for the test images ## 5 and 13 the ratios MSE_{lb}/MSE_{DCT} are about 0.9. This means that the DCT-based filter practically reaches potential efficiency of image denoising for highly textural image especially if the noise intensity is high. However, there remains room for filtering efficiency improving for simpler structure images.

One might also think that modern nonlocal denoising techniques are able to produce results essentially better than the DCT-based filter considered above. For comparison purpose, let us present data for σ_a^2= 65 and R component of color images processed by the state-of-the-art nonlocal BM3D filter [54]. The produced output MSEs are equal to 27.8, 28.3, 45.0, 29.4, and 27.5 for the test images # 5,

Table 3.3 Lower bound and obtained MSEs for R components of some images of the database TID2008

Image #	$\sigma^2 = 65$		$\sigma^2 = 130$	
	MSE_{lb}	MSE_{DCT}	MSE_{lb}	MSE_{DCT}
5	27.5	30.4	48.0	52.4
8	23.2	32.5	40.8	53.7
13	41.0	46.6	72.8	80.5
14	20.7	31.0	35.6	50.5
18	21.4	28.7	37.7	47.1

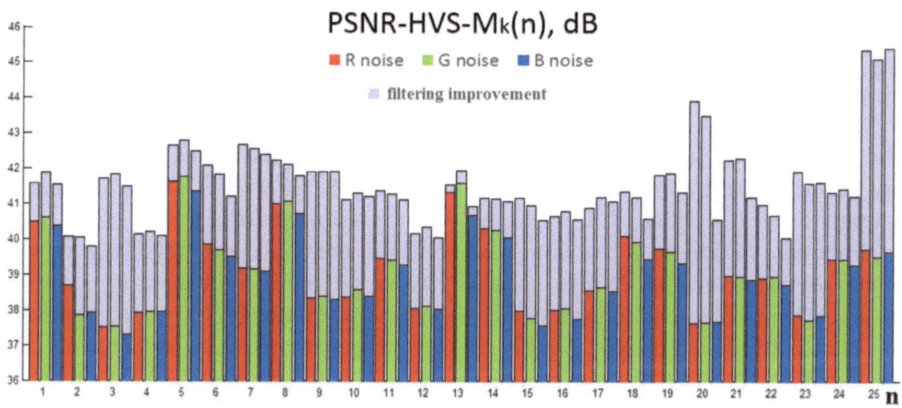

Fig. 3.4 *PSNR-HVS-$M_k(n)$* before and after filtering for $\sigma_a^2 = 25$, AWGN

8, 13, 14, and 18, respectively. Comparing these data to the corresponding data in Table 3.3, we can state that the BM3D is in fact more efficient than the DCT-based filter but only slightly. BM3D produces closer output MSEs to MSE_{lb} in all five cases. However, the difference between MSEs for the DCT and BM3D filters is only about 3 ... 13 %.

We have already mentioned the inadequacy of conventional metrics as output MSE or PSNR to characterize image quality and efficiency of filtering. Because of this, let us also analyze the efficiency of component-wise denoising in terms of visual quality metric, PSNR-HVS-M. The values of PSNR-HVS-M_k for original (noise) and filtered images are represented in Fig. 3.4. Although input PSNR for all these images is the same and equal to 34.2 dB ($\sigma_a^2 = 25$, Fig. 3.4), input PSNR-HVS-M_k (shown by bars of corresponding color) differ by about 4 dB. This is because the analyzed metric takes into account the effects of noise masking by texture. Because of this, for the most textural images ## 1, 5, 8, 13, and 14 the values of PSNR-HVS-M_k for all the color components are over 40 dB. Thus, there is no need to remove the noise in these images at all, since the noise is not visible in them. It is also worth stressing that if DCT-based filtering is applied (for filtered images

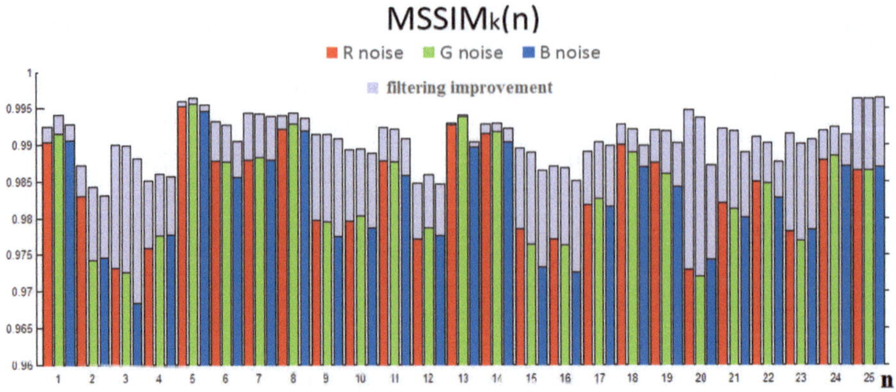

Fig. 3.5 *$MSSIM_k(n)$* before and after filtering for $\sigma_a^2 = 25$, AWGN

the metric values are shown by the same color bars) to these images, their quality characterized by improvement of PSNR-HVS-M_k (IPSNR-HVS-M_k) becomes slightly better (by about 1 dB or even less for the test image # 13).

For other, middle- and low-complexity images, there is a considerable improvement of visual quality. While noise is visible in original images, it becomes practically invisible in denoised ones (the values of PSNR-HVS-M_k exceed 40 dB for almost all filtered images, see Fig. 3.4). More quantitative data and analysis can be found in [42] and [74]. In particular, it is shown that IPSNR-HVS-M can reach 7 . . . 8 dB for simple structure images corrupted by intensive AWGN.

Now consider the same situation (AWGN, $\sigma_a^2 = 25$) but let us analyze the wavelet-based metric MSSIM. The results are presented in Fig. 3.5. The conclusions that can be drawn from analysis of data in Fig. 3.5 are the same as given earlier for results in Fig. 3.4. Note here that the metric values for all three color components are close to each other. Again, there are noisy images for which noise can be hardly noticed ($MSSIM_k > 0.99$, see color bar data in Fig. 3.5). For most images, residual noise becomes invisible after filtering (analyze violet bar data in Fig. 3.5).

Then, there is a question: Is it possible to predict filtering efficiency for a given image and noise variance (assumed known)? This question is tightly connected to other questions—is it worth carrying out filtering for a considered case and is it possible to design an automatic procedure for filtering efficiency prediction and undertaking a decision on expedience of denoising in each particular case.

One step forward to answer these questions was done by Zhu and Milanfar [49]. They have shown that potential efficiency of filtering can be predicted without having noise-free image. However, their approach suffers from two drawbacks. First, one needs prediction that is done quickly (faster than filtering itself) while the algorithm [49] requires intensive computations. Second, the method [49] allows predicting potential efficiency whereas efficiency reachable in practice is desired.

Second step has been motivated by the results in [69] and carried out in [50]. The main idea behind the prediction is the following. Suppose that one has a (statistical) parameter calculated for input (noisy) image and this parameter is strictly

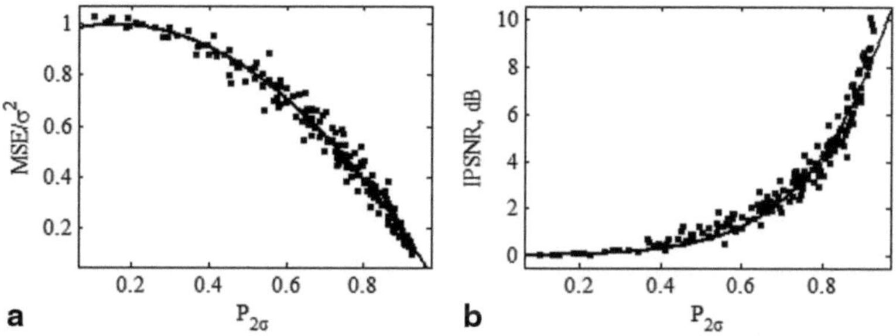

Fig. 3.6 Scatterplots of MSE_{out}/σ^2 (**a**) and *IPSNR* (**b**) on $P_{2\sigma}$ and the fitted curves

connected with an output parameter that is able to characterize filtering efficiency (e.g., IPSNR). Then, calculation of input parameter allows predicting the desired output parameter and to undertake a proper decision.

In fact, prediction can be considered appropriate under the following assumptions that should be valid. First, there are one or a few parameters that can be calculated easily and quickly for a given noisy image. Second, the aforementioned dependence between the considered input and output parameters should be strict and able to adequately characterize filtering efficiency (the dependence is obtained in advance, i.e., off-line).

In [50], it is shown that as input (controlled) parameters it is possible to use some statistics of DCT coefficients in 8 × 8 blocks. One of them is a probability $P_{2\sigma}$ that defines mean probability that AC DCT coefficient absolute value does not exceed 2σ, where σ is a standard deviation of the noise. It has been established in [69] that $P_{2\sigma}$ is strongly connected with the ratio of output MSE to the noise variance. This has allowed obtaining polynomial and other simple dependences of $MSE_{\text{out}}/\sigma^2$ on $P_{2\sigma}$ characterized by high values (about 0.95) of coefficient of determination R^2 [75] (it tends to unity if approximation is perfect). Dependences have been obtained for the DCT-based and BM3D filters. More in detail, the dependences were fitted into scatterplots obtained for a wide set of the test images and noise variances. An example of such scatterplot is presented in Fig. 3.6a. Each point corresponds to one test image corrupted by AWGN with a given variance and processed by a considered filter where horizontal axis relates to the input (calculated) parameter and vertical axis—to the output parameter that describes filtering efficiency and has to be predicted. Having a fitted curve (see examples in Fig. 3.6), it is possible in practice to calculate $P_{2\sigma}$ for an analyzed image and to substitute $P_{2\sigma}$ as argument to the expression characterizing the fitted curve to get the predicted value.

Note that as an output (predicted) parameter it is also possible to use IPSNR expressed in decibel (dB) (see the plot in Fig. 3.6b). As it follows from analysis of the presented plots, it is obviously not worth performing image filtering if $P_{2\sigma}$ is less than 0.45.

Research in this direction continues. It has been recently shown that the proposed approach is, after the corresponding modifications, applicable to processing

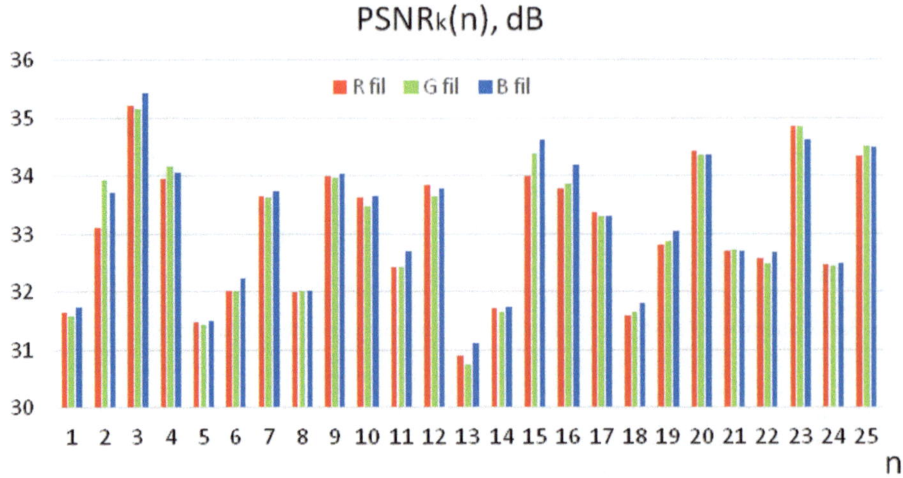

Fig. 3.7 Output $PSNR_k(n)$ for the DCT-based filter with frequency-dependent thresholds applied component-wise, spatially correlated noise with variance 65

images corrupted by model 3 noise [51]. It has been also shown that there is no necessity to calculate $P_{2\sigma}$ in fully overlapping blocks as it was used in [30]. In practice, it is enough to have not less than 300 randomly selected blocks to estimate $P_{2\sigma}$ and to further use it in prediction. This essentially accelerates the image processing performed at the prediction stage and makes this stage considerably faster than DCT-based filtering and by several orders faster than BM3D denoising. Certainly, there are particular tasks needed to be solved, such as what type of curve to fit, is it better to analyze MSE_{out}/σ^2, IPSNR or other parameters, what parameter(s) to use instead of $P_{2\sigma}$, etc.

Now consider the model 2 noise. For the very beginning, let us analyze a dependence $PSNR_k(n)$ presented in Fig. 3.7 for the additive spatially correlated noise with the variance 65. Recalling that PSNR for original images is 30 dB, IPSNR due to filtering is observed for all color components of all test images. However, IPSNR again depends on image complexity and is the smallest for textural images (## 1, 5, 8, 13, 14).

Since noise variance for data in plots of Fig. 3.7 is the same as for AWGN data in Fig. 3.3, it is possible to compare efficiency of removing spatially correlated and white noise. For AWGN, minimal and maximal output PSNRs are equal to 31.3 and 37.2 dB, respectively (see the plots in Fig. 3.3). If the noise is spatially correlated, minimal and maximal output PSNR values are equal to 31.0 and 35.5 dB. This means that suppression of spatially correlated noise is a more difficult task than denoising in the case of AWGN.

Now, let us analyze the DCT-based filter performance in terms of visual quality metrics, namely PSNR-HVS-M. The data for a spatially correlated noise with the variance equal to nine are presented in Fig. 3.8. As it is seen, visual quality of the

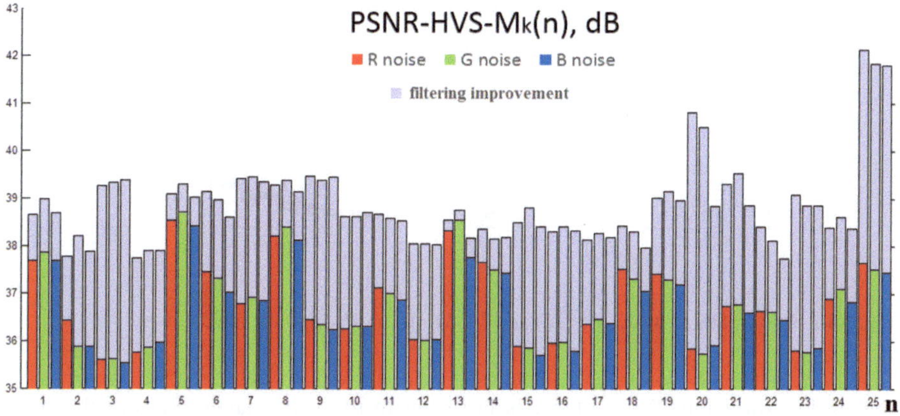

Fig. 3.8 *PSNR-HVS-M$_k$(n)* before and after DCT-based filtering with frequency dependent thresholds for spatially correlated noise with $\sigma_a^2 = 9$

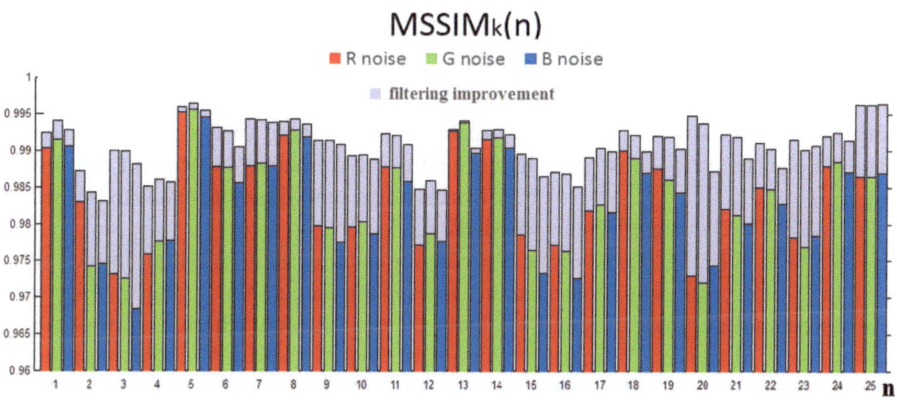

Fig. 3.9 *MSSIM$_k$(n)* before and after DCT-based filtering with frequency-dependent thresholds for spatially correlated noise with $\sigma_a^2 = 9$

noisy images considerably depends on image complexity and noise masking properties. In all test noisy images, spatially correlated noise is visible (PSNR-HVS-M is essentially smaller than 40 dB), although the noise variance is quite small (Fig. 3.8). The data in Fig. 3.8 relate to the denoised component images. It is seen (compare data in Fig. 3.8) that visual quality improves due to filtering for all test images, although there are some test images (e.g., the test image # 13) for which there is practically no benefit. The largest improvement is observed for simple structure images (e.g., the test image # 3 and 23). Improvement of PSNR-HVS-M is about 3 dB that corresponds to considerable enhancement. However, even after filtering PSNR-HVS-M values are mostly smaller than 40 dB, that is, residual noise and distortions due to filtering are visible.

Similar conclusions can be drawn from analysis using the wavelet-based metric MSSIM (see data in [42] and in Fig. 3.9). According to this metric and distortion

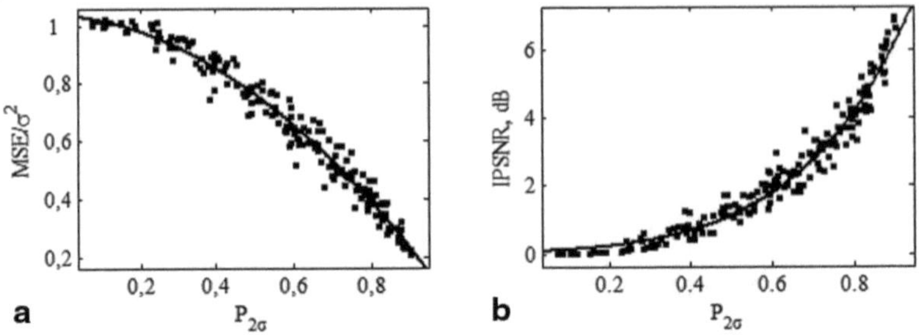

Fig. 3.10 Scatterplots of MSE_{out}/σ^2 (**a**) and *IPSNR* (**b**) on $P_{2\sigma}$ and the fitted curves

invisibility threshold for it, noise is not seen in the test images that are the most textural (## 5, 8, 13 and, with high probability, ## 1 and 14, see color bar data in Fig. 3.9). Residual noise and distortions introduced by the considered filter are (with high probability) invisible in most filtered images except the test images ## 2, 4, 12, and 16 that have many large homogeneous image regions and/or regions with low-contrast textures where noise suppression is problematic [72].

Then, a question appears: Is the prediction possible for a spatially correlated noise. A preliminary answer is "yes," as it follows from the analysis of scatterplots presented in Fig. 3.10. The difference compared to the model 1 case is that the parameter $P_{2\sigma}$ has been determined using frequency-dependent thresholds $T(p,q) = 2\sigma\sqrt{W_{norm}(p,q)}$. In other words, noise spatial spectrum has to be known in advance for filtering efficiency prediction.

The plots in Fig. 3.10 show that a prediction is potentially possible. In fact, for a given noise spectrum it is possible to use a set of test images to simulate a noise with the corresponding properties and to obtain the dependence, which is approximated (fitted) well. However, comparing the plots in Figs. 3.6 and 3.10, it can be noticed that the fitting curves differ a little (not in general behavior but in their parameters). In particular, maximal IPSNR for AWGN is about 10 dB for $P_{2\sigma}$ approaching to unity, while maximal IPSNR for model 2 noise is only about 7 dB. This means that for a good prediction one should be able to encounter the noise spatial correlation, that is, to introduce some spectrum-dependent correcting factors. This task is not solved yet.

Besides, it would be nice to predict some visual quality metrics or their changes due to denoising. One option is to predict IPSNR-HVS-M. To prove the possibility of this, Fig. 3.11 shows the scatterplots of IPSNR-HVS-M on $P_{2\sigma}$ for AWGN and spatially correlated noise. One observation is that the obtained data are clustered but not so well as for IPSNR. Because of this, fitting is not so good and, thus, prediction is not so accurate for IPSNR-HVS-M as it is for IPSNR. The fitted curves differ a little for $P_{2\sigma}$ about 0.6 for AWGN and spatially correlated noise, although maximal values are of the same order. However, analyzing the obtained dependences, we can surely state that it is not worth carrying out image filtering if $P_{2\sigma}$ is smaller than

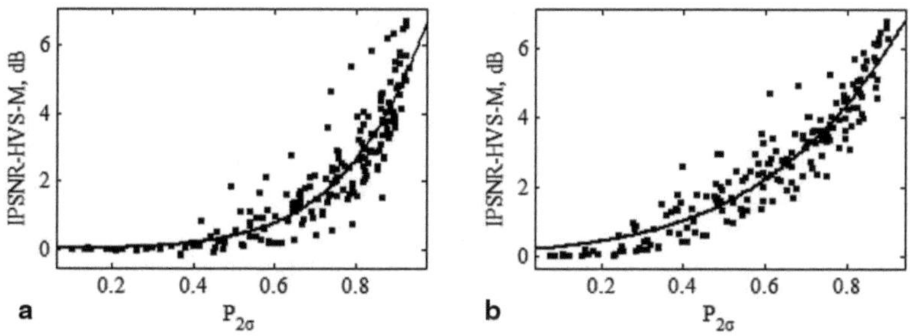

Fig. 3.11 Scatterplots of *IPSNR-HVS-M* on $P_{2\sigma}$ for AWGN (**a**) and spatially correlated (**b**) noise

0.35, then IPSNR-HVS-M is small (close to zero). More detailed study has shown that this, as it can be expected, happens for highly textural images. We hope that accuracy of predicting IPSNR-HVS-M can be improved in the future.

Summarizing the results for component-wise denoising, we can state that there are practical situations (textural images and/or quite low level of the noise) when component-wise denoising is unable to provide image enhancement. Such situations can be predicted.

3.5 3D Denoising of Color Images

Analysis for 3D denoising can be divided into two stages: theoretical and practical (obtained by simulations for a wide set of test images and noise variances). Theory [44] states the following: due to 3D processing of data (instead of component-wise denoising), it is possible to decrease output MSE by 1.6 ... 2.2 times, that is, to increase PSNR by 2 ... 3.5 dB. This is a considerable advantage (benefit) worth implementing in practice.

Let us see what is gained in practice. First, reduction of output MSE by about 1.8 times really takes place for 3D DCT-based filter in the case of AWGN (see simulation data in [40, 42, 45]). This image enhancement in terms of conventional quality metric is supported by improvement of visual quality. Figure 3.12 presents the values of the metric MSSIM for images of TID2008. As it is seen, 3D denoising (3 comp) is always more efficient than component-wise processing (1 comp) that can sometimes lead even to reduction of visual quality (see data for the test image # 13).

Similar conclusions hold for the filters BM3D and C-BM3D. Dependences for them are given in Fig. 3.13 where the metric PSNR-HMA is used. It is worth stressing that PSNR-HMA is considerably improved due to 3D filtering (by C-BM3D) just for textural test images ## 1, 5, 8, 13. The difference is about 3 dB and it is, obviously, sufficient. This means that the use of 3D denoising can be especially expedient for enhancing highly textural images for which component-wise processing has a limited efficiency.

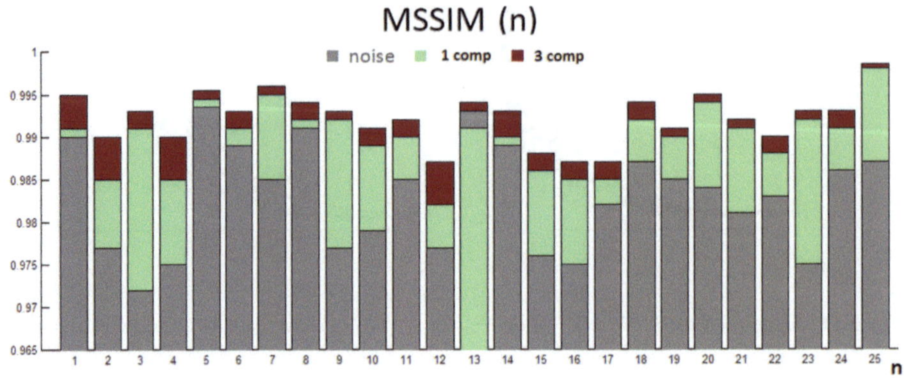

Fig. 3.12 Dependences *MSSIM(n)* for images in TID2008 corrupted by AWGN with variance 65, 2D (component-wise) and 3D DCT-based filters

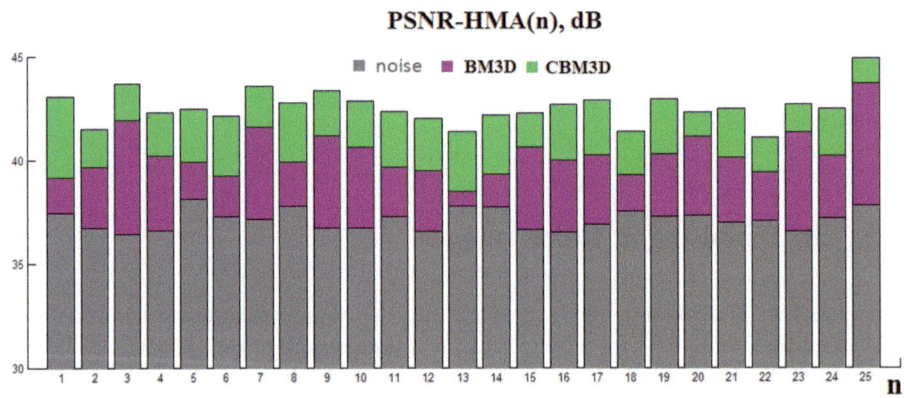

Fig. 3.13 Dependences *PSNR-HMA(n)* for images in TID2008 corrupted by i.i.d. noise with variance 65, BM3D and C-BM3D filters

Similar effects are observed for the model 2 noise. To prove this, Fig. 3.14 presents dependences of PSNR-HVS-M(n) for component-wise (1 comp) and 3D DCT-based filters (3 comp) that both employ frequency-dependent thresholds. Improvement of this metric varies in the limits from 0.5 to 2.5 dB where the largest improvement takes place for the most textural images.

This is a positive moment since component-wise filtering can be not efficient for complex structure images. More details and simulation data can be found in the paper [74].

Efficiency of 3D processing for the model 3 noise is briefly studied in the paper [65]. The obtained results are very similar. 3D processing (after variance stabilizing transform) provides decrease of output MSE by 1.5 … 2 dB compared to component-wise denoising.

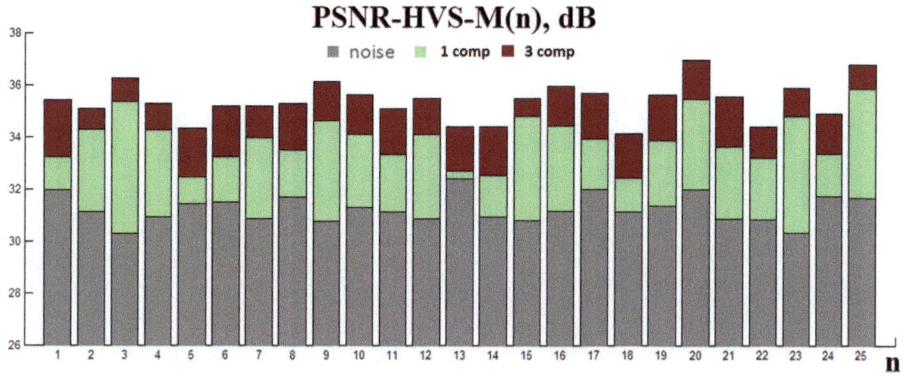

Fig. 3.14 Dependences *PSNR-HVS-M(n)* for images in TID2008 corrupted by spatially correlated noise with variance 65, 2D and 3D DCT-based filters

In general, 3D DCT-based denoising can be applied to processing not necessarily three but also two, four or other number of components of multichannel images (here we mean not color but other kinds of multichannel images as, e.g., mustispectral ones). It can be expected that a larger number of components incorporated in 3D denoising leads to more efficient filtering. To check this expectation, we needed test images with different number of noise-free components for which it was possible to artificially add noise with desired characteristics. Since standard test multichannel images, to the best of our knowledge, are absent, we have carried out the following (slightly artificial) experiment: first, we have processed together only two components (B and G) of color test images exploited above. Second, we have jointly processed all three color components. Third, we have filtered four-component images. They have been formed as follows. First three components of noise-free data were noise-free R, G, and B components of color images and the fourth component was obtained as intensity Y of these images (this component has high correlation with all the color components). Then, AWGN with the same variance has been independently added to all four components.

It occurred that efficiency of filtering depends on what is the position of Y component in 3D data array. For example, the sequence of component positions RGBY is not the best. On the average, the sequence RYGB has shown itself to be the best. So, we present the results below (Fig. 3.15) just for the latter sequence. The plots are given for the metric PSNR-HVS-M (its value averaged for all considered components of multichannel data is presented, the number of jointly processed components is marked in plots) and variance of AWGN equal to 65 for all images of the TID2008.

It can be seen that filtering efficiency increases with increased number of processed components. In particular, for three components the benefit is, on the average, about 1 dB compared to processing two components. This is one more argument in favor of 3D denoising. Note that similar conclusions can be drawn from analysis of the plots for another visual quality metric, FSIMc(n), represented in Fig. 3.16.

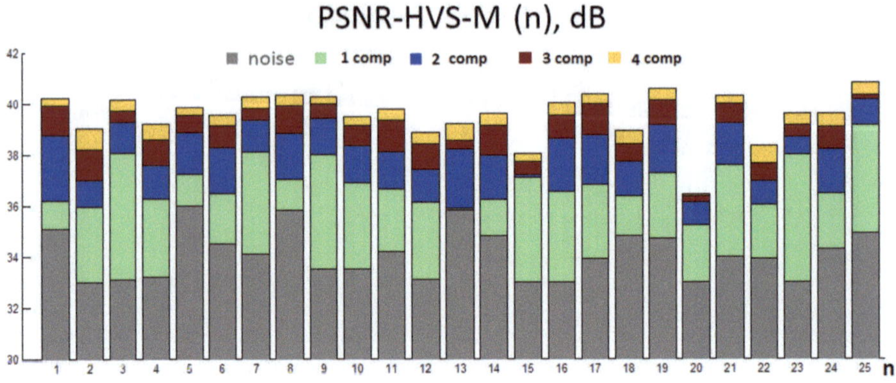

Fig. 3.15 Dependences *PSNR-HVS-M(n)* for different number of processed components, 3D DCT-based filter, $\sigma_k^2 = 65$, AWGN

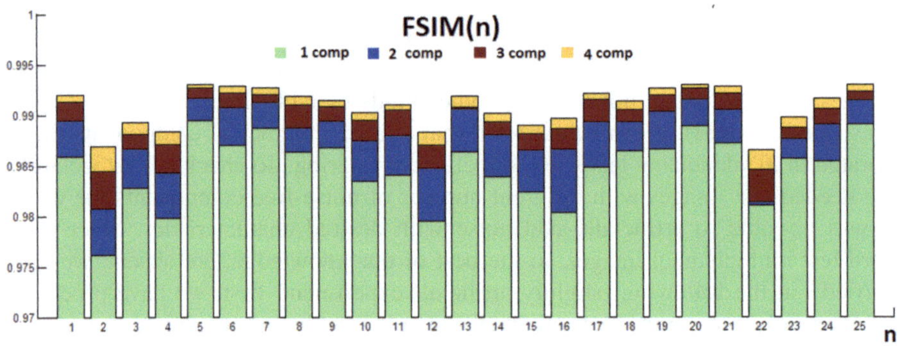

Fig. 3.16 Dependences *FSIMc(n)* for different number of processed components, 3D DCT-based filter, $\sigma_k^2 = 65$, AWGN

The presented results show that efficiency of 3D denoising seems to be roughly predictable as well. In particular, it is possible to determine IPSNR or IPSNR-HVS-M for one component of color image and then to add 2 dB to the obtained value for taking into account better efficiency of 3D denoising compared to 2D counterpart. However, this methodology of prediction needs additional verification.

3.6 Conclusions

The chapter describes several important aspects in enhancement of color images by noise removal. In particular, the following is demonstrated.

For a given noise variance, spatially correlated noise is a more destructive factor than AWGN and it is harder to remove spatially correlated noise than AWGN.

There are highly textural images for which component-wise denoising is unable to enhance images in terms of standard and visual quality metrics; this takes place for good representatives of both transform-based and nonlocal filters.

Such practical situations can be predicted by quite simple and fast preprocessing of noisy images under the condition that noise statistics and spectrum are available.

3D processing obviously provides considerable improvement compared to 2D filtering; thus, 3D denoising can be the only option to enhance textural color images.

References

1. Plataniotis, K.N., Venetsanopoulos, A.N.: Color Image Processing and Applications. Springer-Verlag, NY (2000)
2. Lukac, R. (ed.): Single-Sensor Imaging: Methods and Applications for Digital Cameras (Image Processing Series).CRC Press, Boca Raton (2009)
3. Abbas, Q., Celebi, M.E., Serrano, C., Fondón García, I., Ma, G.: Pattern classification of dermoscopy images: A perceptually uniform model. J. Pattern Recognit. **46**(1), 86–97 (2013)
4. Schowengerdt, R.: Remote sensing: Models and methods for image processing. Academic Press, Orlando (2006)
5. Smolka, B., Plataniotis, K.N., Venetsanopoulos, A.N.: Nonlinear techniques for color image processing. In: Barner, K., Arce, G. (eds.) Nonlinear Signal and Image Processing: Theory, Methods,and Applications, Electrical Engineering & Applied Signal Processing Series. CRC Press, Boca Raton (2003)
6. Morillas, S., Schulte, S., Melange, T., Kerre, E., Gregori, V.: A soft-switching approach to improve visual quality of colour image smoothing filters. In: Blanc-Talon, J., Philips, W., Popescu D., Scheunders P. (eds.) Proceedings of ACIVS, Springer Series on LNCS, vol. 4678, pp. 254–261. Springer, Berlin (2007)
7. Dabov, K., Foi, A., Katkovnik, V., Egiazarian, K.: Color image denoising via sparse 3D collaborative filtering with grouping constraint in luminance-chrominance space. In: Proceedings of IEEE Int. Conf. on Image Process. ICIP 2007, San Antonio, TX, USA, September 2007, pp. 313–316 (2007)
8. Ponomaryov, V., Gallegos-Funes, F., Rosales-Silva, A.: Real-time color image processing using order statistics filters. J. Math. Imaging Vis. **23**(3), 315–319 (2005)
9. Beghdadi, A., Larabi, M.C., Bouzerdoum, A., Iftekharuddin, K.M.: A survey of perceptual image processing methods. Signal. Process Image Commun. **28**(8), 811–831 (2013)
10. Phillips, R.D., Blinn, C.E., Watson, L.T., Wynne, R.H.: An adaptive noise-filtering algorithm for AVIRIS data with implications for classification accuracy. IEEE. Trans. GRS. **47**(9), 3168–3179 (2009)
11. Zhong, P., Wang, R.: Multiple-spectral-band CRFs for denoising junk bands of hyperspectral imagery. IEEE Trans. Geosci. Remote Sens. **51**(4), 2260–2275 (2013)
12. Maggioni, M., Boracchi, G., Foi, A., Egiazarian, K.: Video denoising, deblocking and enhancement through separable 4D nonlocal spatiotemporal transforms. IEEE. Trans. Image Proc. **21**(9), 3952–3966 (2012)
13. Dai, J., Au, O.C., Zou, F., Pang, C.: Generalized multihypothesis motion compensated filter for grayscale and color video denoising. Signal Process **93**(1), 70–85 (2013)
14. Kravchenko, V., Ponomaryov, V., Pustovoit, V.: Filtering of multichannel video sequences distorted by noise, based on the fuzzy-set theory. Dokl. Phys. **58**(10), 447–452 (2013)
15. Astola, J., Haavisto, P., Neuvo, Y.: Vector median filters. Proc. IEEE. **78**, 678–689 (1990)
16. Khriji, L., Gabbouj, M.: Vector median-rational hybrid filters for multichannel image processing. IEEE. Signal Process Lett. **6**(7), 186–190 (1999)
17. Smolka, B.: Adaptive truncated vector median filter. Proc. IEEE. Int. Conf. Comput. Sci. Automation Eng. (CSAE). **4**, 261–266 (2011)

18. Ponomaryov, V.: Real-time 2D-3D filtering using order statistics based algorithms. J. Real-Time Image Proc. **1**(3), 173–194 (2007)
19. Wang, W., Lu, P.: An efficient switching median filter based on local outlier factor. Signal. Process Lett. IEEE. **18**, 551–554 (2011)
20. Jiangtao, X., Wang, L., Shi, Z.: A switching weighted vector median filter based on edge detection. Signal Process **98**, 359–369 (2014)
21. Lebrun, M., Colom, M., Buades, A., Morel, J.M.: Secrets of image denoising cuisine. J. Acta Numer. **21**(1), 475–576 (2012)
22. Pizurica, A., Philips, W., Scheunders, P.: Wavelet domain denoising of single-band and multi-band images adapted to the probability of the presence of features of interest. Proc. SPIE. 591Wavelets XI. **59141**, (2005). doi:1117/619386
23. Colom, M., Buades, A.: Analysis and extension of the percentile method, estimating a noise curve from a single image. Image Process On Line **3**, 332–359 (2013). http://dx.doi.org/5201/ipol.2090
24. Ponomarenko, N.N., Lukin, V.V., Zelensky, A.A., Koivisto, P.T., Egiazarian, K.O.: 3D DCT based filtering of color and multichannel images. J. Telecommun. Radio Eng. **67**, 1369–1392 (2008)
25. Rosales-Silva, A., Gallegos, F.F., Ponomaryov, V.: Fuzzy Directional (FD) Filter for impulse noise reduction in colour video sequences. J. Vis. Commun. Image Represent. **23**(1), 143–149 (2012)
26. Camarena, J., Gregori, V., Morillas, S., Sapena, A.: A simple fuzzy method to remove mixed Gaussian-impulsive noise from colour images. IEEE. Trans. Fuzzy Syst. **21**(5), 971–978 (2013)
27. Schulte, S., Morillas, S., Gregori, V., Kerre, E.E.: A new fuzzy color correlated impulse noise reduction method. IEEE. Trans. Image Process **16**(10), 2565–2575 (2007)
28. Ponomaryov, V., Montenegro-Monroy, H., Nino-de-Rivera, L., Castillejos, H.: Fuzzy filtering method for color videos corrupted by additive noise. Sci. World J. **2014**, 21 (2014) (Article ID 758107)
29. Ponomaryov, V., Rosales, A., Gallegos, F.: 3D filtering of colour video sequences using fuzzy logic and vector order statistics. In: Blanc-Talon, J., Philips, W., Popescu, D., Scheunders, P. (eds.) Proceedings of the 15th International Conference on "Advanced Concepts for Intelligent Vision Systems", vol. LNCS 5807, pp. 210–221. Springer, Berlin (2009)
30. Morillas, S., Gregori, V., Hervas, A.: Fuzzy peer groups for reducing mixed gaussian-impulse noise from color images. IEEE. Trans. Image Process **18**(7), 1452–1466 (2009). doi:1109/TIP.202019305
31. Elad, M.: Sparse and Redundant Representations. From Theory to Applications in Signal and Image Processing. Springer Science + Business Media, LLC (2010)
32. Shreyamsha-Kumar, K.: Image denoising based on non-local means filter and its method noise thresholding. Signal Image Video Process **7**(6), 1159–1172 (2013)
33. Mairal, J., Elad, M., Sapiro, G.: Sparse representation for color image restoration. IEEE Trans. Image Process **17**(1), 53–69 (2008)
34. Mairal, J., Bach, F., Ponce, J., Sapiro, G., Zisserman, A.: Non-local sparse models for image restoration. In: Proc. IEEE 12th International Conference on Computer Vision (ICCV), pp. 2272–2279 (2009)
35. Han, Y., Chen, R.: Efficient video denoising based on dynamic nonlocal means. Image Vis. Comput. **30**(2), 78–85 (2011)
36. Yan, R., Shao, L., Liu, Y.: Nonlocal hierarchical dictionary learning using Wavelets for image denoising. IEEE. Trans. Image Process **22**(12), 4689–4698 (2013)
37. Chandler, D.M.: Seven challenges in image quality assessment: Past, present, and future research. J. ISRN. Signal Process **2013**, 53 (2013). doi:1155/2013/905685 (Article ID 905685)
38. Ponomarenko, N., Ieremeiev, O., Lukin, V., Egiazarian, K., Jin, L., Astola, J., Vozel, B., Chehdi, K., Carli, M., Battisti, F., Jay Kuo, C.-C.: A New color image database TID2013: Innovations and results. In: Proceedings of ACIVS, Poznan, Poland, Oct. 2013, pp. 402–413 (2013)

39. Vansteenkiste, E., Van der Weken, D., Philips, W., Kerre, E.: Perceived image quality measurement of state-of-the-art noise reduction schemes. Lecture Notes Comput. Sci. ACIVS. **4179**, 114–124 (2006)
40. Lukin, V., Abramov, S., Ponomarenko, N., Egiazarian, K., Astola, J.: image filtering: Potential efficiency and current problems. In: Proceedings of ICASSP, May 2011, Prague, Chech Republic, pp. 1433–1436 (2011)
41. Lukin, V., Ponomarenko, N., Egiazarian, K.: HVS-metric-based performance analysis of image denoising algorithms. In: Proceedings of EUVIP, Paris, France, 2011, pp. 156–161 (2011)
42. Fevralev, D., Lukin, V., Ponomarenko, N., Abramov, S., Egiazarian, K., Astola, J.: Efficiency analysis of color image filtering. EURASIP J. Adv. Signal Process **2011**(41), (2011). doi:1186/1687-6180-2011-41
43. Motwani, M.C., Gadiya, M.C., Motwani, R.C., Harris, F.C.: Survey of image denoising techniques. In: Proceedings of GSP 2004, Santa Clara, CA, USA, 27–30 September (2004)
44. Uss, M., Vozel, B., Lukin, V., Chehdi, K.: Potential MSE of color image local filtering in component-wise and vector cases. In: Proceedings of CADSM, February 2011, Ukraine, pp. 91–101 (2011)
45. Rubel, A.S., Lukin, V.V., Egiazarian, K.: Metric performance in similar blocks search and their use in collaborative 3D filtering of grayscale images. In: Proceedings of SPIE, vol. 9019 (2014). doi:1117/2039247
46. Ponomarenko, N.N., Lukin, V.V., Zelensky, A.A., Egiazarian, K.O., Astola, J.T.: Performance evaluation for 2D and 3D filtering methods of noise removal in color images. In: Proceedings SPIE Conference Image Processing: Algorithms and Systems VIII, San Jose, USA, vol. 8295, 12 p. (2012)
47. Lim, S.H.: Characterization of noise in digital photographs for image processing. J. IS & T/SPIE. Electron Imaging **6069**, 1–11 (2008). doi:1117/655915
48. Chatterjee, P., Milanfar, P.: Is denoising dead?. J. IEEE Trans. Image Process **19**(4), 895–911 (2010)
49. Zhu, X., Milanfar, P.: Automatic parameter selection for denoising algorithms using a no-reference measure of image content. J. IEEE. Trans. Image Process **19**(2), 3116–3132 (2010)
50. Abramov, S., Krivenko, S., Roenko, A., Lukin, V., Djurovic, I., Chobanu, M.: Prediction of filtering efficiency for DCT-based image denoising. In: Proceedings of MECO, Budva, Montenegro, June 2013, pp. 97–100 (2013)
51. Krivenko, S., Lukin, V., Vozel, B., Chehdi, K.: Prediction of DCT-based Denoising Efficiency for Images Corrupted by Signal-Dependent Noise. In: Proceedings of ELNANO, Kiev, Ukraine, April (2014)
52. Oktem, R., Egiazarian, K., Lukin, V., Ponomarenko, N., Tsymbal, O.: Locally adaptive DCT filtering for signal-dependent noise removal. EURASIP. J. Adv. Signal Process 10, (2007). doi:1155/2007/42472 (Article ID 42472)
53. Lukin, V.V., Oktem, R., Ponomarenko, N., Egiazarian, K.: Image filtering based on discrete cosine transform. J. Telecommun. Radio Eng. **66**(18), 1685–1701 (2007)
54. Dabov, K., Foi, A., Katkovnik, V., Egiazarian, K.: Image denoising by sparse 3D transform-domain collaborative filtering. J. IEEE. Trans. Image Process **16**(8), 2080–2095 (2007)
55. Ponomarenko, N., Lukin, V., Zelensky, A., Egiazarian, K., Astola, J., Carli, M., Battisti, F.: TID2008–a database for evaluation of full-reference visual quality assessment metrics. Adv. Mod. Radioelectron. **10**, 30–45 (2009) (Moscow)
56. Ponomarenko, N., Ieremeiev, O., Lukin, V., Egiazarian, K., Carli, M.: Modified Image Visual Quality Metrics for Contrast Change and Mean Shift Accounting. In: Proceedings of CADSM, February 2011, Ukraine, pp. 305–311 (2011)
57. Zhang, L., Zhang, Lei., Mou, X., Zhang, D.: FSIM: A feature similarity index for image quality assessment. J. IEEE. Trans. Image Process **20**(8), 2378–2386 (2011)
58. Wang, Z., Simoncelli, E.P., Bovik, A.C.: Multi-scale Structural Similarity for Visual Quality Assessment. Proceedings of the 37th IEEE Asilomar Conference on Signals, Systems and Computers. 2, 1398–1402 (2003)

59. Ponomarenko, N.N., Lukin, V.V., Egiazarian, K., Lepisto, L.: Color image lossy compression based on blind evaluation and prediction of noise characteristics. In: Proceedings of the Conference Image Processing: Algorithms and Systems IX, San Francisco, SPIE Vol. 7870, p. 12 (2011)
60. Theuwissen, A.: Course on Camera System. Lecture Notes, CEU-Europe, pp. 2–5 (2005)
61. Ponomarenko, N.N., Lukin, V.V., Egiazarian, K.O., Astola, J.T.: A method for blind estimation of spatially correlated noise characteristics. In: Proceedings of SPIE Conference Image Processing: Algorithms and Systems VII, San Jose, USA, vol. 7532, 12 p. (2010)
62. Liu, C., Szeliski, R., Kang, S.B., Zitnick, C.L., Freeman, W.T.: Automatic estimation and removal of noise from a single image. J. IEEE. Trans. Pattern Anal. Mach. Intell. **30**(2), 299–314 (2008)
63. Foi, A., Trimeche, M., Katkovnik, V., Egiazarian, K.: Practical poissonian-gaussian noise modeling and fitting for single image raw data. J. IEEE. Trans. Image Process **17**(10), 1737–1754 (2007)
64. Lukin, V., Abramov, S., Ponomarenko, N., Uss, M., Zriakhov, M., Vozel, B., Chehdi, K., Astola, J.: Methods and automatic procedures for processing images based on blind evaluation of noise type and characteristics. SPIE. J. Appl. Remote Sens. **5**(1), 053502 (2011). doi:10.1117/1.3539768
65. Lukin, V.V., Abramov, S.K., Kozhemiakin, R.A., Uss, M.L., Vozel, B., Chehdi, K.: Denoising Efficiency for Multichannel Images Corrupted by Signal-dependent Noise. In: Proceedings of MSMW, Kharkov, Ukraine, June 2013, p. 3 (2013)
66. Lukin, V., Abramov, S., Krivenko, S., Kurekin, A., Pogrebnyak, O.: Analysis of classification accuracy for pre-filtered multichannel remote sensing data. Expert Syst. Appl. **40**(16), 6400–6411 (2013)
67. Starck, J.-L., Murtagh, F.D., Bijaoui, A.: Image Processing and Data Analysis: The Multiscale Approach, p. 297. Cambridge University Press, New York (1998)
68. Abramova, V.V., Abramov, S.K., Lukin, V.V., Egiazarian, K.O., Astola, J.T.: On required accuracy of mixed noise parameter estimation for image enhancement via denoising. EURASIP. J. Image Video Process **2014**(3), (2014). doi:1186/1687-5281-2014-3
69. Pogrebnyak, O., Lukin, V.: Wiener discrete cosine transform-based image filtering. J. Electron Imaging **21**(4), (2012). doi:1117/1.JEI.4.043020 (id. 043020)
70. Ponomarenko, N., Silvestri, F., Egiazarian, K., Carli, M., Astola, J., Lukin, V.: On between-coefficient contrast masking of DCT basis functions. In: Proceedings of the Third International Workshop on Video Processing and Quality Metrics, Scottsdale, USA, January (2007), p. 4
71. Lukin, V., Zriakhov, M., Krivenko, S., Ponomarenko, N., Miao, Z.: Lossy compression of images without visible distortions and its applications. In: CD ROM Proceedings of ICSP, Beijing, October (2010), p. 4
72. Wang, Z., Bovik, A., Sheikh, H., Simoncelli, E.: Image quality assessment: From error visibility to structural similarity. IEEE. Trans. Image Process **13**(4), 600–612 (2004)
73. Rubel, A., Lukin, V., Pogrebniak, O.: Efficiency of DCT-based denoising techniques applied to texture images. In: Proceedings of IAPR, Cancun, Mexico, (2014), p. 10
74. Fevralev, D., Lukin, V., Ponomarenko, N., Abramov, S., Egiazarian, K., Astola, J.: Efficiency analysis of DCT-based filters for color image database. In: Proceedings of SPIE Conference Image Processing: Algorithms and Systems VII, San Francisco, USA, vol. 7870, p. 12 (2011)
75. Cameron, C., Windmeijer, A., Frank, A.G., Gramajo, H., Cane, D.E., Khosla, C.: An R-squared measure of goodness of fit for some common nonlinear regression models. J. Econ. **77**(2), 1790–1792 (1997). doi:1016/S0304-4076(96)01818-0

Chapter 4
Impulsive Noise Filters for Colour Images

Samuel Morillas, Valentín Gregori, Almanzor Sapena, Joan-Gerard Camarena and Bernardino Roig

Noise is often introduced into digital images during the acquisition and transmission processes because of different reasons such as charge-coupled device (CCD) sensor malfunction, transmission errors, storage faults and difficult acquisition conditions. The presence of noise hampers the automatic processing of digital images and also affects their visualization quality. In this context, colour images may be contaminated by several types of noise and they have been widely studied in the literature.

The most common types of noise are Gaussian noise and impulsive noise. Gaussian noise is usually introduced during the acquisition process and is characterized by modifying each pixel in the image, but impulsive noise affects only a portion of the image pixels, replacing their original values with other very different ones. It often appears in digital images as a result of a photoelectronic sensor fault or channel bit errors and can be introduced during the acquisition or transmission of the images. Impulsive noise may be caused by several sources such as ignition systems, industrial machines near the receiver device, lightning in the atmosphere or just by difficulties in the acquisition process [1–3].

In Figs. 4.1 and 4.2, several examples of real images contaminated with impulsive noise during its acquisition or storage are shown. Figure 4.1 (a)–(c) are old images

S. Morillas (✉) · V. Gregori · A. Sapena
Instituto Universitario de Matemática Pura y Aplicada,
Universitat Politècnica de València, Valencia, Spain
e-mail: smorillas@mat.upv.es

V. Gregori
e-mail: vgregori@mat.upv.es

A. Sapena
e-mail: asapena@mat.upv.es

J.-G. Camarena
Conselleria d'Educacio, Valencia, Spain
e-mail: joangerard@gmail.com

B. Roig
Instituto para la Gestión Integral de las zonas Costeras,
Universitat Politècnica de València, Valencia, Spain
e-mail: broig@mat.upv.es

M. E. Celebi et al. (eds.), *Color Image and Video Enhancement*,
DOI 10.1007/978-3-319-09363-5_4

Fig. 4.1 Examples of real images corrupted by impulsive noise during its acquisition or storage. **a** Old image in greyscale 1 (846 × 1200). **b** Old image in greyscale 2 (901 × 1200). **c** Old image in greyscale 3 (1024 × 518). **d** Girl with a Pearl Earring (802 × 923)

Fig. 4.2 Examples of real images corrupted by impulsive noise during its acquisition or storage and corresponding zoom-in images. **a** Galápagos (3264 × 2448). **b** Galápagos zoom-in (359 × 311). **c** Detail example (443×495). **d** Detail example zoom-in (150× 101). **e** Scanned old text (512×512). **f** Scanned old text zoom-in (101 × 101). **g** Real photograph (763 × 667). **h** Real photograph zoom-in (138 × 101)

whose degradation can be seen in some parts as impulsive noise, specially in dark areas. Figure 4.1 (d) is a piece of artwork which is also degraded and impulsive-like faults are observed. Figure 4.2 (a) and (e) are examples of how acquisition conditions can add impulsive noise in the image, such as the presence of floating particles in subaquatic images or texture of the material of a scanned document. Figure 4.2 (c) and (g) contain impulsive noise added by sensor or transmission faults. Figure 4.2 (b), (d), (f) and (h) are zoom-in sections to see with more detail the presence of noise in the original images.

Another type of noise which has been treated in the recent literature, mainly related to medical imaging, is speckle noise, which appears as a random, deterministic interference pattern in an image formed with coherent radiation of a medium containing many sub-resolution scatterers.

In this chapter, we focus our attention on the impulsive noise case and survey a series of classical, relevant and recent techniques for its reduction. The first section of the chapter is devoted to the description of the impulsive noise and its models. In the second section, we review the first approach to the filtering of impulsive noise where the main objective is to solve the problem of how to identify noise-free data in a noisy context. That is, how to process the image in a robust way so that from a noisy image, we can generate a noise-free one. In the third section, we deal with more advanced techniques that, on the basis of the robust filters, try to improve the preservation of the information in the image so that the filtered image is not only noise-free, but also as similar as possible to how the image should be in the absence of noise.

4.1 Modelling Impulsive Noise

There are several basic models of impulsive noise which are used to simulate its presence in colour images [4–7]. The models differ on how it is considered the correlation of noise in the image channels and how the pixel values are affected by noise.

From now on, let us consider the images represented in the red-green-blue (RGB) colour space using 8-bits per pixel component[1]. So, let $\mathbf{F} = \{F_R, F_G, F_B\}$ be the original pixel and let $\mathbf{F}^*$ be the corrupted pixel generated by the modelling process. Suppose that p is the probability of the noise appearance. Then, the different models of impulsive noise are described as follows:

[1] It should be pointed out that there are other colour spaces, different from RGB, where both impulse noise and processing algorithms can be modelled and defined. In fact, several filtering methods have been defined in other colour spaces. However, the RGB colour space is the one most used and both the most classical and best performing methods use RGB representation for filtering impulsive noise. So, in this work we consider the RGB as the reference colour space. Nevertheless, most of the techniques and procedures reviewed in the following can be adapted to work in a different colour space if needed.

I. Fixed value impulsive noise.
All pixels in the colour image are modified following the next scheme:

$$\mathbf{F}^* = \begin{cases} \{d_1, F_G, F_B\} & \text{with probability } p \cdot p_1\,, \\ \{F_R, d_2, F_B\} & \text{with probability } p \cdot p_2\,, \\ \{F_R, F_G, d_3\} & \text{with probability } p \cdot p_3\,, \\ \{d_1, d_2, d_3\} & \text{with probability } p \cdot \left(1 - \sum_{i=1}^{3} p_i\right)\,, \end{cases} \tag{4.1}$$

where d_1, d_2, d_3 are independent and take the values 0 or 255 with the same probability, and p_i for $i = 1, 2, 3$ determines the probability of noise appearance in the image channels.

II. Correlated random value impulsive noise.
$\mathbf{F}^* = \{d_1, d_2, d_3\}$ with probability p, where d_1, d_2, d_3 are random integer values uniformly distributed in the interval $[0, 255]$.

III. Uncorrelated random value impulsive noise.
All pixels in the colour image are modified following the next scheme:

$$\mathbf{F}^* = \{F_R^*, F_G^*, F_B^*\}, \tag{4.2}$$

where

$$F_R^* = \begin{cases} d_R & \text{with probability } p_R \\ F_R & \text{with probability } 1 - p_R \end{cases}$$

$$F_G^* = \begin{cases} d_G & \text{with probability } p_G \\ F_G & \text{with probability } 1 - p_G \end{cases}$$

$$F_B^* = \begin{cases} d_B & \text{with probability } p_B \\ F_B & \text{with probability } 1 - p_B \end{cases},$$

where d_R, d_G and d_B are random values in the interval $[0, 255]$ and p_R, p_G and p_B determine the probability of noise appearance in the red, green and blue channel, respectively. For simplicity, we assume $p_R = p_G = p_B = p$.

IV. α-stable distribution
An α-stable distribution can also be used to model impulsive noise [5–7]. A symmetric α-stable (SαS) random variable is only described by its characteristic function

$$\varphi(t) = exp(j\theta t - \gamma|t|^{\alpha}),$$

where $j \in \mathbb{C}$ is the imaginary unit, $\theta \in \mathbb{R}$ is the location parameter (centrality), $\gamma \in \mathbb{R}$ is the dispersion of the distribution and $\alpha \in (0, 2]$, which controls the heaviness of the tails, is a parameter that controls the degree of impulsiveness so that impulsiveness increases as α decreases. The Gaussian (α=2) and the Cauchy (α=1) distributions are the only symmetric α-stable distributions that have closed-form probability density functions.

4.1.1 Measuring the Quality of a Filter

The above considerations imply that the noise reduction task, also known as image filtering, is a fundamental step in any computer vision system. The objective of the noise reduction techniques is to suppress the noise while preserving the integrity of edges, details and texture in the images.

The performance of image filters is assessed by visual inspection of filtered images and also by objective quality measures that compare the filters' output with the original noise-free image which was contaminated with noise before filtering. The most classical techniques to objectively measure the quality of filtering are based on the computation of pixelwise differences between the original and the output image and most of the measures are evaluated in the RGB colour system which is usual in engineering applications [4]. Objective measures should evaluate noise suppression and detail and colour preservation in the image. In this sense, the usual quality measures are the mean absolute error (MAE) which evaluates detail preservation, the root mean squared error (RMSE), the normalized mean square error (NMSE) and the peak signal-to-noise ratio (PSNR) which measure noise suppression, and the normalized colour difference (NCD) which is better to evaluate colour preservation since it uses the lab colour space which models better colour difference human perception than RGB. A more advanced method is the colour multiscale structural similarity index (CMSSIM) which is intended to take into account three main factors in the similarity in order to better agree with perceptual evaluation: luminance, contrast and structure similarity [8].

In particular, we will use MAE, PSNR [4] and CMSSIM [8] which are defined as follows:

- MAE:

$$MAE = \frac{\sum_{i=1}^{N}\sum_{j=1}^{M}\sum_{q=1}^{Q}\left|F^{q}(i,j) - \hat{F}^{q}(i,j)\right|}{N \cdot M \cdot Q}, \tag{4.3}$$

  where M and N are the image dimensions, Q is the channel number of the image ($Q = 3$ for colour images) and $F^{q}(i, j)$ and $\hat{F}^{q}(i, j)$ denote the q-th component of a vector in the original image in the pixel situated in position (i, j) in the image, respectively.

- PSNR:

$$PSNR = 20\log\left(\frac{2^k - 1}{\sqrt{\frac{1}{NMQ}\sum_{i=1}^{N}\sum_{j=1}^{M}\sum_{q=1}^{Q}\left(F^{q}(i,j) - \hat{F}^{q}(i,j)\right)^2}}\right), \tag{4.4}$$

  where we took $k = 8$ which corresponds with the 8-bits per channel case.

- CMSSIM: The images are divided into different patches of varying size (different scales) and the global similarity is obtained as

$$CMSSIM = (Clr(x, y))^{\delta}(l_M(x, y))^{\alpha_M} \prod^{M} (C_i(x, y))^{\beta_i} \cdot (S_i(x, y))^{\gamma_i}, \quad (4.5)$$

where Clr is a colour similarity factor, l_M is the luminance factor, contrast and structure similarity of scale i are C_i and S_i, respectively and $\beta_1 = \gamma_1 = 0.04448$; $\beta_2 = \gamma_2 = 0.2856$; $\beta_3 = \gamma_3 = 0.3001$; $\beta_4 = \gamma_4 = 0.2363$; $\alpha_5 = \beta_5 = \gamma_5 = 0.1333$ and $\delta = 0.7$. CMSSIM takes values in the interval [0, 1]. To better observe the performance differences, we will use the value of DIS defined as

$$DIS^k = (1 - CMSSIM) \cdot 10^k.$$

4.2 Robust Filters For Impulsive Noise

Given that digital images are of a noisy nature and that there is a need to suppress noise for further use of the images, the first approaches in the field of image filtering focus their efforts in trying to obtain noise-free data from noisy inputs from a robust point of view. That is, they try to process the input noisy image to generate an output image which is free of noise as much as possible, regardless of other considerations such as feature preservation, which is incorporated in more advanced techniques. In this section of the chapter, we review the most important methods for filtering images from this robust point of view and the more advanced techniques are seen in the following section.

Linear approaches were the first choice for reducing noise due to their simplicity and easy implementation. The most well-known, such as the arithmetic mean filter (AMF) and its variants [9], modify the value of each pixel by averaging with surrounding neighbours in a filtering window. However, in this way, these methods are unable to completely reduce the impulse noise since the differences between noise and data are so large that it cannot be totally reduced. In addition, noise in a neighbour pixel usually affects the value of the processed pixel in an undesired way and they tend to blur edges and details significantly [10].

As a consequence, non-linear methods became of interest [11]. In particular, for scalar data contaminated with impulse noise, the median statistical operator gained interest due to its ability to select noise-free samples among noisy data. If there is an impulse in the chosen set, after ordering, the impulse has a higher probability to lie at lower or higher ranks of the ordered set, and so the median becomes the data which is free of noise with the highest probability. The method of replacing each image pixel by the median of its neighbours in a filtering window is known as the median filter.

This method is suitable for univariate data but its generalization to higher dimensional data is not straightforward. In the case of processing colour images contaminated with impulse noise, the simplest approach is to apply the median filter independently in each colour channel, which is called the marginal median filter (MMF). This filter has the advantage of exhibiting the highest noise reduction capability, but since colour channels are processed independently and its correlation is not taken into account, it often produces the so-called colour artefacts: filter output colours which were not present in the original image, to which both the human visual system and other image processing tasks are very sensitive. This fact limits its practical application and motivates the development of alternative methods to generalize the median operator to multivariate data [4, 12].

However, this task is not easy. The main difficulty lies in the fact that the ordering of data is a fundamental point within the median filter and ordering among multivariate data is not defined in a unique way [13]. The most popular alternatives for ordering multivariate data are based on four sub-ordering principles: Marginal ordering (M-ordering), Conditional ordering (C-ordering), Partial ordering (P-ordering) and Reduced ordering (R-ordering) [4]:

- M-ordering: multivariate samples are ordered along each channel of the multichannel signal independently, which is indeed what MMF does.
- C-ordering: multivariate samples are ordered conditional on one of the marginal sets of observations and the other components of each vector are listed (not ordered) according to the position of their ranked component. In this ordering scheme, only information in one channel is used for ordering, so a considerable loss of useful information may occur. It could be useful to reduce noise when it can be isolated in one channel, which seldom occurs.
- P-ordering: subsets of data are grouped together forming minimum convex hulls. The first convex hull is formed such that the perimeter contains a minimum number of points and the resulting hull contains all other points inside it. The points on the perimeter are denoted C-order group 1. These points are discarded and the process is repeated. This method provides no ordering within the groups and the determination of the convex hull is difficult, especially in higher dimensional data. Hence, this method is rarely used.
- R-ordering: multivariate observation is reduced to a scalar value by means of some combination of the component sample values. The resulting scalar values are automatically ordered. The choice of the scalar value is of paramount importance and it is application-dependent.

The R-ordering was found to be the most interesting sub-ordering principle, and the combination of R-ordering and robust statistics produced the most popular vector filter for colour images which is still a reference in the field: the vector median filter (VMF) [14–16]. According to the theory of robust statistics, given a set of multivariate noisy data, the vector in the set which minimizes the accumulated distance to all the other samples in the set is the one which has the higher probability to be free of noise. The VMF proposes to determine the vector median of a colour vector population as the vector that minimizes the accumulated distance (usually measured with the Euclidean distance or, alternatively, other Minkowski metrics) to all the other colour vectors in a filtering window. So, each colour vector in the image should be replaced by the vector median of its neighbourhood. The VMF is the first method able to determine noise-free data in a colour image context, taking into account the correlation among the colour channels since it follows a vector approach. This filter was found to exhibit a robust performance, similar to the MMF but with the advantage that it does not introduce colour artefacts in the output image. However, it is also interesting to note that the noise reduction capability of the VMF can be improved. In fact, the MMF exhibits a higher capability to reduce noise. Consider a vector population where all vectors have one noisy component. In this case, the VMF will always select a noisy vector as output. In order to improve the performance of the VMF, different alternatives have been studied. It should be pointed out that nowadays, VMF is seldom used by itself to process a colour image given that there are other more advanced techniques. However, the VMF operation is commonly used in most of the advanced methods to reduce noise when detected so the improvement of its noise reduction capability is interesting.

On the one hand, there are several filters that follow the same procedure of the VMF but use a different metric to explore different performance possibilities. For instance, the basic vector directional filter (BVDF) ranks the colour vectors based on the orientation difference, by using the angle between two colour vectors as the distance criterion. Since the direction of a colour vector in RGB space is related to its chromaticity, this filter is intended to improve the chromaticity preservation [17, 18]. The generalized vector directional filter (GVDF) is a generalization of the BVDF in which its output is a superset of the single BVDF output. First, the vectors are ranked according to the angular distance criterion and a set of low rank vectors are selected as input to an additional filter that produces a single output vector. In the second step, only the magnitudes of the vectors are considered [18]. The directional distance filter (DDF) is a combination of the VMF and the BVDF. The motivation behind it is to incorporate information about both vector magnitude (brightness) and its direction (chromaticity) in the calculation of the distance metric [19, 20]. Besides, versions of the VMF, BVDF and DDF have been developed using fuzzy metrics instead of the Euclidean and angular metrics [22, 23], and some advantages have been observed. Also, the classical order-statistics-based filters (VMF, BVDF and DDF) have fuzzy and hybrid extensions mixing different filters. They have been investigated to provide a bridge between linear and non-linear techniques suitable for filtering multichannel signals [24, 25].

In a recent research [12], an approach that combine results from robust statistics R-ordering and M-ordering has been developed with the aim of improving the noise reduction capability of VMF without introducing colour artefacts. Two more advanced techniques in this sense are proposed in [26], where the modified vector median filter (MVMF) and the robust vector median filter (RMVF) are based on replacing some components of the vector median in a given set of colour vectors with other more reliable ones to obtain a more robust filter output. The processing is made taking into account channel correlation so that no colour artefacts are introduced. In particular, the replacement is based on information extracted from a robustness analysis of the values in each colour channel, and the introduction of colour artefacts is avoided by using a fuzzy similarity-based ordering of the vectors with respect to the vector median.

Fig. 4.3 Test images and details used for the experimentation. **a** Baboon (512 × 512). **e** Baboon detail (70 × 70). **b** Boats (720 × 576). **f** Boats detail (101 × 101). **c** Goldhill (512 × 512). **g** Goldhill detail (100 × 100). **d** Lenna (256 × 256) and **h** Lenna detail (90 × 90)

4.2.1 Comparison of Robust Methods

The noise models in Sect. 4.1 have been used to contaminate the test Baboon, Boats, Goldhill and Lenna images in Figs. 4.3 (a)–(d). In particular, we have used a detail of each image Figs. 4.3 (e)–(h) to better appreciate the performance differences. We have considered different intensities of noise in the image. For types I, II and III, we considered noise probability equal to 0.1, 0.3 and 0.5, which is a high intensity but interesting to evaluate robustness. For noise type IV, we set α equal to 0.7, 0.5 and 0.4. We have used the objective quality measures MAE, PSNR and DIS^k to objectively compare the performance of a selected group of filters. The contaminated images have been filtered using the filters in Table 4.1 which are a representative selection of the most popular robust methods. In all cases, we have used a 3×3 filtering window since larger filtering windows do not provide significant relative performance differences among the considered methods. We have filtered the image only once to better appreciate the performance differences. Notice that when some noise remains in the image, the resulting image should be filtered once more or until no changes in the image are obtained, or using a larger filtering window. Numerical results are shown in Tables 4.3–4.18 where we have highlighted the best results in each column (optimal in bold, suboptimal in italic). Besides, Fig. 4.4 shows a few filtered images for visual comparison.

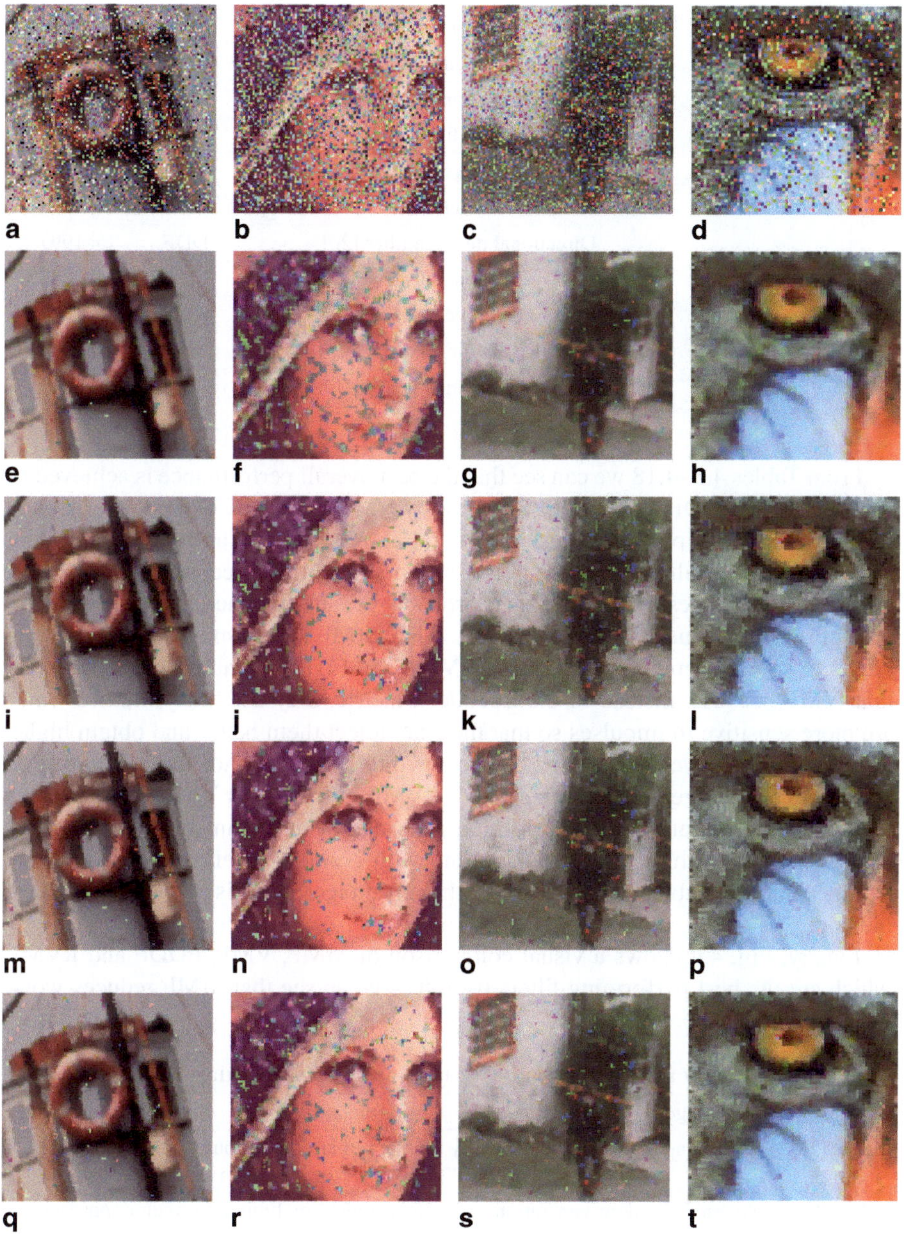

Fig. 4.4 Outputs for filter performance comparison. **a** Boats corrupted with 50 % of noise model I. **b** Lenna corrupted 50 % of noise model II. **c** Goldhill corrupted with 30 % of noise model III. **d** Baboon corrupted with $\alpha = 0.4$ of noise model IV. **e–h** Output of MMF. **i–l** Output of VMF. **m–p** Output of FDDF. **q–t** Output of RVMF

Table 4.1 Filters used in the comparisons

Filter name / reference	Short name	Year
Arithmetic mean filter [27]	AMF	1978
Marginal median filter [28]	MMF	1990
Vector median filter [14]	VMF	1990
Basic vector directional filter [17]	BVDF	1993
Generalized vector directional filter [18]	GVDF	1996
Directional distance filter [20]	DDF	1997
Fuzzy directional distance filter [21]	FDDF	2002
Fuzzy vector median filter [21]	FVMF	2002
Modified vector median filter [26]	MVMF	2011
Robust vector median filter [26]	RVMF	2011

From Tables 4.3–4.18 we can see that the best overall performance is achieved by MMF, FVMF, FDDF and RVMF followed by VMF and DDF. Directional distance methods and AMF perform worse in general. As commented above, MMF obtains high numerical results for its high noise suppression ability excepting for noise type II where its robustness is lower. This is probably because the median operator in this type of random-value noise is less robust. Also, MMF is known to introduce colour artefacts, so its practical use is limited. FVMF and FDDF are able to perform better than VMF because, instead of the Euclidean metric, they use fuzzy metrics which are more sensitive to impulses so that they can reject them better and obtain higher robustness. Moreover, since they are vector filters, they cannot introduce artefact colours in the image. On the other hand, RVMF outperforms VMF by combining marginal ordering and reduced ordering information to obtain the filter output in a reliable way: with better robustness and with a colour artefact generation control mechanism. Table 4.2 summarizes the advantages and disadvantages of these methods.

Finally, Fig. 4.4 shows a visual comparison of MMF, VMF, FDDF and RVMF, which are the best performing filters overall. We can see that MMF reduces worse

Table 4.2 Summary of advantages and disadvantages of best robust filtering methods

Filter	Advantages	Disadvantages
MMF [28]	Highest noise reduction capability	Generates artefact colours Lower performance in random value noise
VMF [14]	No artefact colour generation	Noise reduction limited by vector approach
DDF [20]	Includes directional distance	Noise reduction limited by vector approach
FDDF [21]	Includes directional distance Metrics sensitive to noise	Noise reduction limited by vector approach
FVMF [21]	Metric sensitive to noise	Noise reduction limited by vector approach
RVMF [26]	Noise reduction higher than VMF No artefact colour generation	

Table 4.3 Performance comparison in terms of MAE, PSNR and DIS^k using the detail of the Lenna image contaminated with different densities of noise model I

Filter	No added noise			$p = 10\,\%$			$p = 30\,\%$			$p = 50\,\%$		
	MAE	PSNR	DIS^2	MAE	PSNR	DIS^2	MAE	PSNR	DIS^2	MAE	PSNR	DIS^1
None	0.00	∞	0.00	4.95	19.13	0.07	14.63	14.50	0.21	24.24	12.35	0.35
AMF	6.00	28.13	2.49	9.24	25.12	37.85	14.29	22.02	71.21	18.71	20.03	8.23
MMF	3.99	**29.91**	1.78	*4.28*	*29.51*	3.79	*4.96*	*28.22*	11.73	**5.79**	**26.61**	2.09
VMF	*3.89*	*29.81*	*1.73*	4.38	29.23	*1.78*	5.62	27.40	*3.44*	7.07	24.97	*1.22*
BVDF	3.99	29.45	**1.41**	4.55	28.72	2.20	6.43	25.73	5.89	9.32	21.68	1.98
GVDF	*3.92*	29.70	*1.75*	4.49	28.85	2.33	6.22	25.77	6.44	8.87	21.81	2.08
DDF	*3.82*	29.73	*1.56*	*4.29*	29.21	*2.01*	5.63	27.28	*3.65*	7.19	24.65	*1.35*
FDDF	**3.82**	*29.89*	*1.72*	**4.16**	**29.51**	**1.73**	*4.91*	*28.16*	**3.15**	*6.19*	*25.28*	*1.25*
FVMF	3.96	29.78	1.90	*4.24*	*29.45*	*1.95*	**4.88**	**28.33**	*3.15*	*5.96*	*25.90*	**1.12**
MVMF	*3.91*	*29.82*	1.81	4.35	*29.29*	*2.00*	*5.38*	*27.68*	*4.42*	*6.54*	*25.34*	*1.54*
RVMF	3.95	*29.89*	1.81	*4.34*	*29.37*	2.91	*5.23*	*27.84*	6.49	*6.37*	*25.47*	1.77

Table 4.4 Performance comparison in terms of MAE, PSNR and DIS^k using the detail of the Baboon image contaminated with different densities of noise model I

Filter	No added noise			$p = 10\,\%$			$p = 30\,\%$			$p = 50\,\%$		
	MAE	PSNR	DIS^2	MAE	PSNR	DIS^2	MAE	PSNR	DIS^2	MAE	PSNR	DIS^1
None	0.00	∞	0.00	4.68	19.44	0.07	13.73	14.77	0.21	24.47	12.27	0.37
AMF	11.95	23.42	**1.67**	13.99	22.30	36.82	17.77	20.51	71.47	22.06	18.80	7.87
MMF	10.31	*23.58*	3.61	*10.58*	*23.41*	8.24	*11.09*	**23.01**	20.57	**12.13**	**21.97**	3.53
VMF	*10.10*	*23.58*	*2.06*	10.61	*23.30*	*2.94*	11.57	22.76	*5.35*	13,35	21.19	*1.58*
BVDF	12.01	21.50	*2.00*	12.63	21.35	*2.16*	14.71	20.02	6.29	17.99	17.62	*2.00*
GVDF	11.12	22.30	2.47	11.62	22.10	*2.57*	12.87	21.57	*5.61*	15.65	18.79	2.02
DDF	**9.88**	23.49	*1.69*	*10.49*	23.20	**2.06**	11.55	22.62	*5.10*	13.40	21.09	*1.54*
FDDF	*9.98*	*23.56*	*2.31*	**10.28**	**23.41**	*2.61*	**10.89**	*22.97*	**4.27**	*12.45*	*21.35*	**1.50**
FVMF	10.27	23.51	3.96	*10.50*	*23.37*	4.61	*11.01*	*22.89*	*6.16*	*12.36*	*21.45*	*1.64*
MVMF	*10.16*	*23.56*	2.98	10.62	23.29	4.16	*11.38*	*22.80*	7.14	*12.75*	*21.41*	2.06
RVMF	*10.18*	**23.59**	3.39	*10.55*	*23.34*	5.63	*11.23*	*22.87*	10.22	*12.57*	*21.48*	2.52

noise type II than the rest. Also, it introduces many colour artefacts in the Boats and Baboon images. VMF and FDDF perform similarly, but FDDF is able to reduce more noise than VMF, specially for noise type III and IV. On the other hand, RVMF reduces more noise than VMF and FDDF. We see that remaining clusters of noise are smaller for RVMF, specially for noise type I and III.

Table 4.5 Performance comparison in terms of MAE, PSNR and DIS^k using the detail of the Boats image contaminated with different densities of noise model I

Filter	No added noise			$p = 10\,\%$			$p = 30\,\%$			$p = 50\,\%$		
	MAE	PSNR	DIS^2	MAE	PSNR	DIS^2	MAE	PSNR	DIS^2	MAE	PSNR	DIS^1
None	0.00	∞	0.00	4.67	19.74	0.08	14.93	14.62	0.24	24.57	12.46	0.40
AMF	5.23	30.36	**0.23**	8.14	26.58	34.67	13.66	22.51	70.24	17.95	20.43	8.01
MMF	*3.37*	*32.11*	0.60	**3.65**	**31.56**	2.23	*4.35*	*29.44*	9.22	**5.28**	**27.56**	2.11
VMF	*3.35*	*32.11*	0.45	3.90	31.07	*0.74*	5.00	28.84	*2.72*	6.90	25.11	*1.21*
BVDF	6.37	27.05	*0.35*	6.64	26.67	*0.81*	7.75	24.45	4.09	11.63	19.20	1.92
GVDF	5.09	28.63	0.36	5.39	28.15	*0.71*	6.47	25.69	*3.59*	9.15	21.21	2.01
DDF	3.88	31.14	*0.24*	4.23	30.41	*0.75*	5.21	28.40	*2.66*	7.07	24.76	*1.24*
FDDF	3.53	31.73	*0.27*	*3.77*	*31.29*	**0.61**	*4.47*	*29.35*	**2.15**	*5.84*	*25.37*	*1.23*
FVMF	*3.42*	*31.87*	0.68	*3.68*	*31.40*	1.06	**4.31**	**29.45**	*2.34*	*5.45*	*26.61*	**0.98**
MVMF	*3.35*	*32.10*	*0.35*	*3.83*	*31.15*	0.81	*4.75*	*29.17*	3.68	*6.21*	*25.61*	*1.51*
RVMF	**3.31**	**32.16**	0.39	*3.71*	*31.33*	1.31	*4.59*	*29.35*	5.34	*5.96*	*25.76*	1.79

Table 4.6 Performance comparison in terms of MAE, PSNR and DIS^k using the detail of the Goldhill image contaminated with different densities of noise model I

Filter	No added noise			$p = 10\,\%$			$p = 30\,\%$			$p = 50\,\%$		
	MAE	PSNR	DIS^2	MAE	PSNR	DIS^2	MAE	PSNR	DIS^2	MAE	PSNR	DIS^1
None	0.00	∞	0.00	5.16	19.20	0.08	15.13	14.47	0.23	24.91	12.37	0.38
AMF	5.82	29.46	**0.51**	8.96	25.91	38.33	14.30	22.27	68.93	18.47	20.31	7.92
MMF	*4.68*	**30.18**	*0.85*	**4.86**	**29.90**	1.93	**5.27**	**28.95**	5.26	**5.99**	**27.00**	1.47
VMF	*4.75*	*30.02*	0.95	5.10	29.51	*1.15*	5.97	28.09	*2.51*	7.39	25.13	*1.09*
BVDF	7.22	26.78	*0.92*	7.41	26.56	**0.88**	8.44	23.89	3.61	12.17	18.95	1.76
GVDF	5.85	28.29	0.96	6.09	27.91	*0.99*	6.93	25.99	3.41	9.55	20.97	1.82
DDF	4.99	29.67	*0.90*	5.30	29.17	*0.95*	6.09	27.87	2.82	7.53	24.84	*1.14*
FDDF	4.76	*30.00*	*0.86*	*4.97*	*29.67*	*0.95*	*5.42*	*28.77*	**2.28**	*6.58*	*25.27*	*1.14*
FVMF	*4.76*	29.95	1.20	*4.97*	*29.62*	1.42	*5.39*	*28.73*	*2.31*	*6.30*	*26.11*	**0.84**
MVMF	*4.72*	*30.05*	1.11	*5.02*	*29.59*	1.26	*5.69*	*28.38*	*2.93*	*6.77*	*25.53*	*1.23*
RVMF	**4.67**	*30.13*	1.01	*4.94*	*29.70*	1.44	*5.52*	*28.57*	3.38	*6.59*	*25.66*	1.33

4.3 Adaptive Filtering

The robust vector filters described in the previous section are efficient in reducing impulse noise, but their signal-preserving capability is deficient. In these filters, the filtering operation is applied to each image pixel regardless of whether it is noisy or not, and the filtering operation applied is the same independent of the degree in which impulse noise affected each pixel. The operations made in any image location are fixed so they do not adapt to local image features either. It has been widely

Table 4.7 Performance comparison in terms of MAE, PSNR and DIS^k using the detail of the Lenna image contaminated with different densities of noise model II

Filter	No added noise			$p = 10\,\%$			$p = 30\,\%$			$p = 50\,\%$		
	MAE	PSNR	DIS^2	MAE	PSNR	DIS^2	MAE	PSNR	DIS^2	MAE	PSNR	DIS^1
None	0.00	∞	0.00	7.12	18.95	0.07	22.60	13.86	0.21	37.72	11.62	0.34
AMF	6.00	28.13	2.49	10.38	24.56	48.26	20.05	19.59	86.77	29.27	16.62	9.31
MMF	3.99	**29.91**	1.78	4.75	28.74	6.90	7.75	24.53	34.98	15.30	19.14	7.04
VMF	*3.89*	*29.81*	*1.73*	*4.57*	*28.82*	*2.07*	*6.81*	*25.59*	*11.70*	12.88	19.90	*3.98*
BVDF	3.99	29.45	**1.41**	4.80	28.17	*2.30*	8.22	23.12	16.14	16.57	17.49	4.73
GVDF	*3.92*	29.70	*1.75*	4.77	28.21	2.67	7.37	24.42	*14.60*	12.76	19.26	*3.97*
DDF	*3.82*	29.73	*1.56*	*4.50*	28.81	**1.68**	*6.72*	25.56	*12.19*	*12.57*	*20.00*	*4.02*
FDDF	**3.82**	*29.89*	*1.72*	**4.42**	**29.05**	*1.88*	**6.26**	**26.14**	**8.28**	**11.31**	**20.56**	**3.47**
FVMF	3.96	29.78	1.90	*4.55*	*28.96*	*2.25*	*6.66*	*25.60*	*11.40*	*12.17*	*20.00*	*3.90*
MVMF	*3.91*	*29.82*	1.81	*4.57*	*28.88*	2.51	*6.72*	*25.72*	14.94	*12.40*	*20.18*	4.46
RVMF	3.95	*29.89*	1.81	4.63	*28.89*	3.53	6.82	*25.67*	20.70	*12.38*	*20.26*	5.05

Table 4.8 Performance comparison in terms of MAE, PSNR and DIS^k using the detail of the Baboon image contaminated with different densities of noise model II

Filter	No added noise			$p = 10\,\%$			$p = 30\,\%$			$p = 50\,\%$		
	MAE	PSNR	DIS^2	MAE	PSNR	DIS^2	MAE	PSNR	DIS^2	MAE	PSNR	DIS^1
None	0.00	∞	0.00	6.52	19.32	0.06	22.58	13.95	0.21	37.29	11.76	0.35
AMF	11.95	23.42	**1.67**	14.95	21.93	37.90	22.61	18.77	76.33	30.51	16.38	8.53
MMF	10.31	*23.58*	3.61	10.90	*23.22*	12.37	13.95	21.28	43.76	20.31	18.17	7.16
VMF	*10.10*	*23.58*	*2.06*	*10.72*	*23.24*	*4.02*	13.20	*21.73*	12.08	18.80	18.51	*3.43*
BVDF	12.01	21.50	*2.00*	12.72	21.18	*2.61*	15.64	19.65	*11.84*	21.57	16.66	*3.19*
GVDF	11.12	22.30	2.47	11.58	22.07	*3.55*	13.52	21.17	*10.78*	*17.64*	18.50	**2.64**
DDF	**9.88**	23.49	*1.69*	*10.55*	23.16	**2.59**	*13.01*	21.71	*10.63*	*17.86*	*18.87*	2.90
FDDF	*9.98*	*23.56*	*2.31*	**10.47**	**23.30**	2.78	**12.51**	**21.96**	**8.08**	**17.02**	**19.18**	2.73
FVMF	10.27	23.51	3.96	10.76	23.21	5.14	*12.91*	*21.79*	*11.94*	*18.24*	*18.54*	3.85
MVMF	*10.16*	23.56	2.98	*10.72*	*23.24*	4.90	*13.09*	*21.79*	17.41	18.44	*18.65*	4.52
RVMF	*10.18*	**23.59**	3.39	*10.74*	*23.25*	7.14	*13.13*	*21.75*	26.53	*18.35*	*18.69*	5.52

observed that non-adaptive processing usually results in blurred edges and image details. Indeed, it can be assumed that impulse noise in digital images affects only a small portion of pixels and only for these pixels is the operation of robust methods necessary. Despite the fact that the changes these methods apply to the image are robust, they are unnecessary in many instances, and so they degrade the quality of fine details, edges and texture. This fact limits its practical use and motivates the development of advanced adaptive techniques which we review in the following section.

Table 4.9 Performance comparison in terms of MAE, PSNR and DIS^k using the detail of the Boats image contaminated with different densities of noise model II

Filter	No added noise			$p = 10\,\%$			$p = 30\,\%$			$p = 50\,\%$		
	MAE	PSNR	DIS^2	MAE	PSNR	DIS^2	MAE	PSNR	DIS^2	MAE	PSNR	DIS^1
None	0.00	∞	0.00	6.77	19.49	0.07	21.37	14.45	0.22	35.12	12.28	0.36
AMF	5.23	30.36	**0.23**	9.20	25.89	39.23	17.16	20.83	73.47	24.41	17.85	8.34
MMF	*3.37*	*32.11*	0.60	4.13	*30.77*	3.71	6.86	26.00	26.64	12.86	20.40	5.34
VMF	*3.35*	*32.11*	0.45	*4.10*	*30.73*	*0.94*	6.35	26.66	*8.18*	11.40	20.61	*2.70*
BVDF	6.37	27.05	*0.35*	6.76	26.44	0.99	8.43	23.65	8.46	14.46	18.14	*2.77*
GVDF	5.09	28.63	0.36	5.54	27.91	*0.93*	6.81	25.98	*7.13*	11.11	20.08	*2.44*
DDF	3.88	31.14	*0.24*	4.38	30.21	**0.73**	*6.31*	*26.82*	*7.13*	*10.63*	*21.14*	*2.43*
FDDF	3.53	31.73	*0.27*	*4.10*	*30.70*	*0.73*	**5.87**	**27.35**	**5.57**	**10.04**	**21.31**	**2.34**
FVMF	*3.42*	*31.87*	0.68	*4.10*	30.66	1.14	*6.04*	*26.92*	*7.64*	*10.89*	*20.71*	2.90
MVMF	*3.35*	*32.10*	*0.35*	4.08	*30.74*	*0.98*	*6.21*	*26.88*	10.76	*10.97*	*20.99*	3.23
RVMF	**3.31**	**32.16**	0.39	**4.06**	**30.79**	1.57	*6.23*	*26.89*	15.95	*10.99*	*21.04*	3.76

Table 4.10 Performance comparison in terms of MAE, PSNR and DIS^k using the detail of the Goldhill image contaminated with different densities of noise model II

Filter	No added noise			$p = 10\,\%$			$p = 30\,\%$			$p = 50\,\%$		
	MAE	PSNR	DIS^2	MAE	PSNR	DIS^2	MAE	PSNR	DIS^2	MAE	PSNR	DIS^1
None	0.00	∞	0.00	7.15	19.13	0.07	22.76	14.02	0.21	36.68	12.02	0.34
AMF	5.82	29.46	**0.51**	9.74	25.43	36.53	19.17	20.11	71.25	27.12	17.43	8.15
MMF	*4.68*	**30.18**	*0.85*	5.14	*29.40*	3.40	7.63	25.20	23.67	13.70	20.29	5.64
VMF	*4.75*	*30.02*	0.95	*5.13*	*29.37*	*1.19*	7.13	25.80	*6.03*	11.82	20.91	*2.45*
BVDF	7.22	26.78	*0.92*	7.51	26.13	*1.32*	8.78	23.59	*5.26*	13.26	19.13	*2.03*
GVDF	5.85	28.29	0.96	6.11	27.84	*1.22*	7.00	*26.16*	**3.66**	**10.02**	*21.59*	**1.67**
DDF	4.99	29.67	*0.90*	5.34	29.03	*1.16*	*6.86*	*26.54*	*4.26*	*10.59*	*21.73*	*1.85*
FDDF	4.76	*30.00*	*0.86*	*5.13*	*29.36*	**1.06**	**6.48**	**26.81**	*4.46*	*10.23*	**21.83**	*1.88*
FVMF	*4.76*	29.95	1.20	*5.14*	29.30	1.53	*6.82*	26.02	6.90	*11.38*	20.96	2.82
MVMF	*4.72*	*30.05*	1.11	*5.10*	*29.41*	1.55	*6.96*	*26.10*	8.44	11.51	*21.13*	3.20
RVMF	**4.67**	*30.13*	1.01	**5.09**	**29.43**	1.87	*6.92*	*26.11*	11.20	*11.43*	*21.21*	3.70

A huge number of adaptive filters have been introduced in the literature with the aim to accommodate to varying image characteristics and noise statistics, and so to obtain good performance in real-life applications such as microarray image processing, virtual restoration of artworks, colour video processing and digital image enhancement.

Table 4.11 Performance comparison in terms of MAE, PSNR and DIS^k using the detail of the Lenna image contaminated with different densities of noise model III

Filter	No added noise			$p = 10\,\%$			$p = 30\,\%$			$p = 50\,\%$		
	MAE	PSNR	DIS^2	MAE	PSNR	DIS^2	MAE	PSNR	DIS^2	MAE	PSNR	DIS^1
None	0.00	∞	0.00	7.45	18.62	0.11	22.80	13.87	0.30	37.14	11.72	0.44
AMF	6.00	28.13	2.49	10.70	24.29	58.80	19.86	19.61	87.86	28.64	16.81	9.35
MMF	3.99	**29.91**	1.78	*4.82*	*28.61*	10.58	*8.03*	**24.15**	42.04	**14.63**	**19.47**	*7.39*
VMF	*3.89*	*29.81*	*1.73*	5.20	28.15	*3.19*	10.14	22.48	39.36	19.07	17.70	7.70
BVDF	3.99	29.45	**1.41**	5.82	26.90	*4.16*	12.35	20.78	46.37	23.54	15.85	8.21
GVDF	*3.92*	29.70	*1.75*	5.57	27.26	4.17	10.45	21.74	*37.05*	20.06	16.86	7.62
DDF	*3.82*	29.73	*1.56*	5.13	28.16	*2.68*	10.29	22.30	39.70	19.30	17.60	7.93
FDDF	**3.82**	*29.89*	*1.72*	**4.63**	**28.87**	**2.05**	*8.29*	*23.54*	*29.46*	*16.25*	*18.36*	*7.08*
FVMF	3.96	29.78	1.90	*4.68*	*28.78*	*3.14*	**8.01**	*23.69*	**28.79**	*15.59*	*18.56*	**6.73**
MVMF	*3.91*	*29.82*	1.81	*5.04*	*28.34*	4.32	*8.73*	*23.27*	*37.15*	*15.90*	*18.57*	*7.32*
RVMF	3.95	*29.89*	1.81	*4.94*	*28.45*	6.16	*8.46*	*23.43*	*38.37*	*15.56*	*18.70*	*7.30*

Table 4.12 Performance comparison in terms of MAE, PSNR and DIS^k using the detail of the Baboon image contaminated with different densities of noise model III

Filter	No added noise			$p = 10\,\%$			$p = 30\,\%$			$p = 50\,\%$		
	MAE	PSNR	DIS^2	MAE	PSNR	DIS^2	MAE	PSNR	DIS^2	MAE	PSNR	DIS^1
None	0.00	∞	0.00	7.71	18.51	0.12	22.29	13.98	0.32	37.16	11.79	0.47
AMF	11.95	23.42	**1.67**	15.55	21.67	55.78	22.46	18.81	79.84	30.06	16.52	8.56
MMF	10.31	*23.58*	3.61	*11.22*	*23.07*	19.80	**13.79**	**21.37**	54.63	**20.06**	**18.29**	7.71
VMF	*10.10*	*23.58*	*2.06*	11.57	22.80	8.82	15.60	20.36	36.88	24.10	16.83	7.30
BVDF	12.01	21.50	*2.00*	13.79	20.90	*6.14*	18.77	18.52	*34.27*	28.31	15.15	*7.07*
GVDF	11.12	22.30	2.47	12.60	21.67	*6.31*	15.84	19.92	*30.84*	25.59	15.93	*7.14*
DDF	**9.88**	23.49	*1.69*	11.45	22.73	*6.16*	15.56	20.42	*32.49*	23.66	16.93	*7.07*
FDDF	*9.98*	*23.56*	*2.31*	**10.91**	*23.05*	**5.18**	*13.88*	*21.08*	**26.57**	*21.14*	*17.58*	**6.83**
FVMF	10.27	23.51	3.96	*11.01*	**23.07**	*8.20*	*13.87*	*21.04*	*32.51*	*21.24*	*17.50*	*7.08*
MVMF	*10.16*	*23.56*	2.98	*11.38*	22.89	9.84	*14.47*	*20.77*	42.04	*21.70*	*17.46*	7.27
RVMF	*10.18*	**23.59**	3.39	*11.32*	*22.90*	12.14	*14.22*	*20.89*	46.16	*21.25*	*17.61*	7.49

These filters may be classified into the following main categories:

1. Partition-based filters which classify each pixel to be processed into several signal activity categories which, in turn, are associated to appropriate processing methods.
2. Switching filters which replace only the corrupted pixels. This type of filter works by classifying pixels as corrupted or non-corrupted and so they will be processed by an appropriate denoising filter or left unchanged, respectively. As a consequence, the quality of a switching filter depends on both impulse detection and restoration. This family includes a large amount of filters given that they are intuitive, easy to implement and exhibit a low computational cost.

Table 4.13 Performance comparison in terms of MAE, PSNR and DIS^k using the detail of the Boats image contaminated with different densities of noise model III

Filter	No added noise			$p = 10\,\%$			$p = 30\,\%$			$p = 50\,\%$		
	MAE	PSNR	DIS^2	MAE	PSNR	DIS^2	MAE	PSNR	DIS^2	MAE	PSNR	DIS^1
None	0.00	∞	0.00	7.28	19.06	0.13	21.47	14.44	0.33	36.14	12.13	0.49
AMF	5.23	30.36	**0.23**	9.57	25.38	52.54	17.17	20.83	76.94	25.05	17.68	8.46
MMF	*3.37*	*32.11*	0.60	*4.25*	*30.32*	9.63	**6.81**	**26.11**	35.79	**13.25**	**20.33**	*6.03*
VMF	*3.35*	*32.11*	0.45	4.76	29.58	*3.13*	8.96	23.87	30.84	17.75	18.24	6.50
BVDF	6.37	27.05	*0.35*	7.36	25.94	*3.15*	11.22	21.53	31.86	21.92	15.94	6.88
GVDF	5.09	28.63	0.36	6.17	27.24	*2.40*	9.19	23.14	*27.62*	19.96	16.77	6.74
DDF	3.88	31.14	*0.24*	4.97	29.27	*2.71*	8.89	24.02	*28.99*	17.20	18.40	6.44
FDDF	3.53	31.73	*0.27*	*4.30*	*30.29*	**1.98**	*7.24*	*25.27*	**23.72**	*14.95*	*19.09*	*5.86*
FVMF	*3.42*	*31.87*	0.68	**4.17**	**30.41**	3.18	*7.02*	*25.45*	*24.60*	*14.69*	*19.18*	**5.78**
MVMF	*3.35*	*32.10*	*0.35*	*4.52*	*29.91*	3.59	*7.62*	*24.90*	*30.68*	*14.85*	*19.14*	*6.08*
RVMF	**3.31**	**32.16**	0.39	*4.38*	*30.05*	4.99	*7.34*	*25.12*	31.98	*14.42*	*19.32*	*6.09*

Table 4.14 Performance comparison in terms of MAE, PSNR and DIS^k using the detail of the Goldhill image contaminated with different densities of noise model III

Filter	No added noise			$p = 10\,\%$			$p = 30\,\%$			$p = 50\,\%$		
	MAE	PSNR	DIS^2	MAE	PSNR	DIS^2	MAE	PSNR	DIS^2	MAE	PSNR	DIS^1
None	0.00	∞	0.00	7.09	19.17	0.11	21.68	14.32	0.31	36.42	12.04	0.47
AMF	5.82	29.46	**0.51**	9.85	25.36	48.52	18.08	20.62	77.60	26.88	17.47	8.37
MMF	*4.68*	**30.18**	*0.85*	**5.20**	**29.33**	4.51	**7.22**	**26.03**	28.60	**13.44**	**20.49**	*6.51*
VMF	*4.75*	*30.02*	0.95	5.57	28.78	*1.68*	9.22	23.96	26.22	18.05	18.37	6.93
BVDF	7.22	26.78	*0.92*	7.73	26.01	*1.67*	10.88	22.32	25.56	20.52	16.73	6.96
GVDF	5.85	28.29	0.96	6.56	27.23	*1.71*	8.88	23.97	*21.00*	18.78	17.53	6.91
DDF	4.99	29.67	*0.90*	5.73	28.56	**1.55**	9.06	24.23	*24.77*	16.85	18.84	6.76
FDDF	4.76	*30.00*	*0.86*	*5.26*	*29.21*	*1.55*	*7.58*	*25.34*	**17.58**	*14.95*	*19.40*	*6.36*
FVMF	*4.76*	29.95	1.20	*5.25*	*29.14*	2.05	*7.47*	*25.45*	*18.02*	*14.89*	*19.34*	**6.24**
MVMF	*4.72*	*30.05*	1.11	*5.39*	*28.99*	1.86	*7.88*	*25.01*	*24.60*	*15.06*	*19.31*	*6.59*
RVMF	**4.67**	*30.13*	1.01	*5.31*	*29.09*	2.16	*7.65*	*25.20*	25.59	*14.63*	*19.50*	*6.60*

3. Filters using weighting coefficients which make use of different weights in a non-linear fashion to preserve the original signal structures, such as edges and fine details while removing noise.
4. Fuzzy filters which make use of fuzzy logic tools to provide adaptiveness to local features.
5. Regularization filters which follow a regularization approach based on the minimization of appropriate energy functions by means of partial differential equations.

Table 4.15 Performance comparison in terms of MAE, PSNR and DIS^k using the detail of the Lenna image contaminated with different densities of noise model IV

Filter	No added noise			$\alpha = 0.7$			$\alpha = 0.5$			$\alpha = 0.4$		
	MAE	PSNR	DIS^2	MAE	PSNR	DIS^2	MAE	PSNR	DIS^2	MAE	PSNR	DIS^1
None	0.00	∞	0.00	7.67	20.20	0.12	14.39	16.37	0.21	20.18	14.32	0.30
AMF	6.00	28.13	2.49	8.89	25.73	42.33	12.22	23.46	65.83	14.95	21.90	7.76
MMF	3.99	**29.91**	1.78	**5.11**	**28.88**	9.49	**5.80**	**27.96**	17.75	**6.24**	**27.19**	2.35
VMF	*3.89*	*29.81*	*1.73*	5.61	28.26	*4.38*	6.69	27.04	*10.79*	7.62	25.80	*1.77*
BVDF	3.99	29.45	**1.41**	6.87	26.58	5.53	8.28	24.88	12.05	9.79	23.19	2.08
GVDF	*3.92*	29.70	*1.75*	6.19	27.31	5.58	7.35	25.78	*11.05*	8.59	24.33	1.87
DDF	*3.82*	29.73	*1.56*	5.67	28.17	*4.15*	6.74	26.93	*9.93*	7.80	25.58	*1.72*
FDDF	**3.82**	*29.89*	*1.72*	*5.29*	*28.67*	**4.02**	*6.08*	*27.69*	**7.65**	*6.78*	*26.64*	**1.38**
FVMF	3.96	29.78	1.90	*5.31*	*28.63*	*4.86*	*6.03*	*27.74*	*9.81*	*6.65*	*26.74*	*1.44*
MVMF	*3.91*	*29.82*	1.81	*5.40*	*28.47*	*5.04*	*6.22*	*27.45*	11.12	*6.83*	*26.49*	*1.84*
RVMF	3.95	*29.89*	1.81	*5.28*	*28.62*	6.19	*6.05*	*27.62*	13.26	*6.64*	*26.64*	1.97

Table 4.16 Performance comparison in terms of MAE, PSNR and DIS^k using the detail of the Baboon image contaminated with different densities of noise model IV

Filter	No added noise			$\alpha = 0.7$			$\alpha = 0.5$			$\alpha = 0.4$		
	MAE	PSNR	DIS^2	MAE	PSNR	DIS^2	MAE	PSNR	DIS^2	MAE	PSNR	DIS^1
None	0.00	∞	0.00	7.63	20.29	0.13	14.08	16.52	0.24	19.67	14.53	0.33
AMF	11.95	23.42	**1.67**	13.84	22.46	42.92	15.88	21.43	63.14	18.34	20.32	7.25
MMF	10.31	*23.58*	3.61	**11.23**	**23.22**	16.41	**11.76**	**22.91**	27.16	**12.34**	**22.46**	3.80
VMF	*10.10*	*23.58*	*2.06*	11.55	23.06	*7.22*	12.48	22.50	*11.27*	13.41	21.81	2.50
BVDF	12.01	21.50	*2.00*	15.43	20.29	*7.33*	16.36	19.92	*10.35*	17.67	19.18	*2.44*
GVDF	11.12	22.30	2.47	13.55	21.42	*6.55*	14.23	21.30	*11.33*	15.18	20.61	*2.09*
DDF	**9.88**	23.49	*1.69*	11.69	22.87	*6.33*	12.62	22.39	*9.45*	13.64	21.61	2.22
FDDF	*9.98*	*23.56*	*2.31*	*11.36*	*23.08*	**5.67**	*12.04*	*22.69*	**8.86**	*12.94*	*21.96*	**2.05**
FVMF	10.27	23.51	3.96	*11.40*	*23.08*	10.06	*12.10*	*22.66*	14.08	*12.87*	*22.05*	*2.31*
MVMF	*10.16*	*23.56*	2.98	*11.41*	*23.11*	10.49	*12.16*	*22.61*	14.45	*12.88*	*22.07*	2.75
RVMF	*10.18*	**23.59**	3.39	*11.30*	*23.15*	13.00	*11.97*	*22.70*	18.43	*12.74*	*22.12*	2.99

In the following section, we review these main categories and the different possibilities studied in each one to improve performance.

4.3.1 Partition Based Filters

The main idea behind these methods is that any image presents very different features in different locations and that each location should be processed in a different

Table 4.17 Performance comparison in terms of MAE, PSNR and DIS^k using the detail of the Boats image contaminated with different densities of noise model IV

Filter	No added noise			$\alpha = 0.7$			$\alpha = 0.5$			$\alpha = 0.4$		
	MAE	PSNR	DIS^2	MAE	PSNR	DIS^2	MAE	PSNR	DIS^2	MAE	PSNR	DIS^1
None	0.00	∞	0.00	7.51	20.57	0.14	14.30	16.56	0.26	20.05	14.56	0.35
AMF	5.23	30.36	**0.23**	7.96	27.17	40.72	11.19	24.34	67.46	13.80	22.67	7.50
MMF	*3.37*	*32.11*	0.60	**4.46**	**30.83**	6.85	**5.13**	**29.59**	17.59	**5.57**	**28.80**	2.35
VMF	*3.35*	*32.11*	0.45	4.89	30.16	4.56	6.12	28.23	*10.27*	7.07	26.87	*1.81*
BVDF	6.37	27.05	*0.35*	8.09	25.46	*3.81*	9.24	24.28	10.93	10.00	23.46	2.01
GVDF	5.09	28.63	0.36	6.69	27.02	*3.95*	7.45	26.12	*8.46*	8.32	24.90	1.82
DDF	3.88	31.14	*0.24*	5.16	29.71	*3.71*	6.35	27.89	*9.70*	7.35	26.55	*1.70*
FDDF	3.53	31.73	*0.27*	*4.80*	*30.25*	**3.28**	*5.65*	*28.86*	**7.49**	*6.30*	*27.77*	**1.33**
FVMF	*3.42*	*31.87*	0.68	*4.68*	*30.39*	4.73	*5.49*	*29.01*	*9.20*	*6.11*	*28.04*	*1.48*
MVMF	*3.35*	*32.10*	*0.35*	*4.71*	*30.42*	*4.36*	*5.63*	*28.81*	10.54	*6.28*	*27.75*	*1.74*
RVMF	**3.31**	**32.16**	0.39	*4.60*	*30.59*	5.54	*5.45*	*29.04*	11.99	*6.05*	*28.01*	1.88

Table 4.18 Performance comparison in terms of MAE, PSNR and DIS^k using the detail of the Goldhill image contaminated with different densities of noise model IV

Filter	No added noise			$\alpha = 0.7$			$\alpha = 0.5$			$\alpha = 0.4$		
	MAE	PSNR	DIS^2	MAE	PSNR	DIS^2	MAE	PSNR	DIS^2	MAE	PSNR	DIS^1
None	0.00	∞	0.00	7.63	20.41	0.13	14.79	16.35	0.25	20.41	14.47	0.34
AMF	5.82	29.46	**0.51**	8.43	26.61	39.32	11.66	23.99	66.03	14.26	22.48	7.38
MMF	*4.68*	**30.18**	*0.85*	**5.42**	**29.46**	4.75	**6.03**	**28.49**	11.28	**6.29**	**28.12**	1.74
VMF	*4.75*	*30.02*	0.95	5.88	28.91	*2.82*	6.90	27.49	8.50	7.70	26.40	*1.60*
BVDF	7.22	26.78	*0.92*	8.51	25.58	*2.79*	9.22	24.72	9.04	10.08	23.59	2.02
GVDF	5.85	28.29	0.96	7.03	27.05	3.25	7.73	26.18	*8.05*	8.56	24.78	1.77
DDF	4.99	29.67	*0.90*	6.06	28.60	2.52	7.11	27.19	8.30	7.87	26.15	1.66
FDDF	4.76	*30.00*	*0.86*	*5.75*	*29.05*	**2.41**	*6.49*	*27.93*	**6.14**	*7.02*	*26.94*	*1.20*
FVMF	*4.76*	29.95	1.20	*5.72*	*29.05*	3.37	*6.41*	*27.99*	*7.04*	*6.83*	*27.30*	**1.13**
MVMF	*4.72*	*30.05*	1.11	*5.68*	*29.11*	*2.85*	*6.42*	*27.97*	*7.72*	*6.86*	*27.27*	*1.39*
RVMF	**4.67**	*30.13*	1.01	*5.56*	*29.27*	3.49	*6.28*	*28.13*	*8.01*	*6.66*	*27.48*	*1.40*

way, including the noisy ones. So, they classify the pixels in the image into different categories, explicitly or implicitly, and provide an appropriate processing for each one.

A first approach was made by means of vector median–rational hybrid filters in [29, 30]. The method in [31] analyses similarities between the neighbouring colour vectors in a two-step impulse detection procedure. A three-step procedure including robust estimation, vector partition and weighted filtering has been recently introduced in [32]. Other methods propose to simultaneously use several sub-filters in each image location, and the filter output is computed by choosing the most appropriate sub-filter output [33] or by fusing the sub-filter outputs using a genetic algorithm [34]. Further improvements were presented in [35], [36] and [37], where the use of fuzzy logic was included. On the other hand, the work in [38] uses partition and weighted filters.

4.3.2 Switching Filters

Following the intuitive idea that only noisy pixels should be modified and the rest should be left unchanged, switching filters are based on a classification scheme followed by the appropriate denoising operation when necessary. This idea has inspired a large amount of methods for its simplicity in implementation and low computational cost. The first filters were based on robust estimators to perform the classification and later, other different alternatives were explored to improve performance. So, we may classify the filters in this family as follows.

4.3.2.1 Based on Robust Statistics

A robust statistic is used to characterize the noisiness of each image pixel and decide whether it is noisy or not. These methods are capable of removing even strong noise, while preserving fine image details.

In order to detect the noisy pixels, [39] checks the cluster membership of each pixel after a cluster analysis of its neighbourhood. The work in [40] performs a deviation test with respect to the set of a few robust vectors. The methods in [41,42] use approximation of the multivariate dispersion, and the work in [43] combines a robust statistic and a two-step detection. The technique in [44] is based on computing the confidence limits extracted from the neighbourhood assuming a multi-normal distribution of the colour vectors, and the method in [45] uses quaternion rotation theory to distinguish atypical colours. In [46–48], a special vector ordering procedure that increases the probability of the filter window central pixel to be the filter output is used. This reduces the number of unnecessary substitutions and improves the detail-preserving ability of the filtering.

4.3.2.2 Based on Peer Groups

These filters analyse the neighbourhood of each pixel building its peer group and depending on this, the classification is done. The *peer group* of a given pixel (central pixel) is a set constituted by this pixel and those of its neighbours which are *similar* to it. Depending on this set, the pixels are classified as noise-free or declared as outliers, which should be replaced by a suitable robust filter.

The filters introduced in [49, 50] use the difference between the *peer group* of the pixel under consideration and other *peer groups* in its neighbourhood to form the detection rule. The work in [51] proposes to use windows of different size to determine the peer region of each pixel and then checks the peer region size and shape. The approach introduced in [52] proposes that a pixel can be declared as *noise-free* if and only if its *peer group* size is larger than a determined threshold. The methods in [53,54] introduce several modifications to [52] to improve efficiency and detection where the *peer group* concept is defined in the fuzzy metric context. The filter introduced in [55] divides the detection process into two stages, namely: (i) robust detection and (ii) refinement. The works in [32, 33] extend the filter defined in [50] to the directional domain and use it as a sub-filter in a hybrid structure.

More recently, given that the *similarity* between two colour pixels is not easily expressed in a crisp way, a fuzzy representation introducing the *fuzzy peer group* concept has shown some advantages [56].

4.3.2.3 Using Fuzzy Metrics

Inspired by the performance of the fuzzy metrics [57] when replacing classical metrics in robust vector filters [58, 59], a series of works have tried to design adaptive filters using fuzzy metrics. In fact, these metrics include a parameter in its definition which is used for adaptive processing.

Some examples of fuzzy metrics have been used to measure magnitude differences [47,59] which have been used to develop a local self-adaptive filter [48]. Other works measure angular-colorimetric distances and hybrid magnitude-colorimetric differences [23] or magnitude and spatial distances simultaneously [60].

4.3.2.4 Using Quaternions

Another approach is the use of the quaternion notion to improve the evaluation of the colour dissimilarity. Although the quaternion theory was established in 1843 by Hamilton [61], it had not been applied to colour image filtering until the last two decades.

The choice of the imaginary part to represent the pixel values is fairly obvious, and is supported by the coincidence between the three-space imaginary part of the quaternion and the three-space of the RGB colour values [62]. When we represent colour images in this way, a three-space rotation will always map a pixel value in three-space into another pixel value in three-space, but not necessarily within the bounds of the RGB colour space. A few works have explored the use of quaternions within switching filters in [63–66].

4.3.2.5 Noise Detection in Several Steps

Given the difficulty in distinguishing from noise and image features, some filters tried to improve noise detection performance by using several steps in the detection. First an initial detection is done and later steps are used for refinement. Several steps are used in a few partition filters above [31, 32]. An iterative classification method is proposed in [55]. A two-step procedure for fine detections in used in [43, 67] and a second step to remove noise clusters is proposed in [56].

4.3.2.6 Based on Morphological Operations

Another group of impulsive noise reduction techniques is based on the methods derived from the mathematical morphology [68–70]. The main difficulty of the application of morphological methods into colour image processing lies in the required vector ordering scheme [71, 72].

Actually, impulsive noise can be suppressed through min/max filtering in greyscale images, and so, ordering noisy pixels to be the maximum or minimum ones within the operation window is the key issue in morphological filter design. In the literature, there are many proposals which have focused on morphological filters [73–81], but the extension to colour images is not straightforward.

The definitions of the basic mathematical morphology operations were generalized in various ways, so that they can work on colour images and can be applied for impulsive noise suppression [82–88].

4.3.3 *Filters Based on Weighted Coefficients*

The use of non-linear weighted coefficients is one of the first and better known methods to adapt processing. The difference among the methods in this family, which are many, lies in the way to determine the weights used.

Averaging by weighted filtering is a very well-known efficient technique for reducing Gaussian noise [89], but difficult to generalize to impulse noise in colour images. The approach in [90] added a third term based on a robust estimator to be able to reduce impulse noise also . Alternatively, [91, 92] use robust weights and fuzzy functions, [93–95] use fuzzy logic to compute the weights and [38] partition and weighted sum filters. On the other hand, weights have been used within vector median operations. For instance, the centre weighted vector median filters [4] or more recently, in the restoration operation in [96] or in the computation of accumulated dissimilarities in [97].

4.3.4 Fuzzy Filters

Given the non-linearity of images and the difficult in distinguishing between noise and data, fuzzy logic provides an adaptive framework where multiple criteria can be modelled and implemented in a simple way. A series of concepts used in image processing can be efficiently modelled using fuzzy tools (similarity, neighbourhood,..), which has motivated the development of many fuzzy filters.

The fuzzy inference rule-based filters in [98] have good performance but they cannot properly remove the noise present at the edges. This fact allows to present alternative approaches [56, 99–101]. Nevertheless, although the effectiveness of the rules work well under some constrained environment, there is a high dependence on them and also the number, complexity and variety of such rules makes the design task difficult.

Different fuzzy versions of the vector median filter are given in [102–104]. The scalar and vector median operations are extended to fuzzy numbers in [105]. The techniques in [96, 106–108] use a fuzzy rule-based system to determine the filter output. The vector median and some fuzzy distance/similarity measures are used in [93–95] for calculating the fuzzy coefficients to determine the output as a weighted average of the inputs. The work in [89] computes the fuzzy coefficients, taking into account spatial and value nearness relations between pixels, and the inclusion of an impulse detector in this procedure is addressed in [90]. A robust fuzzy vector filtering framework is introduced in [91, 92]. Further works in [109, 110] use fuzzy coefficients to determine the filter output by selecting the most representative input vector or as the combination of the vectors inside the filter window.

Fuzzy logic and neural networks have been used for impulse noise filtering [111–113]. These methods provide a good performance, but with the cost of higher computational complexity.

Besides, fuzzy cellular automata have been recently used. Cellular automata (CA) are discrete dynamic systems whose behaviour is completely specified in terms of a local relation [114]. Space is represented by a uniform grid, with each site or cell containing a few bits of data. Time advances in discrete steps and the laws of the universe are expressed by a single recipe. At each step, every cell computes its new state from that of its close neighbours. Thus, the laws of the system are local and uniform. CA can be used as a very fast and efficient tool for image processing purposes [115–121]. For example, in [120], CA and fuzzy logic was used as a filtering tool for different type of impulse noise.

4.3.5 Regularization Filters

Another family of promising performance but difficult to extend to colour images are the regularization filters. The minimization of appropriate energy functions by means of partial differential equations has provided a nice performance in removing impulse noise from greyscale images [122–124]. Later, different extensions to

Table 4.19 Filters used in the comparisons

Filter name / reference	Short name	Year
Fast peer group filter [52]	FPGF	2005
Sigma vector median filter [42]	SVMF	2006
Impulse noise reduction [96]	INR	2007
Local self-adaptive fuzzy magnitude impulse vector filter [48]	LSAFSVF	2008
Iterative peer group switching vector filter [55]	IPGSVF	2009
Fuzzy ROD filter [43]	FRF	2010
Quaternion representation vector filter [65]	QRVF	2011
Rank weighted adaptive switching filter [97]	RWVMF	2012
Simple fuzzy rule filter [108]	SFRF	2013

colour images were introduced: [125, 126] use correlation terms, [127] and [128] are based on a vector approach, [129] and [130] use graph models.

4.3.6 Comparison of Advanced Adaptive Methods

In this section, several images have been considered to evaluate the performance of a set of representative filters. We consider both images artificially corrupted with the four types of noise in Sect. 4.1 and real noisy images. The test images Baboon, Boats, Goldhill and Lenna are shown in Fig. 4.3 (a)–(d). We have selected a small detail of each one (Fig. 4.3 (e)–(h)) to better appreciate performance differences. The real noisy images used are shown in Fig. 4.9.

We have considered different intensities of noise in the image. For types I, II and III, we consider noise probability equal to 0.1, 0.2 and 0.3. For noise type IV, we set α equal to 0.7, 0.6 and 0.5. We have used the objective quality measures MAE, PSNR and DIS^k to objectively compare the performance. From the state-of-the art, we selected a set of recent representative filters from the different families which are shown in Table 4.19, ordered by year of publication. In all cases, we have used a 3×3 filtering window since larger filtering windows do not provide significant relative performance differences among the considered methods. We have filtered the image only once to better appreciate the performance differences. Notice that when some noise remains in the image, the resulting image should be filtered once more or until no changes in the image are obtained, or using a larger filtering window. Numerical results are shown in Tables 4.20–4.41 where we have highlighted the best results in each column (optimal in bold, suboptimal in italic). Besides, Figs. 4.5–4.8 show a few filtered images for visual comparison.

From Tables 4.20 to 4.25, we can see that for noise model I (fixed value impulsive noise), the best performing filters are INR and IPGSV, followed by FRF and QRVF and finally RWVMF, SVMF and FPGF. This ranking is closely related to the fact that in this noise model, the formation of noise clusters is frequent and the ability

Table 4.20 Performance comparison in terms of MAE, PSNR and DIS^k using the detail of the Baboon image contaminated with different densities of noise model I

Filter	No added noise			$p = 10\,\%$			$p = 20\,\%$			$p = 30\,\%$		
	MAE	PSNR	DIS^2	MAE	PSNR	DIS^2	MAE	PSNR	DIS^1	MAE	PSNR	DIS^1
None	0.00	∞	0.00	4.96	19.19	38.14	10.62	15.87	5.91	14.55	14.52	7.21
FPGF	*1.93*	*27.66*	*0.53*	*3.48*	*26.04*	2.76	5.74	23.97	0.55	7.47	22.73	1.11
SVMF	4.42	25.53	6.20	4.54	25.26	8.82	*5.37*	*24.42*	1.27	*6.31*	23.04	2.29
INR	*2.01*	*30.20*	1.73	*2.53*	*29.11*	*2.73*	*3.21*	**28.28**	*0.30*	**3.73**	**27.70**	*0.63*
LSAFSVF	3.44	25.20	*0.80*	4.58	24.65	4.08	5.93	23.71	0.64	6.98	*23.06*	1.17
IPGSVF	**0.07**	*40.45*	*0.16*	**1.48**	**30.62**	*1.88*	**3.19**	*27.63*	**0.23**	*6.42*	*24.40*	*0.47*
FRF	*0.31*	*34.42*	*0.18*	*2.01*	*28.56*	*2.88*	*4.53*	*25.09*	*0.36*	*6.95*	*23.15*	*0.80*
QRVF	*0.12*	**43.34**	**0.04**	*1.96*	*27.07*	**0.92**	*4.84*	22.26	*0.31*	*6.34*	21.96	**0.34**
RWVMF	4.00	25.61	0.82	4.54	25.12	2.90	6.77	23.71	*0.27*	8.42	22.88	*0.64*
SFRF	5.09	27.64	1.14	7.47	26.94	5.18	10.75	*24.33*	1.11	12.29	*23.10*	2.22

Table 4.21 Performance comparison in terms of MAE, PSNR and DIS^k using the detail of the Boats image contaminated with different densities of noise model I

Filter	No added noise			$p = 10\,\%$			$p = 20\,\%$			$p = 30\,\%$		
	MAE	PSNR	DIS^2	MAE	PSNR	DIS^2	MAE	PSNR	DIS^1	MAE	PSNR	DIS^1
None	0.00	∞	0.00	4.87	19.56	38.41	10.42	16.15	6.18	15.54	14.45	7.32
FPGF	0.76	36.01	1.78	*0.76*	*36.01*	*1.78*	*1.73*	*30.60*	0.58	3.28	26.11	1.58
SVMF	0.55	37.84	3.09	0.94	*34.67*	4.90	1.74	30.21	0.89	3.13	25.46	2.19
INR	0.39	*39.82*	0.18	**0.63**	**38.15**	**0.30**	**0.96**	**36.06**	**0.12**	**1.38**	**34.24**	*0.30*
LSAFSVF	*0.09*	41.87	**0.00**	*0.83*	34.63	2.15	*1.58*	*31.51*	*0.51*	*2.71*	*27.72*	*1.22*
IPGSVF	**0.00**	∞	**0.00**	*0.70*	*34.71*	*1.44*	*1.26*	*34.36*	*0.14*	*2.31*	*30.82*	**0.21**
FRF	**0.00**	∞	**0.00**	*0.73*	*34.92*	2.01	*1.57*	*31.62*	*0.32*	*3.00*	*27.38*	*0.77*
QRVF	*0.02*	*53.24*	*0.05*	1.45	27.10	*0.60*	3.15	23.47	*0.23*	4.75	21.85	*0.41*
RWVMF	*0.26*	*41.14*	*0.02*	0.91	34.31	1.98	5.39	30.16	2.03	*2.96*	*27.60*	0.89
SFRF	1.06	38.68	*0.11*	4.16	32.51	*1.21*	5.91	28.64	0.65	7.03	26.77	1.67

of each method to solve this handicap is essential. Nevertheless, as it is proposed in [96], a second pass of the filter with a parameter setting able to reduce the remaining noise clusters could be used to improve performance from this point of view. The QRVF filter works well in the Baboon and Lenna images, but is worse in others. This may indicate that QRVF lacks enough adaptiveness to image features. From Fig. 4.5, the visual comparison reveals that INR provides the best trade-off between noise reduction and detail preservation. IPGSVF seems better in preserving details, but misses the detection of some noise clusters and impulse noise when it is similar to the background where it appears. FRF and QRVF also leave noise clusters in the image. QRVF preserves image details pretty well, whereas FRF introduces a little blur in the Baboon image.

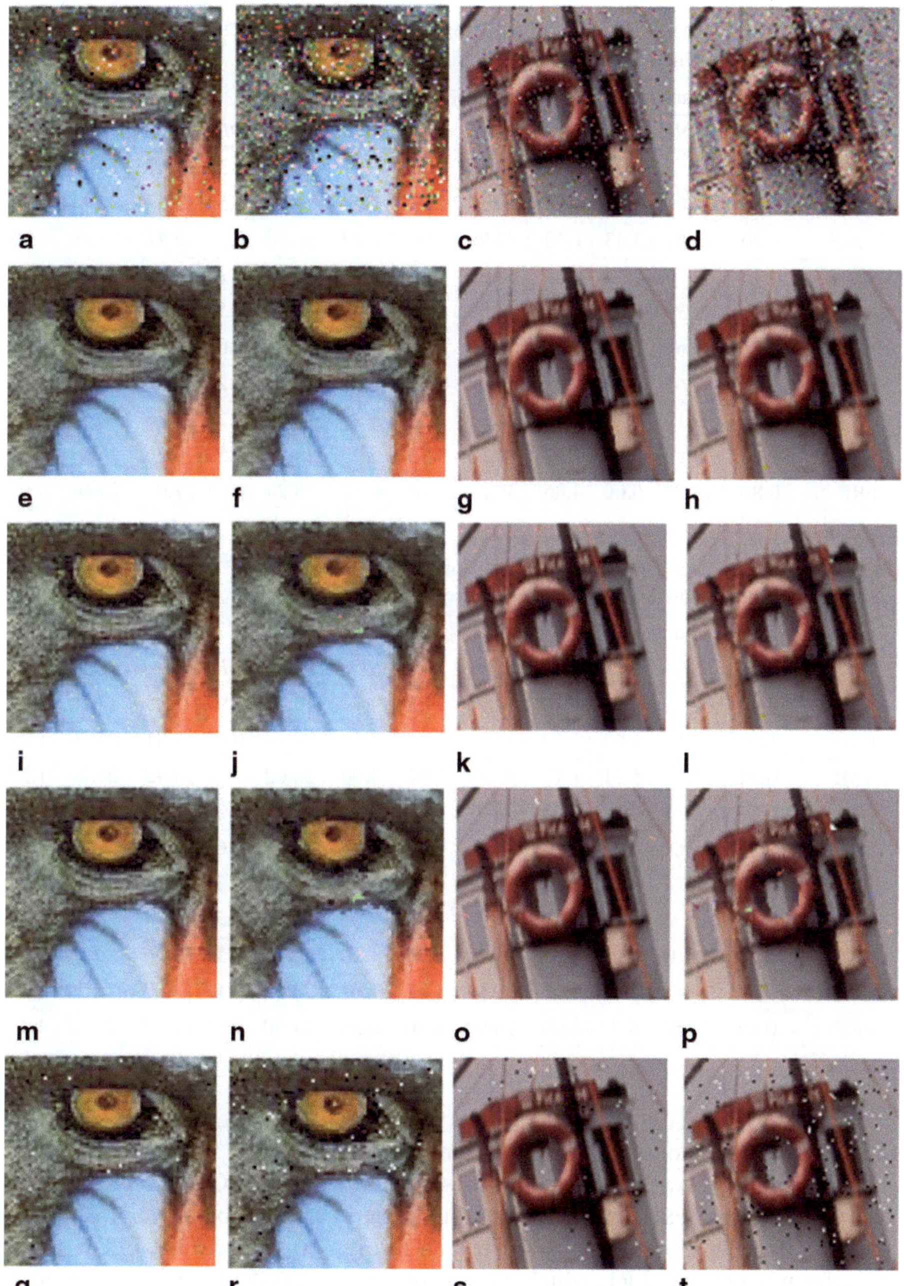

Fig. 4.5 Outputs for filter performance comparison of noise model I. **a, b** Baboon corrupted with $p = 10\,\%$ and $p = 30\,\%$. **c, d** Boats corrupted with $p = 10\,\%$ and $p = 30\,\%$. **e–h** Output of INR. **i–l** Output of IPGSVF. **m–p** Output of FRF. **q–t** Output of QRVF

Table 4.22 Performance comparison in terms of MAE, PSNR and DIS^k using the detail of the Goldhill image contaminated with different densities of noise model I

Filter	No added noise			$p = 10\,\%$			$p = 20\,\%$			$p = 30\,\%$		
	MAE	PSNR	DIS^2	MAE	PSNR	DIS^2	MAE	PSNR	DIS^1	MAE	PSNR	DIS^1
None	0.00	∞	0.00	5.37	19.02	41.63	10.27	16.24	6.25	15.75	14.29	7.29
FPGF	*0.07*	*43.90*	*0.01*	*0.93*	*34.41*	2.03	2.02	29.86	0.56	3.47	26.11	1.40
SVMF	1.20	34.41	3.43	1.30	*33.96*	4.06	*1.99*	*30.30*	0.79	*3.41*	25.89	1.75
INR	0.43	37.76	0.72	**0.70**	**36.68**	*0.84*	**1.04**	**34.93**	*0.13*	**1.41**	**33.66**	**0.15**
LSAFSVF	*0.16*	*39.30*	*0.01*	*1.04*	33.53	2.46	*1.95*	*30.51*	*0.51*	*3.01*	*28.02*	1.09
IPGSVF	**0.00**	**60.66**	**0.00**	*0.87*	33.79	*0.57*	*1.59*	*32.05*	**0.07**	*2.57*	*30.89*	*0.16*
FRF	*0.01*	*52.66*	**0.00**	*0.81*	*36.02*	1.30	*1.76*	*31.74*	*0.27*	*3.07*	*28.06*	*0.64*
QRVF	*0.26*	39.19	0.94	2.30	24.61	*1.06*	3.81	22.80	*0.25*	5.86	20.56	*0.35*
RWVMF	0.81	35.83	0.09	1.12	*34.50*	**0.53**	6.02	29.24	1.53	6.76	*28.01*	1.81
SFRF	0.86	*43.35*	**0.00**	4.86	31.29	*1.16*	6.32	28.24	0.63	7.07	27.03	*0.91*

Table 4.23 Performance comparison in terms of MAE, PSNR and DIS^k using the detail of the Lenna image contaminated with different densities of noise model I

Filter	No added noise			$p = 10\,\%$			$p = 20\,\%$			$p = 30\,\%$		
	MAE	PSNR	DIS^2	MAE	PSNR	DIS^2	MAE	PSNR	DIS^1	MAE	PSNR	DIS^1
None	0.00	∞	0.00	5.18	18.93	39.54	9.79	16.26	5.97	15.30	14.30	7.32
FPGF	1.10	32.55	1.64	1.10	32.55	*1.64*	2.30	28.44	0.62	3.81	25.79	1.31
SVMF	0.94	34.15	5.11	1.17	*33.34*	5.67	2.06	28.62	1.12	*3.42*	25.20	1.98
INR	0.65	*36.38*	*0.27*	*0.94*	*35.07*	*0.88*	*1.32*	*33.05*	0.26	**1.74**	**31.78**	*0.34*
LSAFSVF	*0.36*	35.60	0.42	*1.05*	32.85	1.69	2.05	*29.54*	0.61	*3.08*	*27.41*	1.03
IPGSVF	**0.00**	∞	**0.00**	*0.81*	*33.18*	*1.60*	*1.60*	*31.40*	*0.38*	*2.86*	*29.55*	*0.34*
FRF	*0.04*	*46.30*	**0.00**	*0.79*	*34.96*	*1.31*	*2.03*	29.71	*0.39*	3.44	26.80	*0.70*
QRVF	**0.00**	*56.04*	**0.00**	**0.67**	**37.34**	**0.20**	*1.58*	*32.92*	**0.13**	*2.56*	*30.38*	**0.24**
RWVMF	*0.37*	*37.61*	0.31	4.98	30.94	15.17	**1.01**	**33.63**	**0.13**	3.57	*27.59*	*0.57*
SFRF	0.86	41.54	*0.07*	4.09	29.92	4.30	6.32	27.50	0.91	7.30	26.17	1.41

For noise model II, (correlated random value impulsive noise) (Tables 4.26–4.31) the probability of noise clusters appearance decreases. Here, the best performing methods are those with better detail preservation ability: FPGF and FRF are the best followed by IPSVF and QRVF, which again performs better for Lenna and Baboon. By comparing visually the results in Fig. 4.6, we see that FPGF seems to perform better than FRF, which misses some noise. IPGSVF again misses noise which appears with a colour similar to background.

Tables 4.32–4.37 show the performance comparison for noise model III (uncorrelated random value impulsive noise). In this case, the best performance is exhibited by INR and QRVF, followed by RWVMF and IPGSVF. Given that the appearance

Table 4.24 Performance comparison in terms of MAE, PSNR and DIS^k using the Goldhill image contaminated with different densities of noise model I

Filter	No added noise			$p = 10\,\%$			$p = 20\,\%$			$p = 30\,\%$		
	MAE	PSNR	DIS^2	MAE	PSNR	DIS^2	MAE	PSNR	DIS^1	MAE	PSNR	DIS^1
None	0.00	∞	0.00	5.07	19.33	41.31	10.20	16.27	6.41	15.24	14.53	7.59
FPGF	*0.06*	*44.23*	**0.00**	*0.81*	*35.24*	1.27	*1.76*	30.63	0.49	2.99	27.00	1.23
SVMF	1.16	34.75	0.60	1.22	34.45	1.12	1.81	30.82	0.46	2.99	26.60	1.36
INR	*0.24*	41.06	0.13	**0.49**	**38.77**	**0.20**	**0.78**	**36.97**	**0.04**	**1.12**	**35.34**	*0.08*
LSAFSVF	0.14	39.43	**0.00**	*0.87*	34.54	1.58	*1.69*	*31.53*	0.43	*2.63*	*28.82*	0.91
IPGSVF	**0.00**	**74.99**	**0.00**	*0.74*	*34.98*	*0.54*	*1.35*	*33.79*	*0.04*	*2.13*	*32.19*	**0.06**
FRF	**0.00**	*56.09*	**0.00**	*0.72*	*36.45*	*1.02*	*1.56*	*32.39*	*0.25*	*2.63*	*29.10*	*0.46*
QRVF	*0.05*	*46.67*	0.10	1.48	26.78	*0.48*	2.94	23.77	*0.09*	4.41	22.12	*0.16*
RWVMF	0.54	37.91	*0.01*	1.02	*34.85*	*0.56*	1.92	*31.31*	*0.20*	2.92	*28.68*	*0.47*
SFRF	0.68	*45.20*	**0.00**	4.37	30.76	2.89	6.19	27.10	0.95	6.44	27.73	0.80

Table 4.25 Performance comparison in terms of MAE, PSNR and DIS^k using the Boats image contaminated with different of noise model I

Filter	No added noise			$p = 10\,\%$			$p = 20\,\%$			$p = 30\,\%$		
	MAE	PSNR	DIS^2	MAE	PSNR	DIS^2	MAE	PSNR	DIS^1	MAE	PSNR	DIS^1
None	0.00	∞	0.00	5.15	19.28	39.49	10.19	16.31	6.13	15.32	14.55	7.34
FPGF	*0.02*	*48.67*	**0.00**	*0.59*	*36.74*	1.35	1.33	31.33	0.52	2.47	27.21	1.28
SVMF	0.69	37.99	0.74	0.79	*36.34*	1.56	*1.28*	32.00	0.47	2.35	26.89	1.28
INR	0.23	41.27	0.23	**0.44**	**39.06**	**0.74**	**0.67**	**37.42**	*0.17*	*0.96*	**35.61**	*0.31*
LSAFSVF	*0.08*	41.62	*0.01*	0.59	*36.66*	1.31	*1.17*	*33.24*	*0.37*	*1.94*	*29.77*	0.78
IPGSVF	**0.00**	**79.21**	**0.00**	0.68	33.28	*1.28*	0.97	*34.82*	**0.13**	*1.49*	*33.95*	**0.11**
FRF	**0.00**	*62.03*	**0.00**	0.52	*37.82*	*1.22*	*1.14*	*33.29*	*0.32*	2.07	*29.42*	*0.60*
QRVF	*0.01*	*53.93*	0.15	1.22	27.36	*1.23*	2.46	24.33	*0.27*	3.73	22.66	*0.40*
RWVMF	0.22	43.53	0.06	0.69	36.22	*0.98*	1.33	*32.63*	0.26	*2.13*	*29.20*	*0.59*
SFRF	0.48	*46.22*	*0.01*	3.25	30.92	3.91	5.07	26.84	1.15	5.82	26.09	1.35

of this noise is not correlated among the colour channels, the INR and QRVF filters, which approach the correlation modelling from a different point of view than the classical vectorial filters, perform better. Besides, from Fig. 4.7, INR and QRVF also provide a more visually pleasant image. QRVF seems to preserve texture better but is a little worse in edges and misses some noise detection. RWVMF seems to introduce some blur in the image and leaves some noise, whereas IPGSVF fails in some noise detection for processing the image as a vector field.

Noise model IV (α-stable distribution noise), assessed in Tables 4.38–4.43, appears in the image in a non-correlated fashion and differs from the rest because in

Table 4.26 Performance comparison in terms of MAE, PSNR and DIS^k using the detail of the Baboon image contaminated with different densities of noise model II

Filter	No added noise			$p = 10\,\%$			$p = 20\,\%$			$p = 30\,\%$		
	MAE	PSNR	DIS^2	MAE	PSNR	DIS^2	MAE	PSNR	DIS^1	MAE	PSNR	DIS^1
None	0.00	∞	0.00	6.91	19.06	39.00	14.88	15.79	6.26	23.92	13.69	7.50
FPGF	*1.93*	*27.66*	*0.53*	*3.48*	26.16	*0.88*	*5.19*	*24.88*	**0.14**	*7.51*	*23.33*	**0.32**
SVMF	4.42	25.53	6.20	4.66	25.31	6.67	5.63	24.04	1.42	8.73	21.12	2.71
INR	*2.01*	*30.20*	1.73	*3.50*	*27.63*	2.67	*5.16*	*25.92*	0.42	*7.48*	*23.80*	0.92
LSAFSVF	3.44	25.20	*0.80*	4.42	24.71	*1.20*	5.73	24.13	*0.41*	8.10	22.30	0.98
IPGSVF	**0.07**	*40.45*	*0.16*	*1.75*	*29.14*	*1.31*	**3.45**	**27.13**	*0.17*	*8.05*	*23.07*	*0.77*
FRF	*0.31*	*34.42*	*0.18*	*1.85*	29.29	*0.94*	*4.01*	26.00	**0.14**	*6.98*	*23.55*	*0.40*
QRVF	*0.12*	**43.34**	**0.04**	**1.58**	**30.59**	**0.59**	*3.56*	*26.76*	*0.19*	**6.18**	**23.97**	*0.62*
RWVMF	4.00	25.61	0.82	5.16	24.68	1.35	7.33	23.56	*0.21*	10.46	22.12	*0.51*
SFRF	5.09	27.64	1.14	7.71	*26.61*	4.86	11.21	24.09	1.36	14.84	21.33	3.24

Table 4.27 Performance comparison in terms of MAE, PSNR and DIS^k using the detail of the Boats image contaminated with different densities of noise model II

Filter	No added noise			$p = 10\,\%$			$p = 20\,\%$			$p = 30\,\%$		
	MAE	PSNR	DIS^2	MAE	PSNR	DIS^2	MAE	PSNR	DIS^1	MAE	PSNR	DIS^1
None	0.00	∞	0.00	7.04	19.31	41.01	15.05	15.99	6.50	22.25	14.27	7.57
FPGF	0.76	36.01	1.78	*0.74*	**37.03**	*0.55*	*1.70*	**32.79**	*0.09*	**2.61**	**30.89**	*0.19*
SVMF	0.55	37.84	3.09	1.01	*34.54*	3.66	2.25	29.09	0.90	4.44	24.24	1.94
INR	0.39	39.82	0.18	1.16	*34.82*	**0.36**	2.32	*31.02*	*0.21*	3.49	*28.94*	*0.49*
LSAFSVF	*0.09*	*41.87*	**0.00**	*0.81*	*35.38*	0.75	*1.88*	*31.04*	0.30	*3.20*	28.06	0.63
IPGSVF	**0.00**	∞	**0.00**	*0.89*	34.00	*0.55*	*1.80*	*31.78*	*0.12*	*3.29*	*28.61*	*0.44*
FRF	**0.00**	∞	**0.00**	**0.69**	*36.86*	*0.53*	**1.66**	*32.44*	**0.06**	*2.68*	*30.36*	**0.13**
QRVF	*0.02*	*53.24*	*0.05*	*0.86*	34.08	*0.60*	2.02	29.98	*0.15*	*3.38*	27.23	*0.30*
RWVMF	*0.26*	*41.14*	*0.02*	4.71	31.85	18.57	5.50	29.89	1.85	6.35	*28.34*	1.92
SFRF	1.06	38.68	0.11	4.78	31.03	2.59	6.53	27.98	1.06	8.96	24.65	2.34

addition to the pixels affected by impulses, other pixels are affected with less intensity. Again, INR and QRVF offer nice performance for the non-correlated nature of the noise, but now other filters which are able to correct the pixels affected with a lower noise intensity as SFRF and RWVMF are also competitive. The visual comparison in Fig. 4.8 shows that INR and QRVF only reduce well the noise of high intensity impulses. SFRF reduces more noise but misses some noise clusters and blur texture a little. RWVMF provides the best balance between noise reduction and detail preservation, specially in low noise. It also misses some clusters for high noise.

Table 4.28 Performance comparison in terms of MAE, PSNR and DIS^k using the detail of the Goldhill image contaminated with different densities of noise model II

Filter	No added noise			$p = 10\,\%$			$p = 20\,\%$			$p = 30\,\%$		
	MAE	PSNR	DIS^2	MAE	PSNR	DIS^2	MAE	PSNR	DIS^1	MAE	PSNR	DIS^1
None	0.00	∞	0.00	7.45	18.95	39.47	15.15	15.86	6.27	23.70	13.84	7.38
FPGF	*0.07*	*43.90*	*0.01*	*0.96*	*35.01*	**0.20**	*1.79*	*32.00*	*0.10*	**2.70**	**30.75**	*0.14*
SVMF	1.20	34.41	3.43	1.39	33.36	4.26	2.39	29.02	0.76	5.08	23.40	1.89
INR	0.43	37.76	0.72	1.30	*33.44*	0.78	*2.27*	*31.34*	*0.17*	*3.46*	*29.32*	*0.40*
LSAFSVF	*0.16*	*39.30*	*0.01*	*0.99*	*34.08*	*0.47*	*1.86*	*31.66*	*0.17*	*3.29*	*27.75*	0.56
IPGSVF	**0.00**	**60.66**	**0.00**	*1.00*	*33.66*	*0.44*	*1.86*	*31.68*	*0.08*	*3.43*	*27.90*	*0.46*
FRF	*0.01*	*52.66*	**0.00**	**0.87**	**35.36**	*0.49*	**1.66**	**33.00**	**0.05**	*2.84*	*29.95*	**0.13**
QRVF	*0.26*	39.19	0.94	1.35	30.58	1.14	*2.35*	28.99	*0.20*	3.87	25.94	*0.32*
RWVMF	0.81	35.83	0.09	*1.25*	33.68	*0.34*	6.13	29.11	1.46	7.13	27.57	1.64
SFRF	0.86	*43.35*	**0.00**	4.99	30.99	1.21	7.10	27.05	0.85	9.27	24.04	1.76

Table 4.29 Performance comparison in terms of MAE, PSNR and DIS^k using the detail of the Lenna image contaminated with different densities of noise model II

Filter	No added noise			$p = 10\,\%$			$p = 20\,\%$			$p = 30\,\%$		
	MAE	PSNR	DIS^2	MAE	PSNR	DIS^2	MAE	PSNR	DIS^1	MAE	PSNR	DIS^1
None	0.00	∞	0.00	7.45	18.76	45.59	15.71	15.40	7.37	23.64	13.66	8.34
FPGF	1.10	32.55	1.64	*1.03*	*33.69*	*0.88*	2.04	*30.57*	*0.16*	**3.20**	**28.56**	*0.38*
SVMF	0.94	34.15	5.11	1.33	31.85	6.00	2.50	27.85	1.09	5.26	22.95	2.54
INR	0.65	36.38	0.27	1.52	31.90	*0.78*	2.67	29.26	0.24	4.04	*27.36*	*0.61*
LSAFSVF	*0.36*	35.60	0.42	*1.00*	*33.31*	*0.75*	2.04	*29.92*	0.27	*3.59*	26.94	0.72
IPGSVF	**0.00**	∞	**0.00**	*1.09*	31.87	1.28	*1.82*	*30.67*	*0.15*	*3.93*	*26.98*	0.71
FRF	*0.04*	*46.30*	**0.00**	**0.83**	**34.80**	*0.36*	**1.80**	**30.90**	*0.14*	*3.33*	*27.79*	**0.33**
QRVF	**0.00**	*56.04*	**0.00**	**0.83**	*34.06*	**0.28**	*1.82*	*30.39*	**0.11**	*3.35*	*27.36*	*0.37*
RWVMF	*0.37*	*37.61*	0.31	1.16	*32.88*	0.99	2.54	29.66	*0.17*	4.52	26.93	*0.41*
SFRF	0.86	*41.54*	0.07	4.50	30.97	2.10	6.81	27.09	0.99	9.90	23.28	3.10

Table 4.44 summarizes the main advantages and disadvantages of each filter as shown by the experimental results.

Finally, the results of filtering a few images contaminated with real impulse noise are shown in Fig. 4.9. We compare the results of the filters INR, IPGSVF, RWVMF and QRVF which provide the best performance overall. We see that INR reduces well the real noise and produces nice output images, specially for the Atlas and Stars images. IPGSVF performs similarly, but preserves some edges a little better, such as the letters in the Atlas image. However, it has a lower noise reduction capability for the Hair image which has some noise similar to background. RWVMF reduces the noise and provides smooth images, but performs a little worse than INR and

Table 4.30 Performance comparison in terms of MAE, PSNR and DIS^k using the Goldhill image contaminated with different densities of noise model II

Filter	No added noise			$p = 10\,\%$			$p = 20\,\%$			$p = 30\,\%$		
	MAE	PSNR	DIS^2	MAE	PSNR	DIS^2	MAE	PSNR	DIS^1	MAE	PSNR	DIS^1
None	0.00	∞	0.00	7.50	18.92	41.94	14.95	15.93	6.54	22.54	14.14	7.77
FPGF	*0.06*	*44.23*	**0.00**	0.77	*36.36*	**0.15**	*1.51*	*33.58*	*0.04*	*2.36*	**31.34**	*0.10*
SVMF	1.16	34.75	0.60	1.29	33.78	1.05	2.14	29.55	0.42	4.31	24.39	1.38
INR	*0.24*	41.06	0.13	*1.02*	*35.21*	*0.26*	*1.88*	*32.46*	*0.07*	2.90	*30.38*	*0.19*
LSAFSVF	*0.14*	39.43	**0.00**	*0.85*	*35.07*	0.34	*1.68*	*32.09*	0.13	*2.81*	29.00	0.39
IPGSVF	**0.00**	**74.99**	**0.00**	*0.91*	*34.05*	0.40	*1.61*	*32.63*	*0.04*	*2.74*	*29.78*	*0.19*
FRF	**0.00**	*56.09*	**0.00**	**0.72**	**36.75**	*0.22*	**1.45**	**33.83**	**0.03**	*2.40*	*31.05*	**0.06**
QRVF	*0.05*	*46.67*	0.10	*0.95*	32.41	*0.19*	1.92	29.26	*0.04*	3.04	27.32	*0.10*
RWVMF	0.54	37.91	0.01	1.09	*34.71*	*0.18*	1.98	31.96	0.05	3.06	*29.39*	*0.14*
SFRF	0.68	*45.20*	*0.00*	5.31	28.34	6.18	7.17	26.43	1.03	9.17	23.70	2.15

Table 4.31 Performance comparison in terms of MAE, PSNR and DIS^k using the Boats image contaminated with different densities of noise model II

Filter	No added noise			$p = 10\,\%$			$p = 20\,\%$			$p = 30\,\%$		
	MAE	PSNR	DIS^2	MAE	PSNR	DIS^2	MAE	PSNR	DIS^1	MAE	PSNR	DIS^1
None	0.00	∞	0.00	7.41	19.03	41.99	14.96	15.97	6.49	22.36	14.23	7.66
FPGF	*0.02*	*48.67*	**0.00**	*0.57*	*37.99*	**0.16**	*1.16*	*34.73*	*0.06*	**1.85**	**32.33**	*0.12*
SVMF	0.69	37.99	0.74	0.85	35.64	1.36	1.69	29.64	0.52	3.69	24.41	1.48
INR	0.23	41.27	0.23	0.82	*35.77*	*0.44*	1.51	*33.05*	*0.10*	2.37	*30.77*	*0.25*
LSAFSVF	*0.08*	41.62	*0.01*	*0.62*	*36.51*	*0.42*	*1.32*	32.77	0.16	*2.31*	29.24	0.44
IPGSVF	**0.00**	**79.21**	**0.00**	0.77	34.32	0.53	*1.33*	*32.89*	*0.10*	*2.25*	*29.78*	0.35
FRF	**0.00**	*62.03*	**0.00**	**0.52**	**38.47**	*0.22*	**1.09**	**34.93**	**0.04**	*1.86*	*31.70*	**0.08**
QRVF	*0.01*	*53.93*	0.15	*0.69*	33.60	0.52	1.49	30.22	0.11	2.41	28.13	*0.20*
RWVMF	0.22	43.53	0.06	*0.72*	*36.61*	*0.38*	*1.39*	*32.88*	*0.10*	2.23	*29.60*	*0.27*
SFRF	0.48	*46.22*	*0.01*	4.58	27.91	8.25	6.47	25.85	1.47	8.79	22.96	2.77

IPGSVF in edges. QRVF seems to introduce some blur in the Atlas image and leaves some noise in the rest of the images. Also, it can be seen that the Real image has a larger amount of noise than the rest and no method removes the noise completely. This could be solved by filtering the image once more.

From all this comparison, we can conclude that the desired capabilities for the *perfect* filter would be: reducing noise, specially when clusters appear, preserving image details, appropriate correlated processing of colour images improving vector approach which fails in non-correlated noise and correcting both pixels affected by high impulses as well as those affected by lower ones. Despite the state-of-the-art in this field being so extensive and the existence of promising approaches from each

Fig. 4.6 Outputs for filter performance comparison of noise model II. **a**, **b** Lenna corrupted with $p = 10\,\%$ and $p = 30\,\%$. **c**, **d** Goldhill corrupted with $p = 10\,\%$ and $p = 30\,\%$. **e–h** Output of FPGF. **i–l** Output of FRF. **m–p** Output of IPGSVF. **q–t** Output of QRVF

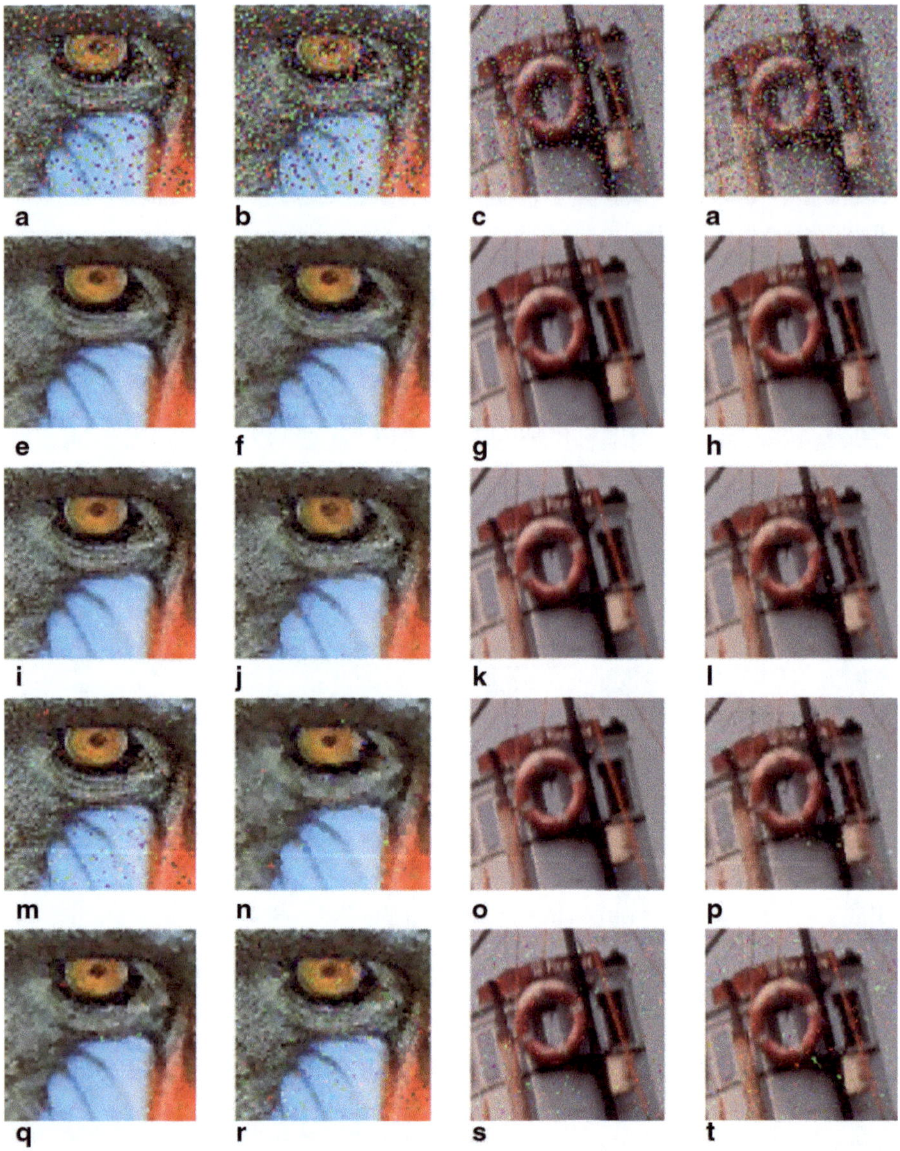

Fig. 4.7 Outputs for filter performance comparison of noise model III. **a**, **b** Baboon corrupted with $p = 10\,\%$ and $p = 20\,\%$. **b**, **c** Boats corrupted with $p = 10\,\%$ and $p = 20\,\%$. **e–h** Output of INR. **i–l** Output of QRVF. **m–p** Output of RWVMF. **q–t** Output of IPGSVF

Table 4.32 Performance comparison in terms of MAE, PSNR and DIS^k using the detail of the Baboon image contaminated with different densities of noise model III

Filter	No added noise			$p = 10\,\%$			$p = 20\,\%$			$p = 30\,\%$		
	MAE	PSNR	DIS^2	MAE	PSNR	DIS^2	MAE	PSNR	DIS^1	MAE	PSNR	DIS^1
None	0.00	∞	1.00	8.17	18.26	0.41	15.52	15.54	0.26	23.62	13.73	0.19
FPGF	*1.93*	*27.66*	*0.99*	*5.93*	24.31	*0.93*	9.74	*22.18*	*0.82*	*14.10*	20.24	0.64
SVMF	4.42	25.53	**0.94**	6.11	23.88	**0.80**	9.03	21.30	**0.61**	14.21	18.14	**0.35**
INR	*2.01*	*30.20*	*0.98*	**3.52**	**28.12**	0.94	**5.36**	**26.05**	0.87	**7.88**	**23.96**	0.74
LSAFSVF	3.44	25.20	*0.99*	6.63	23.12	*0.87*	9.32	21.73	*0.72*	*12.82*	*19.90*	*0.52*
IPGSVF	**0.07**	*40.45*	1.00	*3.91*	*25.66*	*0.86*	*7.72*	*23.35*	*0.81*	*13.05*	*21.09*	*0.70*
FRF	*0.31*	*34.42*	1.00	*4.26*	*25.58*	*0.90*	*8.71*	22.33	*0.84*	15.67	18.90	*0.65*
QRVF	*0.12*	**43.34**	1.00	3.79	27.86	0.96	7.71	24.41	0.87	12.69	21.55	0.72
RWVMF	4.00	25.61	*0.99*	7.73	23.36	0.95	12.09	21.74	0.88	14.81	20.87	0.75
SFRF	5.09	27.64	*0.99*	8.90	*25.77*	0.85	12.75	*22.53*	**0.61**	15.61	*21.18*	*0.47*

Table 4.33 Performance comparison in terms of MAE, PSNR and DIS^k using the detail of the Boats image contaminated with different densities of noise model III

Filter	No added noise			$p = 10\,\%$			$p = 20\,\%$			$p = 30\,\%$		
	MAE	PSNR	DIS^2	MAE	PSNR	DIS^2	MAE	PSNR	DIS^1	MAE	PSNR	DIS^1
None	0.00	∞	0.00	7.57	18.89	56.84	15.13	15.89	7.37	22.35	14.27	8.04
FPGF	0.76	36.01	1.78	2.29	*30.27*	*5.46*	*4.81*	*26.06*	*2.06*	8.25	23.07	3.89
SVMF	0.55	37.84	3.09	*2.25*	29.52	10.36	5.49	23.58	3.28	10.36	19.76	5.63
INR	0.39	39.82	0.18	**1.28**	**34.77**	**1.49**	**2.34**	**31.72**	**0.67**	**3.87**	**29.25**	*1.70*
LSAFSVF	*0.09*	*41.87*	**0.00**	2.43	28.78	9.37	5.06	24.78	2.65	*8.21*	22.06	4.33
IPGSVF	**0.00**	∞	**0.00**	2.85	26.78	12.11	*4.70*	25.60	*1.79*	*6.64*	*25.32*	*2.41*
FRF	**0.00**	∞	**0.00**	*2.17*	*29.60*	*7.23*	*4.53*	*26.41*	*1.49*	*8.14*	22.82	*2.88*
QRVF	*0.02*	*53.24*	*0.05*	*1.85*	*33.27*	*1.52*	*4.07*	*28.99*	*0.78*	*7.00*	*25.66*	**1.83**
RWVMF	*0.26*	*41.14*	*0.02*	*2.24*	*31.14*	*2.36*	7.12	*27.60*	2.55	9.45	25.09	*3.77*
SFRF	1.06	38.68	0.11	5.44	29.13	11.15	7.73	*25.82*	2.62	10.16	*24.14*	4.50

singular point of view, the filter exhibiting these features altogether has not been devised yet, which means that impulse noise reduction in colour images is still an open problem.

Table 4.34 Performance comparison in terms of MAE, PSNR and DIS^k using the detail of the Goldhill image contaminated with different densities of noise model III

Filter	No added noise			$p = 10\,\%$			$p = 20\,\%$			$p = 30\,\%$		
	MAE	PSNR	DIS^2	MAE	PSNR	DIS^2	MAE	PSNR	DIS^1	MAE	PSNR	DIS^1
None	0.00	∞	0.00	7.61	18.85	53.57	15.26	15.80	7.15	22.89	14.00	8.03
FPGF	*0.07*	*43.90*	*0.01*	*2.49*	*29.78*	*6.02*	*4.85*	*26.36*	1.74	*7.80*	23.56	3.46
SVMF	1.20	34.41	3.43	*2.43*	*29.64*	8.06	5.63	23.84	3.03	10.43	19.83	5.62
INR	0.43	37.76	0.72	**1.37**	**33.73**	*1.65*	**2.51**	**31.21**	**0.47**	**3.99**	**28.80**	*1.25*
LSAFSVF	*0.16*	*39.30*	*0.01*	2.69	28.34	7.94	5.14	25.01	2.21	7.98	22.44	3.89
IPGSVF	**0.00**	**60.66**	**0.00**	2.76	27.11	10.95	*4.55*	26.21	1.43	*6.55*	*25.60*	*1.67*
FRF	*0.01*	*52.66*	**0.00**	*2.38*	29.36	*6.78*	*4.64*	*26.39*	1.28	*7.84*	23.16	*2.13*
QRVF	*0.26*	39.19	0.94	*2.23*	*31.95*	*2.33*	*4.25*	*28.80*	**0.47**	*6.55*	*26.23*	**1.12**
RWVMF	0.81	35.83	0.09	2.67	*30.42*	**1.32**	7.40	*27.44*	2.02	8.58	*26.07*	*2.32*
SFRF	0.86	*43.35*	**0.00**	5.72	27.49	13.14	8.02	25.58	2.31	10.05	*24.36*	4.10

Table 4.35 Performance comparison in terms of MAE, PSNR and DIS^k using the detail of the Lenna image contaminated with different densities of noise model III

Filter	No added noise			$p = 10\,\%$			$p = 20\,\%$			$p = 30\,\%$		
	MAE	PSNR	DIS^2	MAE	PSNR	DIS^2	MAE	PSNR	DIS^1	MAE	PSNR	DIS^1
None	0.00	∞	0.00	8.12	18.33	60.15	15.77	15.46	7.93	23.85	13.63	8.71
FPGF	1.10	32.55	1.64	*2.85*	28.33	*6.70*	6.01	24.47	*2.00*	9.27	21.89	3.59
SVMF	0.94	34.15	5.11	*2.42*	*28.80*	11.01	5.90	23.01	3.57	11.58	18.86	6.34
INR	0.65	36.38	*0.27*	**1.55**	**32.59**	*1.68*	**2.63**	**30.57**	*0.79*	**4.44**	**27.57**	**1.57**
LSAFSVF	*0.36*	35.60	0.42	2.88	27.65	9.10	*5.47*	24.34	2.26	*8.69*	21.43	4.29
IPGSVF	**0.00**	∞	**0.00**	2.89	26.75	12.37	*5.06*	*24.97*	2.11	*7.53*	*23.72*	*2.74*
FRF	*0.04*	*46.30*	**0.00**	*2.41*	*29.06*	*6.65*	*5.17*	*25.34*	*1.35*	9.53	21.10	*2.81*
QRVF	**0.00**	*56.04*	**0.00**	*1.99*	*32.05*	**1.32**	*4.29*	*28.24*	**0.65**	*7.31*	*24.56*	*1.61*
RWVMF	*0.37*	*37.61*	0.31	2.60	*29.76*	*2.21*	5.37	*26.39*	*0.86*	*8.40*	*23.89*	*2.11*
SFRF	0.86	*41.54*	*0.07*	5.81	26.97	17.33	8.49	24.38	3.13	11.87	*22.14*	5.56

Table 4.36 Performance comparison in terms of MAE, PSNR and DIS^k using the Goldhill image contaminated with different densities of noise model III

Filter	No added noise			$p = 10\,\%$			$p = 20\,\%$			$p = 30\,\%$		
	MAE	PSNR	DIS^2	MAE	PSNR	DIS^2	MAE	PSNR	DIS^1	MAE	PSNR	DIS^1
None	0.00	∞	0.00	7.53	18.91	56.26	15.13	15.88	7.58	22.70	14.11	8.38
FPGF	*0.06*	*44.23*	**0.00**	*2.19*	*30.39*	*4.60*	4.55	*26.47*	1.76	*7.26*	*23.88*	3.52
SVMF	1.16	34.75	0.60	*2.24*	*30.25*	*5.06*	5.26	24.11	2.59	10.16	19.86	5.35
INR	0.24	41.06	0.13	**1.08**	**35.40**	*0.72*	**2.16**	**32.32**	*0.29*	**3.54**	**29.80**	*0.91*
LSAFSVF	*0.14*	39.43	**0.00**	2.42	29.01	6.86	4.83	25.37	2.13	7.64	22.57	3.94
IPGSVF	**0.00**	**74.99**	**0.00**	2.74	27.04	11.00	*4.36*	26.17	*1.45*	*5.93*	*26.03*	*1.58*
FRF	**0.00**	*56.09*	**0.00**	*2.13*	30.05	5.70	*4.26*	*26.79*	*1.22*	*7.26*	23.43	*2.35*
QRVF	*0.05*	*46.67*	0.10	*1.79*	*33.47*	**0.65**	*3.61*	*29.97*	**0.28**	*5.71*	*27.43*	**0.84**
RWVMF	0.54	37.91	0.01	2.26	*31.39*	*1.59*	*4.23*	*27.49*	*0.79*	*6.77*	*24.37*	*2.13*
SFRF	0.68	*45.20*	**0.00**	5.74	28.32	9.97	8.03	24.71	2.90	10.13	23.66	4.48

Table 4.37 Performance comparison in terms of MAE, PSNR and DIS^k using the Boats image contaminated with different densities of noise model III

Filter	No added noise			$p = 10\,\%$			$p = 20\,\%$			$p = 30\,\%$		
	MAE	PSNR	DIS^2	MAE	PSNR	DIS^2	MAE	PSNR	DIS^1	MAE	PSNR	DIS^1
None	0.00	∞	0.00	7.49	18.97	52.02	14.96	15.97	7.22	22.40	14.23	8.12
FPGF	*0.02*	*48.67*	**0.00**	1.84	*31.13*	*5.10*	*3.96*	*26.87*	*1.66*	6.66	23.90	3.29
SVMF	0.69	37.99	0.74	*1.68*	*31.14*	6.49	4.47	24.45	2.42	9.28	19.99	4.76
INR	0.23	41.27	0.23	**0.86**	**36.57**	**1.14**	**1.70**	**33.54**	*0.39*	**2.89**	**30.74**	*1.01*
LSAFSVF	*0.08*	41.62	*0.01*	2.07	29.47	7.77	4.20	25.71	2.01	6.85	22.72	3.57
IPGSVF	**0.00**	**79.21**	**0.00**	2.79	26.45	12.15	4.23	25.60	1.73	*4.93*	*26.23*	*1.74*
FRF	**0.00**	*62.03*	**0.00**	*1.76*	30.63	*5.75*	*3.55*	*27.22*	*1.16*	*6.34*	23.43	2.24
QRVF	*0.01*	*53.93*	0.15	*1.32*	*35.24*	*1.15*	2.75	*31.26*	**0.37**	*4.63*	*28.01*	**0.97**
RWVMF	0.22	43.53	0.06	*1.57*	*32.68*	*2.65*	*3.19*	*28.11*	*0.95*	*5.56*	*24.22*	*2.32*
SFRF	0.48	*46.22*	*0.01*	4.76	28.08	14.10	6.74	25.23	2.79	8.76	*24.42*	3.86

Table 4.38 Performance comparison in terms of MAE, PSNR and DIS^k using the detail of the Baboon image contaminated with different densities of noise model IV

Filter	No added noise			$\alpha = 0.7$			$\alpha = 0.6$			$\alpha = 0.5$		
	MAE	PSNR	DIS^2	MAE	PSNR	DIS^2	MAE	PSNR	DIS^1	MAE	PSNR	DIS^1
None	0.00	∞	0.00	8.08	20.04	46.57	10.70	18.26	5.92	14.92	16.27	6.81
FPGF	*1.93*	*27.66*	*0.53*	7.58	23.88	*11.76*	8.73	23.24	1.76	10.80	22.23	2.38
SVMF	4.42	25.53	6.20	*7.34*	24.27	17.27	*8.24*	23.57	2.48	*9.32*	22.41	3.43
INR	*2.01*	*30.20*	1.73	**5.27**	*27.39*	*8.47*	**5.96**	**26.55**	*1.34*	**6.63**	**26.14**	*1.51*
LSAFSVF	3.44	25.20	*0.80*	8.05	23.29	12.90	8.87	22.89	2.11	*10.36*	22.08	2.54
IPGSVF	**0.07**	*40.45*	*0.16*	*5.67*	*25.77*	15.33	*6.77*	*25.20*	*1.44*	*10.73*	*22.95*	*1.39*
FRF	*0.31*	*34.42*	*0.18*	*6.09*	*25.21*	*11.63*	*8.01*	*23.74*	*1.49*	10.97	21.95	*1.70*
QRVF	*0.12*	**43.34**	**0.04**	*5.37*	**27.41**	**6.43**	*6.36*	*26.54*	*0.91*	*7.95*	*25.08*	**1.15**
RWVMF	4.00	25.61	0.82	11.88	22.99	27.67	9.64	23.13	**0.89**	11.77	*22.14*	*1.29*
SFRF	5.09	*27.64*	1.14	8.59	*26.29*	*10.88*	10.81	*24.50*	1.64	12.29	*23.48*	2.23

Table 4.39 Performance comparison in terms of MAE, PSNR and DIS^k using the detail of the Boats image contaminated with different densities of noise model IV

Filter	No added noise			$\alpha = 0.7$			$\alpha = 0.6$			$\alpha = 0.5$		
	MAE	PSNR	DIS^2	MAE	PSNR	DIS^2	MAE	PSNR	DIS^1	MAE	PSNR	DIS^1
None	0.00	∞	0.00	7.82	20.40	46.85	10.81	18.24	6.06	14.88	16.39	7.22
FPGF	0.76	36.01	1.78	4.14	29.37	11.97	4.74	*28.58*	1.57	5.83	26.80	2.21
SVMF	0.55	37.84	3.09	*3.72*	*30.10*	*11.75*	*4.53*	28.20	1.72	6.06	25.14	2.99
INR	0.39	*39.82*	0.18	**3.18**	**32.47**	**5.34**	**3.60**	**31.57**	*0.82*	**4.11**	**30.37**	*1.28*
LSAFSVF	*0.09*	41.87	**0.00**	*4.33*	28.32	14.24	5.02	27.41	1.74	6.04	25.88	2.47
IPGSVF	**0.00**	∞	**0.00**	4.55	26.99	16.46	*4.70*	28.04	*1.20*	*5.15*	*28.75*	**1.00**
FRF	**0.00**	∞	**0.00**	*4.09*	29.09	11.91	*4.68*	*28.51*	*1.28*	*5.80*	27.10	*1.58*
QRVF	*0.02*	*53.24*	*0.05*	*3.56*	*31.65*	*6.08*	*4.16*	*30.45*	**0.76**	*5.01*	*29.13*	*1.14*
RWVMF	*0.26*	*41.14*	*0.02*	5.45	*30.69*	21.08	5.91	*29.65*	2.12	*5.31*	*27.47*	*1.42*
SFRF	1.06	38.68	0.11	4.85	*30.69*	*7.34*	6.32	28.34	*1.30*	6.91	*27.77*	1.77

Table 4.40 Performance comparison in terms of MAE, PSNR and DIS^k using the detail of the Goldhill image contaminated with different densities of noise model IV

Filter	No added noise			$\alpha = 0.7$			$\alpha = 0.6$			$\alpha = 0.5$		
	MAE	PSNR	DIS^2	MAE	PSNR	DIS^2	MAE	PSNR	DIS^1	MAE	PSNR	DIS^1
None	0.00	∞	0.00	7.94	20.24	46.47	10.75	18.23	5.81	15.40	16.17	7.18
FPGF	*0.07*	*43.90*	*0.01*	*4.34*	28.97	*10.33*	5.04	27.92	1.51	6.19	26.39	2.18
SVMF	1.20	34.41	3.43	*4.09*	*29.64*	*9.86*	*4.82*	27.87	1.50	6.35	25.25	2.68
INR	0.43	37.76	0.72	**3.29**	**31.82**	*4.35*	**3.66**	**31.08**	**0.58**	**4.19**	**30.10**	**0.74**
LSAFSVF	*0.16*	*39.30*	*0.01*	4.58	27.97	12.06	5.25	27.08	1.63	6.49	25.59	2.29
IPGSVF	**0.00**	**60.66**	**0.00**	4.60	27.25	12.56	*4.83*	27.76	*1.07*	*5.48*	*28.14*	*0.71*
FRF	*0.01*	*52.66*	**0.00**	*4.24*	28.82	10.60	*4.85*	*28.21*	*1.17*	*6.12*	26.72	*1.31*
QRVF	*0.26*	39.19	0.94	*3.96*	*30.43*	*5.23*	*4.50*	*29.66*	*0.71*	*5.54*	*27.83*	*0.90*
RWVMF	0.81	35.83	0.09	6.24	*29.27*	16.08	6.55	*28.73*	1.72	*5.91*	*27.45*	*0.87*
SFRF	0.86	*43.35*	**0.00**	5.34	*30.41*	**4.13**	6.65	*28.16*	*0.88*	7.20	*27.39*	1.34

Table 4.41 Performance comparison in terms of MAE, PSNR and DIS^k using the detail of the Lenna image contaminated with different densities of noise model IV

Filter	No added noise			$\alpha = 0.7$			$\alpha = 0.6$			$\alpha = 0.5$		
	MAE	PSNR	DIS^2	MAE	PSNR	DIS^2	MAE	PSNR	DIS^1	MAE	PSNR	DIS^1
None	0.00	∞	0.00	8.03	20.00	46.31	10.86	18.10	5.77	15.05	16.17	7.04
FPGF	1.10	32.55	1.64	4.57	28.45	*9.78*	5.49	26.90	1.39	6.73	25.37	2.15
SVMF	0.94	34.15	5.11	*4.09*	*28.85*	13.36	*4.93*	*27.28*	1.85	6.36	25.00	2.77
INR	0.65	36.38	*0.27*	**3.51**	**31.28**	*7.32*	**3.94**	*30.16*	*0.90*	**4.46**	**29.29**	*1.21*
LSAFSVF	*0.36*	35.60	0.42	4.75	27.47	12.16	5.50	26.58	1.51	6.48	25.42	2.14
IPGSVF	**0.00**	∞	**0.00**	4.92	26.40	16.10	5.35	26.76	1.51	*5.94*	*27.16*	*1.13*
FRF	*0.04*	*46.30*	**0.00**	*4.36*	28.46	10.10	*5.32*	27.05	*1.24*	*6.53*	25.55	*1.50*
QRVF	**0.00**	*56.04*	**0.00**	*3.75*	*30.78*	**5.00**	*4.42*	**29.61**	**0.64**	*5.17*	*28.50*	**0.84**
RWVMF	*0.37*	*37.61*	0.31	*4.02*	*29.69*	*6.04*	*4.70*	*28.31*	*0.86*	*5.64*	*27.04*	*1.24*
SFRF	0.86	*41.54*	*0.07*	4.93	*30.42*	*7.15*	6.34	*28.35*	*0.91*	7.26	*27.06*	1.76

Table 4.42 Performance comparison in terms of MAE, PSNR and DIS^k using the Goldhill image contaminated with different densities of noise model IV

Filter	No added noise			$\alpha = 0.7$			$\alpha = 0.6$			$\alpha = 0.5$		
	MAE	PSNR	DIS^2	MAE	PSNR	DIS^2	MAE	PSNR	DIS^1	MAE	PSNR	DIS^1
None	0.00	∞	0.00	7.96	20.20	47.61	10.96	18.17	6.21	15.24	16.20	7.39
FPGF	*0.06*	*44.23*	**0.00**	4.18	*29.24*	8.58	4.89	28.10	1.31	5.83	26.75	1.97
SVMF	1.16	34.75	0.60	*3.88*	*30.09*	*5.64*	*4.64*	28.29	1.12	5.97	25.70	2.24
INR	*0.24*	41.06	0.13	**3.11**	**32.55**	**2.42**	**3.43**	**31.85**	**0.36**	**3.85**	**31.01**	*0.56*
LSAFSVF	0.14	39.43	**0.00**	4.43	28.27	10.48	5.16	27.23	1.47	6.06	26.08	2.05
IPGSVF	**0.00**	**74.99**	**0.00**	4.53	27.37	12.41	4.78	27.86	*0.96*	*5.00*	*28.93*	**0.54**
FRF	**0.00**	*56.09*	**0.00**	*4.12*	29.15	*8.75*	*4.69*	*28.55*	0.99	*5.62*	*27.25*	1.23
QRVF	*0.05*	*46.67*	0.10	*3.62*	*31.39*	*2.87*	*4.15*	*30.53*	*0.42*	*4.84*	*29.45*	*0.66*
RWVMF	0.54	37.91	*0.01*	*3.81*	*30.40*	*3.81*	*4.43*	*29.36*	*0.56*	*5.28*	*27.73*	*1.04*
SFRF	0.68	*45.20*	**0.00**	5.32	28.70	9.61	6.06	*28.51*	*0.84*	6.78	27.16	*1.61*

Table 4.43 Performance comparison in terms of MAE, PSNR and DIS^k using the Boats image contaminated with different densities of noise model IV

Filter	No added noise			$\alpha = 0.7$			$\alpha = 0.6$			$\alpha = 0.5$		
	MAE	PSNR	DIS^2	MAE	PSNR	DIS^2	MAE	PSNR	DIS^1	MAE	PSNR	DIS^1
None	0.00	∞	0.00	7.97	20.19	46.22	10.92	18.19	5.99	15.24	16.20	7.22
FPGF	*0.02*	*48.67*	**0.00**	3.86	29.88	*10.28*	4.42	28.78	1.36	5.23	27.23	1.93
SVMF	0.69	37.99	0.74	*3.31*	*31.11*	*8.33*	*4.02*	29.09	1.28	5.22	26.27	2.13
INR	0.23	41.27	0.23	**2.93**	**33.13**	**6.31**	**3.19**	**32.47**	*0.79*	**3.52**	**31.67**	*0.99*
LSAFSVF	*0.08*	41.62	*0.01*	4.06	28.86	11.49	4.64	27.85	1.47	5.35	26.66	1.92
IPGSVF	**0.00**	**79.21**	**0.00**	4.66	26.48	15.98	4.68	27.28	1.38	*4.27*	*29.74*	**0.76**
FRF	**0.00**	*62.03*	**0.00**	*3.80*	*29.69*	*10.55*	*4.18*	*29.15*	*1.15*	*4.91*	27.70	*1.38*
QRVF	*0.01*	*53.93*	0.15	*3.22*	*32.39*	*6.35*	*3.60*	*31.55*	**0.76**	*4.07*	*30.60*	*0.94*
RWVMF	0.22	43.53	0.06	*3.19*	*31.61*	*6.68*	*3.71*	*30.12*	*0.93*	*4.29*	*28.80*	*1.21*
SFRF	0.48	*46.22*	*0.01*	4.64	27.77	13.45	4.70	*29.62*	*0.97*	5.47	*28.04*	1.53

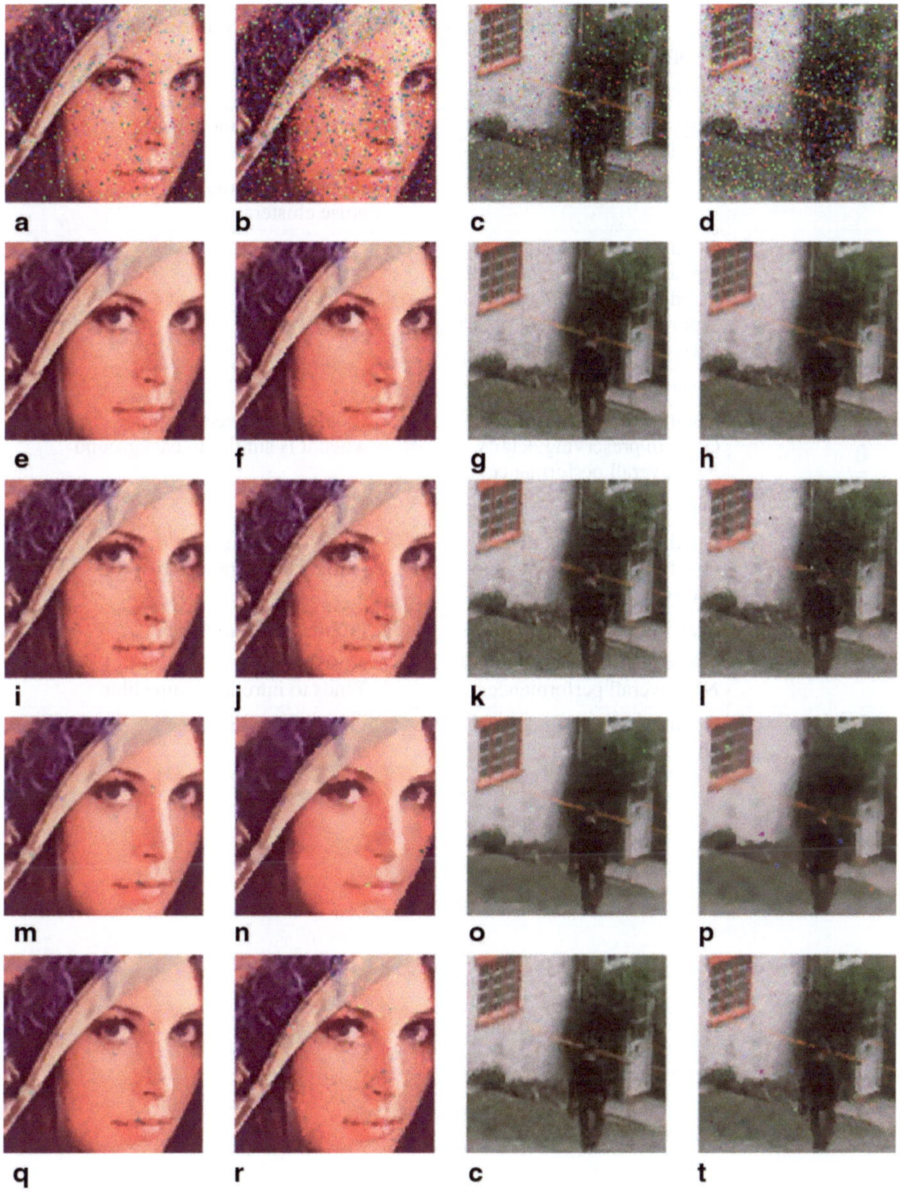

Fig. 4.8 Outputs for filter performance comparison of noise model IV. **a**, **b** Lenna corrupted with $\alpha = 0.7$ and $\alpha = 0.5$. **c**, **d** Goldhill corrupted with $\alpha = 0.7$ and $\alpha = 0.5$. **e–h** Output of INR. **i–l** Output of QRVF; **m–p** Output of SFRF. **q–t** Output of RWVMF

Table 4.44 Summary of main advantages and disadvantages of best performing filtering methods in the experimental comparison

Filter	Advantages	Disadvantages
FPGF [52]	Very efficient Good in preserving details	Problems in removing noise clusters
SVMF [42]		Problems in removing noise clusters
INR [96]	Good for noise clusters Good for non-correlated noise Good trade-off between noise suppression and detail preservation Nice overall performance	
LSAFSVF [48]	No adaptive parameters	
IPGSVF [55]	Good for noise clusters Good in preserving details Nice overall performance	Problems in detecting noise when it is similar to background
FRF [43]	Very efficient	
QRVF [65]	Good in preserving details Good for non correlated noise Nice overall performance	Poor adaptiveness to particular image features
RWVMF [97]	Nice in filtering pixels affected by low intensity noise Nice overall performance	Problems in removing noise clusters Tends to introduce some blur
SFRF [108]	Good in preserving details Nice in filtering pixels affected by low intensity noise	Problems in removing noise clusters

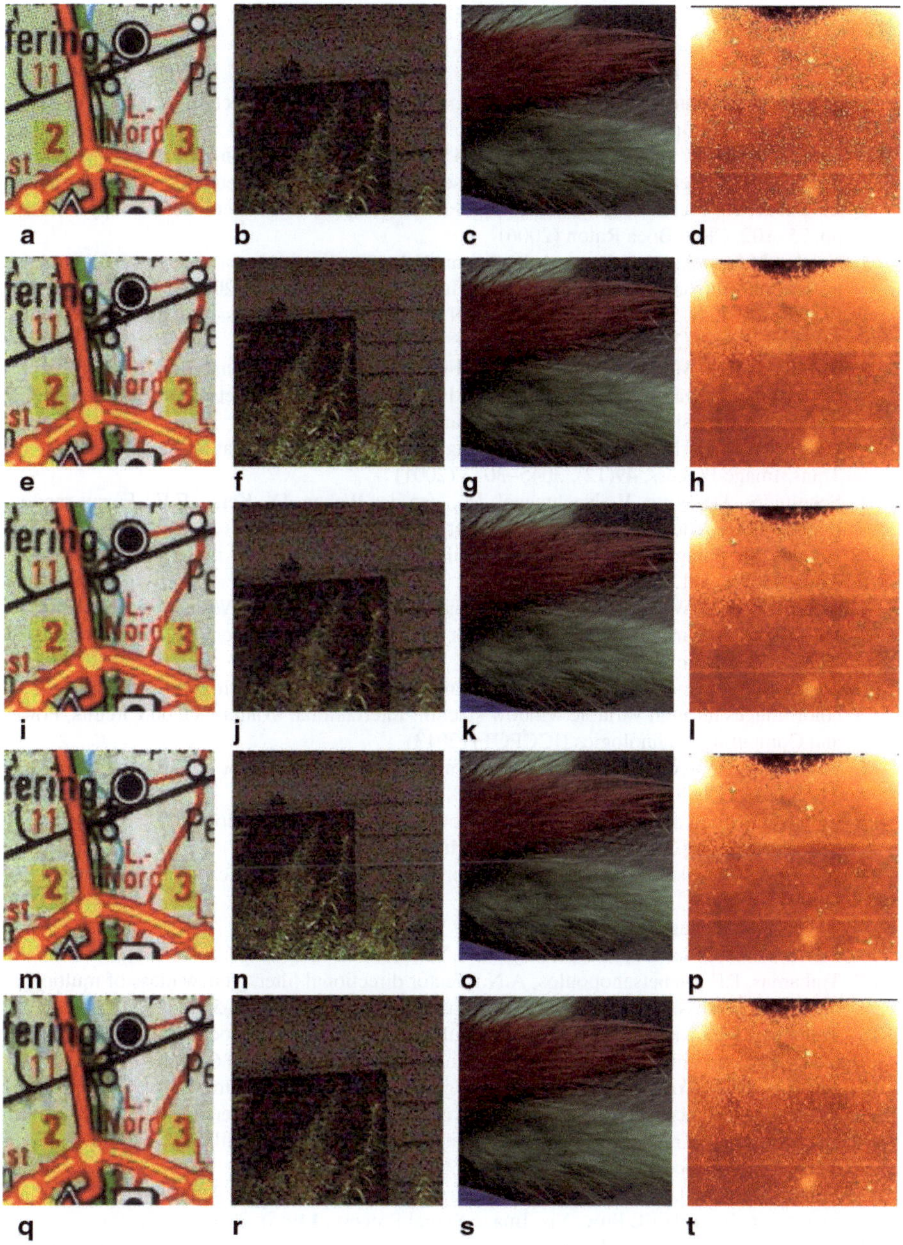

Fig. 4.9 Outputs for filter performance comparison of real noise. **a** Atlas (141 × 140). **b** Detail of real photograph (241 × 241). **e** Hair (443 × 495) and **d** NotStars (200 × 200). **e–h** Output of INR. **i–l** Output of IPGSVF. **m–p** Output of RWVMF. **q–t** Output of QRVF

References

1. Smolka, B., Plataniotis, K.N., Venetsanopoulos, A.N.: Nonlinear techniques for color image processing. In: Barner, K.E., Arce, G.R. (eds.) Nonlinear Signal and Image Processing: Theory, Methods, and Applications, pp. 445–505. CRC, Boca Raton (2004)
2. Smolka, B., Venetsanopoulos, A.N.: Noise reduction and edge detection in color images. In: Lukac, R., Plataniotis, K.N. (eds.) Color Image Processing: Methods and Applications, pp. 75–102. CRC, Boca Raton (2006)
3. Celebi, M.E., Kingravi, H.A., Aslandogan, Y.A.: Nonlinear vector filtering for impulsive noise removal from color images. J. Elec. Imaging **16**(3), 033008 (2007)
4. Plataniotis, K.N., Venetsanopoulos, A.N.: Color Image Processing and Applications. Springer, Berlin (2000)
5. Kuruoglu, E.E., Molina, C., Godsill, S.J., Fitzgerald, W.J.: A new analytic representation for the α-stable probability density function. In: The Fifth World Meeting of the International Society for Bayesian Analysis (ISBA), Istanbul, August 1997
6. Hamza, A.B., Krim, H.: Image denoising: a nonlinear robust statistical approach. IEEE Trans. Image Process. **49**(12), 3045–3053 (2001)
7. Schulte, S., De Witte, V., Nachtegael, M., Van der Weken, D., Kerre, E.E.: Fuzzy random impulse noise reduction method. Fuzzy Set. Syst. **158**(3), 270–283 (2007)
8. Hassan, M., Bhagvati, C.: Structural similarity measure for color images. Int. J. Comput. Appl. **43**(14), 7–12 (2012)
9. Gonzalez, R.C., Woods, R.E.: Digital Image Processing. Addison Wesley, Reading (1992)
10. Pitas, I., Venetsanopoulos, A.N.: Nonlinear Digital Filters: Principles and Applications. Kluwer Academic, Dordrecht (1990)
11. Laskar, R.H., Banerjee, K., Basak, D.: Removal of high density salt and pepper noise from color images through variable window size. In: International Conference on Circuits, Power and Computing Technologies (ICCPCT) (2013)
12. Morillas, S., Gregori, V.: Adaptive vector marginal median filter for colour images. Sensors **11**, 3205–3213 (2011)
13. Lukac, R., Smolka, B., Martin, K., Plataniotis, K.N., Venetsanopoulos, A.N.: Vector filtering for color imaging. IEEE Signal Process. Mag. **22**(1), 74–86 (2005)
14. Astola, J., Haavisto, P., Neuvo, Y.: Vector median filters. In: Proc. IEEE **78**(4), 678–689 (1990)
15. Huber, P.S.: Robust Statistics. Wiley, New York (1981)
16. David, H.A.: Order Statistics. Wiley, New York (1981)
17. Trahanias, P.E., Venetsanopoulos, A.N.: Vector directional filters: a new class of multichannel image processing filters. IEEE Trans. Image Process. **2**(4), 528–534 (1993)
18. Trahanias, P.E., Karakos, D., Venetsanopoulos, A.N.: Directional processing of color images: theory and experimental results. IEEE Trans. Image Process. **5**(6), 868–880 (1996)
19. Karakos, D.G., Trahanias, P.E.: Combining vector median and vector directional filters: the directional distance filters. In: Proceedings of the IEEE ICIP Conference, pp. 171–174 (1995)
20. Karakos, D.G., Trahanias, P.E.: Generalized multichannel image filtering structures. IEEE Trans. Image Process. **6**(7), 1038–1045 (1997)
21. Lorenzo-Ginori, J.V., Plataniotis, K.N., Venetsanopoulos, A.N.: Nonlinear filtering for phase image denoising. IEEE Proc. Vis. Image Signal Process. **149**(5), 290–296 (2002)
22. Morillas, S., Gregori, V., Peris-Fajarnés, G., Latorre, P.: A fast impulsive noise color image filter using fuzzy metrics. Real-Time Imaging **11**(5–6), 417–428 (2005)
23. Morillas, S., Gregori, V., Riquelme, J., Defez, B., Peris-Fajarnes, G.: Fuzzy directional-distance vector filter. Lect. Notes Comput. Sci. **4578**, 355–361 (2007)
24. Khriji, L., Gabbouj, M.: Adaptive fuzzy order statistics-rational hybrid filters for color image processing. Fuzzy Set. Syst. **128**, 35–46 (2002)
25. Plataniotis, K.N., Androutsos, D., Venetsanopoulos, A.N.: Adaptive fuzzy systems for multichannel signal processing. In: Proc. IEEE **87**, 1601–1622 (1999)

26. Morillas, S., Gregori, V.: Robustifying vector median filter. Sensors **11**, 8115–8126 (2011)
27. Pratt, W.: Digital Image Processing (4th ed.). Wiley, New York (2007)
28. Pitas, I.: Marginal order statistics in color image filtering. Opt. Eng. **29**(5), 495–503 (1990)
29. Khriji, L., Gabbouj, M.: Vector median-rational hybrid filters for multichannel image processing. IEEE Signal Proc. Let. **6**(7), 186–190 (1999)
30. Khriji, L., Gabbouj, M.: Adaptive fuzzy order statistics-rational hybrid filters for color image processing. Fuzzy Set. Syst. **128**, 35–46 (2002)
31. Ma, Z., Feng, D., Wu, H.R.: A neighborhood evaluated adaptive vector filter for suppression of impulsive noise in color images. Real-Time Imaging **11**(5–6), 403–416 (2005)
32. Ma, Z., Wu, H.R., Qiu, B.: A window adaptive hybrid vector filter for color image restoration. In: Proceedings of International Conference on Acoustics, Speech and Signal Processing ICASSP'04. **3**, 205–208 (2004)
33. Ma, Z., Wu, H.R., Qiu, B.: A robust structure-adaptive hybrid vector filter for color image restoration. IEEE T. Image Proc. **14**(12), 1990–2001 (2005)
34. Tsai, H., Yu, P.T.: Adaptive fuzzy hybrid multichannel filters for removal of impulsive noise from color images. Signal Process. **74**, 127–151 (1999)
35. Lin, T.C., Yu, P.T.: Partition fuzzy median filter based on fuzzy rules for image restoration. Fuzzy Set. Syst. **147**(1), 75–97 (2004)
36. Ma, Z., Wu, H.R., Feng, D.: Partition based vector filtering technique for noise suppression in digital color images. IEEE Trans. Image Process. **15**(8), 2324–2342 (2006)
37. Ma, Z., Wu, H.R., Feng, D.: Fuzzy vector partition filtering technique for color image restoration. Comput. Vis. Image Underst. **107**, 26–37 (2007)
38. Shao, M., Barner, K.E.: Optimization of partition-based weighted sum filters and their application to image denoising. IEEE Trans. Image Process. **15**(7), 1900–1915 (2006)
39. Allende, H., Galbiati, J.: A non-parametric filter for image restoration using cluster analysis. Pattern Recognit. Lett. **25**(8), 841–847 (2004)
40. Lukac, R.: Adaptive vector median filtering. Pattern Recognit. Lett. **24**(12), 1889–1899 (2003)
41. Lukac, R., Plataniotis, K.N., Venetsanopoulos, A.N., Smolka, B.: A statistically-switched adaptive vector median filter. J. Intell. Robot. Syst. **42**(4), 361–391 (2005)
42. Lukac, R., Smolka, B., Plataniotis, K.N., Venetsanopoulos, A.N.: Vector sigma filters for noise detection and removal in color images. J. Vis. Commun. Image R. **17**(1), 1–26 (2006)
43. Camarena, J.G., Gregori, V., Morillas, S., Sapena, A.: Two-step fuzzy logic-based method for impulse noise detection in colour images. Pattern Recognit. Lett. **31**(13), 1842–1849 (2010)
44. Camacho, J., Morillas, S., Latorre, P.: Efficient impulsive noise suppression based on statistical confidence limits. J. Imaging Sci. Techn. **50**(5), 427–436 (2006)
45. Jin, L., Li, D.: A switching vector median filter based on the CIELAB color space for color image restoration. Signal Process. **87**(6), 1345–1354 (2007)
46. Smolka, B., Plataniotis, K.N., Chydzinski, A., Szczepanski, M., Venetsanopoulos, A.N., Wojciechowski, K.: Self-adaptive algorithm of impulsive noise reduction in color images. Pattern Recognit. **35**(8), 1771–1784 (2002)
47. Morillas, S., Gregori, V., Peris-Fajarnés, G., Latorre, P.: A fast impulsive noise color image filter using fuzzy metrics. Real-Time Imaging **11**, 417–428 (2005)
48. Morillas, S., Gregori, V., Peris-Fajarnés, G., Sapena, A.: Local self-adaptive fuzzy filter for impulsive noise removal in color images. Signal Process. **88**(2), 390–398 (2008)
49. Deng, Y., Kenney, C., Moore, M.S., Manjunath, B.S.: Peer group filtering and perceptutal color image quantization. In: Proceedings of the IEEE International Symposium on Circuits and Systems **4**, 21–24 (1999)
50. Kenney, C., Deng, Y., Manjunath, B.S., Hewer, G.: Peer group image enhancement. IEEE Trans. Image Process. **10**(2), 326–334 (2001)
51. Ho, J.Y.F.: Peer region determination based impulsive noise detection. In: Proceedings of the International Conference on Acoustics, Speech and Signal Processing ICASSP'03. **3**, 713–716 (2003)

52. Smolka, B., Chydzinski, A.: Fast detection and impulsive noise removal in color images. Real-Time Imaging **11**(5–6), 389–402 (2005)
53. Camarena, J.G., Gregori, V., Morillas, S., Sapena, A.: Fast detection and removal of impulsive noise using peer groups and fuzzy metrics. J. Vis. Commun. Image R. **19**(1), 20–29 (2008)
54. Camarena, J.G., Gregori, V., Morillas, S., Sapena, A.: Some improvements for image filtering using peer group techniques. Image Vis. Comput. **28**(1), 188–201 (2010)
55. Morillas, S., Gregori, V., Peris-Fajarnes, G.: Isolating impulsive noise pixels in color images by peer group techniques. Comput. Vis. Image Underst. **110**(1), 102–116 (2008)
56. Morillas, S., Gregori, V., Hervás, A.: Fuzzy peer groups for reducing mixed Gaussian-impulse noise from color images. IEEE Trans. Image Process. **18**, 1452–1466 (2009)
57. George, A., Veeramani, P.: On some results in fuzzy metric spaces. Fuzzy Set. Syst. **64**, 395–399 (1994)
58. Celebi, M.E.: Distance measures for reduced ordering-based vector filters. IET Image Process. **3**, 249–260 (2009)
59. Morillas, S., Gregori, V., Peris-Fajarnés, G., Latorre, P.: A new vector median filter based on fuzzy metrics. In: Proceedings of the International Conference on Image Analysis and Recognition, ICIAR 2005, LNCS vol. 3656, pp. 81–90 (2005)
60. Morillas, S., Gregori, V., Peris-Fajarnés, G., Sapena, A.: New adaptive vector filter using fuzzy metrics. J. Electron. Imaging **16**(3), 033007 (2007)
61. Hamilton, W.R.: Elements of Quaternions. Ginn & Company, Boston (1887)
62. Sangwine, S.J., Ell, T.A.: Colour image filters based on hypercomplex convolution. IEEE Proc. Vis. Image Signal Process. **147**(2), 89–93 (2000)
63. Evans, C.J., Sangwine, S.J., Ell, T.A.: Hypercomplexcolor-sensitive smoothing filters. In: Proceedings of the Seventh IEEE International Conference on Image Processing (ICIP2000). **1**, 541–544 (2000)
64. Denis, P., Carre, P., Fernandez-Maloigne, C.: Spatial and spectral quaternionic approaches for colour images. Comput. Vis. Image Underst. **107**(1–2), 74–87 (2007)
65. Jin, L., Liu, H., Xu, X., Song, E.: Color impulsive noise removal based on quaternion representation and directional vector order-statistics. Signal Process. **91**, 1249–1261 (2011)
66. Geng, X., Hu, X., Xiao, J.: Quaternion switching filter for impulse noise reduction in color image. Signal Process. **92**, 150–162 (2012)
67. Schulte, S., De Witte, V., Nachtegael, M., Van der Weken, D., Kerre, E.E.: Fuzzy two-step filter for impulse noise reduction from color images. IEEE Trans. Image Process. **15**(11), 3567–3578 (2006)
68. Serra, J.: Image Analysis and Mathematical Morphology. Academic, New York (1982)
69. Soille, P.: Morphological Image Analysis, Principles and Applications. Springer, New York (1999)
70. Stevenson, R.L., Arce, G.R.: Morphological filters: statistics and further syntactic properties. IEEE T. Circ. Syst. **CAS-34**(11), 1292–1305 (1987)
71. Barnett, V.: The ordering of multivariate data. J. Roy. Stat. Soc. Ser. A **139**(3), 318–355 (1976)
72. Hanbury, A., Serra, J.: Mathematical morphology in the HLS colour space. In: Proceedings of the British Machine Vision Conference. 451–460 (2001)
73. Crespo, J., Serra, J., Schafer, R.W.: Theoretical aspects of morphological filters by reconstruction. Signal Process. **47**, 201–225 (1995)
74. Crespo, J., Maojo, V.: New results on the theory of morphological filters by reconstruction. Pattern Recognit. **31**(4), 419–429 (1998)
75. Terol-Villalobos, I.R., Cruz-Mandujano, J.A.: Contrast enhancement and image segmentation using a class of morphological nonincreasing filters. J. Electron. Imaging **7**(3), 641–654 (1998)
76. Pessoa, L.F.C., Maragos, P.: MRL-filters: A general class of nonlinear systems and their optimal design for image processing. IEEE Trans. Image Process. **7**(7), 966–978 (1998)

77. Cheng, F., Venetsanopoulos, A.N.: An adaptive morphological filter for image processing. IEEE Trans. Image Process. **1**(4), 533–539 (1992)
78. Deng-Wong, P., Cheng, F., Venetsanopoulos, A.N.: Adaptive morphological filters for color image enhancement. J. Intell. Robot. Syst. **15**, 181–207 (1996)
79. Cheng, F., Venetsanopoulos, A.N.: Adaptive morphological operators, fast algorithms and their applications. Pattern Recognit. **33**, 917–933 (2000)
80. Harvey, N.R., Marshall, S.: The use of genetic algorithms in morphological filter design. Signal Process: Image Commun. **8**(1), 55–71 (1996)
81. Oh, J., Chaparro, L.F.: Adaptive fuzzy morphological filtering of impulse noise in images. Multidimens. Syst. Signal Process. **11**, 233–256 (2000)
82. Comer, M.L., Delp, E.J.: Morphological operations for color image processing. J. Electron. Imaging **8**(3), 279–289 (1999)
83. Louverdis, G., Vardavoulia, M.I., Andreadis, I., Tsalides, Ph.: A new approach to morphological color image processing. Pattern Recognit. **35**(8), 1733–1741 (2002)
84. Louverdis, G., Andreadis, I.: Soft morphological filtering using a fuzzy model and its application to colour image processing. Pattern Anal. Appl. **6**, 257–268 (2004)
85. Angulo, J.: Morphological colour operators in totally ordered lattices based on distances: application to image filtering, enhancement and analysis. Comput. Vis. Image Underst. **107**(1–2), 56–73 (2007)
86. Zhou, H., Mao, K.Z.: An impulsive noise color image filter using learning-based color morphological operations. Digit. Signal Process. **18**(3), 406–421 (2008).
87. Soleymani, M., Kasaei, S.: An FPCA-based color morphological filtering for noise removal. Sci. Iran. **16**(1), 8–18 (2009)
88. Khan, N., Pandey, P., Sahoo, A., Suchi, R., Srivastava, M.: Color image restoration using morphological detectors and adaptive filter. In: Contemporary Computing. Communications in Computer and Information Science, vol. 40, pp. 381–388 Springer, Berlin (2009)
89. Tomasi, C., Manduchi R.: Bilateral filter for gray and color images. In: Proceedings of the IEEE International Conference Computer Vision, pp. 839–846 (1998)
90. Garnett, R., Huegerich, T., Chui, C., He, W.: A universal noise removal algorithm with an impulse detector. IEEE Trans. Image Process. **14**(11), 1747–1754 (2005)
91. Plataniotis, K.N., Androutsos, D., Venetsanopoulos, A.N.: Multichannel filters for image processing. Signal Process.: Image Commun. **9**(2), 143–158 (1997)
92. Plataniotis, K.N., Androutsos, D., Venetsanopoulos, A.N.: Adaptive fuzzy systems for multichannel signal processing. In: Proc. IEEE **87**(9), 1601–1622 (1999)
93. Shen, Y., Barner, K.E.: Fuzzy vector median based surface smoothing. IEEE T. Vis. Comput. Gr. **10**(3), 252–265 (2004)
94. Shen, Y., Barner, K.E.: Fast adaptive optimization of weighted vector median filters. IEEE T. Signal Process. **54**(7), 2497–2510 (2006)
95. Shen, Y., Barner, K.E.: Optimization of fuzzy vector median filters. In: Proceedings of the 38th Annual Conference of Information Sciences & Systems, Princeton (2004)
96. Schulte, S., Morillas, S., Gregori, V., Kerre, E.E.: A new fuzzy color correlated impulsive noise reduction method. IEEE Trans. Image Process. **16**(10), 2565–2575 (2007)
97. Smolka, B., Malik, K., Malik, D.: Adaptive rank weighted switching filter for impulsive noise removal in color images. J. Real-Time Image Proc. **10**(2), 289–311 (2012). doi: 10.1007/s11554-012-0307-0
98. Russo, F., Ramponi, G.: A fuzzy filter for images corrupted by impulse noise. IEEE Signal Process. Lett. **3**, 168–170 (1996)
99. Toprak, A., Guler, I.: Impulse noise reduction in medical images with the use of switch mode fuzzy adaptive median filter. Digit. Signal Process. **17**, 711–723 (2007)
100. Toprak, A., Guler, I.: Angiograph imageres to ration with the use of rule base fuzzy 2-D Kalman filter. Exp. Syst. Appl. **35**, 1752–1761 (2008)
101. Toprak, A., Ozerdem, M.S., Guler, I.: Suppression of impulse noise in MR images using artificial intelligent based neuro-fuzzy adaptive median filter. Digit. Signal Process. **18**, 391–405 (2008)

102. Luo, W.: Efficient removal of impulse noise from digital images. IEEE Trans. Consum. Elec. **52**, 523–527 (2006)
103. Toh, K.K.V., Ibrahim, H., Mahyuddin, M.N.: Salt-and-pepper noise detection and reduction using fuzzy switching median filter. IEEE Trans. Consum. Elec. **54**, 1956–1961 (2008)
104. Meher, S.K.: Recursive and noise-exclusive fuzzy switching median filter for impulse noise reduction. Eng. Appl. Artif. Intel. **30**, 145–154 (2014)
105. Chatzis, V., Pitas, I.: Fuzzy scalar and vector median filters based on fuzzy distances. IEEE Trans. Image Process. **8**(5), 731–734 (1999)
106. Arakawa, K.: Median filter based on fuzzy rules and its application to image restoration. Fuzzy Set. Syst. **77**(1), 3–13 (1996)
107. Schulte, S., De Witte, V., Nachtegael, M., Van der Weken, D., Kerre, E.E.: Histogram-based fuzzy colour filter for image restoration. Image Vis. Comput. **25**(9), 1377–1390 (2007)
108. Camarena, J.G., Gregori, V., Morillas, S., Sapena, A.: A simple fuzzy method to remove mixed Gaussian-impulsive noise from color images. IEEE Trans. Fuzzy Syst. **21**(5), 971–978 (2013)
109. Lukac, R., Plataniotis, K.N., Smolka, B., Venetsanopoulos, A.N.: cDNA microarray image processing using fuzzy vector filtering framework. Fuzzy Set. Syst.: Special Issue on Fuzzy Sets and Systems in Bioinformatics. **152**(1), 17–35 (2005)
110. Lukac, R., Plataniotis, K.N., Smolka, B., Venetsanopoulos, A.N.: A multichannel order-statistic technique for cDNA microarray image processing. IEEE T. Nanobiosci. **3**(4), 272–285 (2004)
111. Civicioglu, P.: Using uncorrupted neighborhoods of the pixels for impulsive noise suppression with ANFIS. IEEE Trans. Image Process. **16**(3), 759–773 (2007)
112. Qin, H., Yang, S.X.: Adaptive neuro-fuzzy inference systems based approach to nonlinear noise cancellation for images. Fuzzy Set. Syst. **158**(10), 1036–1063 (2007)
113. Yuksel, M.E.: A hybrid neuro-fuzzy for edge preserving restoration of images corrupted by impulse noise. IEEE Trans. Image Process. **15**(4), 928–936 (2006)
114. Wolfram, S.: Universality and complexity in cellular automata. Physica **10**(1–2), 1–35 (1984)
115. Selvapeter, P.J., Hordijk, H.: Cellular automata for image noise filtering. In: World Congress on Nature Biologically Inspired Computing (NaBIC). pp. 193–197 (2009)
116. Popovici, A., Popovici, D.: Cellular automata in image processing. In: Proceedings of the 15th International Symposium on the Mathematical Theory of Networks and Systems. (2002)
117. Rosin, P.L.: Training cellular automata for image processing. IEEE Trans. Image Process. **15**(7), 2076–2087 (2006)
118. Rosin, P.L.: Image processing using 3-state cellular automata. Comput. Vis. Image Underst. **114**(7), 790–802 (2010)
119. Songtao, L., Chen, H., Yang, S.: An effective filtering algorithm for image salt-pepper noises based on cellular automata. In: Congress on Image and Signal Processing, vol. 3, pp. 294–297 (2008)
120. Sadeghi, S., Rezvanian, A., Kamrani, E.: An efficient method for impulse noise reduction from images using fuzzy cellular automata. Int. J. Electron. Commun. (AEU) **66**(9), 772–779 (2012)
121. Sahin, U., Uguz, S., Sahin, F.: Salt and pepper noise filtering with fuzzy-cellular automata. Comput. Electr. Eng. **40**, 59–69 (2014)
122. Nikolova, M.: A variational approach to remove outliers and impulse noise. J. Math. Imag. Vis. **20**, 99–120 (2004)
123. Chan, R.H., Ho, C., Nikolova, M.: Salt-and-pepper noise removal by median-type noise detectors and detail-preserving regularization. IEEE Trans. Image Process. **14**(10), 1479–1485 (2005)
124. Plonka, G., Ma, J.: Nonlinear regularized reaction-diffusion filters for denoising of images with textures. IEEE Trans. Image Process. **17**(8), 1283–1294 (2007)
125. Keren, D., Gotlib, A.: Denoising color images using regularization and correlation terms. J. Vis. Commun. Image R. **9**(4), 352–365 (1998)

126. Li, X.: On modeling interchannel dependency for color image denoising. Int. J. Imag. Syst. Tech. **17**(3), 163–173 (2007)
127. Blomgren, P., Chan, T.: Color TV: total variation methods for restoration of vector-valued images. IEEE Trans. Image Process. **7**(3), 304–309 (1998)
128. Tschumperl, D., Deriche, R.: Vector-valued image regularization with PDEs: a common framework from different applications. IEEE Trans. Pattern Anal. Mach. Intell. **27**(4), 506–517 (2005)
129. Lezoray, O., Elmoataz, A., Bougleux, S.: Graph regularization for color image processing. Comput. Vis. Image Underst. **107**(1–2), 38–55 (2007)
130. Elmoataz, A., Lezoray, O., Bougleux, S.: Nonlocal discrete regularization on weighted graphs: a framework for image and manifold processing. IEEE Trans. Image Process. **17**(7), 1047–1060 (2008)

Chapter 5
Spatial and Frequency-Based Variational Methods for Perceptually Inspired Color and Contrast Enhancement of Digital Images

Edoardo Provenzi

5.1 Introduction

Variational principles amount to defining a suitable linear functional, that is, a scalar-valued linear function defined on a suitable functional space, so that its minima (ideally, its *unique* minimum) provide the optimal solution of the problem under analysis.

In this chapter, we consider the problem of color correction inspired by human perception. More specifically, given any digital image, acquired under general illumination conditions, we try to modify its chromatic attributes so that the resulting image is as close as possible to the human perception of the scene photographed in the picture.

Useful applications of this kind of algorithms can be found in several fields, such as interior architecture, flickering reduction, calibration, tracking in large camera environments, photorealism, and computational photography, to quote but a few, see for example [2, 8, 13, 15].

There exists a plethora of perceptually inspired color correction models, for example [6, 10, 11], to quote but a few, so it is impossible to perform a thorough analysis of all of them here.

For this reason, we are concerned only with those called Retinex-like, that is, the models that follow two basic principles stated by E. Land and J.J. McCann in their seminal work [12]. In this paper, the authors studied 2D physical configurations made of juxtaposed patches which they called "Mondrian" pictures. Their experiments have proven that color perception of each patch depends much more on the reflectance of the surrounding patches than on the illumination. This remarked that perceived color is an inherently local property, which can only be specified "in context" and it is not an absolute property of physical surfaces. They proposed a strategy to recover the perceived color of Mondrian patches via a quite complicated computation called ratio-threshold-reset.

E. Provenzi (✉)
Laboratoire MAP5 (UMR CNRS 8145), Université Paris Descartes, Sorbonne Paris Cité,
45 rue des saint Pères, 75006, Paris, France
e-mail: edoardo.provenzi@parisdescartes.fr

M. E. Celebi et al. (eds.), *Color Image and Video Enhancement*,
DOI 10.1007/978-3-319-09363-5_5

Without entering into the quite involved details of this computation, we can say that the core of the original Retinex computation are the two following principles: the first is that color correction can be performed separately on the three chromatic channels RGB, the second is *locality*, that is, the fact that the correction of the RGB intensity of a pixel should be driven by the local context, that is, the intensity distribution of the surrounding pixels. The way Land and McCann implemented locality is by comparing pixel intensities over paths travelling around the Mondrian pictures. A chain of intensity ratios is propagated through the path until a very bright pixel is found. When this condition is reached the reset mechanism comes into play and the computation starts back. In this way, the bright pixel encountered in the path is assumed as being the local reference for white. For this reason the original model of Retinex is said to fall in the category of local white patch models. The formalization of these operations and their interpretation can be found in [19, 20].

Variational principles allowed building a general framework in which existing Retinex-like color correction algorithms can be embedded and compared, see for example [3, 4, 16], a quite noticeable result, considering the fact that the direct equations of these algorithms seem very distant from each other.

The presentation that we give here does not follow the actual chronological development of the variational model as it can be found in the literature, but it has the advantage of being more logical and easier to follow.

The chapter is structured as the following. In Sect. 5.2 we introduce the fundamental result of [22] about the variational interpretation of histogram equalization. In Sect. 5.3, we briefly introduce the basic Human Visual System (HVS) properties that we use in Sect. 5.4 to propose a set of axioms to be satisfied by a perceptual functional and highlight the relationship between these functionals and the one that induces histogram equalization. Examples on natural images are shown and analyzed. The wavelet analogue of the perceptual functionals is discussed in Sect. 5.5. We conclude with some remarks about future perspectives to further improve the models presented in this chapter.

5.2 Variational Interpretation of Histogram Equalization

In this section we introduce a basic result for the development of the rest of this work, which consists in the variational interpretation of histogram equalization.

It is worthwhile starting with some notation. The functional space that we consider here is that of RGB continuous image functions. To introduce these functions we first denote the *spatial domain* of a digital image with $\Omega \subset \mathbb{R}^2$, its area with $|\Omega|$, and the coordinates of two arbitrary pixels with Ω $x \equiv (x_1, x_2)$ and $y \equiv (y_1, y_2)$. We always consider a *normalized dynamic range* in $[0, 1]$, so that an *RGB image function* will be denoted with

$$\begin{aligned} \mathbf{I} : \Omega &\longrightarrow [0,1]\times[0,1]\times[0,1] \\ x &\mapsto (I_R(x),\ I_G(x),\ I_B(x)) \end{aligned}$$

where each scalar component $I_c(x)$ defines the intensity level of the pixel $x \in \Omega$ in the Red, Green, and Blue channel, respectively.

We stress that we will perform every computation on the scalar components of the image, thus *treating each chromatic component separately* as in the original Retinex paper [12]. Therefore, we avoid the subscript c and write simply $I(x)$ to denote the intensity of the pixel x in a given chromatic channel.

From the point of view of functional analysis, we implicitly consider the space of image functions as a subspace of $L^2(\Omega)$, the space of square-integrable (finite-energy) functions from Ω to [0, 1].

Let us also recall very briefly what histogram equalization is. Let $\lambda \in [0, 1]$ be a generic intensity level, then the *histogram* of I computed in λ is:

$$h(\lambda) = \frac{1}{|\Omega|}\,\text{Area}\{x \in \Omega \mid I(x) = \lambda\} \qquad \lambda \in [0, 1], \tag{5.1}$$

that is, the *occurrence probability* of the level λ in the image.

The *cumulative histogram* of I computed in λ, $H(\lambda)$, is:

$$H(\lambda) = \frac{1}{|\Omega|}\,\text{Area}\{x \in \Omega \mid I(x) \le \lambda\} \qquad \lambda \in [0, 1], \tag{5.2}$$

that is, the *probability to find a pixel with intensity less than* λ.

Of course, the relationship between h and H is:

$$H(\lambda) = \int_0^{\lambda} h(t)\,dt, \quad H'(\lambda) = h(\lambda), \tag{5.3}$$

that is, *H is the integral function of h* in the interval [0, 1] and the first derivative of H in each level gives the histogram of that level.

It will be useful for later purposes to notice that the relationship $H(\lambda) = \int_0^{\lambda} h(t)\,dt$ can be rewritten as follows:

$$H(\lambda) = \int_0^{\lambda} h(t)\,dt = \frac{1}{|\Omega|}\int_0^1 \text{sign}^+(\lambda - I(t))\,dt \tag{5.4}$$

where

$$\text{sign}^+(s) = \begin{cases} 1 & \text{if } s \ge 0; \\ 0 & \text{if } s < 0. \end{cases}$$

and its "spatial version"

$$H(I(x)) = \frac{1}{|\Omega|}\int_{\Omega} \text{sign}^+(I(x) - I(y))\,dy. \tag{5.5}$$

An image is said to be *equalized* if *each level has the same occurrence probability*, that is, if $h(\lambda) \equiv 1$ (recall that the histogram is normalized) $\forall\lambda$, which of course can be translated to the condition $H(\lambda) = \lambda \ \forall\lambda$ on the cumulative histogram.

Classical histogram equalization consists in the transformation that modifies the level distribution of an image in such a way that its histogram is as homogeneous as possible. It is easy to proof (see e.g., [7]) that the transformation $\lambda \longmapsto H(\lambda)$, that is, the application of the cumulative histogram, seen as a function from [0, 1] to itself, is the easiest modification that implements histogram equalization.

However, this is not the only histogram equalization transformation available in literature. In particular, a variational interpretation has been provided in the paper [22] and it is the basis for what we will discuss in Sect. 5.4. This interpretation is discussed in the following theorem.

Theorem 5.1. *Given the functional*

$$E_{\text{hist eq}}(I) \equiv 2\int_{\Omega}\left(I(x)-\frac{1}{2}\right)^2 dx - \frac{1}{|\Omega|}\iint_{\Omega^2}|I(x)-I(y)|\,dxdy \qquad (5.6)$$

if $I^* = \text{argmin}_I\, E_{\text{hist eq}}(I)$*, then* I^* *has equalized histogram, that is,* $H(I^*(x)) = I^*(x)$ *for all* $x \in \Omega$.

Moreover, if I_0 *is the original image, then the initial value problem for the gradient descent*

$$\begin{cases} \partial_t I = -\delta E_{\text{hist eq}}(I) \\ I(0) = I_0 \end{cases}$$

has only one solution, where t *is the evolution parameter of the iterative gradient descent scheme and the symbol* δ *represents the first variation of the functional.*

The proof of the first part of this theorem is reported in Appendix 1. Here we prefer to stress the meaning of the functional that appears in the theorem, that will be called "Caselles–Sapiro functional" from now on.

To understand the meaning of the energy functional $E_{\text{hist eq}}(I)$, let us write the histogram equalization energy functional as follows: $E_{\text{hist eq}}(I) = D_{\frac{1}{2}}(I) - C(I)$, where the two functional terms $D_{\frac{1}{2}}(I)$ and $C(I)$ are

$$D_{\frac{1}{2}}(I) \equiv 2\int_{\Omega}\left(I(x)-\frac{1}{2}\right)^2 dx$$

and

$$C(I) \equiv \frac{1}{|\Omega|}\iint_{\Omega^2}|I(x)-I(y)|\,dxdy.$$

The minimization of $E_{\text{hist eq}}(I) = D_{\frac{1}{2}}(I) - C(I)$ is achieved through the *minimization of* $D(I)$ and the *maximization of* $C(I)$, because $C(I)$ has a minus sign in front of it, so it becomes more negative as we increase it.

Let us discuss the meaning of the two functional terms:

- $D_{\frac{1}{2}}(I)$ is called *global quadratic dispersion* term around the middle gray level $1/2$ and it is minimized when $I(x) \equiv 1/2$ for all $x \in \Omega$, that is the minimization of this term tends to turn I into a uniform gray image
- $C(I)$ is called *global contrast* term and its maximization corresponds to the maximization of the global contrast of the image I, measured through the absolute differences $|I(x) - I(y)|$

Thus, the argmin of the Caselles–Sapiro functional is the image corresponding to the optimal balance between *two opposite effects*: on one side the minimization of $D_{\frac{1}{2}}(I)$ tends to *set all the levels to the average gray* $1/2$ but, on the other side, the minimization of $-C(I)$ tends to *spread the intensity levels apart, as far as possible from each other*. So, the intrinsical meaning of Theorem 5.1 is that *the equilibrium between two conflicting actions, dispersion control around the middle gray and contrast enhancement, induces histogram equalization.*

One practical consequence of this result is that, applying for example the gradient descent technique to minimize $E_{\text{hist eq}}(I)$, one can stop the minimization process before reaching the complete equalization, thus realizing a *partial equalization* that can nonetheless be useful to avoid the typical over-enhancement of low-key images (see [7] for more details).

However, for the purposes of color image processing, the most important consequences of Theorem 5.1 are theoretical: in fact, as we will see in the next section, we can modify the functional $E_{\text{hist eq}}(I)$ in such a way that the basic principle of histogram equalization, that is, the balance between dispersion control and contrast enhancement, is preserved but we can change the analytical form of the terms $D_{\frac{1}{2}}(I)$ and $C(I)$ taking inspiration from human visual perceptual features. The argmin image I^* of the modified functional is a color-corrected version of the original image driven by perceptual properties of the HVS.

5.3 A Basic Set of HVS Properties

The HVS is an extraordinarily complex system, nowadays still far from being fully understood. As it would be impossible to provide an even approximate description, here we just recall the *four basic phenomenological mechanisms of color vision* that will be used to select perceptual functionals. As a general reference for the following section, the reader can refer to the complete book [26].

5.3.1 Adaptation to the Average Luminance Level

The primary event that allows vision consists in a visible photon penetrating the eyes and hitting the retina. The photon may be absorbed by a photoreceptor, one of the three *cones* L, M, S or a *rod*, depending on the intensity and spectral content of the photon energy. The absorbtion activates a photochemical transduction from the electromagnetic energy of the photon to the difference of electric potential of the photoreceptor.

More precisely, it has been measured (see e.g., [23]) that when a photoreceptor absorbs a photon of luminance L, the electric potential of its membrane changes according to an empirical law known as *Michaelis–Menten's equation* (or *Naka–Rushton's equation* when the coefficient n is unitary):

$$r(L) = \frac{\Delta V}{\Delta V_{\max}} = \frac{L^n}{L^n + L_S^n}, \tag{5.7}$$

where $\Delta V_{\max}$ is the highest difference of potential that can be generated, n is a constant (measured as 0.74 for the rhesus monkey), and L_S is the luminance at which the photoreceptor response is half maximal, called the *semisaturation level*, and which is usually associated with the *level of adaptation*. Each type of cone L, M, S is most sensitive in a particular waveband and the semisaturation constant depends on the amount of light in the particular waveband that is absorbed, not on the global luminance of the light source.

The interpretation of the sigmoid-like transformation (5.7) is the following: the photoreceptors response has an intensity *threshold*, below which the receptors do not respond, then they *respond fairly linearly* and, finally, the response *saturates* as we approach large values of luminance.

The "center" of this behavior is the semisaturation value L_S, in this sense the retinal cells "adapt" to each scene average luminance, allowing us to perceive modulations of light intensity around L_S. This adaptive capacity of retinal cells is of paramount importance for human vision, without it we simply would not be able to see when passing from a dim-illuminated environment to a very intensely illuminated one, and vice versa.

5.3.2 Local Contrast Enhancement

The HVS operates a local contrast enhancement, with the primary (but not the only) effect of improving edge perception, as proven by visual effects as simultaneous contrast or Mach bands, reported in Figs. 5.1 and 5.2, respectively.

Edges contain the most important part of the visual information, so it is not surprising that the HVS has developed a way to improve their perception.

The most important feature that must be highlighted here is that edge enhancement is *local*, that is, it depends on the local distribution of light intensity around each point.

Fig. 5.1 *Top*: Colored simultaneous contrast. *Bottom*: Grayscale simultaneous contrast. In both pictures, the *inner gray squares* have exactly the same *physical* luminance, however, their *perceived* luminance is very different

5.3.3 Color Constancy

If light adaptation is the HVS ability to adapt to different light *intensities*, *color constancy* may be described at first glance as the ability to adapt to different light *spectral content*, to perceive colors as constant as possible as the illumination changes.

This feature is so intrinsically hardwired in our visual system that we often take it for granted. However, whenever we take two pictures of the same scene with two different illuminants, we see that we must perform a very careful white balance to obtain two images that produce similar color sensations.

The psychophysiological details that allow color constancy are beyond the scopes of this work, for more details see for example, [9].

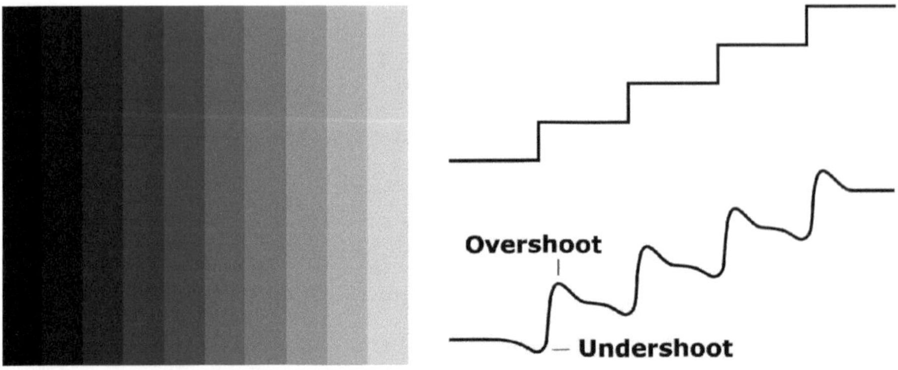

Fig. 5.2 *Left*: Mach bands effect. *Right*: real and apparent luminance pattern, revealing the typical perceptual undershoots and overshoots, that is, nonphysical jumps of light intensity around the edges

Color constancy will be essential for the selection of the analytical form of the contrast enhancement term of perceptual functionals in Sect. 5.4.

5.3.4 Weber's Law

The German physicist E. Weber in the second half of the nineteenth century developed some psychophysical experiments to test contrast perception: a dark-adapted human observer was put in a dim room in front of a white screen on which a narrow beam of light was thrown in the center of the visual field. The luminous intensity I of the beam was increased very slowly and the observer was asked to tell whether he/she could perceive an intensity change. Nowadays, we call the least perceptible intensity change ΔI the *JND* for *Just Noticeable Difference.*

Weber found out that the JND increased proportionally with the luminous intensity[1], that is, $\Delta I = K \cdot I$, or, as it is presented more often:

$$\frac{\Delta I}{I} \simeq K, \tag{5.8}$$

Eq. (5.8) is called *Weber's law* and K Weber's constant. Weber's law says that, as we increase the background light I, the difference ΔI must increase proportionally to be able to appreciate $I + \Delta I$ as different from I. This partially explains why we are more sensitive to noise in dark areas of a visual scene and thus why it is more important to perform a good denoising in dark areas of digital images rather than in bright ones.

[1] Weber's law is approximately valid not only for the visual sense, but also for all the other senses, with different values of Weber's constant.

5.4 Variational Perceptually Inspired Color Enhancement of Digital Images

We are now going to use the perceptual features recalled in the previous section to transform the Caselles–Sapiro functional into a perceptual one by quoting the major result of [16].

Theorem 5.2. *The only class of energy functionals complying with all the four HVS features discussed in Sect. 5.3 is the following:*

$$E_{\mu,I_0,w,\varphi}(I) = D_{\mu,I_0}(I) + C_w(I), \tag{5.9}$$

where

$$D_{\mu,I_0}(I) = \int_\Omega \left[\alpha \left(\mu \log \frac{\mu}{I(x)} - (\mu - I(x)) \right) + \beta \left(I_0 \log \frac{I_0}{I(x)} - (I_0 - I(x)) \right) \right] dx, \tag{5.10}$$

and

$$C_{w,\varphi}(I) = \iint_{\Omega^2} w(x,y)\, \varphi \left(\frac{\min\{I(x), I(y)\}}{\max\{I(x), I(y)\}} \right) dxdy \tag{5.11}$$

where,

- I_0 *is a given channel of the original image function;*
- μ *is the average value of* I_0*;*
- $\alpha, \beta > 0$ *are real coefficients that control the attachment to* μ *and to the original image function values* I_0*, respectively;*
- $w{:}\Omega \times \Omega \to \mathbb{R}^+$ *is a spatial kernel, which depends only the Euclidean distance* $\|x - y\|$ *between two generic pixels* $x, y \in \Omega$ *and it is monotonically decreasing with the distance itself, that is,* $w(x, y){:} = g(\|x - y\|)$*, where* $g{:}\mathbb{R}^+ \to \mathbb{R}$ *is monotonically decreasing;*
- $\varphi{:}[0, 1] \to \mathbb{R}$ *is a monotonically increasing differentiable function.*

Instead of reporting the (long and complex) proof of this theorem (which of course can be found in [16]), we consider to discuss the meaning of the analytical form of the functional terms appearing in the perceptual energy functional much more interesting.

To understand the motivation for the analytical shape of the contrast term $C_{w,\varphi}(I)$ let us consider the basic image formation model, that is, $I(x) = \rho(x) \cdot \lambda$, where $\rho(x)$ represents the reflectance of a point x and λ represents the illuminant (supposed to be constant all over the scene). Since $C_{w,\varphi}(I)$ is written in terms of a ratio, it is evident that it is independent with respect to illuminant changes, thus it is coherent with the color constancy feature. Moreover, as proven in [16], this analytical form is the only one in which color constancy and Weber's law can be combined.

The function φ represents a degree of freedom and it is chosen to be monotonically increasing not to reverse contrast between pixels.

Finally, notice that the minimization of $C_{w,\varphi}(I)$ indeed induces a contrast enhancement. In fact, the ratio $\min\{I(x), I(y)\}/\max\{I(x), I(y)\}$ is minimized when the lowest value between two pixels is decreased and the highest is increased, which, of course, corresponds to an intensification of contrast. The enhancement is spatially local due to the presence of the weighting function w. Typically w is a Gaussian kernel with center in x, its standard deviation σ can be set by a user to increase or decrease the locality of contrast enhancement. Small values of σ push the effect toward *sharpening*, large values of σ instead push toward a global enhancement.

Let us now discuss the dispersion term. Its choice has been guided by dimensional coherence with the contrast term, which has dimension 0 with respect to I. The easiest meaningful candidate is the *entropy functional* D_{μ,I_0}, whose minimization produces a reduction of entropy, that is, disorder, around the average value μ (which can be different in each chromatic channel) and around the original image I_0. This last attachment is introduced to avoid an excessive departure from the original intensity values and can be modulated via the coefficients ratio α/β.

If we compare the Caselles–Sapiro functional and the perceptual ones, we can infer that *a perceptually inspired Retinex-like color correction can be interpreted as a suitable local and nonlinear version of histogram equalization.*

5.4.1 Stability of the Numerical Scheme for the Minimization of Perceptual Functionals and Reduction of Computational Complexity

The Euler–Lagrange equations of $E_{\mu,I_0,w,\varphi}(I)$, that is, $\delta E_{\mu,I_0,w,\varphi}(I) = 0$ are integral equations that cannot be solved analytically. Moreover, the perceptual functionals are not convex, so that we cannot easily find a global minimum. However, in [16], it has been proved that if we substitute the null values of the input image with $1/255$ (the smallest nonzero value in the normalized dynamic range), and we consider $\alpha, \beta > 1$, then the gradient descent scheme with respect to $\log I$, that is,

$$\partial_t \log I = -\delta E_{\mu,I_0,w,\varphi}(I) \tag{5.12}$$

converges to a unique image function, that we can consider the perceptually inspired color-corrected one. The gradient descent is written in terms of $\log I$ because the logarithmic derivative, that is, $\partial_t \log I = \frac{1}{I}\partial_t I$ allows rewriting the gradient descent equation as follows:

$$\partial_t I = -I \cdot \delta E_{\mu,I_0,w,\varphi}(I), \tag{5.13}$$

the right-hand side of this equation has dimension 0 (which is the correct perceptual dimension), since $\delta E_{\mu,I_0,w,\varphi}(I)$ has dimension[2] -1 in terms of I. Moreover, notice

[2] In general, the first variation of a homogeneous functional of degree n is a homogeneous functional of degree $n-1$, in our case $E_{\mu,I_0,w,\varphi}(I)$ has degree 0, so its first variation has degree -1.

that, as proven in [1], the effect of taking the logarithmic image is simply to change the speed of convergence of the gradient descent scheme and not its final result.

The generic step of the iterative scheme is given by the following equation [16]:

$$I_{k+1}(x) = \frac{I_k(x) + \Delta t \left(\alpha\mu + \beta I_0(x) + \frac{1}{2} C^{\varphi}_{I_k}(x) \right)}{1 + \Delta t(\alpha + \beta)}, \tag{5.14}$$

where

$$C^{\varphi}_{I_k}(x) = \int_{\Omega} w(\|x - y\|)\varphi \left(\frac{\min\{I_k(x), I_k(y)\}}{\max\{I_k(x), I_k(y)\}} \right) dy. \tag{5.15}$$

The computational complexity of these algorithms is $\mathcal{O}(N^2)$, N being the number of image pixels. This high complexity is due to the contrast term: its first variation generates the integral term $C^{\varphi}_{I_k}(x)$ in the Euler–Lagrange equations.

In [16], an approximation technique has been developed that is able to reduce the computational complexity to $\mathcal{O}(N \log N)$ that we are going to describe, avoiding the iterative subscript k to simplify the notation and help the comprehension of the approximation strategy. Let p be a generic polynomial of order n of the variables $I(x), I(y)$ and define $\tilde{\varphi}_n(I(x), I(y)) = \operatorname{argmin}_p \|p - \varphi\|_2$, that is, $\tilde{\varphi}_n$ is the polynomial of order n with *minimal quadratic distance* with respect to the function φ. It is convenient to write $\tilde{\varphi}_n$ as follows

$$\tilde{\varphi}_n(I(x), I(y)) = \sum_{j=0}^{n} f_j(I(x)) I(y)^j, \tag{5.16}$$

where $f_j(I(x)) = \sum_{m=0}^{n-j} p_{m,j} I(x)^m$. The numerical coefficients $p_{m,j}$ depend on the approximation order n, but we will not make explicit this dependence for the sake of a more readable mathematical notation.

Introducing this expression of $\tilde{\varphi}_n$ in $C\varphi_I(x)$ instead of φ and noticing that f_j does not depend on y, one gets

$$\tilde{C}^{\varphi}_{w,n}(x) = \sum_{j=0}^{n} f_j(I(x)) \int_{\Omega} w(\|x - y\|) I(y)^j dy, \tag{5.17}$$

but $\int_{\Omega} w(\|x - y\|) I(y)^j dy = (w * I^j)(x)$, that is the convolution between the kernel w and the j-th power of the image I, hence

$$\tilde{C}^{\varphi}_{w,n}(x) = \sum_{j=0}^{n} \sum_{m=0}^{n-j} p_{m,j} I(x)^m (w * I^j)(x). \tag{5.18}$$

The n convolutions $w * I^j$ can be *precomputed* through the FFT (Fast Fourier Transform), which has computational complexity $\mathcal{O}(N \log N)$, thus drastically decreasing the computational time. The degree n of the polynomial is a parameter that controls the precision of the polynomial approximation.

5.4.2 *Relationship with Existing Perceptually Inspired Color Correction Models*

Another major result allowed by the variational framework is the possibility to recover in this setting two well-known models for perceptually inspired color correction. This can be done simply by selecting two different analytical expressions of the function φ.

In [3] and [4], respectively, it has been proven that:

- If $\varphi \equiv \log$, then the minimization of $E_{\mu,I_0,w,\log}(I)$ leads to the so-called ACE algorithm formulated in [21];
- If $\varphi \equiv \text{id}$, then the minimization of $E_{\mu,I_0,w,\text{id}}(I)$ leads to a continuous symmetrized version of the original Retinex algorithm [12].

To have an example of the action of these methods on the color of digital images, three results of the method obtained by the gradient-descent minimization of $E_{\mu,I_0,w,\text{id}}(I)$ are shown in Fig. 5.3. It can be seen that contrast is enhanced both in under- and over-exposed images and color cast is removed. In Fig. 5.4 the enhancement is quantified via the difference between the original and the enhanced histograms.

5.5 Wavelet-based Implementation of Variational Perceptually Inspired Color Correction

Disclaimer: the reader unfamiliar with the wavelet framework can read a brief summary in Appendix 2.

As we have discussed in Sect. 5.4, the perceptual color correction proposed in [16] is performed through a local contrast enhancement balanced by the action of a dispersion term around the average value plus a conservative term that avoids over-enhancement. Wavelet detail coefficients are related to local contrast; let us see how this fact has been used in [18] to provide a wavelet-based framework for perceptually inspired color correction.

The general scheme of the algorithm is the following:

1. We consider the coarsest approximation coefficients $\{a_{J,k},\ k \in \Omega\}$, and we modify them to implement adaptation to the average gray level in the wavelet domain (according to Eq. (5.19) of Sect. 5.5.1);
2. We fix these new approximation coefficients and use them along with $\{d^{\ell}_{J,k},\ k \in \Omega\}$, $\ell = H, V, D$, to modify the horizontal, vertical, and diagonal detail coefficients according to the equations of Proposition 5.3 of Sect. 5.5.2. This will implement local contrast enhancement in the wavelet domain at the scale 2^J;
3. We then pass to the scale 2^{J-1} and we compute the approximation coefficients by summing the approximation and detail coefficients just computed at the scale 2^J;

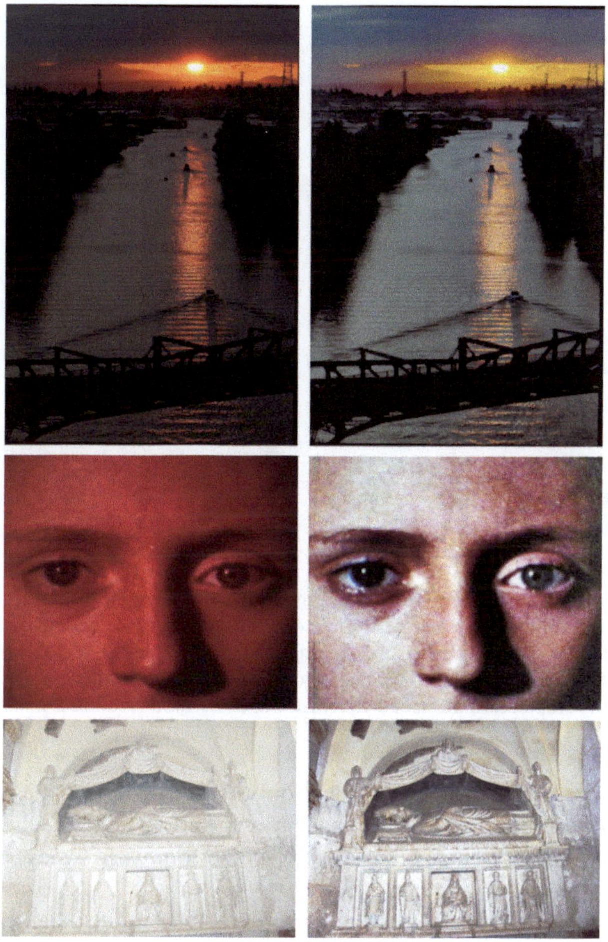

Fig. 5.3 *Images on the left*: Originals. *Images on the right*: Filtered versions obtained by minimizing $E_{\mu,I_0,w,\mathrm{id}}(I)$ with a gradient-descent technique. w is a Gaussian kernel with a standard deviation equal to one forth of the image diagonal. First image: courtesy of P. Greenspun. Middle image: courtesy of A. Pardo, this image is a frame of a very corrupted old film, thus the artifacts shown by the *middle image*, *on the right*, are present in the original film frame and not introduced by the algorithm

4. We fix these approximation coefficients and we repeat step 2 at the scale 2^{J-1};
5. We iterate this scheme until reaching the finest scale.

The equations quoted in the scheme above are presented and analyzed in detail in the next two sections, but before doing that it is convenient to make some assumptions that will greatly simplify the exposition.

First of all, since changing the sign of a wavelet coefficient can result in drastic modifications of an image, we modify only the *absolute value* of the wavelet coefficients, restoring the original sign at the end of the computation.

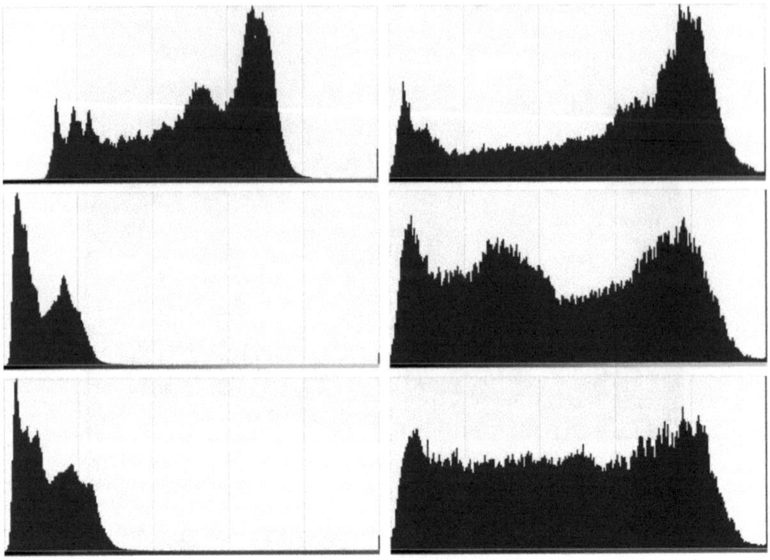

Fig. 5.4 *Left column*: Originals RGB histograms of the image in the second row of Fig. 5.3 (in every picture, the horizontal axis ranges from 0 to 255, the vertical axis ranges from 0 to 1). *Right column*: enhanced RGB histograms corresponding to the image in the second row of Fig. 5.3. It can be seen that the perceptual enhancement algorithms tends toward histogram equalization, but they automatically stop before the actual equalization, that would result in a washed-out image

Moreover, in every scale 2^j, we deal only with coefficient magnitudes bigger than a fixed threshold $T_j > 0$, leaving the other coefficients unchanged to avoid intensification of noise. Thus, we deal with *positive, bounded, and finite* sequences of wavelet coefficient magnitudes.

To simplify the notation, we avoid the superscript ℓ in the detail coefficients, by making the implicit assumption that the operations are repeated on the horizontal (H), vertical (V), and diagonal (D) detail coefficients.

The variability of the scale coefficient 2^j is confined within the scales 2^L and 2^J, $L, J \in \mathbb{Z}$, $L < J$, 2^L being the finest and 2^J the coarsest. Typically, $J - L$ ranges between 2 and 10, depending on the image dimension. Finally, with the notation $k \in \Omega$, we implicitly assume a column-wise ordering of Ω, the spatial support of the image, so that Ω can be seen as a finite subset of $\mathbb{Z}$.

5.5.1 Adjustment to the Average Value in the Wavelet Domain

Let us first consider the effect of adaptation to the average level. If we were dealing with Fourier transforms, the average image intensity value μ would be represented by the zero-order Fourier coefficient. In the wavelet domain there is no such direct correspondence. However, since the coefficients $\{a_{j,k},\ k \in \Omega\}$ represent the image approximation at the scale 2^j, a natural analogue of the average value μ in the wavelet framework at the scale 2^j is represented by $\overline{a}_j \equiv \frac{1}{|\Omega|}\sum_{k\in\Omega} a_{j,k}$, that is, the average approximation coefficient.

We also stress that we only need to modify the approximation coefficients of the coarsest scale, since this modification will be propagated to finer scales. Thus, we implement the adaptation to the average value at the coarser scale balanced by the adjustment to the original values through this convex linear combination

$$a_{J,k} \equiv \alpha\overline{a}_J + (1-\alpha)a^0_{J,k}, \tag{5.19}$$

where $\{a^0_{J,k},\ k \in \Omega\}$ is the original sequence of approximation coefficients at the scale 2^J, and $\alpha \in [0,1]$ is a suitable weight coefficient. The bigger the value of α, the stronger the adjustment to the average value $\overline{a}_J$, and vice versa.

5.5.2 Local Contrast Enhancement in the Wavelet Domain

Let us now consider local contrast enhancement. We have remarked that the two most important features of the contrast functional $E_{\mu,I_0,w,\varphi}(I)$ are the fact that it enhances contrast *locally* and in an *illuminant-independent* way. To maintain these characteristics also in the wavelet domain, in [18] the following local contrast enhancement functional at the scale 2^j has been proposed:

$$\mathcal{C}_{w_j,\varphi,\{a_{j,k}\}}(\{d_{j,k}\}) \equiv \sum_{k\in\Omega} w_j\,\varphi\left(\frac{a_{j,k}}{d_{j,k}}\right), \qquad 2^J \geq 2^j \geq 2^{L+1}, \tag{5.20}$$

where w_j are positive coefficients that permit to differentiate the contrast enhancement action depending on the scale 2^j, and φ:$[0,\infty) \to [0,\infty)$ is a differentiable monotonically increasing function, such that $\varphi(r) \to +\infty$ as $r \to \infty$.

We stress that the approximation coefficients are passed to the functional $\mathcal{C}$ as fixed parameters in every scale, from the coarsest to the finest. In particular, the approximation coefficients used in the coarsest scale are those defined by Eq. (5.19).

$\mathcal{C}_{w_j,\varphi,\{a_{j,k}\}}$ is minimized when the ratio between the approximation and detail coefficients decreases, but since the approximation coefficients are fixed at each scale, the minimization of $\mathcal{C}_{w_j,\varphi,\{a_{j,k}\}}$ corresponds to an intensification of the coefficients $d_{j,k}$. This implies a local and multiscale contrast enhancement of the image. The locality depends both on the mother wavelet chosen (because different mother wavelets have different shape) and on the scale 2^j at which one operates: the finer the scale, the more local is contrast enhancement, and vice versa.

Furthermore, the invariance with respect to global illumination changes, in the sense of the von Kries model [25], is guaranteed by the fact that also the functional $\mathcal{C}_{w_j,\varphi,\{a_{j,k}\}}$ is *homogenous of degree 0*, that is, so the transformation $I \mapsto \lambda I$, $\lambda > 0$, will have no effects on $\mathcal{C}_{w_j,\varphi,\{a_{j,k}\}}$ and its Euler–Lagrange equations.

To prevent an excessive magnification of the original detail coefficients, whose absolute value is denoted by $d^0_{j,k}$, a conservative term should be introduced. To maintain dimensional coherence with $\mathcal{C}_{w_j,\varphi,\{a_{j,k}\}}$, an entropic dispersion functional is a suitable choice:

$$\mathcal{D}_{d^0_{j,k}}(\{d_{j,k}\}) \equiv \sum_{k\in\Omega}\left[d^0_{j,k}\log\frac{d^0_{j,k}}{d_{j,k}} - \left(d^0_{j,k}-d_{j,k}\right)\right], \qquad 2^J \geq 2^j \geq 2^{L+1}. \tag{5.21}$$

Combining these two effects, one can define the energy functional that realizes local contrast enhancement as $\mathcal{E}_{w_j,\varphi,\{a_{j,k}\},d^0_{j,k}} = \mathcal{C}_{w_j,\varphi,\{a_{j,k}\}} + \mathcal{D}_{d^0_{j,k}}$, that is,

$$\mathcal{E}_{w_j,\varphi,\{a_{j,k}\},d^0_{j,k}} \equiv \sum_{k\in\Omega}\left[w_j\varphi\left(\frac{a_{j,k}}{d_{j,k}}\right) + d^0_{j,k}\log\frac{d^0_{j,k}}{d_{j,k}} - \left(d^0_{j,k}-d_{j,k}\right)\right], \tag{5.22}$$

with $2^J \geq 2^j \geq 2^{L+1}$. The following theorem ensures the existence of a minimum of $\mathcal{E}_{w_j,\varphi,\{a_{j,k}\},d^0_{j,k}}$ and determines the corresponding Euler–Lagrange equations. Its proof can be found in [18].

Theorem 5.3. *There exists a minimum of the functional $\mathcal{E}_{w_j,\varphi,\{a_{j,k}\},d^0_{j,k}}$. Moreover, the Euler–Lagrange equations for the detail coefficients are:*

$$\frac{\partial\mathcal{E}_{w_j,\varphi,\{a_{j,k}\},d^0_{j,k}}}{\partial\{d_{j,k}\}}(d_{j,k}) = 0 \iff d_{j,k} = d^0_{j,k} + w_j\varphi'\left(\frac{a_{j,k}}{d_{j,k}}\right)\frac{a_{j,k}}{d_{j,k}}, \tag{5.23}$$

where φ' denotes the derivative of φ. In particular, when $\varphi \equiv id$,

$$\frac{\partial\mathcal{E}_{w_j,\{a_{j,k}\},d^0_{j,k}}}{\partial\{d_{j,k}\}}(d_{j,k}) = 0 \iff d_{j,k} = d^0_{j,k} + w_j\frac{a_{j,k}}{d_{j,k}}. \tag{5.24}$$

In Theorem 5.3 the role of the identity has been highlighted because it is the easiest choice for φ. Equation (5.24) is an implicit equation that has to be solved using a numerical method. In [18] Newton–Raphson's method has been used to find the zero of the function $F(d_{j,k}) \equiv d_{j,k} - d^0_{j,k} - w_j\frac{a_{j,k}}{d_{j,k}}$ by iteratively solving the equation

$$d^n_{j,k} = d^{n-1}_{j,k} - \frac{F(d^{n-1}_{j,k})}{F'(d^{n-1}_{j,k})}, \tag{5.25}$$

$n \geq 1$. Since the solution is not expected to differ too much from the original magnitude $d^0_{j,k}$, Newton–Raphson's algorithm is initialized with $d^0_{j,k}$.

A standard result guarantees the convergence of Newton–Raphson's algorithm as long as the initial condition $d^0_{j,k}$ is sufficiently near the solution, $F'(d^0_{j,k})$ is small enough, $F''(d_{j,k})$ varies smoothly, and the inverse of $F''(d_{j,k})$ is bounded near the solution, see for example, [5]. In particular, these conditions imply that one cannot take the weights w_j to be too big, otherwise Newton's algorithm can oscillate. In [17], it has been proven that if the identity function is substituted by the

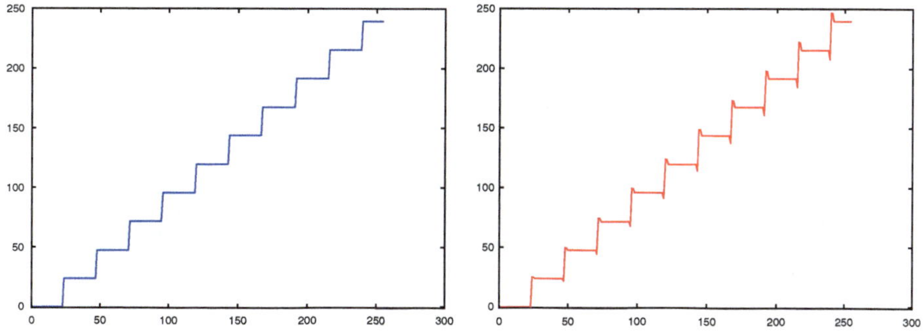

Fig. 5.5 *From left to right*: a scan-line of the Mach-bands image before and after the wavelet algorithm, respectively. The undershoot and overshoot effects are typical HVS features that the wavelet model is able to reproduce. The wavelet algorithm was applied with following parameters: the mother wavelet is the biorthogonal wavelet with two vanishing moments, the computation is performed over the maximum number of scales allowed for each image, $w_j = 0.5$, $T_j = \frac{\max_{k \in \Omega}\{d_{j,k}\}}{2.5}$ for each scale, and $\alpha = 0.1$

gamma function, then the stability of Newton–Raphson's algorithm in this setting is improved.

Notice that, since Newton–Raphson's method is initialized with the coefficients $d^0_{j,k}$, which are bounded from below, the solution of Eq. (5.24) belongs to an open neighborhood of $d^0_{j,k}$ and the numerical method converges to a positive solution of Eq. (5.24).

To stress the perceptual nature of the model, in Fig. 5.5 it is reported a scan-line of the classical Mach-bands picture: the wavelet algorithm is able to reproduce the well-known undershoots and overshoots typical of the HVS behavior.

The wavelet-based method just described can run in real time and it is as efficient as the spatially based one. In Fig. 5.6, the action of the wavelet algorithm can be seen on three images affected by distinct problems: underexposure, color cast, and overexposure. As can be seen, the wavelet algorithm is able to perform a radiometric adjustment of the nonoptimally exposed pictures and to strongly reduce the color cast, as proven in Fig. 5.7.

5.6 Conclusions and Future Perspectives

In this work, we have discussed in an organic way the major results of a recent variational formulation of perceptually inspired color correction algorithms.

We have shown that the variational framework provides a unified setting for such models, which allows understanding their action in terms of optimal balance between local, and illumination-invariant, contrast enhancement and entropy-like adjustment to the average luminance.

Fig. 5.6 *Images on the left*: original ones. *Images on the right*: enhanced versions after the wavelet algorithm: details appear in originally underexposed and overexposed areas, and the pink color cast in the "Lena" image is removed. The filtering parameters are the following: the mother wavelet is the Daubechies wavelet with two vanishing moments, the computation is performed over the maximum number of scales allowed for each image, $w_j = 0.5$, and $T_j = \frac{\max_{k\in\mathfrak{J}}\{d_{j,k}\}}{10}$ for each scale 2^j

The variational framework can be formulated both in the spatial and in the wavelet domain. The latter has the advantage of providing a much faster implementation and it distributes the computation over different scales.

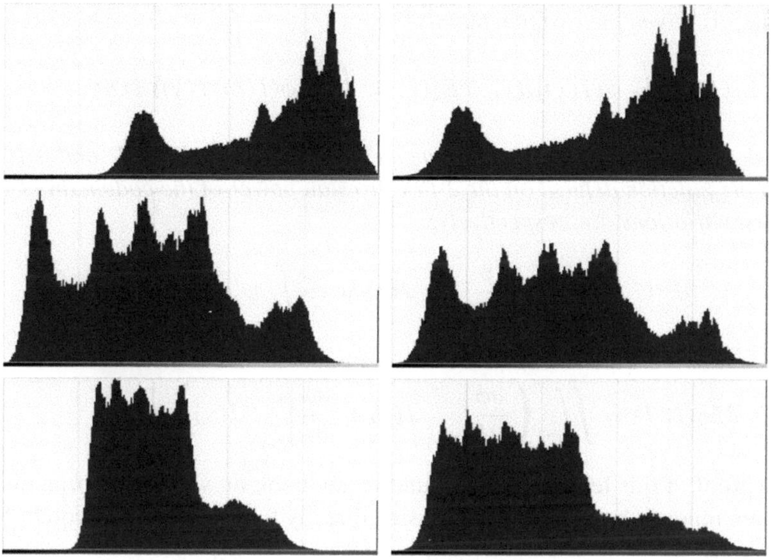

Fig. 5.7 *Left column*: Originals RGB histograms of the image in the second row of figure 5.6 (in every picture, the *horizontal* axis ranges from 0 to 255, the *vertical* axis ranges from 0 to 1). *Right column*: enhanced RGB histograms corresponding to the image in the second row of Fig. 5.6. Again, notice the histogram stretching without reaching a complete equalization, that would have been inappropriate

From the point of view of human vision, the major drawback of these models is that they consider the three chromatic channels independently, while the HVS operates a recombination of the LMS cones into luminance plus color-opponent channels, see for example, [9]. A natural evolution of these models is an implementation in color opponent spaces.

Acknowledgements The author would like to dedicate this work to the memory of Prof. Vicent Caselles with infinite gratitude for sharing his deep intelligence, culture, passion for science, and, last but not least, honesty.

Appendix 1

We provide here the proof of Theorem 5.1. We think that this proof is instructive because it explicitly shows the link between the variations of the functionals appearing in Theorem 5.1 and histogram equalization, which is far from being intuitive at the first sight.

Let us start with a useful lemma.

Lemma 5.1. *Given the two functionals*

$$E_1(I) = \int_\Omega \psi(I(x))\,dx, \quad E_2(I) = \iint_{\Omega^2} \phi(I(x), I(y))\,dxdy, \tag{5.26}$$

where ψ is a differentiable function defined on the codomain of I, and ϕ is a differentiable function defined on the 2-th Cartesian power of the codomain of I, then their first variations are, respectively:

$$\delta E_1(I, J) = \int_\Omega \left.\frac{\partial \psi}{\partial I}\right|_{I(x)} J(x)dx \equiv \int_\Omega \psi'(I(x))\,J(x)\,dx \tag{5.27}$$

and

$$\delta E_2(I, J) = \iint_{\Omega^2} \left(\left.\frac{\partial \phi}{\partial I}\right|_{I(x)} J(x) + \left.\frac{\partial \phi}{\partial I}\right|_{I(y)} J(y) \right) dxdy. \tag{5.28}$$

The proof of this lemma can be found in any book on variational principles and we do not report it here, preferring passing directly to the proof of Theorem 5.1.

Proof. By linearity, we can compute the first variation of the two terms of the energy functional separately and then add the results. For that, let us recall that we have written:

$$D_{\frac{1}{2}}(I) = 2\int_\Omega \left(I(x) - \frac{1}{2}\right)^2 dx; \tag{5.29}$$

$$C(I) = \frac{1}{|\Omega|} \iint_{\Omega^2} |I(x) - I(y)|\,dxdy. \tag{5.30}$$

By virtue of formula (5.27), we have

$$\delta D_{\frac{1}{2}}(I, J) = \int_\Omega 4\left(I(x) - \frac{1}{2}\right) J(x)\,dx, \tag{5.31}$$

and by virtue of formula (5.28), we have

$$\begin{aligned} \delta C(I, J) &= \frac{1}{|\Omega|} \iint_{\Omega^2} \left[\operatorname{sign}(I(x) - I(y))J(x) - \operatorname{sign}(I(x) - I(y))J(y)\right] dxdy \\ &= \frac{1}{|\Omega|} \iint_{\Omega^2} \operatorname{sign}(I(x) - I(y))J(x)\,dxdy + \\ &- \frac{1}{|\Omega|} \iint_{\Omega^2} \operatorname{sign}(I(x) - I(y))J(y)\,dxdy. \end{aligned} \tag{5.32}$$

Now, interchanging the role of the "mute" variables x and y in the second integral of the last step, we have that

$$\frac{1}{|\Omega|} \iint_{\Omega^2} \operatorname{sign}(I(x) - I(y))J(y)\,dxdy = \frac{1}{|\Omega|} \iint_{\Omega^2} \operatorname{sign}(I(y) - I(x))J(x)\,dydx \tag{5.33}$$

but then, using the oddness of the sign function,

$$\frac{1}{|\Omega|}\iint_{\Omega^2} \operatorname{sign}(I(x)-I(y))J(y)\,dxdy = -\frac{1}{|\Omega|}\iint_{\Omega^2} \operatorname{sign}(I(x)-I(y))J(x)\,dydx. \tag{5.34}$$

Hence, we can write

$$\begin{aligned}\delta C(I,J) &= \frac{1}{|\Omega|}\iint_{\Omega^2} \operatorname{sign}(I(x)-I(y))J(x)\,dydx + \\ &+ \frac{1}{|\Omega|}\iint_{\Omega^2} \operatorname{sign}(I(x)-I(y))J(x)\,dydx = \\ &= \frac{2}{|\Omega|}\iint_{\Omega^2} \operatorname{sign}(I(x)-I(y))J(x)\,dydx\end{aligned} \tag{5.35}$$

that can be conveniently rearranged as follows:

$$\delta C(I,J) = \int_\Omega \left(\frac{2}{|\Omega|}\int_\Omega \operatorname{sign}(I(x)-I(y))\,dy\right) J(x)\,dx. \tag{5.36}$$

Now, since $\delta E_{\text{hist eq}}(I,J) = \delta D_{\frac{1}{2}}(I,J) - \delta C(I,J)$, by using formulas (5.31) and (5.36) we have

$$\delta E_{\text{hist eq}}(I,J) = \int_\Omega 4\left(I(x)-\frac{1}{2}\right) J(x)\,dx - \int_\Omega \left(\frac{2}{|\Omega|}\int_\Omega \operatorname{sign}(I(x)-I(y))\,dy\right) J(x)\,dx \tag{5.37}$$

that is,

$$\delta E_{\text{hist eq}}(I,J) = \int_\Omega \left[4\left(I(x)-\frac{1}{2}\right) - \frac{2}{|\Omega|}\int_\Omega \operatorname{sign}(I(x)-I(y))\,dy\right] J(x)\,dx. \tag{5.38}$$

The stationary condition $\delta E_{\text{hist eq}}(I,J) = 0,\ \forall J$, implies that the expression in the square bracket must be zero, that is,

$$\delta E_{\text{hist eq}}(I,J) = 0 \iff 4\left(I(x)-\frac{1}{2}\right) - \frac{2}{|\Omega|}\int_\Omega \operatorname{sign}(I(x)-I(y))\,dy = 0, \tag{5.39}$$

so that the Euler–Lagrange equation relative to the energy functional $E_{\text{hist eq}}$ is the following implicit equation

$$2\left(I(x)-\frac{1}{2}\right) - \frac{1}{|\Omega|}\int_\Omega \operatorname{sign}(I(x)-I(y))\,dy = 0, \tag{5.40}$$

that can be suitably rewritten as

$$\frac{1}{|\Omega|}\int_\Omega \operatorname{sign}(I(x)-I(y))\,dy = 2I(x) - 1. \tag{5.41}$$

Now, using the identity sign(t) $= 2\text{sign}^+(t) - 1$, we can express the left-hand side of the Euler–Lagrange equation as

$$\begin{aligned}\frac{1}{|\Omega|}\int_\Omega (2\text{sign}^+(I(x)-I(y))-1)\,dy &= \frac{2}{|\Omega|}\int_\Omega \text{sign}^+(I(x)-I(y))\,dy - \frac{\int_\Omega dy}{|\Omega|}\\ &= 2H(I(x))-1,\end{aligned} \tag{5.42}$$

where we have used the fact that $\frac{1}{|\Omega|}\int_\Omega \text{sign}^+(I(x)-I(y))\,dy$ is the spatial version of the cumulative histogram $H(I(x))$, as noticed in Eq. (5.5).

Thus, the Euler–Lagrange Eq. (5.41) is equivalent to $2H(I(x))-1 = 2I(x)-1$, that is, to $H(I(x)) = I(x)$, but then

$$\delta E_{\text{hist eq}}(I, J) = 0 \quad \Longleftrightarrow \quad H(I(x)) = I(x), \ \forall x \in \Omega, \tag{5.43}$$

which means that the image function I which satisfies the Euler–Lagrange equations of the functional $E_{\text{hist eq}}(I)$ has an equalized histogram.

The proof of existence and uniqueness of the solution of the gradient descent scheme written in Theorem 5.1 is quite long and technical and can be found in [22]. □

Appendix 2

Let us start this section by briefly recalling the basic information about wavelet theory in 1D, then we will extend the discussion to 2D wavelets, the main reference for all the results quoted hereafter is [14]. A 1D (mother) wavelet $\psi \in L^2(\mathbb{R})$ is a unit norm and null-mean function. Of course this is possible only if ψ oscillates, but, unlike infinite waves, wavelets can have compact support. The ψ-wavelet transform $W_\psi f$ of $f \in L^2(\mathbb{R})$ in the point ξ at the scale s is given by the inner product $W_\psi f(\xi, s) = \int_\mathbb{R} f(x)\frac{1}{\sqrt{s}}\overline{\psi}\left(\frac{x-\xi}{s}\right)dx$. $W_\psi f$ gives a "measure of similarity" between f and ψ around the point ξ at the scale s. So, if a signal is constant or do not vary "too much" in the support of a wavelet, then its wavelet transform will be zero or very small, this is how wavelets provide a multiscale information about the *local contrast* of a signal.

The set $\{\psi_{j,k}\}_{(j,k)\in\mathbb{Z}^2} \subset L^2(\mathbb{R})$ given by $\psi_{j,k}(x) \equiv \frac{1}{\sqrt{2^j}}\psi\left(\frac{x-2^j k}{2^j}\right)$ is a complete orthonormal system of $L^2(\mathbb{R})$. Moreover, $L^2(\mathbb{R})$ can be recovered by the closure of the union of a sequence of nested closed subspaces $V_j \subset V_{j-1}$ with suitable properties (see Mallat's book [14] for more details). The orthogonal projections of $f \in L^2(\mathbb{R})$ onto V_j and V_{j-1} give the approximation of f at the scales 2^j and 2^{j-1}, respectively. The 2^j-approximation is coarser and the missing details with respect to the finer 2^{j-1}-approximation are contained in the orthogonal complement W_j of V_j in V_{j-1}: $V_{j-1} = V_j \oplus W_j$. W_j is called the j-th *detail space* and it can be proven that the orthogonal projection of f on W_j is given by $P_{W_j} f = \sum_{k\in\mathbb{Z}} \langle f, \psi_{j,k}\rangle \psi_{j,k} \equiv \sum_{k\in\mathbb{Z}} d_{j,k}\psi_{j,k}$. The coefficients $d_{j,k}$ are called *detail coefficients of f at the scale* 2^j. Fine-scales detail coefficients at fine scale are sparse, in fact, they are non-null only when the support of $\psi_{j,k}$ intersects a high contrast zone, that is, around sharp edges.

Finally, let us recall that every wavelet ψ is related to a mirror filter h and to a function ϕ, called *scale function*, through the following equation that involves their Fourier transforms: $\hat{\psi}(2\omega) = \frac{1}{\sqrt{2}} e^{-i\omega} \hat{h}^* (\omega + \pi) \hat{\phi}(\omega)$, see [24] for a complete and detailed description of how to generate wavelets using the filter design methodology. ϕ appears in the orthogonal projection of a signal f onto the approximation space V_j, in fact it can be proven that $P_{V_j} f = \sum_{k\in\mathbb{Z}} \langle f, \phi_{j,k}\rangle \phi_{j,k} \equiv \sum_{k\in\mathbb{Z}} a_{j,k}\phi_{j,k}$, where $\phi_{j,k}(x) = \frac{1}{\sqrt{2^j}} \phi\left(\frac{x - 2^j k}{2^j}\right)$ and $a_{j,k}$ are called *approximation coefficients* at the scale 2^j. It follows that $P_{V_{j-1}} f = P_{V_j} f + P_{W_j} f = \sum_{k\in\mathbb{Z}} a_{j,k}\phi_{j,k} + \sum_{k\in\mathbb{Z}} d_{j,k}\psi_{j,k}$.

In practical applications one is interested in a multiresolution analysis between two fixed scales 2^L and 2^J, $L, J \in \mathbb{Z}, L < J$. In this case $V_{J-1} = V_J \oplus W_J$, $V_{J-2} = V_{J-1} \oplus W_{J-1} = V_J \oplus W_J \oplus W_{J-1}$ and so on, thus $V_L = V_J \oplus \bigoplus_{2^J \geq 2^j \geq 2^{L+1}} W_j$. For this reason, following [14], we say that a discrete orthogonal wavelet multiresolution representation of a 1D signal f between two fixed scales 2^L and 2^J, $L, J \in \mathbb{Z}$, $L < J$, is given by the collection of detail coefficients $\{d_{j,k}\}$ at all scales, completed by the approximation coefficients at the coarser scale, that is, $\{a_{J,k}\}$.

When we deal with 2D signals, as images, we have to consider a multiresolution analysis of $L^2(\mathbb{R}^2)$. Multidimensional wavelet bases can be generated with tensor products of separable basis functions defined along each dimension. In this case, an orthogonal wavelet multiresolution representation between two scales 2^L and 2^J, $L, J \in \mathbb{Z}$, $L < J$, is given by three sets of detail coefficients $\{d^H_{j,k}, d^V_{j,k}, d^D_{j,k}\}$ at all scales, which correspond to the *horizontal, vertical, and diagonal detail coefficients*, respectively, completed by the approximation coefficients at the coarser scale, that is, $\{a_{J,k}\}$.

References

1. Ambrosio, L., Gigli, N., Savaré, G.: Gradient flows in metric spaces and in the space of probability measures. Lectures in Mathematics, Birkhauser (2005)
2. Banich, N., Loncaric, S.: Light random sprays retinex: Exploiting the noisy illumination estimation. IEEE Signal Process. Lett. **20**, 1240–1243 (2013)
3. Bertalmío, M., Caselles, V., Provenzi, E., Rizzi, A.: Perceptual color correction through variational techniques. IEEE Trans. Image Process. **16**, 1058–1072 (2007)
4. Bertalmío, M., Caselles, V., Provenzi, E.: Issues about the retinex theory and contrast enhancement. Int. J. Comput. Vision. **83**,101–119 (2009)
5. Ciarlet, P.: Introduction to numerical linear algebra and optimisation. Cambridge University Press, Melbourne (1989)
6. Frankle, J., McCann, JJ.: Method and apparatus for lightness imaging. United States Patent, US 4,348,336. (1983)
7. Gonzales, R., Woods, R.: Digital image processing. Prentice Hall, Upper Saddle River (2002)

8. Han, Y., Zheng, D., Baciu, G., Feng, X., Li, M.: Fuzzy region competition-based auto-color-theme design for textile images. Textile Res. J. **83**(6), 638–650 (2013)
9. Hubel, D.: Eye, brain, and vision. Scientific American Library, New York (1995)
10. Kimmel, R., Elad, M., Shaked, D., Keshet, R., Sobel, I.: A variational framework for retinex. Intern. J. Comput. Vis. **52**(1), 07–23 (2003)
11. Kuang, J., Johnson, G.M., Fairchild, M.D.: iCAM06: A refined image appearance model for HDR image rendering. J. Vis. Commun. Image R. **18**, 406–414 (2007)
12. Land, E., McCann, J.: Lightness and retinex theory. J. Opt. Soc. Am. **61**(1), 1–11 (1971)
13. Li, H., Zhang, L., Shen, H.: A perceptually inspired variational method for the uneven intensity correction of remote sensing images. IEEE Trans. Geosci. Remote Sens. **50**(8), 3053–3065 (2012)
14. Mallat, S.: A wavelet tour of signal processing, 3rd ed. Academic Press, San Diego (1999)
15. Montagna, R., Finlayson, G.D.: Constrained pseudo-Brownian motion and its application to image enhancement. J. Opt. Soc. Am. A. **28**(8), 1677–1688 (2011)
16. Palma-Amestoy, R., Provenzi, E., Bertalmío, M., Caselles, V.: A perceptually inspired variational framework for color enhancement. IEEE Trans. Pattern Anal. Mach. Intell. **31**(3), 458–474 (2009)
17. Provenzi, E.: Boosting the stability of wavelet-based contrast enhancement of color images through gamma transformations. J. Modern Opt. **60**, 1145–1150 (2013)
18. Provenzi, E., Caselles, V.: A wavelet perspective on variational perceptually-inspired color enhancement. Int. J. Comput. Vis. **106**, 153–171 (2014)
19. Provenzi, E., De Carli, L., Rizzi, A., Marini, D.: Mathematical definition and analysis of the Retinex algorithm. J. Opt. Soc. Am. A. **22**(12), 2613–2621 (2005)
20. Provenzi, E., Fierro, M., Rizzi, A., De Carli, L., Gadia, D., Marini, D.: Random spray retinex: A new retinex implementation to investigate the local properties of the model. IEEE Trans. Image Process. **16**(1), 162–171 (2007)
21. Rizzi, A., Gatta, C., Marini, D.: A new algorithm for unsupervised global and local color correction. Pattern Recognit. Lett. **24**, 1663–1677 (2003)
22. Sapiro, G., Caselles, V.: Histogram modification via differential equations. J. Diff. Eq. **135**, 238–266 (1997)
23. Shapley, R., Enroth-Cugell, C.: Visual adaptation and retinal gain controls. Prog. Retin. Res. **3** (Chap 9), 263–346 (1984)
24. Strang, G., Nguyen, T.: Wavelets and filter banks. Wellesley-Cambridge Press, Wellesley (1996)
25. von Kries, J.: Chromatic Adaptation. Festschrift der Albrecht-Ludwigs-Universität 135:145–158 (1902)
26. Wyszecky, G., Stiles, W.S.: Color science: Concepts and methods, quantitative data and formulas. Wiley, New York (1982)

Chapter 6
The Color Logarithmic Image Processing (CoLIP) Antagonist Space

Yann Gavet, Johan Debayle and Jean-Charles Pinoli

6.1 Introduction

The initial goal when addressing the representation of color images was to set up an algebraic mathematical framework that considers color images as vectors in an abstract linear space. The key idea was to develop this framework not only with a mathematical relevance but in accordance with the main laws and characteristics of the human color visual perception.

As noticed by the authors, the current color spaces (e.g., RGB, CIE XYZ, CIE $L^*a^*b^*$, CIECAM02) do not follow an additive superposition law, and therefore fail to obey the linear concept that is of a very high theoretical and practical interest in mathematics and its applications. From an image processing and analysis viewpoint, these color spaces do not thus allow to combine, amplify, and transform color images with the tools outcoming from the abstract linear algebra mathematical branch. Indeed, the classical linear image processing (CLIP) approach is no more valid as the classical additive operation "+" is not relevant for color images.

The color logarithmic image processing (CoLIP) framework is based on the logarithmic image processing (LIP) theory that was developed for the representation and processing of images valued in a bounded intensity range [20, 21, 33]. The LIP theory is physically and psychophysically well justified as it is consistent with the multiplicative image formation model and with several laws and characteristics of human brightness perception (e.g., Weber's law, Fechner's law, saturation effect, brightness range inversion) [34].

The work in the field of color image processing is somehow complexbecause of the variety of the definitions and their interpretations and the variety of the observers: always remember that the standard observer is an average observer.

Y. Gavet (✉) · J. Debayle · J.-C. Pinoli
École Nationale Supérieure des Mines, Saint-Etienne, France
e-mail: gavet@emse.fr

M. E. Celebi et al. (eds.), *Color Image and Video Enhancement*,
DOI 10.1007/978-3-319-09363-5_6

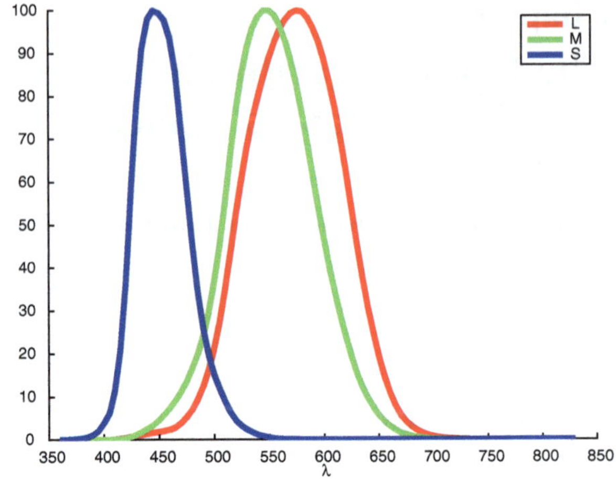

Fig. 6.1 Color-matching functions in LMS [7]. Wavelengths are represented on *abscissa*, in nanometers. These values are normalized to reflect the chromatic adaptation

6.2 Human Color Perception

6.2.1 Trichromacy and Color Photoreception

The trichromacy theory (Young, beginning of the 19th century, [39, 52]) states that humans have three different receptors sensitive to color stimuli. This assertion was later confirmed and the theory was extended by Helmholtz [18, 28] in the middle of the 19th century.

Indeed, the color photoreceptors in the retina, namely the cones, are sensitive to three different wavelength ranges: long, medium, and short, and thus classified into three types of cones: L, M, and S, respectively.

The color-matching functions (see also Sect. 6.5.3) reflect the trichromacy with an average observer, first established by the Commission Internationale de l'Eclairage (CIE) in 1931 and later in 1964. The LMS color-matching functions are represented in Fig. 6.1 for the observer of CIE 1931.

6.2.2 Opponent Process

The opponent-process theory was first proposed by Hering [19], who noticed that particular colors like reddish-green or yellowish-blue would never be observed. He explained it by a bipolar response of the cones. Later in the 20th century, this theory was confirmed by physiological experiments [44].

Schematically, the color information is coded in opponent red-green (denoted rg) and yellow-blue (denoted yb) channels to improve the efficiency of the transmission (reducing the noise) and to decorrelate the LMS channels [7].

6.2.3 *Just Noticeable Differences and Logarithmic Laws*

The psychophysicist Weber argued [50] that the human brightness detection depends on light intensity ratios rather than differences between two light intensity values (called stimuli). The just noticeable difference (JND, [15]) is the relative amount of light necessary to add or subtract to a visual test field of constant intensity value, such that it can be discriminated. JND is also known as the Weber's constant (or ratio, [15, 25], or fraction). The Fechner's brightness perception law [8, 9, 25], which is more general than Weber's law, states that arithmetic steps in perceived brightness (sensation) are produced by geometrical steps in the incident light intensity (stimulus).

From a mathematical continuous viewpoint, this yields a logarithmic relationship between the perceived brightness and the stimulus. For each channel LMS, a Weber's fraction exists, yielding the LMS Fechner's law [43].

6.2.4 *Color Representation Systems and Conversions*

The starting point for all CoLIP operations is the LMS color space [7]. Thus, this section introduces the conversion methods from different color spaces used in this chapter, for example, XYZ, sRGB, and $L^*a^*b^*$.

The numerical calculations are done using the OptProp toolbox[1].

6.2.4.1 XYZ to LMS Conversion

The conversion from XYZ to LMS is done in agreement with the CIECAM02 specifications [7], considering a complete adaptation. The matrix $\mathrm{M_{HPE}}$ (Hunt, Pointer, Estevez) is used to perform the following conversion:

$$\begin{pmatrix} L \\ M \\ S \end{pmatrix} = \mathrm{M_{HPE}} \times \begin{pmatrix} X \\ Y \\ Z \end{pmatrix}, \tag{6.1}$$

[1] by Jerker Wågberg, *More Research* and *DPC*, www.more.se.

with

$$\mathrm{M}_{\mathrm{HPE}} = \begin{pmatrix} 0.38971 & 0.68898 & -0.07868 \\ -0.22981 & 1.18340 & 0.04641 \\ 0.00000 & 0.00000 & 1.00000 \end{pmatrix}. \tag{6.2}$$

6.2.4.2 sRGB to XYZ Conversion

There exist a lot of different conversion matrices, taking into account the illuminant, the observer, and the RGB space used. Here, we chose to keep the conventions of the CIE 1931, i.e., a 2° 1931 observer, with the equal energy illuminant E and the sRGB space, to be consistent with the color-matching functions data presented in Sect. 6.5.3.

6.2.4.3 XYZ to $L^*a^*b^*$ Conversion

These operations are defined in the CIELAB specifications. $X_n Y_n Z_n$ define the tristimulus value of the reference white (E). L^* is defined in the range [0;100]. The $L^*a^*b^*$ space is discussed in Sect. 6.5.2 with regard to the CoLIP space. The conversion operations are presented in Eqs. 6.3 and 6.4:

$$\begin{aligned} L^* &= 116 \cdot f\left(\tfrac{Y}{Y_n}\right) - 16, \\ a^* &= 500 \cdot \left[f\left(\tfrac{X}{X_n}\right) - f\left(\tfrac{Y}{Y_n}\right)\right], \\ b^* &= 200 \cdot \left[f\left(\tfrac{Y}{Y_n}\right) - f\left(\tfrac{Z}{Z_n}\right)\right], \end{aligned} \tag{6.3}$$

with

$$f(t) = \begin{cases} t^{1/3} & \text{if } t > 0.008856 \\ 7.787 \cdot t + \frac{16}{116} & \text{if } t \leq 0.008856 \end{cases}. \tag{6.4}$$

6.3 LIP

The LIP theory was introduced in the mid-1980s [33]. It defines an algebraic framework that allows operations on images in a bounded range [20]. This model is mathematically well defined as well as physically consistent with the transmitted light imaging process. Moreover, it satisfies the Weber–Fechner human visual law (see Sect. 6.2.3).

This chapter mainly focuses on color image processing, and does not consist in a review of nonlinear frameworks. For example, the reader is referred to the multiplicative homomorphic image processing framework [31] or the log-ratio image processing framework [40]. More recent publications try to generalize (generalized LIP, [6]), introduce parameters (parameterized LIP, [11, 32]), different nonlinear laws (power image processing, [12]), and some symmetry (symmetric LIP [47, 48], and [29]). All these frameworks are more or less physically justified, and for several of them no psychophysical justification is reported. No color antagonist color spaces have been reported based on them. In some cases, the choice of specific parameters yields the same mathematical relations as for the LIP framework. Pinoli [35] gives more details of the main approaches. This review is the subject of a future publication.

6.3.1 Gray-Tone Functions

In the LIP theory, a gray-tone function f is associated to an intensity image F. A gray-tone function f is defined on a spatial support $D \subset \mathbb{R}^2$ and has its values in the real-number range $[0;M_0[$, with M_0 being a strictly positive real number. In the context of transmitted light imaging, the value 0 corresponds to the total transparency and M_0 to the total opacity. Thus, the gray-tone function f is defined, for $F_{\max}$ being the saturating light intensity level (glare limit), by:

$$f = M_0 \left(1 - \frac{F}{F_{\max}}\right). \tag{6.5}$$

6.3.2 The Vectorial Structure

The vector space S of gray-tone functions is defined by the following operations of addition, scalar multiplication, opposite, and subtraction:

$$\forall f, g \in S,\ f \triangle\!\!\!+ g = f + g - \frac{fg}{M_0}, \tag{6.6}$$

$$\forall f \in S, \forall \lambda \in \mathbb{R},\ \lambda \triangle\!\!\!\times f = M_0 - M_0 \left(1 - \frac{f}{M_0}\right)^{\lambda}, \tag{6.7}$$

$$\forall f \in S,\ \triangle\!\!\!- f = \frac{-M_0 f}{M_0 - f}, \tag{6.8}$$

$$\forall f, g \in S,\ f \triangle\!\!\!- g = M_0 \frac{f - g}{M_0 - g}. \tag{6.9}$$

The definition of the opposite operation $\triangle\!\!\!-$ extends the gray-tone range to the unbounded real-number range $]-\infty;M_0[$. This vectorial space S is algebraically and topologically isomorphic to the classical vector space of real-valued functions, defined through the following isomorphism φ:

$$\forall f \in S,\ \varphi(f) = -M_0 \ln\left(1 - \frac{f}{M_0}\right).$$

The inverse isomorphism is then defined by φ^{-1}:

$$f = \varphi^{-1}(\varphi(f)) = M_0\left(1 - \exp\left(-\frac{\varphi(f)}{M_0}\right)\right).$$

This isomorphism φ allows the introduction of notions and structures outcoming from functional analysis [24], such as the Euclidean norm:

$$\forall f \in S,\ \| f \|_{\triangle} = |\varphi(f)|_{\mathbb{R}},$$

with $|\cdot|_{\mathbb{R}}$ being the usual absolute value.

6.3.3 *Illustrations and Applications*

The LIP framework is particularly efficient in the case of images presenting small intensity values. This is why the dynamic enhancement method is interesting. For example, Fernandes et al. [10] have illustrated the use of the LIP framework to perform shape-from-focus reconstruction.

This section illustrates the use of the LIP operations with elementary operations first, and then details a method for image enhancement.

6.3.3.1 Elementary Operations Illustrations

The elementary operations are presented in Fig. 6.2. The operation of multiplication is illustrated in Fig. 6.2b (for $2 \triangle\!\!\!\!\triangle f$) and c (for $\frac{1}{2} \triangle\!\!\!\!\triangle f$). The operation of addition of two images f and g, namely $f \triangle\!\!\!\!+ g$, is illustrated in Fig. 6.2e.

6.3.3.2 Dynamic Range Maximization

A gray-tone function denoted f is defined on the spatial support $D \subset \mathbb{R}^2$. Its upper and lower bounds are denoted $f_u = \inf_{x \in D} f(x)$ and $f_l = \inf_{x \in D} f(x)$, assuming that $0 < f_l < f_u < M_0$ for practical reasons.

The dynamic range of f on D is defined by $R(f) = f_u - f_l$. The LIP scalar multiplication of f by a real number $\lambda > 0$ yields the following dynamic :

$$R(\lambda \triangle\!\!\!\!\triangle f) = \lambda \triangle\!\!\!\!\triangle f_u - \lambda \triangle\!\!\!\!\triangle f_l.$$

Fig. 6.2 Basic LIP operation for gray-tone images: illustration of the LIP multiplication $\triangle\!\!\!\!\times$ and addition $\triangle\!\!\!\!+$. **a** Original Manet image denoted f, green channel. **b** LIP scalar multiplication $2 \triangle\!\!\!\!\times f$. **c** LIP scalar multiplication $0.5 \triangle\!\!\!\!\times f$. **d** Original Monet image denoted g, green channel. **e** $f \triangle\!\!\!\!+ (0.5 \triangle\!\!\!\!\times g)$

It has been shown [21] that there exists an optimal value $\lambda_0(f)$ that maximizes the dynamic range, that is,

$$R(\lambda_0(f) \triangle\!\!\!\!\times f) = \max_{\lambda>0} R(\lambda \triangle\!\!\!\!\times f).$$

This value is explicitly defined by:

$$\lambda_0(f) = \frac{\ln\left(\frac{\ln((M_0-f_u)/M_0)}{\ln((M_0-f_l)/M_0)}\right)}{\ln\left(\frac{M_0-f_l}{M_0-f_u}\right)}.$$

6.4 CoLIP

The previous section introduced the LIP theory for gray-tone images. This section presents the color space CoLIP, previously defined in [16,17]. It models the different stages of the human color vision and defines a vector space for color images. In the literature, the only publication that can be found on LIP with color images is LIPC [22]. However, this model does not have the same objectives as the CoLIP, as it does not involve the Hering opponent-process theory, nor the human cone response.

6.4.1 *From Cone Intensities to Achromatic and Chromatic Tones*

6.4.1.1 Color Tones

The trichromacy theory that results from the works of Maxwell, Young, and Helmholtz has established that three types of color receptors must be present in the human eye. Indeed, there exist three types of cones, sensitive to long, medium, and short wavelengths (red, green, and blue regions of the spectrum, see [7]). In the CoLIP framework, the chromatic tones are defined from the cone intensities L, M, and S as:

$$\forall c \in \{l, m, s\}, C \in \{L, M, S\}, c = M_0 \left(1 - \frac{C}{C_0}\right), \tag{6.10}$$

where C_0 is the maximal transmitted intensity value. M_0 is arbitrarily chosen at normalized value 100. Notice that $C \in]0;C_0]$ and $c \in [0;M_0[$.

6.4.1.2 Color Logarithmic Response

The logarithmic response of the cones, as in the LIP theory, is modeled through the isomorphism φ:

$$\text{for } c \in \{l, m, s\},\ \tilde{c} = \varphi(c) = -M_0 \ln\left(1 - \frac{c}{M_0}\right), \tag{6.11}$$

where $(\tilde{l}, \tilde{m}, \tilde{s})$ are called the logarithmic chromatic tones.

6.4.1.3 Opponent Process

The opponent-colors theory comes from the acknowledgment that certain hues never occur together [19]. More recently, electrophysiological experiments corroborated this theory [44]. Thus, the three logarithmic color channels $(\tilde{l}, \tilde{m}, \tilde{s})$ are represented by a logarithmic achromatic tone $\tilde{a}$, and two logarithmic chromatic tones $\tilde{rg}$ and $\tilde{yb}$, where $\tilde{rg}$ opposes red and green and $\tilde{yb}$ opposes yellow and blue. The conversion is obtained by the following relation:

$$\begin{pmatrix} \tilde{a} \\ \tilde{rg} \\ \tilde{yb} \end{pmatrix} = P \times \begin{pmatrix} \tilde{l} \\ \tilde{m} \\ \tilde{s} \end{pmatrix}, \tag{6.12}$$

with the transfer matrix P given by:

$$P = \begin{pmatrix} \alpha & \beta & \gamma \\ \alpha' & \beta' & \gamma' \\ \alpha'' & \beta'' & \gamma'' \end{pmatrix}. \tag{6.13}$$

Notice that after applying the isomorphism φ (i.e., when considering the logarithmic tones), the classical mathematical operations are used. The relation between the parameters of P reflects the decorrelation of the colors. In the case of achromatic images, $L = M = S$ implies that the relation $\tilde{l} = \tilde{m} = \tilde{s}$ is valid. In this case, the coefficients of the matrix P verify the following relations:

$$\begin{cases} \alpha & +\beta & +\gamma & = 1, \\ \alpha' & +\beta' & +\gamma' & = 0, \\ \alpha'' & +\beta'' & +\gamma'' & = 0. \end{cases} \tag{6.14}$$

An equivalent relation, in the case of (a, rg, yb) tones, holds:

$$\begin{cases} a = & M_0 - & (M_0 - l)^{\alpha} & (M_0 - m)^{\beta} & (M_0 - s)^{\gamma}, \\ rg = & M_0 - & (M_0 - l)^{\alpha'} & (M_0 - m)^{\beta'} & (M_0 - s)^{\gamma'}, \\ yb = & M_0 - & (M_0 - l)^{\alpha''} & (M_0 - m)^{\beta''} & (M_0 - s)^{\gamma''}. \end{cases} \tag{6.15}$$

Using a matrix notation, we get:

$$\begin{pmatrix} a \\ rg \\ yb \end{pmatrix} = P \triangle scale = 1.2 \begin{pmatrix} l \\ m \\ s \end{pmatrix}. \tag{6.16}$$

The antagonist transformation matrix P is defined according to the CIECAM02 specifications in the following way [7]. The achromatic channel is computed considering a ratio 40:20:1 in red, green, and blue sensibility of the eye [49]. The antagonist channels are defined like CIECAM02 opponent-color responses. The numerical values used are proposed in the following definition:

$$P = \begin{pmatrix} 40/61 & 20/61 & 1/61 \\ 1 & -12/11 & 1/11 \\ 1/9 & 1/9 & -2/9 \end{pmatrix}. \tag{6.17}$$

6.4.2 The Trichromatic Antagonist Vectorial Structure

A color-tone function, denoted f, is defined on a compact set $D \subset \mathbb{R}^2$, with values in $]-\infty; M_0[^3$, by:

$$x \in D,\ F(x) = \begin{pmatrix} L(x) \\ M(x) \\ S(x) \end{pmatrix} \mapsto f(x) = \begin{pmatrix} a_f(x) \\ rg_f(x) \\ yb_f(x) \end{pmatrix}. \tag{6.18}$$

Thus, the operations of addition, scalar multiplication, and subtraction can be defined in Eqs. 6.19, 6.20, and 6.22 as follows:

$$f \mathbin{\blacktriangle} g = \begin{pmatrix} a_f \mathbin{\blacktriangle} a_g \\ rg_f \mathbin{\blacktriangle} rg_g \\ yb_f \mathbin{\blacktriangle} yb_g \end{pmatrix}, \tag{6.19}$$

$$\lambda \mathbin{\blacktriangle} f = \begin{pmatrix} \lambda \mathbin{\blacktriangle} a_f \\ \lambda \mathbin{\blacktriangle} rg_f \\ \lambda \mathbin{\blacktriangle} yb_f \end{pmatrix}, \tag{6.20}$$

$$\triangle g = \begin{pmatrix} \triangle a_g \\ \triangle rg_g \\ \triangle yb_g \end{pmatrix}, \tag{6.21}$$

$$f \triangle g = \begin{pmatrix} a_f \triangle a_g \\ rg_f \triangle rg_g \\ yb_f \triangle yb_g \end{pmatrix}. \tag{6.22}$$

The set I of color-tone functions defined on D and valued in $]-\infty; M_0[^3$, with the operations of multiplication $\mathbin{\blacktriangle}$ and internal addition $\mathbin{\blacktriangle}$, is a real vector space. With the logarithmic color-tone functions, $\tilde{f} = \varphi(f)$, the classical operations are used, and thus straightforwardly,

$$\tilde{f} + \tilde{g} = \begin{pmatrix} \tilde{a_f} + \tilde{a_g} \\ \tilde{rg_f} + \tilde{rg_g} \\ \tilde{yb_f} + \tilde{yb_g} \end{pmatrix}, \tag{6.23}$$

$$\lambda \times \tilde{f} = \begin{pmatrix} \lambda \times \tilde{a_f} \\ \lambda \times \tilde{rg_f} \\ \lambda \times \tilde{yb_f} \end{pmatrix}, \tag{6.24}$$

$$-\tilde{g} = \begin{pmatrix} -\tilde{a_g} \\ -\tilde{rg_g} \\ -\tilde{yb_g} \end{pmatrix}, \tag{6.25}$$

$$\tilde{f} - \tilde{g} = \begin{pmatrix} \tilde{a_f} - \tilde{a_g} \\ \tilde{rg_f} - \tilde{rg_g} \\ \tilde{yb_f} - \tilde{yb_g} \end{pmatrix}. \tag{6.26}$$

6.4.3 Bounded Vector Space

The vector space I defines a framework for manipulating color-tone functions with values in $]-\infty;M_0[^3$. The opponent channels rg and yb are thus not symmetric, which can cause problems for computational manipulation and storage, or for representation.

To handle this, it is proposed to introduce the three channels $\hat{f} = (\hat{a}, \hat{rg}, \hat{yb})$ defined by:

$$\hat{a} = a \tag{6.27}$$

$$\hat{rg} = \begin{cases} rg & \text{if } rg \geq 0 \\ -\triangle rg & \text{if } rg < 0 \end{cases} \tag{6.28}$$

$$\hat{yb} = \begin{cases} yb & \text{if } yb \geq 0 \\ -\triangle yb & \text{if } yb < 0. \end{cases} \tag{6.29}$$

This representation is illustrated in the next sections.

6.4.4 Summary

The different conversion operations are presented in the following synoptic diagram:

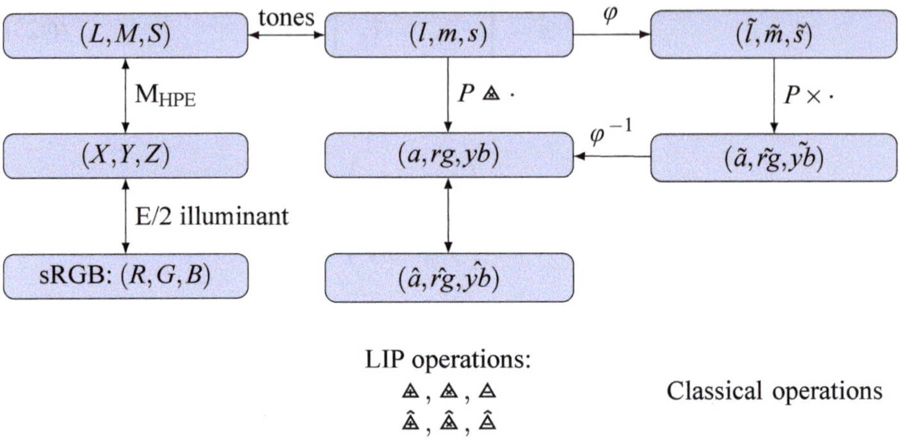

Now that the formal definitions are introduced, the next section illustrates the connections with psychophysical theories.

6.5 Psychophysical Connections

6.5.1 Color Combination

The trichromacy theory was established by the works of Young and Grassmann who proposed several laws for mixing colors [18]. They state that every color is a combination of three primary colors (RGB). Their generalization [30,51] is summarized by the definition of a three-dimensional vector space for colors, as for the CoLIP framework.

The first law states that any color C can be described by a combination of red, green, and blue (with a vector notation in bold capital letters):

$$C = R_C\mathbf{R} + G_C\mathbf{G} + B_C\mathbf{B},$$

where R_C, G_C, and B_C quantify the light power. The equal sign means a matching (the perception of identical colors) and not a strict equality.

The second law states that the mixture of any two colors (C_1 and C_2) corresponds to the addition of their decomposition in the RGB primaries:

$$C_3 = C_1 + C_2 = (R_1 + R_2)\mathbf{R} + (G_1 + G_2)\mathbf{G} + (B_1 + B_2)\mathbf{B}.$$

The third law states that the first two laws remain valid under different illuminance conditions (α being a real-number constant value):

$$\alpha C = \alpha R_C \mathbf{R} + \alpha G_C \mathbf{G} + \alpha B_C \mathbf{B}.$$

6.5.2 Comparison with the $L^*a^*b^*$ Space

The $L^*a^*b^*$ space is intended to code the opponent-process theory from Hering (a^* is equivalent to red-green opposition, b^* is equivalent to yellow-blue opposition). Psychophysically, the main difference lies in the modelization of the nonlinearity of the intensity perception. $L^*a^*b^*$ follows a Stevens-like visual law (i.e., a cubic-root law), whereas CoLIP follows a Weber–Fechner law (i.e., logarithmic). The coefficients used for the $L^*a^*b^*$ conversion are not fully scientifically justified and it involves ad-hoc adapted coefficients, so that a power law can resemble a logarithmic law. Mathematically, contrary to the $L^*a^*b^*$ antagonist color space, the CoLIP antagonist space allows to manipulate and combine color images by means of algebraic operations (addition, subtraction, and convolution), and so to apply color image transformations and measurements that are psychophysically justified.

6.5.3 Color-Matching Functions

The standard observer of the CIE (in 1931) was introduced to average the human response to colors. Three color-matching functions define this standard observer, $\bar{x}$, $\bar{y}$, and $\bar{z}$ (see also Sect. 6.2.1). In 1964, the CIE defined a second standard observer. The data presented here use the 1931 standard observer. If λ is the wavelength of a monochromatic light (λ is restricted to the visible spectrum wavelengths $\mathcal{V}$), and if the observed color has the spectrum distribution $I(\lambda)$, the tristimulus values are :

$$X = \int_{\lambda \in \mathcal{V}} I(\lambda)\bar{x}(\lambda)d\lambda, \tag{6.30}$$

$$Y = \int_{\lambda \in \mathcal{V}} I(\lambda)\bar{y}(\lambda)d\lambda, \text{ and} \tag{6.31}$$

$$Z = \int_{\lambda \in \mathcal{V}} I(\lambda)\bar{z}(\lambda)d\lambda. \tag{6.32}$$

After a conversion from XYZ into LMS space, the color-matching functions are illustrated in Fig. 6.3 in the LMS and $(\hat{a}, \hat{rg}, \hat{yb})$ spaces. The physical meaning of $(\hat{a}, \hat{rg}, \hat{yb})$ is the same as (a, rg, yb). The achromatic channels a and $\hat{a}$ are identical. The chromatic channels $\hat{rg}$ and $\hat{yb}$ represent the opponent colors red-green and yellow-blue. This "hat" antagonist space is set up to get a symmetric mathematical representation for both the negative and positive components of the red-green and the yellow-blue channels, respectively.

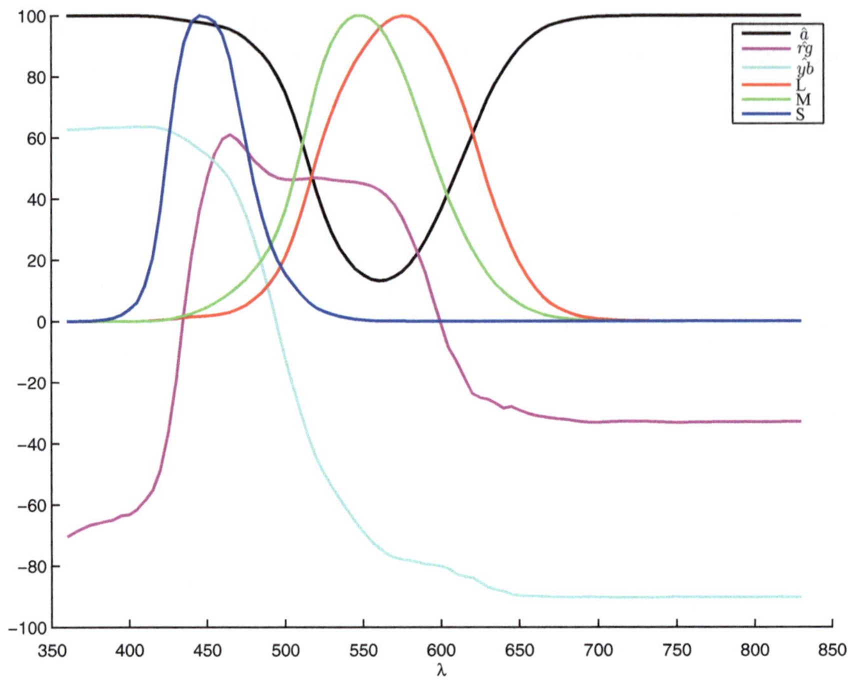

Fig. 6.3 Color-matching functions for the visible wavelengths $\lambda \in \mathcal{V}$ in nanometers, in the LMS and $(\hat{a}, \hat{rg}, \hat{yb})$ spaces

6.5.4 Chromaticity Diagram and the Maxwell Triangle

From the color-matching functions in $(\hat{a}, \hat{rg}, \hat{yb})$, the chromaticity diagram is represented in a projection into $(\hat{rg}, \hat{yb})$ in Fig. 6.4 and in three dimensions in Fig. 6.5.

The purple line is obtained by first converting the values into the xyY space, taking points on a line segment from the extreme values, and then translating these values into the XYZ, LMS, and $(\hat{a}, \hat{rg}, \hat{yb})$ spaces.

Regarding the representation of the RGB cube in $(\hat{a}, \hat{rg}, \hat{yb})$, each triplet of values is taken and converted to $(\hat{a}, \hat{rg}, \hat{yb})$.

6.6 Application to Image Processing and Analysis

6.6.1 Contrast Enhancement

The dynamic enhancement has been introduced previously in Sect. 6.3.3.2. The dynamic enhancement in color images is somehow different, because three color channels are considered.

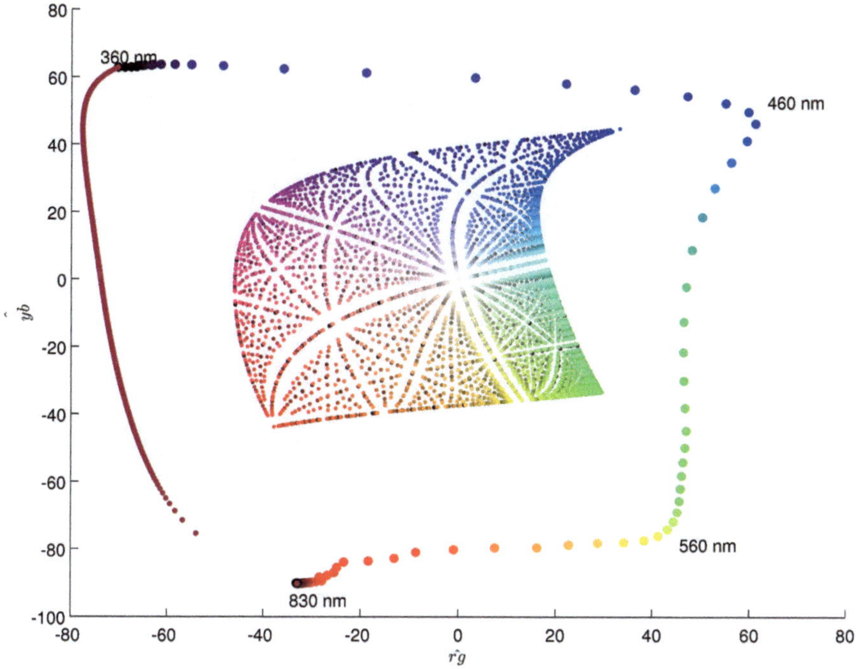

Fig. 6.4 Chromaticity diagram in the plane $(\hat{rg}, \hat{yb})$. The wavelength (in nanometers) appears on the diagram *border*. The RGB *square* is converted into $(\hat{a}, \hat{rg}, \hat{yb})$ and is represented in this diagram. The *purple curve* is also represented, but not as a line segment in this space

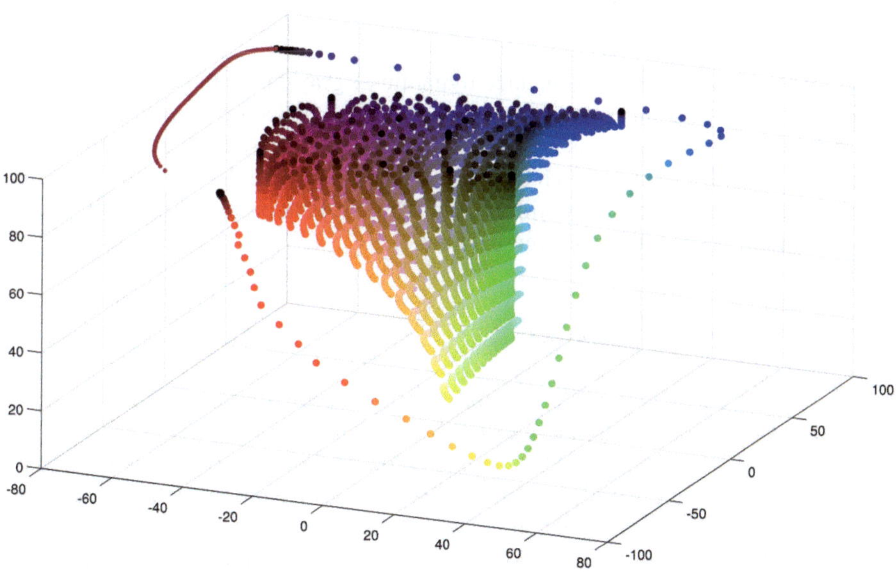

Fig. 6.5 Chromaticity diagram in three dimensions in the $(\hat{a}, \hat{rg}, \hat{yb})$ space, with the RGB *cube* and the *purple curve*

Fig. 6.6 Histogram equalization on the achromatic channel a for the CoLIP space and the luminance channel L for $L^*a^*b^*$ space (**c** and **d**). The small barge in the *center* appears more contrasted in **c** than in **d**. The dynamic enhancement in the $(\tilde{a}, \tilde{rg}, \tilde{yb})$ space is presented in Fig. 6.8b. **a** *Impression, soleil levant* by Claude Monet, 1872. Original image used in the different test. **b** Histogram equalization on the three channels of the RGB image. **c** Histogram equalization on achromatic channel $\tilde{a}$ in the $(\tilde{a}, \tilde{rg}, \tilde{yb})$ space. **d** Histogram equalization on channel L in the $L^*a^*b^*$ space

Some classical methods to enhance dynamic of gray-tone images are histogram stretching and histogram equalization. If applied to RGB color images, these methods introduce false colors. It is common in the literature to see enhancement of one single channel (the luminance channel) in color spaces like $L^*a^*b^*$. If applied on a^* et b^* channels, false colors still appear. Thus, it is appropriate to apply such a transformation only on the achromatic channel.

6.6.1.1 Histogram Equalization

The histogram equalization is a method that transforms the intensities of a given original image so that the cumulative distribution function of intensities of the resulting image is close to be linear [13]. This method applies only on pixels intensities. There also exist some locally adaptive histogram equalization methods [23], but they are not illustrated in this chapter.

The results are shown in Figs. 6.6 and 6.7. The original image is converted into $L^*a^*b^*$ and $(\tilde{a}, \tilde{rg}, \tilde{yb})$ spaces, and the histogram equalization is applied on the luminance and achromatic channels, respectively, before being converted back to the

Fig. 6.7 Histogram equalization on the achromatic channel a for the CoLIP space and on luminance channel L for the $L^*a^*b^*$ space. The small barge in the *center* appears more contrasted in **c** than in **d**. **a** Original image of a car race. **b** Histogram equalization on the three channels of the RGB image. **c** Histogram equalization on achromatic channel a in the (a, rg, yb) space. **d** Histogram equalization on channel L in the $L^*a^*b^*$ space

RGB space. The results show that the resulting image appears to be more saturated in the $L^*a^*b^*$ spacethan in (a, rg, yb) space. The histogram equalization in the three channels of RGB is also shown. Although the image looks nice, false colors are present, and it looks really different from the original images.

As a comment, a general drawback of histogram equalization is that extreme values are saturated, which can be observed on all images of Figs. 6.6 and 6.7.

6.6.1.2 Dynamic Enhancement

As previously discussed, the dynamic enhancement consists in finding an optimal scalar value and multiplying (with the LIP operation) a channel by this scalar value.

First, the RGB images, denoted (f_a, f_{rg}, f_{yb}), are converted into the (a, rg, yb) space. Second, the optimal value λ_0 is found by maximizing the dynamic range of the achromatic channel:

$$R(\lambda_0(a) \triangle f_a) = \max_{\lambda > 0}\{R(\lambda \triangle f_a)\},$$

with $R(a) = f_{a,u} - f_{a,l}$ and $f_{a,u} = \inf_{x \in D} f_a(x)$ and $f_{a,l} = \inf_{x \in D} f_a(x)$, with $0 < f_{a,l} < f_{a,u} < M_0$.

Last, the image $(\lambda_0 \triangle f_a, f_{rg}, f_{yb})$ is converted back to the RGB space. The results are presented in Fig. 6.8 for a painting and a real photograph. For the painting 6.8a, the dynamic was already maximal, the image was thus artificially stretched

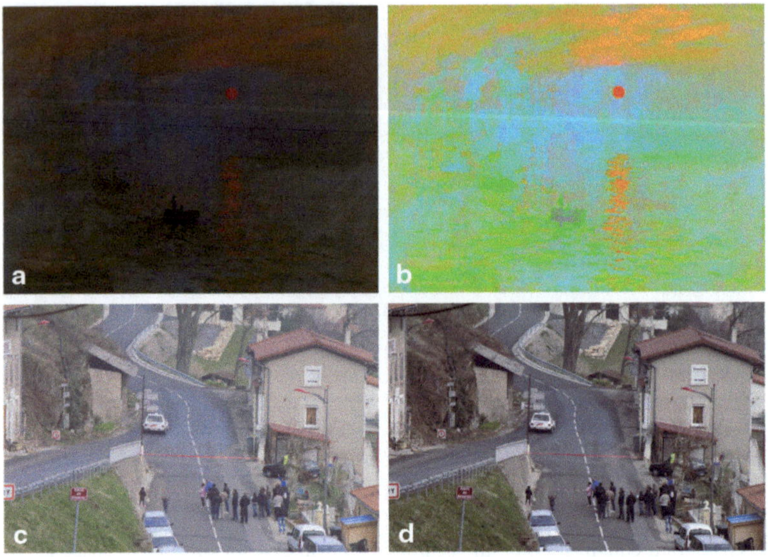

Fig. 6.8 Dynamic enhancement. **a** Image from Fig. 6.6a modified in $L^*a^*b^*$ to test enhancement. L is shrunk to [15;85]. **b** Dynamic enhancement of $\tilde{a}$ of part **a**. **c** Car race image from Fig. 6.7a and dynamic enhancement of $\tilde{a}$. **d** Dynamic stretching in $L^*a^*b^*$ space, for L channel

(by a linear expansion of the values of L into a given interval) the histogram into a narrower band. This was done in the $L^*a^*b^*$ space, and therefore, a dynamic enhancement in $L^*a^*b^*$ by linearly extending the histogram has no sense.

Regarding the real photographs, both operations have been performed in $L^*a^*b^*$ and $(\tilde{a}, \tilde{rg}, \tilde{yb})$ spaces. They are presented in Fig. 6.8c and d.

6.6.2 White Balance Correction

The proposed white balance correction method derives from the von Kries adaptation model [46]. If L, M, and S represent the cone responses and L_a, M_a, and S_a represent the adapted cone responses, this model can be written as (see [7]):

$$L_a = \frac{L}{L_{\text{White}}}, \tag{6.33}$$

$$M_a = \frac{M}{M_{\text{White}}}, \tag{6.34}$$

$$S_a = \frac{S}{S_{\text{White}}}. \tag{6.35}$$

In the CoLIP framework, this is shortly written as:

$$f_a = f \triangle f_{\text{White}},$$

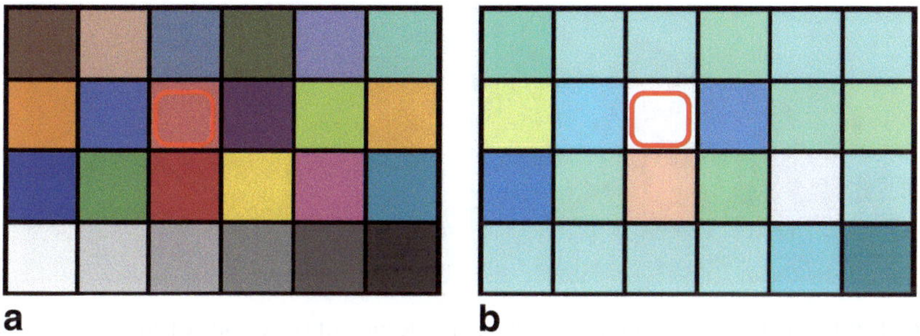

Fig. 6.9 White balance correction with a manual selection of the *white*. **a** Macbeth ColorChecker. The chosen color is delineated with a *red square*. **b** White balance correction for one color. The *red square* designates the white after correction

Fig. 6.10 White balance correction with a manual selection of the white. **a** *Claude Monet in Argenteuil*, by Édouard Manet, 1874. **b** White balance correction for one color manually chosen in the original image **a**

where f_{White} is the color of the illuminant. Practically, the CoLIP addition on rg and yb corresponds to the addition of a color filter (the addition of an illuminant), and the CoLIP subtraction corresponds to the subtraction of a color filter. The addition on the achromatic channel is the gain control.

This is illustrated in Fig. 6.9. The illuminant is chosen as one particular color in Fig. 6.9a and subtracted in Fig. 6.9b. The original chosen color is now represented as white. Notice that the white in the original image is now represented with the opposite color (in the CoLIP space) in the corrected image.

In the second example (Fig. 6.10 below), the color considered as white is manually chosen and subtracted from the original image (Fig. 6.10a). The resulting image is presented in Fig. 6.10b.

The third example presents a photograph taken with the wrong color balance configuration (Fig. 6.11). It presents an area that should be white (some snow). The illuminant color is manually picked up in the snow area and subtracted from the original image (Fig. 6.11a). The resulting image is presented in Fig. 6.11b. The snow is now white, although it, along with other colors, seems a bit saturated.

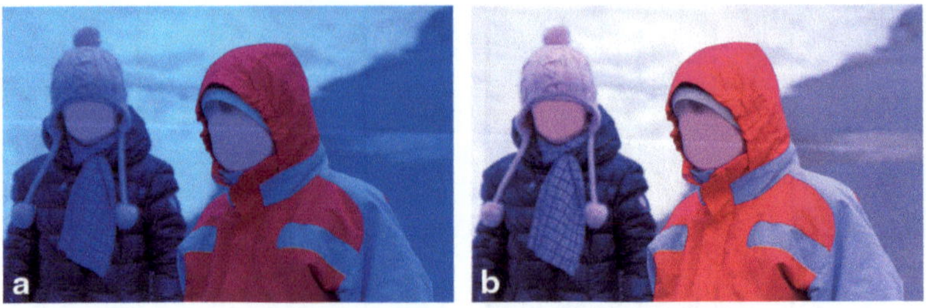

Fig. 6.11 White balance correction with a manual selection of the white. **a** Kids with snow background. The white balance was intentionally poorly defined. **b** White balance correction of **a**, with white point manually chosen in the snow area

6.6.3 Color Transfer

Color transfer is another application of white balance correction. Let us consider two images f_1 and f_2. The following notation is introduced for a given collection of values c: $\mu(c)$ is the mean related to c and $\sigma(c)$ is the standard deviation related to c. For a color image in the CoLIP space,

$$\mu(f) = \begin{pmatrix} \mu(a) \\ \mu(rg) \\ \mu(yb) \end{pmatrix} \tag{6.36}$$

and

$$\sigma(f) = \begin{pmatrix} \sigma(a) \\ \sigma(rg) \\ \sigma(yb) \end{pmatrix}. \tag{6.37}$$

The transfer of colors of image f_1 into image f_2 corresponds to Eq. 6.38, which gives the resulting image f_{new}. This formula centers and normalizes the distribution of colors in the original image f_1, and applies the same color distribution as in image f_2 to the new image f_{new}.

$$f_{\text{new}} = \left(\frac{\sigma(f_1)}{\sigma(f_2)} \mathbin{\triangle\!\!\!\!\times} (f_2 \triangle\!\!\!- \mu(f_2)) \right) \triangle\!\!\!+ \mu(f_1) \tag{6.38}$$

An example is illustrated in Fig. 6.12. f_1 is the image of Fig. 6.6a, f_2 is the image of Fig. 6.12a, and f_{new} is the image of Fig. 6.12b.

6.6.4 K-Means Clustering

The K-means is a clustering technique of a vector space into n clusters [26]. A point of this vector space is associated to a cluster with the nearest mean point. In color

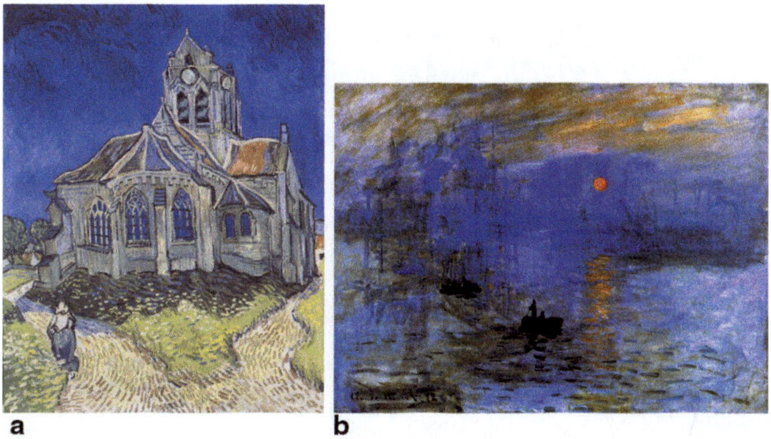

Fig. 6.12 Color transfer: the overall color from **a** is transferred into the original painting from Fig. 6.6a, resulting in **b. a** *The Church in Auvers-sur-Oise, View from the Chevet* by Vincent Van Gogh, 1890. **b** Color transfer

image processing, this is a way to perform an image segmentation based only on color information (and not on spatial information).

An example is shown in Fig. 6.13. The original image "Peppers" (from the Matlab software, see Fig. 6.13a) is employed. The number of clusters is fixed to five for visual convenience. The image is then converted into the CoLIP and $L^*a^*b^*$ color spaces. Finally, the K-means clustering is performed using the Euclidean distance, and each cluster is represented with its mean RGB value.

Figure 6.13 shows that the K-means segmentation in RGB gives poor results. Qualitatively, the results in the $L^*a^*b^*$ space and in the CoLIP space look similar. Some differences can be seen for the orange pepper, in the right background of the image, which gives a slight advantage to the CoLIP space. The whole background appears violet for both spaces, but some violet is present in the middle of the peppers, which is incorrect. It also seems that the segmentation in the CoLIP space presents more "onion-white" labels than in the $L^*a^*b^*$ space, which is in favor of $L^*a^*b^*$.

6.6.5 Mathematical Morphology

Mathematical morphology (MM; [27, 38]) is a theory for image analysis based on set theory and topology. Initially, it was established on binary images but was later extended to gray-level images [42] and color images [1, 2].

MM has been successfully used in several applications. Some examples are remote sensing [3], biomedicine [5], quality control in industry [36], geoscience [45], texture description [14], etc.

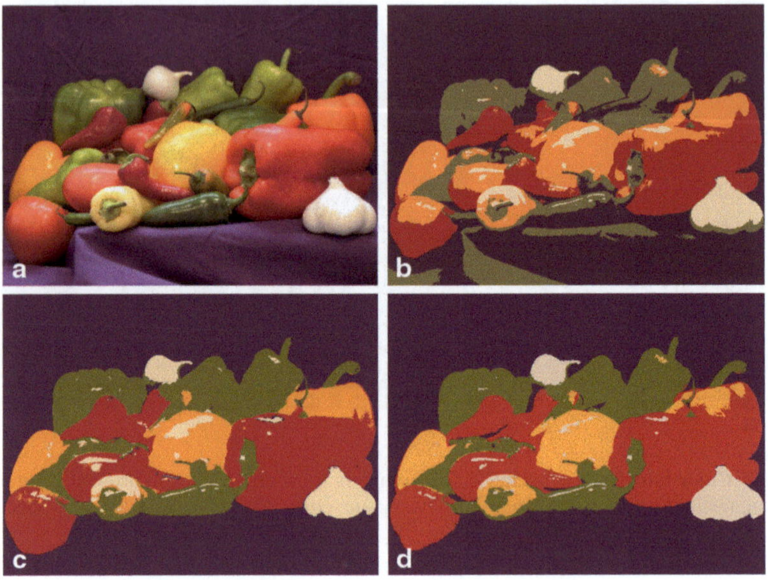

Fig. 6.13 K-means segmentation with five clusters in different color spaces: RGB, $L^*a^*b^*$, and (a,rg,yb). **a** Image peppers. **b** K-means segmentation in RGB using the Euclidean distance. **c** K-means segmentation in (a, rg, yb) using the Euclidean distance. **d** K-means segmentation in $L^*a^*b^*$ using the Euclidean distance

6.6.5.1 Color Morphology

Morphological operators require the definition of a total-order relationship between pixel intensity values. However, in the case of color images this is not straightforward because of the vectorial nature of their points.

In the literature several order relationships have been proposed (i.e., marginal, lexicographical, partial, or reduced ordering) [1, 4]. Among these relationships, the Ω-ordering, denoted $\prec_\Omega$, proposed in [1], presents several advantages.

Let $c_1 = (c_1^A, c_1^B, c_1^C)$ and $c_2 = (c_2^A, c_2^B, c_2^C)$ be two colors which belong to a color space E_{ABC} with its three components A, B, and C.
The Ω-ordering is defined as:

$$c_1 \prec_\Omega c_2 = \begin{cases} d_{ABC}(c_1, c_0) > d_{ABC}(c_2, c_0) \quad \text{or} \\ d_{ABC}(c_1, c_0) = d_{ABC}(c_2, c_0) \quad \text{and} \\ \begin{cases} c_1^A < c_2^A \quad \text{or} \\ c_1^A = c_2^A \quad \text{and} \quad c_1^B < c_2^B \quad \text{or} \\ c_1^A = c_2^A \quad \text{and} \quad c_1^B = c_2^B \quad \text{and} \quad c_1^C < c_2^C \end{cases} \end{cases}, \tag{6.39}$$

where c_0 is the color reference and d_{ABC} is the distance in the color space E_{ABC} such as the L^2 metric.

The resulting color erosion and dilation of an image f at point $x \in D$ by means of the structuring element (SE) $B \subseteq D$ is given respectively by:

$$E_B(f)(x) = \inf_{\prec_\Omega}\{f(w){:}w \in B_x\}, \tag{6.40}$$

$$D_B(f)(x) = \sup_{\prec_\Omega}\{f(w){:}w \in B_x\}, \tag{6.41}$$

where $\sup_{\prec_\Omega}$ and $\inf_{\prec_\Omega}$ stand for the supremum and infimum according to the total Ω-order relationship $\prec_\Omega$, and B_x corresponds to the SE B translated to the point $x \in D$.

Based on these two elementary operators, more advanced morphological transforms could be defined. For example, the opening O_B and closing C_B operators are defined by composition of erosion and dilatation as:

$$O_B(f) = D_{\breve{B}} \circ E_B(f) \tag{6.42}$$

$$C_B(f) = E_{\breve{B}} \circ D_B(f), \tag{6.43}$$

where $\breve{B} = \{-b, b \in B\}$ denotes the reflected SE related to B.

6.6.5.2 CoLIP Morphology

In the case of CoLIP morphology, the original image f is first converted to $\tilde{f}$ within the CoLIP color space $(\tilde{a}, \tilde{rg}, \tilde{yb})$. Then, the morphological operator is applied to $\tilde{f}$ (e.g., $\tilde{g} = E_B(\tilde{f})$). Finally, the resulting image $\tilde{g}$ is converted back to g within the RGB color space.

Figure 6.14 illustrates the operators of CoLIP opening and closing of the original color image "baboon," with a comparison to the RGB and $L^*a^*b^*$ color spaces.

6.6.5.3 CoLIP Image Enhancement

The following example (Fig. 6.15) is focused on contrast enhancement. The considered image enhancement technique is an edge sharpening process: the approach is similar to unsharp masking type enhancement [37] where a high pass portion is added to the original image. The CoLIP contrast enhancement process is a toggle contrast filter [41], whose operator κ_r is defined as follows:

$$\kappa_r(f)(x) = \begin{cases} D_{B_r}(f)(x) \text{ if } d_{\mathrm{CoLIP}}(D_{B_r}(f)(x), f(x)) < d_{\mathrm{CoLIP}}(f(x), E_{B_r}(f)(x)) \\ E_{B_r}(f)(x) \text{ otherwise,} \end{cases} \tag{6.44}$$

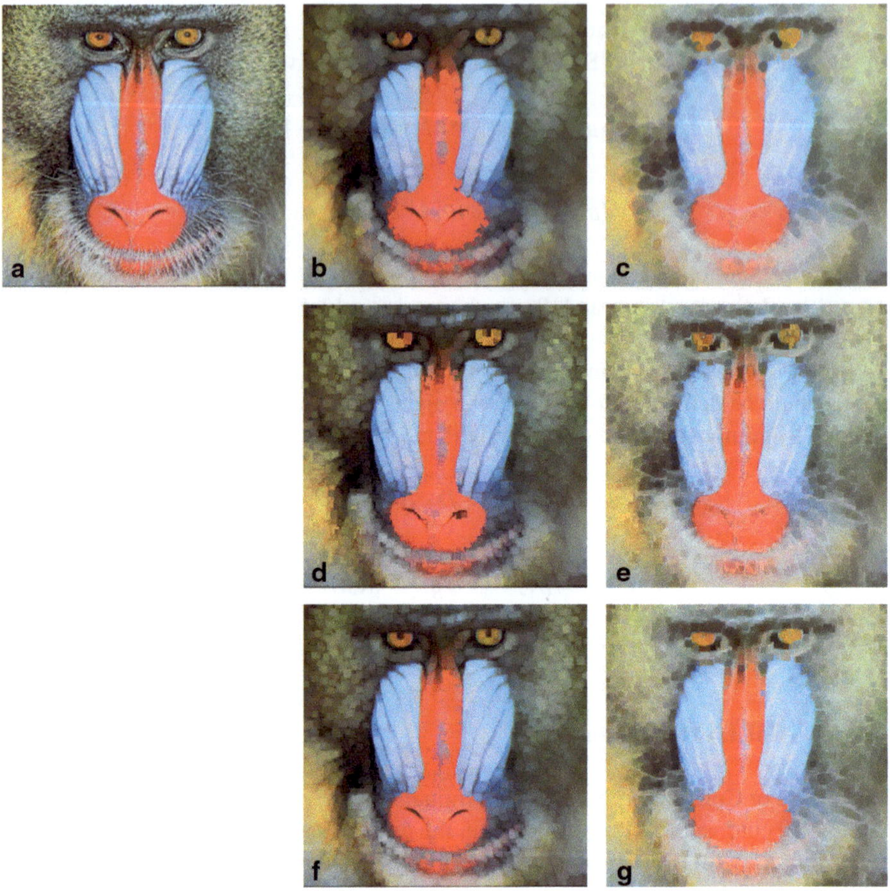

Fig. 6.14 Illustration of CoLIP (**f**, **g**), $L^*a^*b^*$ (**d**, **e**), and RGB (**b**, **c**) morphological operators on the original image "baboon" (**a**). **a** Original image. **b** RGB opening. **c** RGB closing. **d** $L^*a^*b^*$ opening. **e** $L^*a^*b^*$ closing. **f** CoLIP opening. **g** CoLIP closing

where D_{B_r} and E_{B_r} denote the CoLIP dilation and erosion, respectively, using a disk B_r of radius r as structuring element. d_{CoLIP} is the L^2 distance within the CoLIP color space (a, rg, yb).

Figure 6.15 illustrates the enhancement of an image from the painter, artist Gamze Aktan within the RGB, $L^*a^*b^*$, and CoLIP spaces.

Fig. 6.15 Enhancement of an image from the painter, artist Gamze Aktan (**a**) within the RGB (**b**), $L^*a^*b^*$ (**c**), and CoLIP (**d**) spaces. **a** Original image. **b** RGB image enhancement. **c** $L^*a^*b^*$ image enhancement. **d** CoLIP image enhancement

6.7 Concluding Discussion and Perspectives

In this chapter, the CoLIP framework has been presented and illustrated in several applications. It is consistent with several human visual perception laws and characteristics, and is mathematically and computationally relevant.

It has been illustrated with some practical applications of image processing and enhancement techniques (white balance correction, color transfer, K-means clustering, and filtering). The qualitative comparison with the $L^*a^*b^*$ space shows that the (a, rg, yb) space has equivalent properties.

The perspective of this work is to define a color appearance model by providing hue and saturation properties. The distance between colors, based on this model, should be defined. It could be beneficial to the color mathematical morphology applications, and also for the study of the JNDs and the MacAdam ellipses. In short, the objectives are to be defined within the CoLIP framework, an equivalent to the HSV color space.

References

1. Angulo, J.: Morphological colour operators in totally ordered lattices based on distances: Application to image filtering, enhancement and analysis. Comput. Vision Image Underst. **107**(1–2), 56–73 (2007)
2. Aptoula, E., Lefèvre, S.: A comparative study on multivariate mathematical morphology. Pattern Recognit. **40**(11), 2914–2929 (2007)
3. Aytekin, O., Ulusoy, I.: Automatic segmentation of VHR images using type information of local structures acquired by mathematical morphology. Pattern Recognit. Lett. **32**(13), 1618–1625 (2011)
4. Barnett, V.: The ordering of multivariate data. J. R. Stat. Soc. Ser. A. **139**(3), 318–354 (1976)
5. Bouraoui, B., Ronse, C., Baruthio, J., Passat, N., Germain, P.: 3D segmentation of coronary arteries based on advanced mathematical morphology techniques. Comput. Med. Imag. Graph. **34**(5), 377–387 (2010)
6. Deng, G.: A generalized logarithmic image processing model based on the gigavision sensor model. IEEE Trans. Image Process. **21**(3), 1406–1414 (2012)
7. Fairchild, M.D.: Color Appearance Models. Wiley, The Atrium (2013)
8. Fechner, G.T.: Elemente der Psychophysik. Breitkopf und Härtel, Leipzig (1860)
9. Fechner, G.: Elements of Psychophysics, vol. 1. Thoemmes Press, New York (1966). Traduction by E. G. Boring and by H. E. Adler
10. Fernandes, M., Gavet, Y., Pinoli, J.C.: Improving focus measurements using logarithmic image processing. J. Microsc. **242**(3), 228–241 (2011). doi: 10.1111/j.1365-2818.2010.03461.x
11. Florea, C., Florea, L.: Parametric logarithmic type image processing for contrast based auto-focus in extreme lighting conditions. Int. J. Appl. Math. Comput. Sci. **23**(3), 637–648 (2013)
12. Florea, C., Vertan, C., Florea, L.: Logarithmic model-based dynamic range enhancement of hip x-ray images. In: Blanc-Talon, J., Philips, W., Popescu, D., Scheunders, P. (eds.) Advanced Concepts for Intelligent Vision Systems, pp. 587–596. Springer, Berlin (2007)
13. Gonzalez, R.C., Woods, R.E.: Digital Image Processing, 2nd ed. Prentice Hall (2002)
14. González-Castro, V., Debayle, J., Pinoli, J.C.: Color adaptive neighborhood mathematical morphology and its application to pixel-level classification. Pattern Recognit. Lett. **47**:50–62 (2014)
15. Gordon, I.: Theories of Visual Perception, 3rd ed. Psychology Press (2004).
16. Gouinaud, H.: Traitement logarithmique d'images couleur. Ph.D. thesis, École Nationale Supérieure des Mines de Saint-Etienne (2013)
17. Gouinaud, H., Gavet, Y., Debayle, J., Pinoli, J.C.: Color correction in the framework of color logarithmic image processing. In: Proceedings of the 7th IEEE International Symposium on Image and Signal Processing and Analysis (ISISPA), pp. 129–133. Dubrovnik, Croatia (2011)
18. Grassmann, H.: Zur theorie der farbenmischung. Annalen der Physik. **165**(5), 69–84 (1853)
19. Hering, E.: Outlines of a Theory of the Light Sense. Harvard University Press, Cambridge (1964). (Trans. L. M. Hurvich and D. Jameson)
20. Jourlin, M., Pinoli, J.C.: Logarithmic image processing. Acta Stereologica. **6**, 651–656 (1987)
21. Jourlin, M., Pinoli, J.C.: Image dynamic range enhancement and stabilization in the context of the logarithmic image processing model. Signal Process. **41**(2), 225–237 (1995). doi:10.1016/0165-1684(94)00102-6. URL http://dx.doi.org/10.1016/0165-1684(94)00102-6
22. Jourlin, M., Breugnot, J., Itthirad, F., Bouabdellah, M., Closs, B., et al.: Logarithmic image processing for color images. Adv. Imag. Electron. Phys. **168**(2):65–107 (2011)

23. Kaur, M., Kaur, J., Kaur, J.: Survey of contrast enhancement techniques based on histogram equalization. (IJACSA) Int. J. Adv. Comput. Sci. Appl. **2**(7) (2011)
24. Kreyszig, E.: Introductory Functional Analysis with Applications, vol. 81. Wiley New York (1989)
25. Krueger, L.E.: Reconciling fechner and stevens: Toward a unified psychophysical law. Behav. Brain Sci. **12**, 251–267 (1989). doi: 10.1017/S0140525X0004855X
26. MacQueen, J., et al.: Some methods for classification and analysis of multivariate observations. In: Proceedings of the fifth Berkeley symposium on mathematical statistics and probability, vol. 1, pp. 281–297. California, USA (1967)
27. Matheron, G.: Random Sets and Integral Geometry. Wiley (1975)
28. Maxwell, J.C.: Theory of the perception of colors. Trans. R. Scottish Soc. Arts. **4**, 394–400 (1856)
29. Navarro, L., Deng, G., Courbebaisse, G.: The symmetric logarithmic image processing model. Digit. Signal Process. **23**(5), 1337–1343 (2013)
30. Oleari, C., Pavesi, M.: Grassmann's laws and individual color-matching functions for non-spectral primaries evaluated by maximum saturation technique in foveal vision. Color Res. Appl. **33**(4), 271–281 (2008). doi: 10.1002/col.20421. URL http://dx.doi.org/10.1002/col.20421
31. Oppenheim, A.S., Stockham, R., et al.: Nonlinear filtering of multiplied and convolved signals (1968)
32. Panetta, K.A., Wharton, E.J., Agaian, S.S.: Human visual system-based image enhancement and logarithmic contrast measure. IEEE Transac. Syst. Man Cybern. Part B **38**(1), 174–188 (2008)
33. Pinoli, J.C.: Contribution à la modélisation, au traitement et à l'analyse d'image. Ph.D. thesis, Département de Mathématiques, Université de Saint-Etienne, France (1987)
34. Pinoli, J.C.: The logarithmic image processing model: Connections with human brightness perception and contrast estimators. J. Math. Image. Vis. **7**(4), 341–358 (1997)
35. Pinoli, J.C.: Mathematical Foundations of Image Processing and Analysis, vol. 2. Wiley (2014)
36. Priya, S., Kumar, T., Paul, V.: A novel approach to fabric defect detection using digital image processing. In: Signal Processing, Communication, Computing and Networking Technologies (ICSCCN), 2011 International Conference on, pp. 228–232 (2011)
37. Ramponi, G., Strobel, N., Mitra, S.K., Yu, T.H.: Nonlinear unsharp masking methods for image-contrast enhancement. J. Electron. Imag. **5**(3), 353–366 (1996)
38. Serra, J.: Image Analysis and Mathematical Morphology. Academic Press (1982)
39. Sharma, G., Bala, R.: Digital Color Imaging Handbook. CRC press, Boca Raton (2002)
40. Shvayster, H., Peleg, S.: Inversion of picture operators. Pattern Recognit. Lett. **5**(1), 49–61 (1987)
41. Soille, P.: Morphological Image Analysis Principles and Applications. Springer, New York (2003)
42. Sternberg, S.R.: Grayscale morphology. Comput. Vis. Graph. Image Process. **35**(3), 333–355 (1986)
43. Stockman, A., Mollon, J.: The spectral sensitivities of the middle-and long-wavelength cones: an extension of the two-colour threshold technique of ws stiles. Perception. **15**, 729–754 (1986)
44. Svaetichin, G.: Spectral response curves from single cones. Acta. Physiol. Scand. Suppl. **39**(134), 17–46 (1956)
45. Tuia, D., Pacifici, F., Kanevski, M., Emery, W.: Classification of very high spatial resolution imagery using mathematical morphology and support vector machines. IEEE Trans. Geosci. Remote Sens. **47**(11), 3866–3879 (2009)
46. von Kries, J.: Die gesichtsempfindungen. Handbuch der physiologie des menschen. **3**:109–282 (1905)
47. Vorobel, R.A.: Logarithmic image processing. part 1: Basic model. Inf. Extr. Process. **107**(31), 26–35 (2009)

48. Vorobel, R.A.: Logarithmic image processing. part 2: generalized model. Inf. Extr. Process. **107**(31), 36–46 (2009)
49. Vos, J., Walraven, P.: On the derivation of the foveal receptor primaries. Vis. Res. **11**(8), 799–818 (1971). doi: 10.1016/0042-6989(71)90003-4
50. Weber, E.: Der Tastsinn und das Gemeingefühl. Handwörterbuch der Physiologie. **3**(2):481–588 (1846)
51. Wyszecki, G., Stiles, W.S.: Color science, vol. 8. Wiley, New York (1982)
52. Young, T.: The bakerian lecture: On the theory of light and colours. Philosophical transactions of the Royal Society of London. pp. 12–48 (1802)

Chapter 7
Color Management and Virtual Restoration of Artworks

Giuseppe Maino and Mariapaola Monti

7.1 Introduction

Restoration is already a digital or virtual common practice, which is often carried out together with the "physical" work of the restorer. Therefore, a better understanding of the possibilities and limitations of this procedure is necessary as well as the establishment of rules to be followed in order to plan the restoration work. In particular, the color analysis plays a fundamental role in the documentation, preservation, and restoration of archival, archaeological, and artistic objects. Color contrast and enhancement techniques are commonly used to improve the quality of images and to guarantee the best reproduction of the "true" color in the digital format of the considered artifact. Consequently, it is possible to identify defects, retouches, or repaintings, hardly visible by human eyes (even those of the best-trained restorer or art historian), and to make comparisons with previous images of the same object in order to evaluate its present state of preservation. Moreover, images resulting by diagnostic investigations such as multispectral analyses, for instance,—in addition to images in the visible light—may also contribute to a better understanding of the conservation status of the studied artifact, and to prepare the (possibly virtual) restoration. Therefore, virtual restoration is not in competition with the "physical" restoration, but it represents an interesting procedure to support and prepare the physical work. Therefore, the image obtained by suitable numerical color processing can be useful to the final user of the computing work, such as archaeologists, art historians, conservators, and restorers in order to:

- Ensure better use of cultural property
- Perform teaching and training activities
- Use it as a guide to the actual restoration

These are only a few of the possibilities offered by color image processing and virtual restoration, depending on the constituent materials of works of art and on

G. Maino (✉) · M. Monti
University of Bologna, Bologna, Italy
e-mail: giuseppe.maino@enea.it

M. E. Celebi et al. (eds.), *Color Image and Video Enhancement*,
DOI 10.1007/978-3-319-09363-5_7

the reasons that lead to their use. In some fields, like photography or cinema, virtual restoration is a real effective restoration technique, which is recognized as a valid alternative to the traditional one. It preserves the portrayed image or movie and its historical significance, regardless of the material support.

In this chapter, we consider two main applications of image processing, namely, color enhancement and contrast enhancement.

Color enhancement is obviously essential in digital restoration of cultural heritage artifacts and is used for recovering the original color of paintings and sculptures, as well as photographs and movies of historical and aesthetic values, or archaeological objects. Starting from a physical and chemical knowledge of the pigments used by the artists and their degradation processes, it is possible to speculate and suggest suitable hypotheses about original colors and verify their impact on the considered artistic object by means of suitable computing simulations.

Contrast enhancement is also introduced and applied to improve the legibility of ancient archival documents and manuscripts, for instance, illuminated codes or palimpsests where the original but erased text can be recovered by means of infrared (IR) reflectography and contrast improvement. Moreover, contrast enhancement allows one to identify the contours of tesserae in mosaics and the working days in frescoes, and thus to provide important information about the preservation condition and the execution techniques.

However, one must consider that generally the aim of an artistic object is its vision and its aesthetic value. In these operations, subjectivity plays an important role, and the final result of image processing has to be evaluated within this framework. Therefore, in the following section, the role of subjective perception of colors is investigated with reference to cleaning and restoration of paintings. Theories of colors developed in the nineteenth and twentieth centuries starting from Goethe's work on visual perception are briefly addressed and summarized, then applied to different approaches in painting restoration. Then, Sect. 7.3 describes the Retinex theory and the most used approaches to color enhancement. Section 7.4 introduces the concept of artwork digital restoration; applications are presented in the following sections (manuscripts and books in Sect. 7.5; old masters paintings in Sect. 7.6 –7.8; mosaics in Sect. 7.9, and sculptures in Sect. 7.10). Finally, Sect. 7.11 is devoted to concluding remarks.

7.2 Color Perception

The human perception of colors is a complex phenomenon involving, at least, three different aspects: the optical properties of the object to be seen, the lighting conditions and sources, and, finally, the observer. Generally, scientific instruments are available to derive objective information (direct or indirect measurements) on the former two characteristics. Spectrophotometers, etc., are widely used to this aim, and a whole field of research is based upon them, namely, colorimetry, which has also important applications in industrial processes and products.

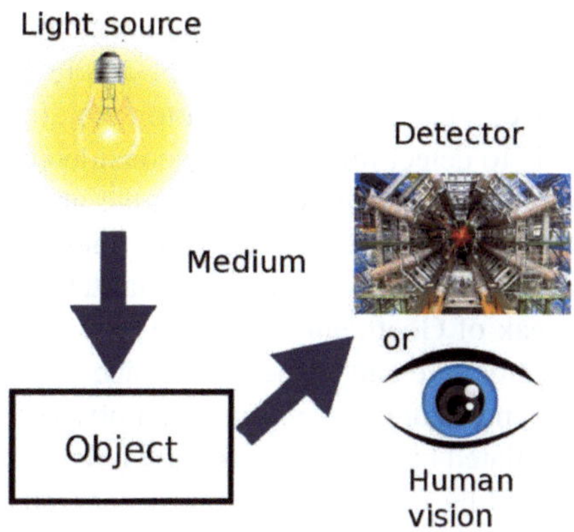

Fig. 7.1 Scheme of visual perception

On the other hand, relatively less attention is usually paid to the observer, whose importance is difficult to underestimate in the whole process of color perception. If it is represented by an experimental equipment such as a spectrophotometer, independent, objective, repeatable information is acquired. A real person has a completely different impact: two identical couples of eyes do not exist, and each person has his/her own bias in identifying and perceiving colors. Extreme examples of people affected by Daltonism or other defects or illnesses are obvious, but in general cases physiological or psychological conditions influence the mechanism of color perception as well as contour conditions or bias due to cultural traditions, personal experiences, and so on.

The visual culture is essential in the evaluation of restoration work since it concerns aesthetic values and strongly depends—as outlined by Cesare Brandi in his fundamental book on the theory of restoration [1] and by other authors—on the visual approach even if one attempts to develop and apply recipes as objective as possible (introduction of rigatino, etc., as suitable means to differentiate restoration intervention with respect to the original painting).

Therefore, in this section, we intend to review the different approaches to visual perception developed from the eighteenth century, starting from the conflict between the Newtonian approach and Goethe's theory of color up to recent models of visual perception developed from Gestalt psychology as well as other approaches that also resulted in suitable algorithms and computing codes for color management, such as the Retinex technique.

In an interview with Roger Lewin [2], Brian Goodwin of Open University, UK, a leading scientist in the field of complexity, says that "ours (science of physical complex systems, ed) is a science of qualities, not quantities, and is therefore a Goethean science."

The process of visual perception can be simply described by Fig. 7.1, where the electromagnetic (e.m.) radiation emitted by a suitable light source passes through a medium (generally, air) and hits on the surface of the considered object. Then, it is partly absorbed, partly reflected, and—at some distance—grabbed by a detector system. Alternatively, an (human) observer captures the reflected radiation. A computer or the brain performs the whole process of storing and elaborating images. The detector can be a spectrophotometer, for instance, or a multispectral imaging system, able to detect images of the considered object in different wavelength ranges [3]. In this way, a perfectly quantitative analysis of the grabbed image can be performed and quantitative data are kept, relevant to image properties. On the other hand, the observer perceives a quanti-qualitative vision of the object, and—in this sense—we speak of Goethean science in opposition or, better, as a complement to Newtonian quantitative and objective science [4, 5].

The visual perception by an observer is therefore an objective-subjective process mediated by the previous knowledge—and prejudices—of the observer her/himself, including cultural and, more generally, social influences. According to Goethe's approach, colors exist only as far as an independent observer perceives them, and where light and color phenomena develop is not the space but the human eye. Colors exist in a perceptive field where they are located and assume variable, always different, characters, constant but never equal to themselves [4]. The role and influence played by society, history as well as culture on the perception of colors are investigated in many books; we quote here the classic text by Portal on color symbolism [6] and the excellent account by Pastoureau of a specific color, blue [7], from the Classic Era to the Middle Ages and modern times when it assumes a dominant role in paintings (by Yves Klein, for instance), photographs, magazines, dresses, and so on. Pastoureau wrote other books dedicated to particular colors (black, green,. . .) or color patterns subsequently to the successful work on blue color.

The main steps in the development of a comprehensive theory of color and visual perception can be summarized in the following dates:

1810 Wolfgang Goethe
1854 Arthur Schopenhauer
1802–1855 Thomas Young–Hermann Helmholtz
1864 Michel-Eugène Chévreul
1910 Gestalt psychology
1930 Neuropsychology
1950 New Look
1952–1963 Joseph Albers
1954 Rudolf Arnheim
1961 Johannes Itten
1970 Edwin Land

In his work, Schopenhauer refers to Goethe's theory of color and puts it in its most radical form pointing out that colors exist only as far as an observer sees them and they are characterized by properties such as intensity, quality, and extension.

To Young (1802) and Helmholtz (1855), we are indebted for the first formulation of the three-chromatic theory according to which every color can be obtained from a suitable mixing of three basic (primary) colors. Recent experiments have shown that the retinal photoreceptors are peaked around 447, 540, and 577 μm, approximately corresponding to blue, green, and red colors, respectively.

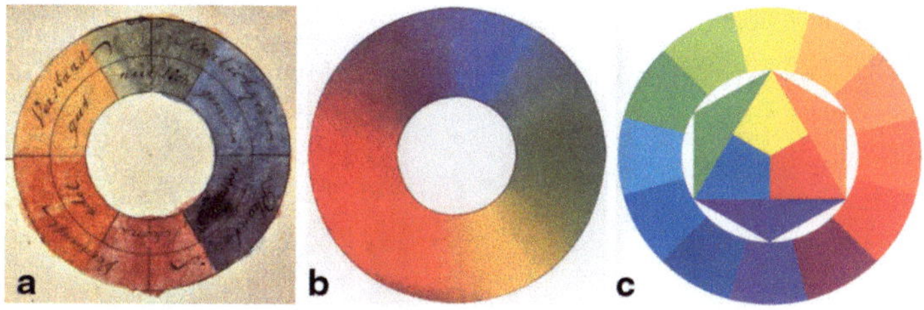

Fig. 7.2 Wheels of Goethe (*left*), Chévreul (*middle*), and Itten (*right*)

Fig. 7.3 Examples of contrasting figures and colors in visual perception according to Chevreul's experiments

By exploiting complementary colors, the French chemist and physicist Chévreul defined the laws governing contrast on scientific yet quantitative grounds, to be used in manufacturing. He discovered that contrasting colors are both enhanced when seen from a distance less than that necessary to mix them on the retina (see Fig. 7.2 in the middle and Fig. 7.3).

At the beginning of the twentieth century, the psychology of Gestalt goes behind the classical physiological theory of Young–Helmoltz and its reductionism. The psychology of the form (Gestalt) opposes atomistic systems and proves that perception is more than the simple addition of elementary components, thus extending the results of Chévreul.

The following school of neuropsychology represents a further improvement with respect to the previous theories that confined themselves to study the relationship between an isolated stimulus and the corresponding response obtained by means of introspection or of suitable indirect measurements. The neuropsychological approach introduces, between stimulus and its response, the organism in its whole complexity of anatomy and functionality [8].

The New Look approach considers the subject (the observer of the scheme in Fig. 7.1) as a "person" and not only as an "organism." As a consequence, the visual perception depends not only on the objective conditions (lighting, optical properties of the medium, and the considered object, etc.) but even on the culture of the observer and her/his previous experience [9, 10].

Fig. 7.4 Contrast of colors according to Albers

As a word assumes a precise meaning only in connection with other words making the phrase, analogously single colors take their expression and meaning only when related to other neighboring colors and including the previous experience of the observer in evaluating them. Moreover, Itten [11] says that the reality of colors takes content and meaning through the retina and brain. The retina and the brain produce an "exact" perception by means of operations of comparison and opposition. To sum up, colors are not defined independently of the observer but, for their existence, the observer plays an active and determinant role (see Fig. 7.2, right). A similar situation occurs in quantum mechanics, where the observer (or the detector) has an essential and an unavoidable role, from which, for instance, the principle of uncertainty derives.

The essay *Interaction of color* [12] collects a series of lessons (1952–1963) given by Albers, where colors have been studied by analyzing mutual relations, influences, and ratios, namely, interactions arising from the visual perception of each one. Albers introduces the concept of perceptive field originating from Gestalt: We can never see an isolated color, not in relations with other colors. Colors represent a continuous flux in relationship with neighboring variable colors and in variable conditions (see Fig. 7.4).

According to Land [5,7], the sense of color is not absolute, arising from a simultaneous comparison of blue, red, and green intensities among all the present objects in the visual field. Land's theory has been named "Retinex" (from "retina" and "cortex"), since Land argued for the existence of a brain region located between the retina and the cortex appointed to compare quantities of radiation of a given color (red, green, or blue) reflected by a surface with quantities reflected by the neighboring surfaces.

Although Land's assumptions have been largely overcome by the successive researches [8]; however, they retain a non-negligible interest with regard to the algorithmic applications of image processing. Furthermore, many studies have been performed to investigate and elucidate the mechanisms of human vision, the relevant brain functions, and the areas concerned with visual perception. New advanced techniques of noninvasive medical imaging have allowed us a more deepened and effective knowledge of how the vision works and, therefore, simulation models have been developed suggesting useful applications to computer science. It is worth mentioning the pioneering work of David Marr [13], while more recent results have been obtained by Semir Zeki et al. [14] as well as other authors. Some references to this research activity can be found in [15].

7.3 The Retinex Approach and Other Techniques of Color Enhancement

The Retinex model, as proposed by Land, is based on two important theories of color vision, namely, the Young–Helmoltz or trichromatic theory and that of the opponent process, founded by Ewald Hering. The latter explains effects such as negative afterimages on the basis of the observation that color is perceived on a scale of opposites. Essentially, Hering states that if a color is seen for a long enough time, the relevant cone receptors tire out and therefore stop the transmission of information to the brain. This lack of information in the process corresponds to the perception of the opposite color.

The Retinex approach is only an approximation to the real human vision of colors, according to the most recent studies. However, this model is commonly used in computer science applications and found some interesting applications also to the reproduction of colors of artistic objects. In particular, significant examples have been carried out in the fields of virtual restoration and augmented reality. Significant results have been obtained in recent years by Alessandro Rizzi and coworkers at the Milan University [16]. They can be found, for instance, in the proceedings of the annual conference of the Gruppo del Colore—Associazione Italiana Colore, which is a scientific-cultural nonprofit entity that aims to promote science and culture regarding color, light, and the related technical-scientific, cultural, and professional fields. The association wants to spread scientific, technological, cultural, and applicative knowledge of the cited disciplines, and support matters of common interest between members in national and international contexts. Since 2014, the association has also been publishing a half-yearly journal, freely available for everyone, in which the most important papers presented in the past editions of the Conferenza del Colore are reported (ISSN 2384-9568). Many of them concern the cultural heritage.

In the last few years, many other approaches and algorithms have been introduced for color image enhancement and applied to the investigation of culture heritage problems. In the following sections, some applications to documents, books, paintings, and mosaics will be described. It is worth mentioning the techniques of histogram equalization, largely used in the literature: an approach aimed to gain maximum brightness preservation, thus avoiding visual deteriorations of colored images such as saturation effects, was presented in Ref. [17], the Minimum Mean Brightness Error Dynamic Histogram Equalization (MMBEHDE).

Another study [18] concerns the color enhancement by scaling the discrete cosine transform (DCT) coefficients, thus allowing the treatment of chromatic components in addition to the processing of the luminance one. According to the comparative investigation performed by the authors, it greatly improves the quality of the image, maintaining a satisfactory computational efficiency.

A critical review of different theoretical approaches to color contrast and enhancement techniques is presented in two recent papers [19, 20] concerning both spatial domain methods and frequency domain methods. Traditional and more recent algorithms are discussed as for their accuracy and performance: gamma corrections, contrast stretching, histogram equalization, contrast-limited adaptive histogram equalization (CLAHE) for the former category, and Retinex, homomorphic and wavelet multiscale procedures for the latter. A conclusion [19] is reached about the good performance of the wavelet transform technique for image denoising, and the use of a combination of this algorithm and the Retinex one is suggested to attain the best results for the color rendition problem, the maximum peak signal to noise ratio (PSNR) and the minimum root mean square error in color enhancement.

In Tewari and Singh [20], color contrast enhancement methods are discussed by taking into account nonuniform illumination images, comparing algorithms in both spatial and compressed domains and stating that the CLAHE method provides better performances in this case, while for true color images the DCT approach significantly improves the color contrast. However, all these techniques are yet to be evaluated as regards possible applications to cultural heritage investigation. In the following, we present some results preliminary to further studies in a field of growing interest also from an economic point of view.

7.4 Color and Painting Restoration

Actually, cleaning and restoration of paintings are often devoted to enhance the chromatic variations of the ancient painting itself, in close analogy with the modern art, where acrylic colors are used (see Fig. 7.5). As a matter of fact, the visual perception of the surrounding objects is intrinsically linked to the prescriptive images assimilated by our cognitive system during previous experiences, both at individual and collective level, and—in this context—one can say that a real perceptive and mental habitus exists. Thus, it is worth remarking that the theories of color perception summarized in the previous sections elucidate the paradigm within which the restorer works and the restoration is performed and then perceived by the public.

Fig. 7.5 A. Kapoor, *As If to Celebrate, I Discovered a Mountain Blooming with Red Flowers*, 1981

Fig. 7.6 *Pietas* by Giovanni Bellini, before (*left*) and after (*right*) cleaning and restoration

Generally, the specific technical and scientific reasons offered to explain and motivate a restoration, at a deepest insight, result in the product of choice arising from the taste of a given time and culture. Of course, the diagnostics images grabbed by a suitable detector provide objective information and are used to identify pigments, put in evidence an underlying drawing (IR reflectography), outline past restorations (ultraviolet (UV) fluorescence), etc.

On the other hand, visual customs are determined by several media and different sources, such as magazines, television, cinema, advertising, photographs, etc. Therefore, we are induced to prefer or favor aspects of the painting that these tools accustomed us. We see paintings without noticing their thickness due to pigments and, therefore, we forget transparencies and opacities of the original painting (see for instance Fig. 7.6).

In the history of painting restoration, one can state that, therefore, cleaning and restoration of a painting are very difficult operations to be performed only in the case of possible damages to the original masterpiece. Moreover, these problems can be partially solved by means of a multidisciplinary approach, where art historians, conservators, and restorers are supported by scientists. Particularly computer science—namely, image processing—can help through well-grounded simulations of the supposed original characteristics of the painting, the so-called virtual or digital restoration.

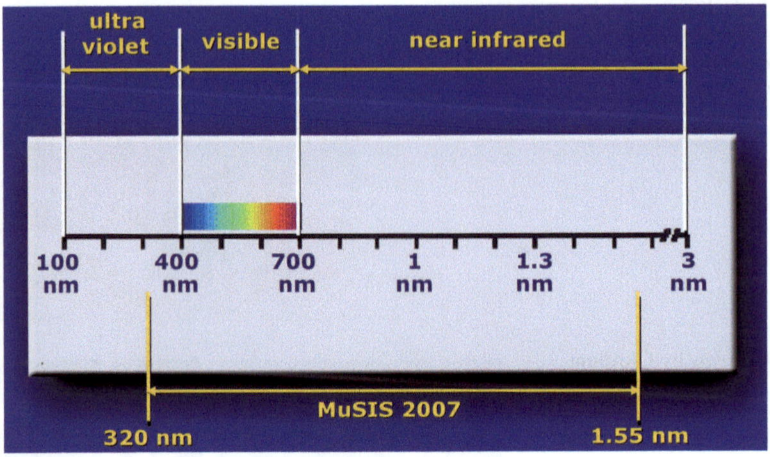

Fig. 7.7 Electromagnetic range of multispectral analysis

7.5 Multispectral Images

The application of multispectral nondestructive techniques—initially developed in the field of materials science—to the study of archaeological and artistic objects as well as books and documents conserved in libraries and historical archives provides the scholar with reliable qualitative and often quantitative information on the considered objects. In this way, it is possible to understand, for instance, the technical and intellectual know-how of the artists or craftsmen of the period under study, and—as for the conservators—to obtain preliminary, often essential information for conservation and restoration of damaged or degraded objects and in many cases to assess the authenticity of artifacts and documents.

On the other side, availability of a portable apparatus represents a strong constraint for many analyses to be performed in situ, since paintings, books, or documents of large dimensions or particularly fragile cannot be easily and/or safely transferred to specific laboratories. IR and UV spectrometry is then a suitable technology for this kind of analysis.

The investigation of many artistic and archaeological works can be thus provided ranging from IR radiation to visible light and UV fluorescence (Fig. 7.7), in order to perform suitable analyses of paintings, sculptures, illuminated codes, parchments, books, and documents in historical archives, etc., preliminary to any restoration or cleaning, and to achieve high-resolution images of the considered objects, to be implemented in a multimedia Internet database.

In many cases, the multispectral analysis allows one to improve the readability of a document or to identify again previous writings or images that have been severely damaged by natural hazards such as fire or flood, or by human interventions, or erased in the past. Finally, the availability of multispectral data in digital format together with suitable high-resolution images of the document itself can be used to

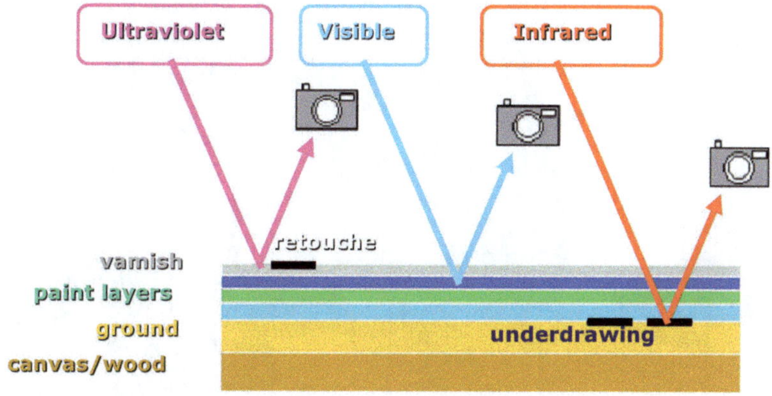

Fig. 7.8 Schematic picture of painting layers and the effects of radiation of different wavelengths

obtain a reliable restoration of the considered historical object without performing operations on the original copy, due to suitable algorithms and physical models of degradation processes.

In particular, IR reflectography allows a deeper penetration in the surface than visible light, thus making the original underdrawing evident, when it is present as in thirteenth–fourteenth-century paintings, for instance. On the other hand, UV fluorescence, which is produced by molecular excitations and decays from the surface of the object, is useful to show repaintings, previous restorations, and so on, due to different aging times of the pigments and therefore to their different microscopic behavior (Fig. 7.8) once exposed to the UV radiation flux.

To summarize, multispectral analyses, including images in visible light, are produced by the radiation–matter interaction in the presence of an observer (person or detector, as previously outlined). Depending on the frequency or the wavelength of the considered radiation, one recovers images from different physical processes, namely, IR reflection, visible light, UV reflection, UV fluorescence, X-ray transmission, and X- and gamma-ray reflection.

An integrated multispectral digital imaging system can provide information for:

- The observation of fabrication techniques
- The examination and analysis of the constituent materials
- The assessment of the effects of time on works of art
- The study of their degradation and/or alteration processes
- The implications of these diagnoses on the state of conservation and on the means of prevention of deterioration and restoration

As for the investigation and digital images further presented in this chapter, the adopted multispectral digital instrument consists of a portable camera with imaging charged-coupled device (CDD) optical sensor, imaging converter, and filter lenses, covering a spectral responsiveness in the wavelength range of 320–1550 nm. Different imaging modes are thus possible with the same experimental apparatus: UV

Fig. 7.9 Experimental setup for grabbing digital images

diffusive reflection, visible diffuse reflection in the red, green, blue (RGB) basis, visible fluorescence, false color IR, and IR diffuse reflection in three spectral bands, namely, in the ranges 750–1000 nm, 980–1150 nm, and 1150–1550 nm. The optics is supplemented by a 25-mm lens and a zoom-macro system with magnification up to 60 $\times$. A portable computer provides the storage and visualization system during the analyses and the relevant digitization of the considered objects or documents.

Finally, the illumination sources are provided by two 30-W halogen lamps and, for UV lightening, an UV lamp. The whole apparatus at work is shown in Fig. 7.9.

More details on the instrumentation, the working planning and conditions, and the adopted procedures can be found in [3].

Figures 7.10 and 7.11 show an example of the results of the digitization activity carried out in the Library of Seminario Arcivescovile Cosentino, one of the few institutions allowed by the Inquisition to conserve prohibited books, such as works of Tommaso Campanella, Gioacchino da Fiore, and Giordano Bruno. It is worth remarking the use of image-processing techniques for incorporating a visible watermarking in the digital image in addition to a hidden one, in order to satisfy the requirements of copyright asked by the owner, the Seminario Arcivescovile Cosentino.

Figure 7.12 refers to the digitization survey performed in the Library of the Osservanza's Friars convent in Bologna where the whole stock of *incunaboli* and *cinquecentine* has been investigated in order to provide useful information on the preservation status of the books and on planning a restoration work. The image on the left side has been acquired in the visible domain, while that on right side in IR reflection spectroscopy (so it is reproduced in a false scale of grays). The hole refers to a road or a worm, while the dark spot is a chromatic alteration of the paper due to exposure to water or presence of humidity.

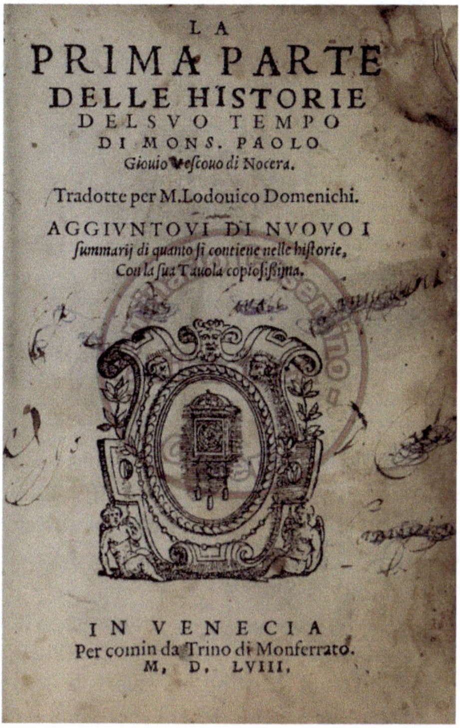

LA
PRIMA PARTE
DELLE HISTORIE
DEL SVO TEMPO
DI MONS. PAOLO
Giouio Vescouo di Nocera.
Tradotte per M.Lodouico Domenichi.
AGGIVNTOVI DI NVOVO I
summarij di quanto si contiene nelle historie,
Con la sua Tauola copiosissima.

IN VENECIA
Per comin da Trino di Monferrato.
M. D. LVIII.

Fig. 7.10 Paolo Giovio, *La prima parte delle historie del suo tempo*, 1558, Biblioteca Arcivescovile del Seminario Cosentino, Paola (Cosenza), codex CS 0144 pos. XXVII A 2

Fig. 7.11 Two pages of books conserved in the Library of the Seminario Arcivescovile Cosentino

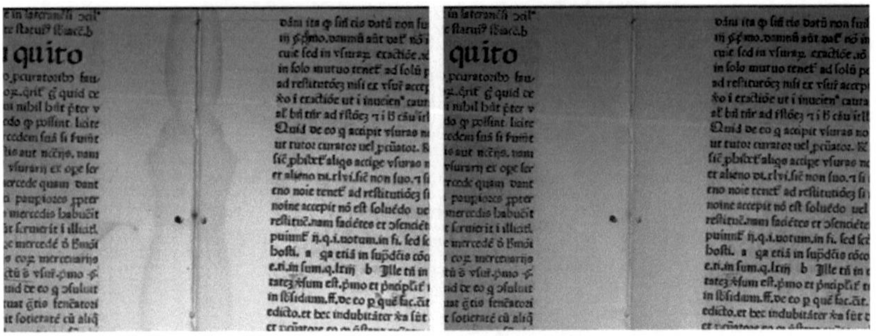

Fig. 7.12 Detail of two pages of an open book from the fifteenth century conserved in the Osservanza's Library of Bologna: visible light (*left*) and IR reflection (*right*)

The same page has been digitized in IR reflectography (see Fig. 7.12, right side), thus avoiding the presence of the humidity spot since IR radiation comes deeper into the paper surface. Therefore, the only visible defect is the existence of the wormhole, and this simple procedure provides a very effective proof of the possibility of "virtual" restoration on the digital images.

Figure 7.13 shows an another significant example, where pages of a book exposed in the past to humidity exhibit the so-called foxing spots, arising from the combination of water and dust and showing the characteristic chromatic alterations of the paper in rust color that, in the future, can more and more deteriorate. While the upper image in Fig. 7.13 is taken in visible light, the other is provided by the nearest IR reflection where the longer wavelength radiation penetrates deeper in the surface and therefore the foxing spots disappear, thus allowing us to see the pages in their original status or chromatic values. Moreover, to enhance the IR image, an adaptive histogram equalization [27] has been applied.

As for UV fluorescence analysis, Fig. 7.14 reports the image of a portion of a page in which a written record has been partly erased, partly covered by dirt, yet is clearly readable in the UV figure. The name (Giuseppe, Joseph) and family name of a possible ancient owner are written, but different actions (erasure, dirt, ink easing) made it impossible to identify at visual investigation.

The proposed study cases—in addition to providing a digital repository of the books themselves to be distributed on the Internet, for instance, while the best preservation conditions are possible for the original material copies—give a useful insight into the past histories of the documents, their preservation status, and the needs and requirements for future restorations. Therefore, all this digital information is suitable for archiving and consulting as a reference example.

The most impressive work, however, is performed on ancient documents and codex severely damaged by humidity, fire, or other accidents, erased and rewritten parchments (palimpsests), and illuminated books. Figures 7.15, 7.16, and 7.17 show, respectively, a page of a palimpsest on parchment, the same image after a color enhancement application, and, finally, the IR image in false gray scale and rotated by 90°, making evident the erased writing that in the other two figures can only

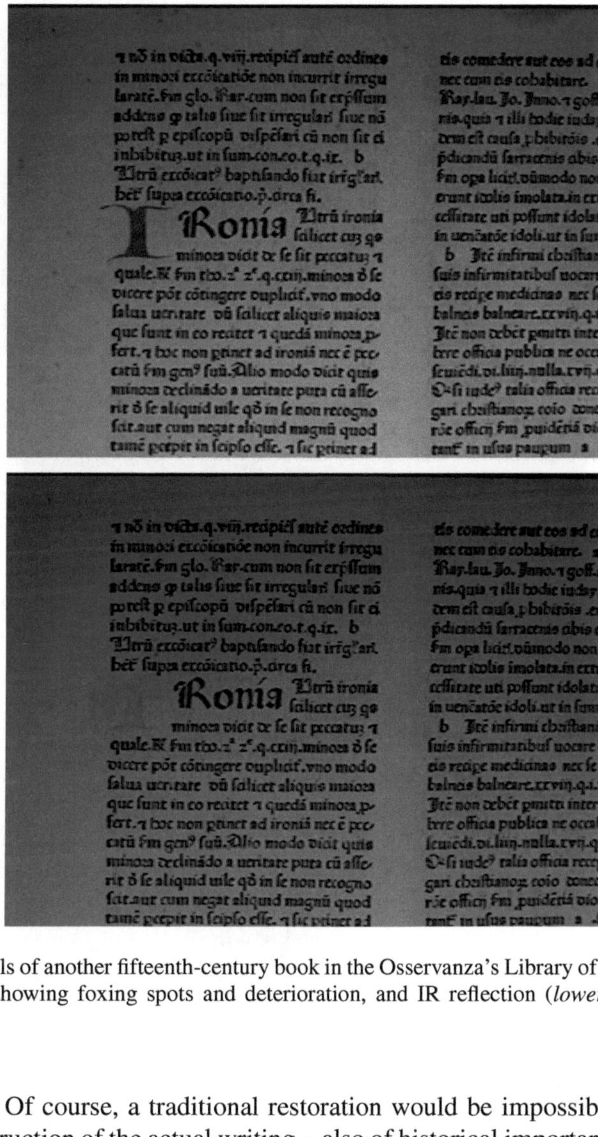

Fig. 7.13 Details of another fifteenth-century book in the Osservanza's Library of Bologna: visible light (*upper*), showing foxing spots and deterioration, and IR reflection (*lower*) with removed defects

be glimpsed. Of course, a traditional restoration would be impossible in this case since the destruction of the actual writing—also of historical importance—is strictly forbidden. Virtual digital restoration—here performed by Fotoscientifica, Parma, Italy—does not act on the original artifacts but is able to improve its legibility and therefore its knowledge.

The combination of images at different wavelengths and contrast enhancement tools is really effective in providing an integrated approach to recover important information from damaged or changed documents. Another example is shown in Figs. 7.18 and 7.19, relevant to a manuscript exposed in the past to a fire and therefore of impossible reading. By means of the described technique, the text can be recovered.

Fig. 7.14 UV fluorescence image of a page where the name of a previous owner of the book can be read, otherwise illegible at visual inspection

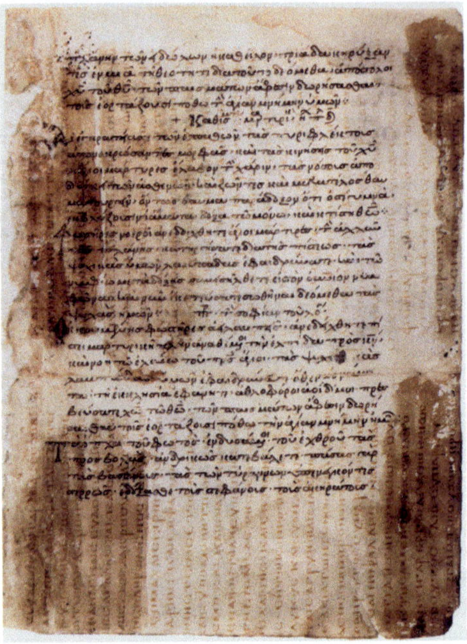

Fig. 7.15 Visible light image of a typical palimpsest

7.6 Applications to Paintings

We introduce the use of IR radiation and color enhancement techniques in the analysis of artistic works by considering, as a significant example, a very controversial and partial restoration carried out on a painting by Pietro di Giovanni Lianori, representing the *Virgin and the Child*, an artist active in the fifteenth century in Bologna, Italy [21]. This work has been conserved in the Museum of Cappuccini in Bologna, Italy, since 1928. In the provincial archives of the Cappuccini Friars of Bologna, the photographs are preserved, documenting the status of the work before (Fig. 7.20)

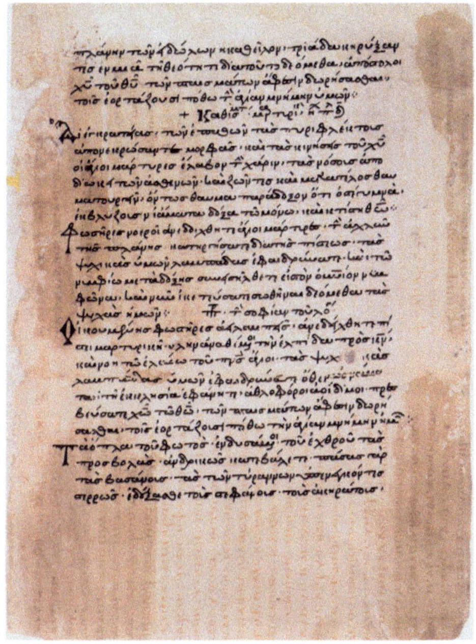

Fig. 7.16 Color contrast enhancement by means of homomorphic filters of the image in Fig. 7.15

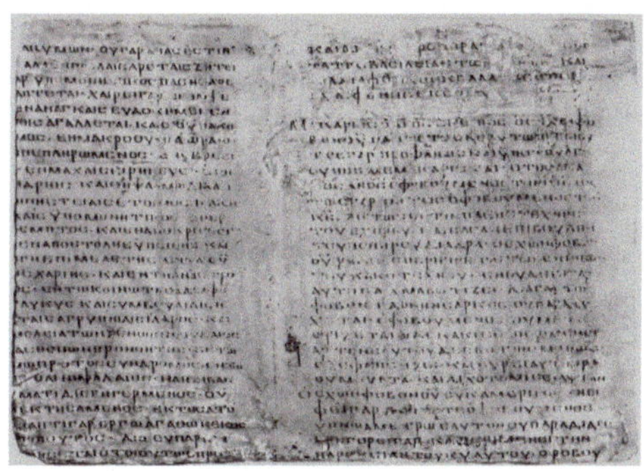

Fig. 7.17 The IR, rotated by 90°, processed image of the palimpsest of Fig. 7.15

and during the restoration that has been interrupted because of the discovery of the original painting below a successive remaking (Fig. 7.21).

In this case, digital restoration is the only way to intervene on an artistic work, namely, when the paintings show successive repaintings, and the second painting cannot be removed since it is an artistic artifact in itself to be preserved. In digital image processing, conversely, one has the freedom to remove or divide the two paintings by techniques that allow retouching with higher magnification to restore the original features, without the work suffering a loss.

Within a digital framework, there is a large freedom of action to create hypotheses for restoration of not only paintings and frescoes but also photographs, architecture, and three-dimensional (3D) objects. Today, the digital (or virtual) restoration is considered an accepted technique for the restoration of old photographs, but it is

Fig. 7.18 Parchment severely damaged by fire

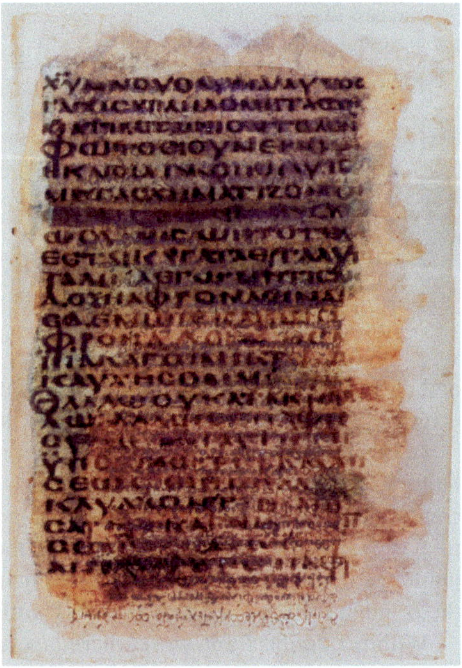

Fig. 7.19 The writing of Fig. 7.18 after digital restoration

Fig. 7.20 Pietro Lianori, *Virgin with the Child*, Cappuccini Museum, Bologna, fifteenth century, 171 × 124 cm, before the restoration

Fig. 7.21 Pietro Lianori, *Virgin with the Child*, image in visible light, current issue

also an excellent opportunity to develop possible interventions on paintings, sculptures, and archaeological artifacts. These images confirmed a very difficult situation for restorers and conservators due to a complex overlapping of both original and repainted layers in order to refresh the painting and to highlight the identity of a new donor.

In fact, the comparison between data recorded from multispectral images in various spectral bands has allowed the identification of quantitative and qualitative differences subsequent to the drafting of the original (Fig. 7.21). The comparison of the recovery image of this panel in visible light before, during, and after the restoration in the seventies of the twentieth century clearly shows that the work had been almost entirely repainted. If the choice to remove, in some areas, repaintings arising from the seventeenth-century modifications made it possible to uncover the original paint surface that is still preserved in the underlying layers, although very impoverished in terms of material, it has also compromised the ability to read all the work in a consistent manner.

The spatial resolution in the acquisition phase has been chosen in such a way that the digital images could reproduce the *craquelures* of the painted surface; this result has been obtained with a sampling frequency corresponding to 6 pixels/mm. Therefore, the whole painting has been digitized by capturing partial images referring to an area of 9×6 cm and, in order to improve the signal to noise ratio, each image consists of an average image of 16 acquired frames for the following recomposition of the whole digital image. The resulting noise on the signal ratio is reduced by a factor $4 = \sqrt{16}$, according to a statistical Poisson error distribution. The complete final digital images of the painting have been obtained by means of suitable algorithms of reassembly and are shown in Figs. 7.21 and 7.22.

The dynamic composition of the painting is monumental, a feature typical of the artist, where the Virgin is represented seated on the throne with her right arm raised and holding the Child, and in her left hand a white rose, symbol of purity. Above the shelves beside the throne, one can see two coats of arms that may lead back to bidders, smaller paintings in the lower part of the work, at the sides of the central throne (Figs. 7.23 and 7.24).

In the latter, there is a scroll with the following inscription: "PETRUS IOHANIS DE LIANORIS PXT ANO 236 GABRIEL DARDUS MED. DOCTOR DONAVIT ANNO DNI 1611," written in gothic letters in the first row and in the second one in Roman letters (Fig. 7.25).

By means of the multispectral investigation coupled with color enhancement procedure, discussed next, it was possible to examine nondestructively the whole surface of the painting and find some significant areas where the penetration of the various utilized radiation wavelengths has highlighted the differences between the underlying drafts and the fifteenth–seventeenth-century modified version as in the case of the two coats of arms.

Fig. 7.22 Pietro Lianori, *Virgin with the Child*, IR image (*on the left*) and image in UV fluorescence (*on the right*)

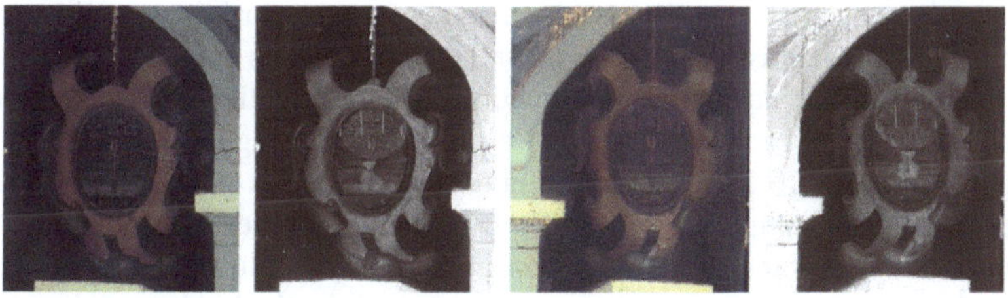

Fig. 7.23 Pietro Lianori, *Virgin with the Child*, details showing the two coats of arms in visible light (*left*) and IR reflectography (*right*) with different heraldic symbols

In order to improve the color contrast, all images have been made radiometrically uniform by means of an equalization procedure, and a mean-centering scaling algorithm was adopted. The multispectral images of Figs. 7.21 and 7.22 can be summed up in a three-way array, G(IxJxK), with I and J indexes labelling row and column, respectively, of each image, while K refers to the different wavelengths (three channels in the visible light corresponding to the RGB basis, one up to three for IR radiation, one for the UV fluorescence region). The method of standard principal component analysis (PCA) applied to G matrix is described in the following section.

Fig. 7.24 Pietro Lianori, *Virgin with the Child*, other details in visible light (*on the left*), IR reflectography (*in the middle*), and UV fluorescence (*on the right*)

The PCA has been performed for the whole panel, starting from these photon spectra, digitally recorded and processed. Once obtained the G matrix, pixel-by-pixel local operations have been carried out by subtracting the content of each pixel in the G matrix corresponding to the RGB image before the restoration.

Here, it is worth noticing that the visualization of PCA model parameters, loading, and score vectors has an important role in interpreting the results of multivariate spectroscopy data analysis. Moreover, taking into account the difference between the G matrix evaluated from the visible image before the partial restoration and that obtained from the multispectral analysis of the painting in its present status, a quantitative estimate of the introduced modifications is obtained allowing a well-grounded digital restoration to be performed pixel-by-pixel in each region of interest.

Figures 7.24 and 7.25 show a few details of Lianori's painting where digital images have been captured in visible RGB basis, in IR reflectography and UV fluorescence to be combined together according to this procedure. Therefore, a digital or virtual restoration of the painting is recovered where the hidden original layer is brought back to pristine condition, as shown in Fig. 7.26.

The concept of "virtual restoration," while its introduction is quite recent, is not without ambiguity of use and meaning. Founded in the field of cultural heritage, it has gradually emancipated widening the scope of its application and getting to lick even the linguistic and literary studies; yet, the term "virtual restoration," even in its variant of "digital restoration," continues to show just the used technical tools rather than a clear instance of methodology.

Recently, someone has seriously proposed abandonment of these expressions in favor of a more comprehensive definition, but still with a good degree of vagueness, such as, for example, "iconological digital restoration."

Our example proves that even a virtual restoration can be performed by respecting the principles stated by Cesare Brandi [1] for every "real" physical restoration and overall in the world adopted, namely, respect for the aesthetic and historical

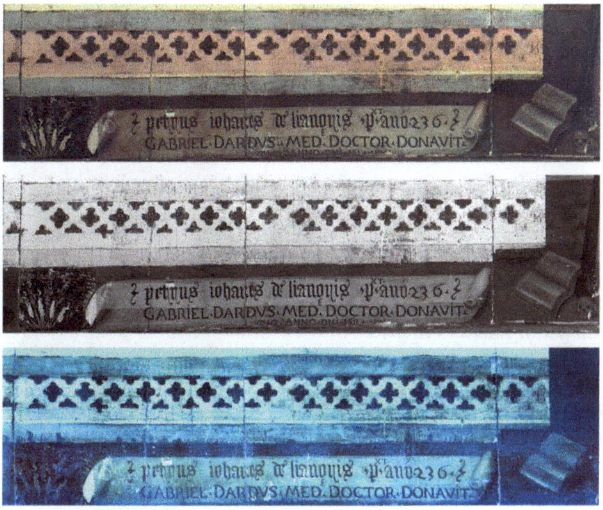

Fig. 7.25 Pietro Lianori, *Virgin with the Child*, details of the base of the throne with the inscription in visible light (*top*), IR reflectography (*middle*), and UV fluorescence (*bottom*)

aspects, recognizability of the intervention, reversibility of the materials, and minimal intervention, while the fourth one, compatibility of materials, is pertinent only to the real restoration.

7.7 Principal Component Analysis

The use of suitable hardware systems for digital imaging grabbing in multispectral analyses allows treatments by means of color enhancement and image processing techniques that increase the precision and reliability of the cultural heritage examination. In particular, a method of numerical processing based on the PCA was used for improving the diagnostic procedure carried out on the conservation status of the Medieval *Crucifix* of San Damiano in Assisi [3], Italy (Fig. 7.27).

The investigated object is a wooden crucifix of the twelfth century, whose surface is painted in tempera and measures 200×150 cm; the whole painting has been digitized by grabbing partial images referring to a painted area of 9×6 cm (see Fig. 7.28) and, in order to improve the signal to noise ratio, each image consists of an average of 16 acquired frames [3]. The complete final digital image of the Crucifix has been obtained by means of suitable standard algorithms of reassembly.

Figure 7.29 shows the results of Fourier transform applied to the single frame image and to the average image; the so-gained better resolution is evident.

In Fig. 7.30, multispectral diagnostic images are shown, obtained—respectively, from the left to the right side—at the wavelengths $700 - 450$ nm (RGB), 900 nm (IR), and 400 nm (UV) for a detail of the Crucifix, while Fig. 7.31 underlines in false

Fig. 7.26 Two details of Lianori's painting during virtual restoration

Fig. 7.27 The *Crucifix* of San Damiano; unknown author, twelfth century, tempera on wood, 200 × 150 cm, Assisi, Basilica di Santa Chiara

colors the regions where damages to the work or repaintings and past restorations are present, as resulting from the IR and UV analyses, respectively.

The evaluation of the conservation state of artistic as well historical works largely requires methods of investigation based on the interpretation of images in multi-spectral bands. The noninvasive nature of this technique represents one of the main factors of its success and diffusion. However, the multispectral image analysis and interpretation are not always an easy duty, namely, when paintings of very large size are studied, and several wavelengths must be included in the investigation.

Fig. 7.28 The process of digital multispectral images acquisition

Fig. 7.29 Comparison of Fourier transforms (*upper parts*) of the single image frame (*left*) and averaged image (*right*), respectively

Therefore, a more objective approach must be introduced than the merely visual one shown in Fig. 7.31.

In fact, a restorer has to consider, first of all, each image separately in order to extract the maximum content of information. Then, by comparing the so-gained data, she/he has to evaluate the state of conservation of the whole object.

The analysis procedure and comparison require long times and considerable efforts; moreover, accuracy and reliability of the relevant diagnosis strongly depend on the experience of the examiner. The main problem of such a traditional approach is the large amount of data that in this case are constituted by pictorial information. A lot of them could be redundant because of their correlation; other information

Fig. 7.30 Detail of the *Crucifix* in visible light (*left*), IR spectroscopy (*middle*), and UV fluorescence (*right*)

Fig. 7.31 The same detail as in Fig. 7.30, where defects are underlined in *yellow* (from IR image) and *violet* (from UV image)

could be hidden by strong signal or noise. A possible way to overcome these limitations is represented by the use of multispectral PCA [22].

This technique of multivariate analysis allows one to rearrange the multispectral image information in a convenient yet unusual manner that makes the interpretation easier and the evaluation of the conservation state of the object less subjective [23–26].

Moreover, by means of PCA, it is possible to perform a multivariate segmentation of images: The original image is partitioned in regions that are homogeneous from the diagnostic point of view, and areas that have been restored or damaged can be grouped together, thus making the restorer's task easier. The study described here has been performed in collaboration with restorers of the COO.BE.C. firm in Spoleto, Italy.

The same detail of the multispectral image obtained as previously described is shown in Fig. 7.32, corresponding to wavelengths, λ, in the range between 400 nm (UV) and 900 nm (IR), and referring to the part of the body of Christ and three people behind him. With respect to Fig. 7.30, the images are presented in order of decreasing wavelength, and the three RGB components have been separated in three different (grey) images. The three images in the middle, namely the RGB components, have decreasing light intensities, so that the blue component is difficult to read. The IR image on the left side emphasizes the contours of the figures and two dark vertical areas at the right top which correspond to past interventions of restoration. The same areas are even more outlined in the UV image at the right side, where it is worth remarking a few scattered spots identifying restorations and repaintings.

In order to improve the image contrast, all images have been made radiometrically uniform by means of an equalization procedure [27], and a mean-centering scaling algorithm was adopted. As shown by Bonifazzi et al. [28], the multispectral images of Fig. 7.32 can be summed up in a three-way array, $G(I \times J \times K)$,

Fig. 7.32 From the left-hand side, multispectral images are shown, obtained, respectively, at the following wavelengths: 900 nm (IR), 700—450 nm (RGB), 400 nm (UV)

with I and J indexes labelling rows and columns, respectively, of each image, and K referring to the different utilized wavelengths.

The results of the PCA applied to the G matrix are summarized by the following equation:

$$\underline{G} = \sum_{a=1}^{A} T_a \times p_a,$$

where A is the rank of a two-way G(I $\times$ J) $\times$ K matrix resulting from the rearrangement of G, and each term in the summation is a multivariate product, $T_a \times p_a$, between the score images, T_a, and the corresponding loading vectors, p_a [22]. Only three principal components (PCs) explain with "sufficient accuracy" the multivariate image G; the correspondent score images, T_a, and loading vectors are shown in Fig. 7.33. As evident in the loading plots, T1 summarizes the information common to all the wavelengths with some emphasis on RGB ones. As a consequence, the physical features of the figures and details such as, for instance, the borders of the haloes and draperies are well outlined. Second- and third-score images, T2 and T3, mainly resume IR and/or UV wavelength concerning the preparatory drawings as well as tonalities present both in the figure of Christ and in the background. Moreover, in these score images the restoration areas are clearly identified. T4 and T5 images (not shown) are responsible for a small amount of the whole multispectral variability, whose importance must be evaluated time by time.

These results prove that the compression of information produced by the PCs allows a simpler and more rational extraction of information than that deduced from visual analysis only of images of scores. They also suggest that through the segmentation of the most important multivariate PCs so obtained, one can proceed to a separation of areas of interest for a study of the conservation status of the painting and the determination of repainting. Similarly, this method of analysis is used in the field of cultural heritage applications in order to extract the information essential to the identification and recognition of handwritten characters in archival documents and illuminated manuscripts, for instance.

Score images can be then used to deduce the multivariate segmentation of the original image [22]. In Fig. 7.34A, scores T1, T2, and T3 have been associated, respectively, to red (R), green (G), and blue (B) channels, thus obtaining a pseudocolor image of the painting. Scatter plots of the score images T3-T1, T3-T2, and

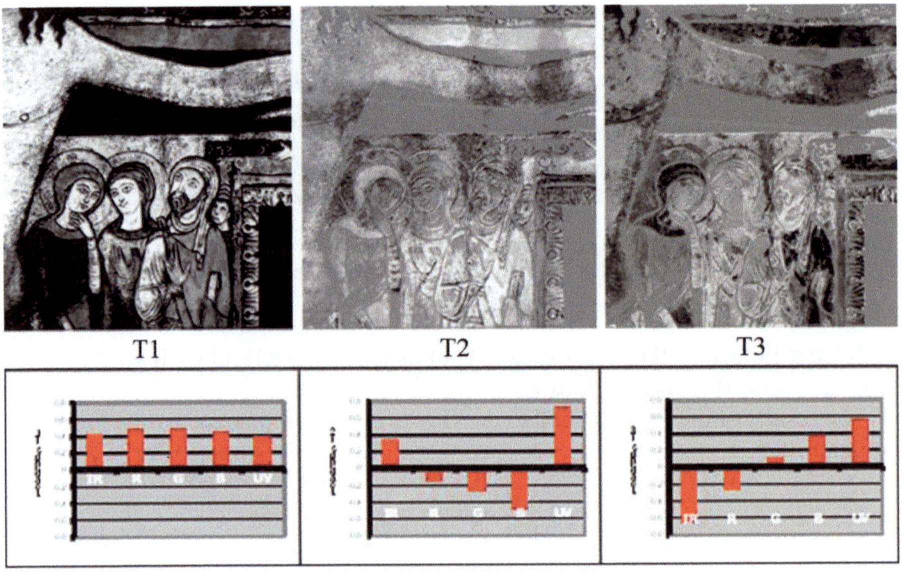

Fig. 7.33 Score images produced by the PCA and loadings plots. The variance accounted for each image is, respectively, T1 = 76 %, T2 = 12 %, and T3 = 7 %

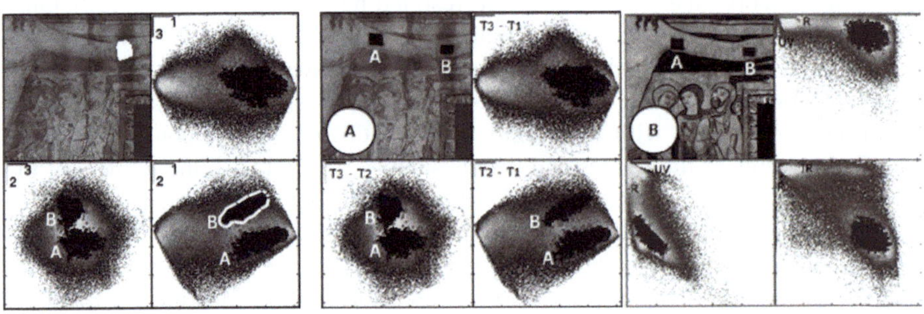

Fig. 7.34 Pseudocolor representation of scores images T1, T2, T3 (*A, left*), and R, IR, UV wavelength images (*B, middle and right*). Scatter plots of the images are shown (see text)

T2-T1 are also shown. Figure 7.34B was obtained associating RGB channels to R, IR, and UV components of the original multivariate image, respectively.

Two regions of interest, A and B, have been determined on both pseudocolor images. Coming from the Christ arm, they are figuratively homogeneous but—since region B is located on a restored area—quite different from a diagnostic point of view. This feature clearly emerges from the multispectral segmentation on both the pseudocolor image of Fig. 7.34A and on the scatter plots, T2-T3 and T2-T1, where two splitted clusters corresponding to A and B regions are evident. On the contrary, from Fig. 7.34B and the related scatter plot, it is not possible to distinguish regions A and B since the original wavelength images do not allow the separation of the figurative and diagnostic features. It is worth remarking that IR, R, and UV maintain a large amount of information about the restored area, as shown in Fig. 7.32.

In conclusion, these results show that the multivariate analysis of digital images at different wavelengths can greatly improve the knowledge of the conservation state of an artistic object, such as the painting on wood here considered. As a matter of fact, since PCA focuses the information on a limited number of score images whose importance is rapidly decreasing, a simple visual analysis of them allows a very immediate, clear, and impressive reading of the surface as well as the history of an artwork. Moreover, as multivariate images provide complementary information, it is possible to determine a multivariate segmentation of the original image on the basis of characteristics such as the local composition of the painting, thus putting the restored or damaged areas in evidence.

Finally, it is also possible to compare the abovementioned analyses with the digitalized radiographies of the painting, whose detail is shown in Fig. 7.35. The yellow contour zone in the upper image is magnified in the lower image and illustrates the high quality of the result since even the texture of the canvas glued to the wooden table as a ground for the preparatory layer and the pigments is clearly visible. Figure 7.36 refers to a comparison between the radiography (superimposed to the relevant area of the painting) and the results of diagnostics images of Fig. 7.31.

It is evident that the main damages occur in the juncture of wooden planks, where thermal and mechanical stresses have been stronger.

7.8 Evaluation of Restoration Work: A Case Study

We discuss here the results of the restoration performed at the end of the last century on a painting of Jean Gossaert. The painting we consider here has been restored in the laboratories of the National Gallery in London, to whose collection it has belonged for a long time. The painting represents the *Virgin and Child* (see Fig. 7.37).

An accurate cleaning has established that this painting, long believed to be a copy, is Gossaert's original [29–31]. It dates from about 1527. Long regarded as merely a late copy after the original, the removal of overpaint and a thick, discolored varnish from the surface of the painting revealed it to be a beautiful and authentic work of Gossaert.

Jean (or Jan) Gossaert, a native of Flanders (active 1503; died 1532), was one of the most startling and accomplished artists of the Northern Renaissance. He is sometimes referred to as "van Mabuse," from Maubeuge, where he was born in about 1478.

Dendrochronological analysis to determine the age of the oak panel support found that the painting could have been made as early as 1509. It would be highly unlikely for a "late copy" to have been painted on an early sixteenth-century panel.

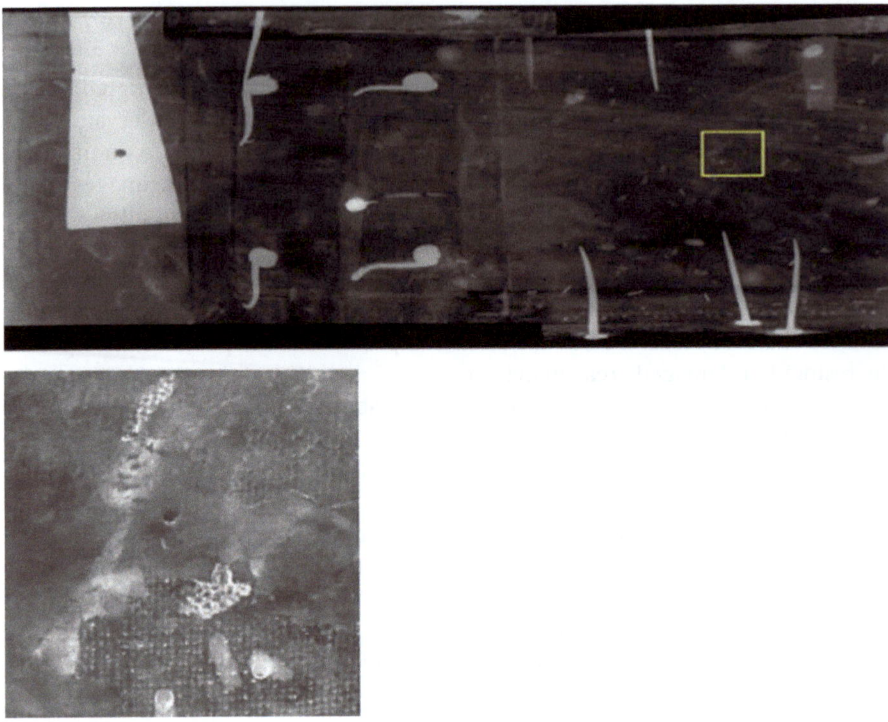

Fig. 7.35 X-ray radiograph of part of the Crucifix (*upper*) where the yellow contour area is magnified below

Fig. 7.36 Comparison among different diagnostic images: visible light (*left*), superimposed X-ray image (*middle*), and inclusion of damaged zones as pointed out by IR and UV analyses (*right*)

Paint samples, analyzed in cross-section, also gave promising results: The ground and priming layers were consistent with sixteenth-century Netherlandish paintings. The disfiguring varnish, which gave the painting the appearance of a much later work, was also studied in the National Gallery scientific laboratory; gas

Fig. 7.37 Jan Gossaert (Jean Gossart)—NG1888—*Virgin and Child*, oil on oak panel, (*arch-topped*), 30.7 × 24.3 cm (before the cleaning)

chromatography–mass spectrometry (GC-MS) analysis identified a particular combination of components that had prevented the thick varnish, probably applied in the nineteenth century, from drying properly, resulting in wrinkles and surface cracks.

Removal of the varnish revealed the extent of the repainting (see Fig. 7.38). A photograph taken under UV light during cleaning shows the left half of the painting still covered with the old varnish, which has a greenish fluorescence. Furthermore, comparison of the X-ray radiograph of *The Virgin and Child* with Crispijn de Passe

Fig. 7.38 Jan Gossaert (Jean Gossart)—NG1888—*Virgin and Child*, after the cleaning

the Elder engraving (dated 1589 and then done from the original painting) showed that the original design beneath the overpaint corresponded precisely with the print.

Our work consisted in providing an objective quantitative method for evaluation of the amount of restoration work. To this aim, the two high-quality images of the Mabuse painting, taken before and after the cleaning in the scientific laboratory of the National Gallery under the same technical and lighting conditions, have to be compared pixel-by-pixel. As discussed in the previous section, the two images were standardized with a radiometrical equalization procedure [4] and a

Fig. 7.39 Details of the painting showing in false colors the differences from before and after the cleaning (see text)

mean-centering scaling algorithm. Then, the differences between the two images were pointed out in false colors, sky blue where the removal of repainting has revealed previously hidden original parts and yellow where losses of color shooting in previous interventions were found (Fig. 7.39).

We point out that this procedure was performed as a blind test, and only after the color enhancement and processing, the results have been presented to the restorer of the National Gallery (J. Dunkerton) and discussed with her, by matching perfectly with the evidence of her work even in the smallest details.

7.9 Applications to Mosaics

Mosaic cartoons are tempera paintings which reproduce ancient mosaics, traced tile by tile, in order to document and make copies of the mosaics themselves. We take into account the cartoons that were created by leading mosaic masters and restorers of Ravenna, such as Alessandro Azzaroni, Zelo Molducci, and Libera Musiani (between 1900 and 1975 ca). Some of them are kept, as teaching materials, at the Art Institute for Mosaic "Gino Severini" in Ravenna.

The virtual restoration of some of these cartoons is part of a complete project for the conservation and enhancement of them [32, 33]. The virtual restoration we discuss here does not aim at foreshadowing a real restoration intervention (which is often not necessary as a conservative intervention may be enough), but at replacing the lacunose paintings with digital images reconstructing their missing parts, in order both to facilitate their use for educational purposes and to protect the original ones from everyday use. As a matter of fact, these cartoons are still regularly used for the didactic activities of the mosaic school.

Intervention attempts may be different and may also be modified subsequently: Each phase of the intervention may be registered at a different level in a photo-editing program such as Adobe Photoshop, Corel PhotoPaint, or GIMP. In a certain sense, this allows for the reversibility of the intervention and immediate comparisons between the different phases and possibly among several operational choices.

The reconstruction hypotheses digitally carried out may also be mimetic, maybe carried out on a different level compared to the original image in order to guarantee their identification. It is important, though, that they are justified by a philological analysis of the piece. In this case, with mosaic cartoons, we do not risk to make arbitrary reconstructions of missing parts since we can refer to the ancient mosaics the cartoons were drawn from.

Prior to the real intervention, an adequate color management system (CMS) should be used. Due to the use of a colorimeter or a spectrophotometer, CMS can coordinate the gamut (the set of all the colors that a device can display), that is, the color spaces of different output devices (such as printer and monitor), in order to have the colors of the scanned images correspondent as much as possible to one another and to those of the original artwork.

The process of virtual restoration on the mosaic cartoons consists of the following phases:

- Digital data acquisition
- Mosaicking
- Elaboration (balance, cleaning, reconstruction, extraction of the grid)
- Electronic filing

Of course, memory of each step and phase of digital processing must be kept in a different level of the program, able to store several overlying images, saving them without compression and therefore without losing quality. The digital acquisition is generally performed using a digital camera, the resolution of which today usually ranges from 8 to 14 Megapixels (millions of pixels). This is a camera in which the traditional photosensitive film is replaced by a CCD or complementary metal-oxide semiconductor sensor (CMOS), able to capture the image, transforming it into an electrical signal. To obtain a higher resolution, we can also use scanners or digital backs.

If the cartoon to be photographed is very large, it can be acquired through several shots, overlapping the edges which then must be reassembled through mosaicking, in order to obtain a resolution of at least 500 or 600 ppi (pixels per inch) on the real size, required to display a screen magnification of all details and to make extremely high-quality prints.

The mosaicking consists in the perspective rectification of several shots that are then reassembled to form the image of the entire object, perfectly matching the margins. Once reassembled, the high-resolution photographs not only allow one to gain a better understanding of the artwork but also help to assess the conservation status and identify any previous restoration work.

A first elaboration of the image obtained in this way is to balance brightness and contrast, and to regulate dominant colors, in order to make the picture as similar as possible to the original one, since each acquisition inevitably causes overexposure

Fig. 7.40 Original photograph of a mosaic cartoon

or underexposure and color toning due to lighting conditions and the characteristics of the used sensors.

In order to adequately perform these corrections, photos must be carried out by placing a color scale next to the cartoon. By doing so, one can have a reliable reference so as to assess the deviation of the color and brightness values of the picture from those of the original one. Otherwise a color-checker chart can be used, an arrangement of standardized color samples which can be used to calibrate and to profile graphic devices, such as digital cameras and scanners.

As an example, we consider the photograph of the painting by Alessandro Azzaroni representing the pair of doves from the Mausoleum of Galla Placidia, produced by TeCoRe (degree course in "Technologies for the Conservation and Restoration of cultural heritage," Bologna University) during the laboratory analysis of the same painting, with a resolution of 200 ppi at actual size (43 $\times$ 71 cm).

In Fig. 7.40 we see that the acquired image appears too pale, blurry, and slightly turned towards a red color. On the contrary, in Fig. 7.41, the same image was balanced on the basis of the colors shown by the original mosaic.

In this way, it was possible to correct not only the imperfections due to image acquisition but also the changes of the color of the tesserae, due to alterations of paper color (photochromic degradation) over time; indeed, the new yellow-orange dominant color of paper can be seen through the thin layer of the tempera.

Then, if necessary, one can perform cleanup operations (removal of stains) and intensification of the faded lines. This is a recovery of information which is not always possible to carry out with a traditional restoration intervention. Due to the use of IR and UV photography—as previously discussed—it is also possible to separate and give back distinctness to any palimpsests.

The reconstruction of missing parts, as shown in Fig. 7.42, consists in the use of the intact parts of the image in order to recreate the shape of the lost areas, only where it is possible to do so with certainty, therefore for very small gaps and also

Fig. 7.41 Same as Fig. 7.40 with balanced colors

Fig. 7.42 Virtual restoration of Fig. 7.40

for much larger ones, if there are other copies of the artwork to be restored. In our case, we keep the original mosaics on which the cartoons are based.

The gold tesserae at the base of the cup took on a greenish color in the cartoon, probably due to an alteration of *porporina*, a pigment consisting of a mixture of metal powders, usually a golden color, easily alterable in contact with air humidity, especially in tempera paintings.

Conversely, the golden band that decorates the cup was made with yellow, orange, and ocher tempera; therefore, it did not suffer any alteration.

The final image (Fig. 7.42) reproduces the mosaic cartoon as it presumably could have looked like originally.

The extraction of the grid can be useful to make copies in mosaic. Indeed, according to the practice used by the mosaic masters, the grid can be printed by impression on fresh plaster on which the mosaic will be reproduced. Hence, the extraction of the grid was carried out on the image of the restored cartoon, transformed into a grey scale; therefore, the contrast was enhanced to place greater emphasis on the

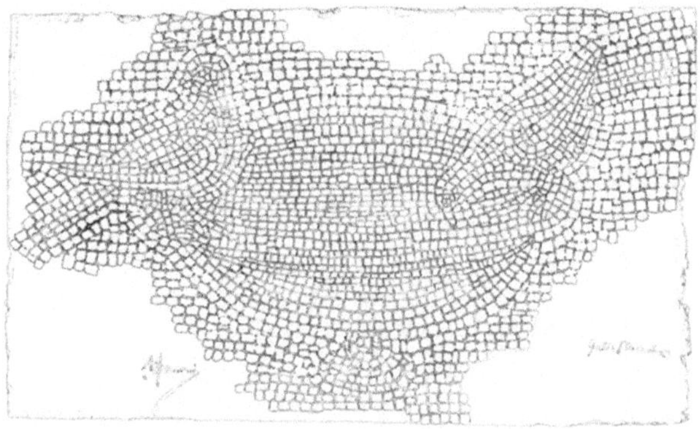

Fig. 7.43 Extraction of the grid from Fig. 7.42

edges of the tesserae, which were then cleaned one by one, as the color variations inside were also highlighted, albeit mildly, using a suitable filter.

The final result is shown in Fig. 7.43.

The last phase of intervention consists in electronic storage of high- and low-resolution images in order to keep the acquired images and all the subsequent elaborations, foreseeing both their consultation and any new virtual restoration based on different principles.

For this reason, we need to save images both in a format that, although occupies a lot of space, does not imply a loss of data ("bmp," "psd," and "tiff" formats that use lossless data compression algorithms), and in compressed format (usually using "jpeg," which uses a lossy compression algorithm), which causes a loss of information at each subsequent save but allows an easier consultation and less waste of space and resources.

Saved high-resolution images may be used to make high-quality prints for periodicals or magazines and for sales. Low-resolution ones, instead, may be used to make online catalogues or user-friendly virtual museums.

Such interventions, possibly with the use of tools that improve resolution, carried out on all cartoons with greater historic value, would allow one to preserve the originals from daily use, without subtracting the information contained inside them, but rather improving and spreading them via prints or accessible online data banks.

Another color enhancement application to mosaics is an experimental project of virtual restoration carried out on a mosaic from the church of San Severo in Classe, near Ravenna. The construction of San Severo dates from the late sixth century; the church was consecrated in 582 and was pulled down and abandoned in the early 1820s. The floor of the church was formed by a rich mosaic, which was only partially found [34, 35].

At the end of the 1966 excavation survey, a mosaic carpet—4.50 $\times$ 2.75 m—was discovered at the center of the main nave. Figure 7.44 shows a scheme of the excavations and a plant of the church; the area where the mosaic was found is indicated in violet. This mosaic showed a grid of rows of tangent pised square containing figures

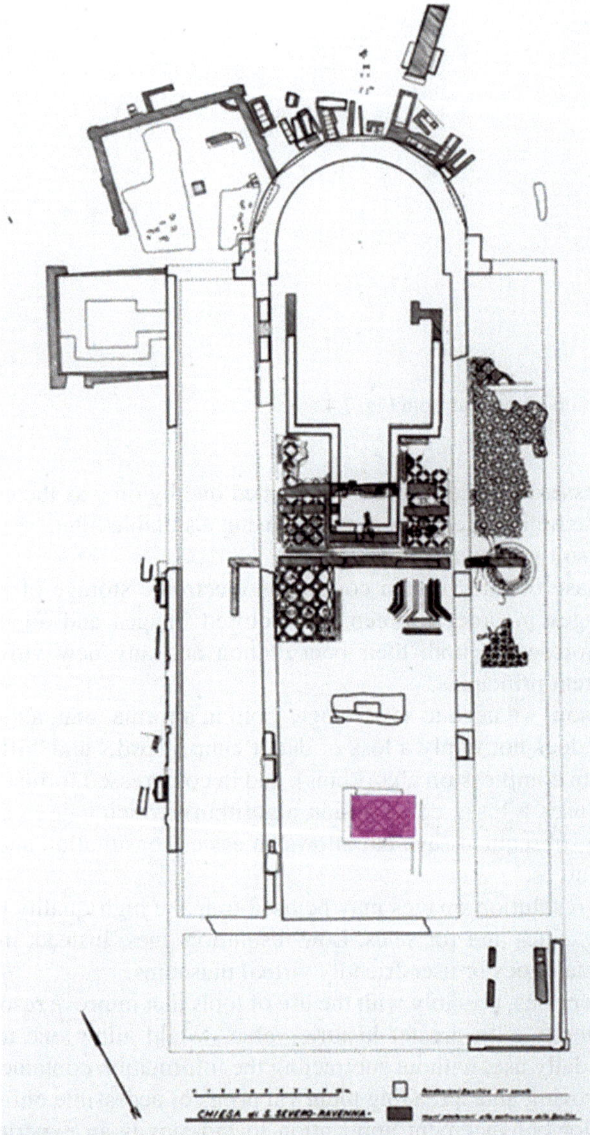

Fig. 7.44 Plan of the San Severo church; the original location of the mosaic considered in this work is represented by the *rectangle* in the lower part of the building, near the entrance

of birds made of glass paste, of great elegance, and extremely naturalistic; around this main scheme runs a shaded three-strand guillocheon on a black ground.

The mosaic was found at a height of 0.40 m beneath the floor of the church, almost in the middle of the main nave, but unlike all the other mosaics, it was not

Fig. 7.45 The actual location of the San Severo mosaics

aligned and oriented in north to south direction. These data have suggested that it might be connected to an ecclesia domestica, an oratory that might have been part of the Roman domus, whose remains are beneath the church. A place of great devotional importance, perhaps, the place itself where the saint had exercised his apostolate.

Today, the floor, lifting and relaiding on cement mortar in the late 1960s, is located at the Museo Nazionale in Ravenna, divided into two sections, placed separately on a wall in a narrow and dark lobby, as shown in Fig. 7.45.

The verticality of the mosaic, the separation of the two sections, decontextualize the mosaic and impoverish it of its original function as an integral part of the internal architecture of the basilica.

Precisely for this reason it was decided to opt for a virtual restoration project, with the intent to provide guidance for future restorations and to revalue this art piece [36].

The characteristics of many mosaics—the symmetry and the repetition of geometric patterns—make them suitable for the use of information technology both for a simulated integration and for the reconstruct by analogy of the geometric pattern. The use of computers and image-processing programs, in fact, offers a preview of the restoration that can support the restorer in the technical operations.

Our virtual restoration project started from the idea of bringing together the two sections. In order to deal with a virtual restoration, it is necessary to have the images scanned at high resolution; the photographic documentation was made with a Nikon D90 at 300 dpi resolution and Kodak color reference band. We have chosen the two images that were considered the best for color reliability and for minor distortion, since the narrowness of the environment did not allow a clear global frontal view of the entire floor.

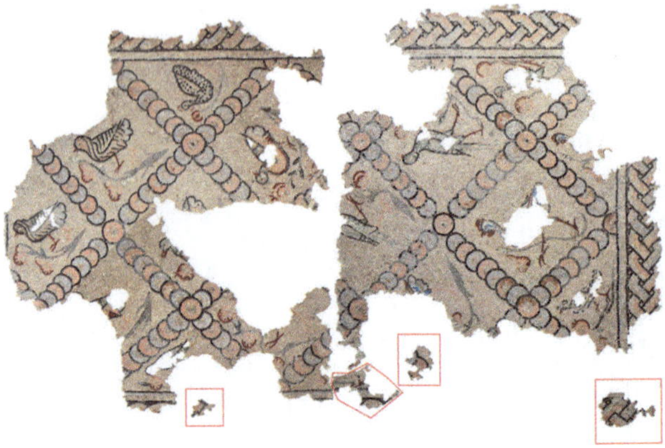

Fig. 7.46 Front image of the two mosaic fragments; the parts inside rectangles refer to fragments mistakenly placed in the museum relocation

It was then necessary to eliminate the perspective and bring all images in 1:10 scale. To this end, we used a simple photo-editing open-source program such as GIMP, which was also used for all subsequent operations. As a matter of fact, we used very simple and easily provided software to perform this work since our purpose was to provide an example of virtual processing useful for conservators and restorers as well as for scholars (archaeologists, art historians, etc.) that can be carried out without specific expertise and computer skills.

After the images were acquired in digital format, we were able to proceed with the actual reconstruction of the floor, taking into account the geometric motifs that make up the mosaic assembly and that made the recomposition the most reliable. It was not possible to use excavation maps as they had no metric references at all.

Once a basic image was obtained, we went on to reconstruct it by analogy: the geometric motif of the floor, which covered both the outer frame with its shaded three-strand guillocheon, as well as the internal allocation of the carpet, formed by a grid of partly overlapping circles, in alternating colors.

After the virtual reconstruction had been completed, some fragments of the mosaic did not fit perfectly, probably because of injuries and of deformations caused by the tearing up and repositioning the mosaic on mortar support; therefore, small corrections were needed, with minimal and calibrated geometric modifications (Fig. 7.46).

In addition to the reconstruction of geometric patterns shown in Fig. 7.47, a proposal was made concerning the integration of the figurative lacunae that are not reinstated, consisting of three tables.

The lacunae have been reintegrated, respectively:

- With a neutral color
- With a deliberately discordant color
- With a camouflage integration

Fig. 7.47 Front image of the two mosaic fragments where the smaller pieces have been placed in the correct positions and geometric patterns reconstructed

The first and third different choices are shown in Figs. 7.48 and 7.49, respectively, and make an evaluation of the obtained results possible from both an aesthetic and a conservative point of view. The virtual restoration offers a number of possibilities in the field of mosaic restoration, although to this date it is not yet so widely adopted. In the case under consideration—the Gallinelle mosaics—it allowed us, at low cost and with great ease, to create a graphic reconstruction of the mosaic floors, so that when a conservative restoration will be carried out, the restorers will have more significant information and different solutions for the integrations. In addition, the handling of mosaic fragments—often very bulky and made heavy by concrete support—certainly does not make delicate operations such as recompositions easy.

We observe that the virtual restoration not only offers to restorers the possibility of recomposing a floor: It also provides the opportunity of simulating on the digital image, with a photo-editing program, the type of integration and the color so that scholars, restorers, and conservators may evaluate the final appearance of the work and the different aesthetic choices.

The virtual restoration is also an essential tool for the revaluation of cultural heritage. The new kind of consumer must in fact find the most suitable means to understand what she/he sees. A mosaic floor with large lacunae may in fact resemble a ruin rather than remind the magnificent mansions that it had been part of. Therefore, a mosaic floor, decontextualized, hung on a wall and with large lacunae deprives the visitor of vital information.

Fig. 7.48 The “restored” mosaic with neutral color integration

Virtual restoration then, at very low cost, may intervene by proposing the reconstruction of the entire floor, thus playing an important didactic role and giving back to the mosaic its readability. In the case of the Gallinelle’s mosaic, split in two fragments, attached on a wall in a dark, narrow room, the visitor cannot understand its importance: She/he, certainly, cannot imagine that same floor within the church where its position probably indicated an important ancient worship area.

In this case, the presence of the digital reconstruction of the floor with camouflage mimetic integration could revalue the importance of this mosaic floor with its delicate birds, made more precious by the glass tesserae, and give back, albeit partially, its integrity and dignity as a work of art, while waiting for the conservative restoration.

This case study highlights the importance of a preliminary study, as a preparation of a restoration before performing surgery. In this way, scholars, conservators, and superintendents can evaluate the final appearance of the work and the different aesthetic choices.

In the case of discordant colors, for instance, we deliberately chose a red-violet great impact in order to emphasize the integration. The color of integration must

Fig. 7.49 The virtually restored mosaic with camouflage integration

be reduced to the background level and should not compose directly with the color distribution of the surface of the work (as seen in Fig. 7.48).

Last but not least, the virtual restoration is an essential tool for the promotion of cultural heritage. In the face of increasing and diverse audiences with different levels of cultural education, museums and archaeological sites seek more and more to create accurate and comprehensive educational courses. The new user should in fact find the most appropriate ways to understand what one sees.

The mosaic floors were, in fact, in the Roman domus as well as later in churches, closely related to architecture and function of rooms: For example, the triclinium, that is, the floor space that would house the beds had no drawing, while at the center an emblem or a representation was placed so as to be watched by the landlord and the guests when they were eating. In every room, then, the drawings of the floor were turned towards the entrance, to be admired by those who entered.

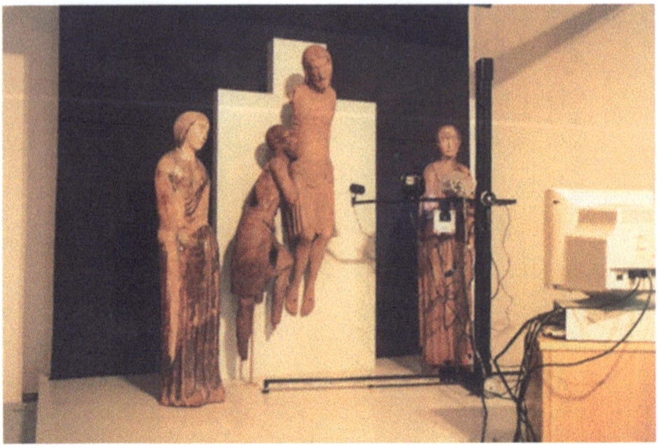

Fig. 7.50 The wooden sculptures of *Deposition of Christ*, in the Museum of Montone (Umbria region, Italy) on the occasion of our survey, after the restoration

7.10 3D Applications

In the last few years, new advanced techniques of image processing have been introduced and applied also to 3D reproduction and reconstruction of cultural heritage artifacts, in addition to well-established studies for historical buildings, monuments, and archaeological sites (see, e.g., [37]). Moreover, color enhancement is increasingly important in this field in order to achieve a closer resemblance and likelihood to the represented object. Generally, this aim is achieved by means of usual two-dimensional (2D) image and color processing applied to suitable textures that will be then implemented on the 3D model.

However, even in these cases, multispectral imaging can be very useful to make clear old restorations, repaintings, and not immediately recognizable details. Figures 7.50, 7.51 and 7.52 refer to the investigation of a group of wooden medieval sculptures representing the *Deposition of Christ*, conserved in the Museum of Montone, a small city of Umbria (Italy) and restored some years ago on the occasion of an important exhibition of Deposition's groups [38].

Figure 7.50 shows the wooden group of sculptures during the multispectral analyses with the used instrument, already adopted for the investigation of St. Damiano Crucifix. Figures 7.51 and 7.52 present two details of the head of two characters, respectively, the Virgin and St. John. The traces of the original colors, the refined decoration of the Virgin's mantle and the tears running down her face are now visible. In the other sculpture, further details and the original colors are clearly visible due to the multispectral shootings.

Very recently, 3D scanning technologies have given impetus to the realization of advanced hardware for rapid prototyping. Now, it is possible to obtain realistic models of cultural heritage objects in different scales up to 1:1 or even bigger and in various materials. Therefore, it is essential also to improve the capabilities of these innovative systems of rapid prototyping in the reproduction of colors. A first step in

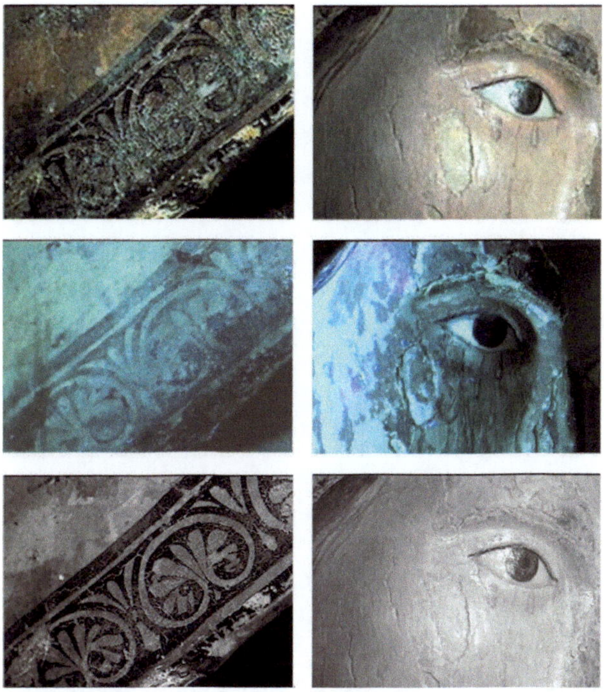

Fig. 7.51 Details of the Virgin's head in visible light (upper images), UV fluorescence (middle images), and IR reflectography (lower images)

this direction was proposed a few years ago in [39], where the adopted techniques allowed the production of 3D models without resorting to manual post-production interventions or hard painting.

The recent advances in rapid prototyping technology suggest driving a strong effort in this direction, and further important improvements in color enhancement procedures for 3D reconstructions are expected.

7.11 Conclusions

The cultural heritage field has undergone profound changes in recent years, because of an increasingly demanding public but also due to the greater awareness that culture is a real and important "good" that can generate wealth and employment. The computer expert that can manage virtual restoration programs has now joined the more traditional professional figure of the restorer, in a still experimental way. However, it is necessary to develop a scientific approach to digital restoration of artworks in order to avoid simplistic applications made with the sole purpose to impress the audience. Moreover, a close collaboration among art historians, conservators, and scientists is needed to produce results of real significance and utility.

Fig. 7.52 Details of St.John's head in visible light (upper images), UV fluorescence (middle images), and IR reflectography (lower images)

We must never forget that a restoration intervention, as in itself always traumatic for the piece of art, should be carried out only in case it is necessary for the survival of the artwork; therefore, when it is necessary to improve the readability of the image, then traditional restoration can be replaced by digital restoration, thereby conserving the integrity of the original materials: In other words, one can place side by side the fragmented piece and its reconstructed image, cleaned up or rebuilt, but only where it can be done without falling into arbitrary solutions.

There are applications that, in this specific field, provide a number of interesting proposals: The virtual restoration does not act on the art work but simulates a visual and aesthetic improvement of the work, thus enhancing it. It also gives the possibility to choose a series of solutions before technical operations are performed.

Electronic or digital restoration can be therefore defined as the set of digital processing of 2D or 3D computer graphics, allowing for visual and aesthetic improvements of the work or a hypothetical reconstruction that is not real, but virtual precisely. Moreover, it is useful to better understand a work of art (both mobile and immobile) or document archives which for serious reasons of physical degradation cannot be easily restored in the traditional way [40].

Color enhancement and contrast techniques are increasingly useful to improve the quality of the virtual restoration or reconstruction, and the cultural heritage has proven to be a valid and interesting field of research to validate the developed methods.

This definition of "virtual restoration" means the technical tools by which the restoration is done, rather than the methodology used. A feature of the virtual restoration is that direct action does not require real work of art but is made on the digital image that represents it. For this fact, virtual restoration is not in competition with the "physical" restoration, but may represent an interesting procedure to support and study it, as described in the abovementioned study cases.

Acknowledgments We are grateful to Prof. Donatella Biagi Maino (University of Bologna), Claudio Bonifazzi and Agostino Tartari (University of Ferrara), and Bruno Toscano (University of Roma III) for useful discussions and valuable suggestions and contributions to this work. We are thankful for the kind hospitality at the National Gallery of London, and collaboration by Jill Dunkerton, John Cupitt, and David Saunders (now Head of the Scientific and Restoration Laboratories of the British Museum). Restorers Bruno Bruni, Bernardino Sperandio, and Rolando Ramaccini (COO.BE.C., Spoleto), and Paolo Virilli (Tecnireco, Spoleto) provided essential advice. Finally, Drs. Stefano Ferriani, Alfio Musumeci, and Daniele Visparelli (ENEA, Bologna), Marco Landi, and Elena Nencini (University of Bologna, Ravenna campus) greatly contributed to our longstanding collaboration.

References

1. Brandi, C.: Teoria del restauro. Edizioni di Storia e Letteratura, Roma (1963)
2. Lewin, R.: Complexity, 2nd edn, p. 41. Orion Books Ltd, London (2001)
3. Maino, G., Bruni, S., Ferriani, S., Musumeci, A., Visparelli, D.: Multispectral analysis of paintings and wooden sculptures. In: Proceedings of II Congresso Nazionale AIAr Scienza e Beni Culturali, pp. 203–214. Patron Editore, Bologna. (2002)
4. Goethe, W.: Zur Farbenlahre. Cotta, Tubingen (1810)
5. Torcellini, D., Biagi Maino, D., Maino, G.: Il dibattito otto-novecentesco sulla percezione visiva del colore per la storia e la teoria del restauro. In: Raggi A., (ed.) Colore e colorimetria: contributi multidisciplinari, Atti della 1° Conferenza Nazionale del Gruppo del Colore, Pescara, October 20–21, 2005, Collana Quaderni di Ottica e Fotonica 13, pp. 203–210. Centro Editoriale Toscano, Firenze (2005)
6. Portal, F.: Des couleurs symboliques dans l'Antiquité, le Moyen Age et le temps modernes, Treuttel et Würtz, Paris (1837)
7. Pastoureau, M.: Bleu. Histoire d'une couleur. Editions du Seuil, Paris (2000)
8. Cesa Bianchi, M., Beretta, A., Luccio, R.: La Percezione. Un'introduzione alla psicologia della visione. Franco Angeli, Milan (1972)
9. Maino, G.: Immagini per la storia. La documentazione nell'età digitale, in L'immagine del Settecento da Luigi Ferdinando Marsili a Benedetto XIV, Biagi Maino,D. (ed.) Archivi di Arte Antica, pp. 129–159. Umberto Allemandi Editore, Torino, (2005)
10. Maino, G.: Igino Benvenuto Supino nella cultura del suo tempo, in Igino Benvenuto Supino 1858–1940. Omaggio ad un padre fondatore. Paola Bassani Pacht, (ed.), pp. 215–229. Edizioni Polistampa, Firenze (2006)
11. Itten, J.: Kunst der Farbe. Otto Meier Verlag, Ravensburg (1961)
12. Albers, J.: Interaction of Color. Yale University Press, New Haven (1971)
13. Marr, D.: Vision. A Computational Investigation into the Human Representation and Processing of Visual Information, W.H. Freman and Company, New York (1982)

14. Zeki, S.: Inner Vision. An Exploration of Art and the Brain. Oxford University Press, Oxford (1999); idem, Splendors and Miseries of the Brain. Love, Creativity, and the Quest for Human Happiness, Wiley-Blackwell (2009)
15. Hérault, J.: Vision: Images, Signals and Neural Networks. Models of Neural Processing in Visual Perception. World Scientific Publishing Co., Singapore (2010)
16. Parraman, C., Rizzi, A.: Spatial image processing for the enhancement and restoration of film, photography and print. In: Proceedings of INFOS 2010, the 7th International Conference on Informatics and Systems, March 28–30, 2010, Cairo, Egypt, and references quoted therein (2010)
17. Hossain, Md.F., Alsharif, M.R., amashita, K.Y.: An approach to color image enhancement using minimum mean brightness error dynamic histogram equalization. Int. J. Innov. Comput. **7**, 827–840 (2011)
18. Ramesh, S.M., Shanmugam, A.: A new technique for enhancement of color images by scaling the discrete cosine transform coefficients. Int. J. Electron. Commun. Technol. **2**, 34–37 (2011)
19. Vishwakarma, A.K., Mishra, A.: Color image enhancement techniques: a critical review. Indian J. Comput. Sci. Eng. **3**, 39–45 (2012)
20. Tiwari, A., Singh, P.: Analysis of color contrast enhancement techniques. Int. J. Emerg. Technol. Adv. Eng. **4**, 185–191 (2014)
21. Biagi, D., Maino, E., Grimaldi, Maino, G.: Analisi multispettrali su un dipinto di Pietro Lianori. Archeomatica. **0**, 16–20 (2009)
22. Geladi, P., Grahn, H.: Multivariate Image Analysis. Wiley, New York (1996)
23. Geladi, P.: Chemometrics in spectroscopy I: classical chemometrics. Spectrochim. Acta B. **58**, 767–782 (2003)
24. Huang, J., Wium, H., Qvist, K., Esbensen, K.H.: Multi-way method in image analysis: Relationships and applications. Chemom. Intell. Lab. Syst. **66**, 141–158 (2003)
25. Jackson, J.E.: A User Guide to Principal Components. Wiley Interscience, New York (1992)
26. Kiers, H.A.L.: Towards a standardized notation and terminology in multiway analysis. J. Chemometr. **14**, 105–122 (2000)
27. Gonzales, C., Woods, E.: Digital Image Processing. Addison-Wesley, New York (1993)
28. Bonifazzi, C., Ferriani, S., Maino, G., Tartari, A.: Multispectral examination of paintings and works of art: A principal component analysis approach. In: Proceedings of CLADAG 2003 Meeting of the Classification and Data Analysis Group of the Italian Statistical Society, pp. 67–70. CLUEB, Bologna (2003)
29. Campbell, L.: 'Jean Gossart, Virgin and Child' published online 2011, from The Sixteenth Century Netherlandish Paintings with French Paintings before 1600, London. http://www.nationalgallery.org.uk/paintings/research/jan-gossaert-virgin-and-child Accessed 23 June 2015.
30. Bomford, D., Dunkerton, J., Wyld, M.: A Closer Look: Conservation of Paintings, pp. 91–94. National Gallery, London (2009)
31. Campbell, L., Dunkerton, J.: A famous Gossaert rediscovered. The Burlington Magazine 138, pp. 164–173. (1996)
32. Monti, M., Maino, G.: L'informatica per il mosaico, tre casi prototipali, Archeomatica III, 1, pp. 22–27. (2012)
33. Monti, M., Maino, G.: Image processing and a virtual restoration hypothesis for mosaics and their cartoons. In: Proceedings of Image Analysis and Processing—ICIAP 2011, Lecture Notes in Computer Science (LNCS 6979),pp. 486–495. Springer, Berlin (2011)
34. Maioli, M.G.: La basilica di San Severo a Classe, scavo e architettura, in Santi, Banchieri, Re. Ravenna e Classe nel VI secolo. San Severo il tempio ritrovato, Milan (2006)
35. Racagni, P.: Del distacco dei mosaici e della loro conservazione, in La basilica ritrovata. I restauri dei mosaici antichi di San Severo a Classe, Ravenna (2010)
36. Nencini, E., Maino, G.: From the physical restoration for preserving to the virtual restoration for enhancing. In: Proceedings of Image Analysis and Processing—ICIAP 2011, Lecture Notes in Computer Science (LNCS 6978), pp. 700–709. Springer, Berlin (2011)

37. Santos, J. jr, Bellon, O., Silva, L., Vrubel, A.: Improving 3D reconstruction for digital art preservation, In: Proceedings of Image Analysis and Processing—ICIAP 2011, Lecture Notes in Computer Science (LNCS 6979), pp. 374–383 Springer, Berlin (2011)
38. Maino, G., Biagi Maino, D.: Indagini sulla caratterizzazione e la datazione di sculture lignee, in La Deposizione lignea in Europa. L'immagine, il culto, la forma, pp. 677–711. Electa Editori Umbri Associati, Perugia (2004)
39. Cignoni, P., Gobbetti, E., Pintus, R., Scopigno, R.: Color enhancement for rapid prototyping. In: Proceedings of 9th International Symposium on Virtual Reality, Archaeology and Cultural Heritage—VAST. (2008)
40. Barni, M., Pelagotti, A., Piva, A.: Image processing for the analysis and conservation of paintings: opportunities and challenges. IEEE Signal Process Mag. **22**, 1–4 (2005)

Chapter 8
A GPU-Accelerated Adaptive Simultaneous Dynamic Range Compression and Local Contrast Enhancement Algorithm for Real-Time Color Image Enhancement

Chi-Yi Tsai and Chih-Hung Huang

8.1 Introduction

Modern robotic vision systems usually require the use of image enhancement algorithms as they play an important preliminary processing role in computer vision applications. Numerous image enhancement methods have been proposed to deal with monochromatic images. However, they often cannot be directly applied on multichannel images, which usually give more information about the environment than monochromatic images. Consequently, color image enhancement has now become an active research issue in image enhancement processing. Moreover, because color images contain much more data to be processed, the improvement on computational efficiency of color image enhancement to achieve real-time performance is still a challenging task.

This chapter presents a parallel implementation of an existing adaptive color image enhancement scheme, which improves visual quality of color images suffered from low dynamic range (LDR) and poor contrast defects. Images captured from a digital camera may be in poor visibility due to the limitation on dynamic range of image sensors. This problem typically can be resolved by compressing the dynamic range of captured images, and this image enhancement process is commonly known as dynamic range compression. Several dynamic range compression techniques have been proposed based on Retinex theory, which is a lightness and color perception model of human vision [1]. Retinex-based techniques [2–7] are general purpose methods, simultaneously achieving dynamic range compression, local-contrast enhancement, and color consistency; however, these algorithms are usually computationally expensive, requiring algorithmic simplification and hardware acceleration to achieve real-time performance [8].

C.-Y. Tsai (✉) · C.-H. Huang
Department of Electrical Engineering, Tamkang University, 151 Ying-zhuan Road, Tamsui District, New Taipei City 25137, Taiwan R.O.C.
e-mail: chiyi_tsai@mail.tku.edu.tw

M. E. Celebi et al. (eds.), *Color Image and Video Enhancement*,
DOI 10.1007/978-3-319-09363-5_8

On the other hand, some dynamic range compression methods are combined with a local-contrast enhancement algorithm to improve visual quality of color images. For example, the authors in [9] proposed an adaptive and integrated neighborhood dependent approach for nonlinear enhancement (AINDANE) algorithm, which consists of two separate processes: adaptive luminance enhancement and adaptive contrast enhancement. The former process compresses dynamic range of the input image, and the latter process enhances local contrast after range compression. The authors also extended their work to an illuminance perception-based nonlinear enhancement algorithm, which provides high quality of enhanced images, robust performance, and fast processing speed, i.e., achieving 26 frames per second (fps) for frame size of 360 × 240 pixels [10]. The authors in [11] proposed a spatially variant dynamic range compression algorithm with local contrast preservation based on the concept of local contrast range transform. This method usually performs well in color image enhancement; however, its enhancement procedure requires high computational costs with a large memory and leads to an inefficient algorithm. To improve computational efficiency, Unaldi et al. proposed a fast and robust wavelet-based dynamic range compression (WDRC) algorithm with local-contrast enhancement [12]. The processing time of WDRC algorithm is notably reduced, since it fully operates in wavelet domain. The authors in [13] proposed a fast dynamic range compression with local-contrast preservation (FDRCLCP) algorithm, which is able to preserve image details during dynamic range compression. An important feature of FDRCLCP algorithm is that, it achieves real-time performance (30 fps for 640 × 480 videos on a 2.4-GHz Intel Core 2 desktop platform) without hardware acceleration.

In a previous work [14], a simultaneous dynamic range compression and local-contrast enhancement (SDRCLCE) algorithm, which is able to enhance image local contrast during dynamic range compression, was developed. One merit of the SDRCLCE algorithm is that, it is amenable to parallel processing on a multi-core processor such as a graphics processing unit (GPU) device. This advantage motivates us to develop a GPU-accelerated SDRCLCE algorithm to improve the real-time performance for color image enhancement. An existing adaptive intensity transfer function, which efficiently handles the enhancement of LDR images, is first presented. The proposed algorithm is then derived by combining the adaptive intensity transfer function with the GPU-accelerated SDRCLCE algorithm, achieving real-time performance in processing high-resolution color images.

The rest of this chapter is organized as follows. Section 2 introduces the existing SDRCLCE method, which is related to the development of the proposed algorithm. Section 3 introduces an existing image-dependent dynamic range compression format with a local-contrast enhancement algorithm to enhance LDR images efficiently. Section 4 presents a GPU acceleration method for the proposed algorithm to achieve better real-time performance. Section 5 presents experimental results to evaluate the effectiveness and efficiency of the proposed GPU-accelerated color image enhancement method. Section 6 concludes the contributions of this work.

8.2 The Existing SDRCLCE Algorithm

This section briefly introduces the existing SDRCLCE algorithm [14] that efficiently achieves the dynamic range compression while enhancing local contrast of color images. The SDRCLCE algorithm [14] provides an efficient way to compress image dynamic range and to enhance image local contrast simultaneously. Let $L_{in} \in [0,1]$ denote the normalized luminance component of an input color image, $T[\bullet] \in C^1$ be an arbitrary monotonically increasing and continuously differentiable intensity transfer function, and $T'[L_{in}(x,y)] = dT[X]/dX\,|_{X=L_{in}(x,y)}$ be the corresponding derivative function of T. Then, the general form of SDRCLCE algorithm is given by

$$L_{out}^{SDRCLCE}(x,y) = \left\{ f_n^{-1}(x,y)\left\{\beta(x,y)L_T(x,y) + [1-\beta(x,y)]L_{lce}(x,y)\right\} \right\}_0^1, \tag{8.1}$$

$$f_n(x,y) = \left\{\beta_{\max}(x,y)T(L_{in}^{\max}) + [1-\beta_{\max}(x,y)][\alpha T'(L_{in}^{\max})L_{in}^{\max}]\right\}_\varepsilon^1, \tag{8.2}$$

$$L_{lce}(x,y) = \alpha T'[L_{in}(x,y)]L_{in}(x,y), \tag{8.3}$$

where $\alpha = \pm 1$ is a two-valued parameter, $L_T(x,y) = T[L_{in}(x,y)]$ is the luminance output obtained from the function T, and $L_{lce}(x,y)$ is the local-contrast enhancement component for each pixel. In Eq. (8.1), the operator $\{x\}_a^b$ means that the value of x is bounded to the range $[a, b]$, and $\beta(x, y)$ is a weighting factor defined as

$$\beta(x,y) = \frac{L_{in}(x,y)+\varepsilon}{\bar{L}_{in}(x,y)+\varepsilon}, \tag{8.4}$$

where ε is a small positive value to avoid dividing by zero, and $\bar{L}_{in}(x,y)$ is the value of local average luminance of the pixels in a neighborhood of specified size. In Eq. (8.2), $f_n \in [\varepsilon, 1]$ is a normalization factor associated with the weighting factor $\beta(x, y)$ such that

$$\beta_{\max}(x,y) = \beta(x,y)|_{L_{in}(x,y)=L_{in}^{\max}}, \tag{8.5}$$

where $L_{in}^{\max}$ is the maximum input luminance value. To compute the luminance image of color images, several methods can be used based on NTSC (National Television System Committee) standard [9], sRGB (standard Red-Green-Blue) standard [15], or the definition of HSV (Hue-Saturation-Value) intensity value [16]. However, according to [10], adopting HSV intensity value is suggested because it achieves color consistency in the RGB color image enhancement without color space conversion. Hence, we use the HSV intensity value as the input luminance $L_{in}(x, y)$ for color image enhancement processing.

The SDRCLCE algorithm requires working with a linear color restoration process to extend to color image enhancement, because it only processes the luminance component of the input image. The linear color restoration process usually works in RGB color space. Let $\mathbf{P}_{in}^{RGB} = \left[R_{in}\ G_{in}\ B_{in}\right]^T$ and $\mathbf{P}_{out}^{RGB} = \left[R_{out}\ G_{out}\ B_{out}\right]^T$, respectively, denote the input and output color value of each pixel in RGB color space. Then, the linear color restoration process for each input RGB color pixel can be expressed as:

$$\mathbf{P}_{out}^{RGB}(x, y) = \rho(x, y)\mathbf{P}_{in}^{RGB}(x, y), \tag{8.6}$$

where $\rho(x, y)$ is a nonnegative color-mapping ratio given by

$$\rho(x, y) = \frac{L_{out}^{SDRCLCE}(x, y) + \varepsilon}{L_{in}(x, y) + \varepsilon}, \tag{8.7}$$

where, again, ε is a small positive value to avoid dividing by zero. Equation (8.6) is able to preserve color information of the original image for minimal color shifts appeared in the enhanced result.

8.3 The Adaptive SDRCLCE Algorithm

This section first presents an existing adaptive intensity transfer function that achieves adaptive dynamic range compression to efficiently deal with LDR images, because selecting a suitable intensity transfer function is an important task when employing the SDRCLCE algorithm. The proposed algorithm is then derived by applying the SDRCLCE formula on the adaptive intensity transfer function.

8.3.1 An Existing Adaptive Intensity Transfer Function

In the authors' previous work [17], an adaptive intensity transfer function was proposed to achieve the image-dependent dynamic range compression, while preventing over enhancement in the dark region of the image. Let $S \in (0,2]$ denote a nonzero positive constant, and $\phi \in [0, \min(1, S)]$ a positive parameter controlling the capability of dynamic range compression. Then the adaptive intensity transfer function is given by

$$T[L_{in}(x, y)] = \frac{1}{2}\Big\{L_{in}^{[(1-\phi)z+\phi]}(x, y) + L_{in}^{(2-z)}(x, y) + S(1-z)L_{in}^{(\phi+1)}(x, y) \\ [1 - L_{in}(x, y)]\Big\}, \tag{8.8}$$

where the parameter z is determined by an intensity level L related to the darkness of the input luminance image such that

$$z = \begin{cases} 0, & for\ L \le 50 \\ \dfrac{L-50}{100}, & for\ 50 < L \le 150\,. \\ 1, & for\ L > 150 \end{cases} \tag{8.9}$$

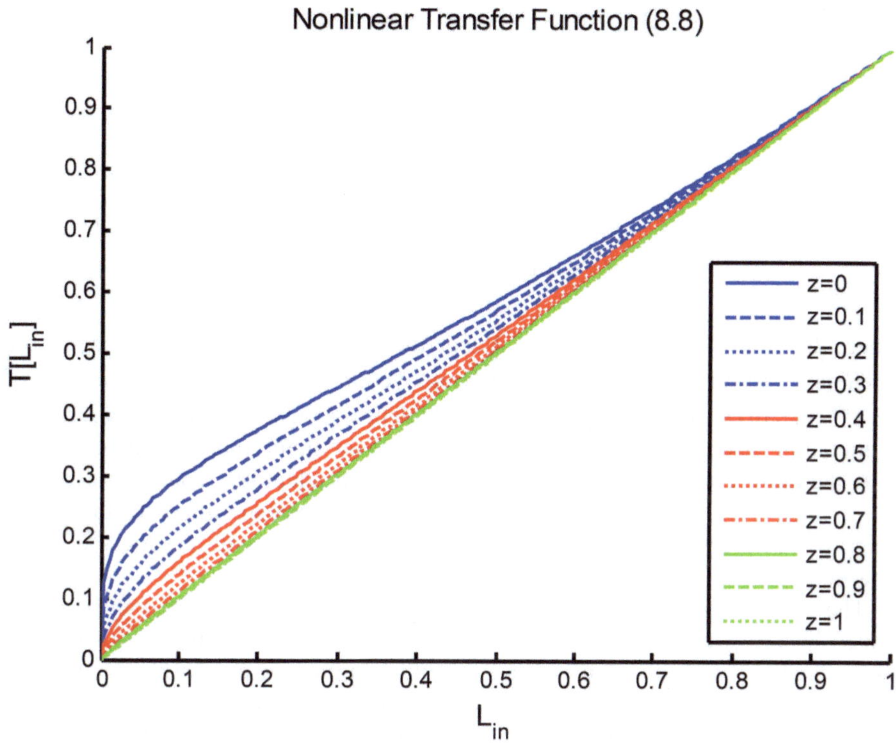

Fig. 8.1 The intensity mapping curve processed by the nonlinear transfer function (8.8) with $S = 0.4$ and $\varphi = 0.25$

Figure 8.1 shows the intensity mapping curve processed by the nonlinear transfer function (8.8) with different values of z. It is clear from Fig. 8.1 that the nonlinear transfer function (8.8) remaps low-intensity pixels to higher intensity pixels and leaves high-intensity pixels as close to unchanged as possible. This feature makes the nonlinear transfer function (8.8) usable to enhance LDR images. Moreover, the adaptive intensity transfer function (8.8) satisfies a zero-input, zero-output condition for all $z \in [0, 1]$. This property helps to achieve an acceptable dynamic range compression result with less color artifacts.

8.3.2 Application of SDRCLCE Formula into the Adaptive Intensity Transfer Function

As the adaptive intensity transfer function (8.8) satisfies the monotonically increasing and continuously differentiable conditions, an image-dependent image enhancement algorithm can be easily derived by substituting this function to the SDRCLCE formula (8.1)–(8.3) such that

$$L_{out\_T}^{SDRCLCE}(x,y) = \left\{ [f_n^T(x,y)]^{-1} \{ \beta(x,y) L_T(x,y) + [1-\beta(x,y)] L_{lce}^T(x,y) \} \right\}_0^1, \tag{8.10}$$

$$f_n^T(x,y) = \{ \beta_{\max}(x,y) T(L_{in}^{\max}) + [1-\beta_{\max}(x,y)][\alpha T'(L_{in}^{\max}) L_{in}^{\max}] \}_\varepsilon^1, \tag{8.11}$$

$$L_{lce}^T(x,y) = \alpha T'[L_{in}(x,y)] L_{in}(x,y), \tag{8.12}$$

$$T'[L_{in}(x,y)] = \frac{1}{2}\left\{ \begin{array}{l} [(1-\phi)z+\phi][L_{in}(x,y)+\varepsilon]^{[(1-\phi)z+\phi-1]} + (2-z)L_{in}^{(1-z)}(x,y) \\ +S(1-z)L_{in}^{\phi}(x,y)\{(\phi+1)[1-L_{in}(x,y)]-L_{in}(x,y)\} \end{array} \right\}, \tag{8.13}$$

where $L_{in}^{\max} = 1$ and $T(L_{in}^{\max}) = 1$. For the enhancement of color images, the linear color restoration process (8.6) is employed, but using $L_{out\_T}^{SDRCLCE}$ instead of $L_{out}^{SDRCLCE}$ in (8.7). Moreover, the value of z depends on an intensity level L related to the cumulative distribution function (CDF) of the input intensity image. Therefore, the range compression property of Eq. (8.10) is adaptively changed according to the intensity histogram of the input image. This advantage improves the range-compression property of the proposed method in dealing with a variety of LDR images.

The contrast-enhancement property of the SDRCLCE algorithm can be determined by the two-valued parameter α. That is, setting α =1 in Eqs. (8.11) and (8.12) leads to preserve local contrast of output luminance during range compression, and $\alpha = -1$ leading to enhance local contrast. Figure 8.2 shows an example to explain this property. From Fig. 8.2, visually comparing the SDRCLCE output (8.10) with the adaptive tone-mapping output (8.8) observes that the SDRCLCE algorithm is able to enhance local contrast of the image without employing an extra contrast enhancement process. This advantage makes the SDRCLCE algorithm more suitable to meet the requirements of real-time applications.

8.3.3 Extension to Video Signal Processing

The proposed color image enhancement algorithm also can be extended to video signal processing. According to digital video standards, YC_bC_r is the most commonly used color space in video rendering, and thus most of the existing video enhancement methods are performed in YC_bC_r color space. However, these methods may result in low saturation images, as they usually leave C_b, C_r components unchanged in the enhancement process. To overcome this problem, a YC_bC_r linear color restoration process, which can preserve the color information of input video signals during the video enhancement process in YC_bC_r color space, was proposed in [14]. Let $\left[Y_{in}\ C_{in}^b\ C_{in}^r \right]$ and $\left[Y_{out}\ C_{out}^b\ C_{out}^r \right]$ denote the input and output color pixel in YC_bC_r color space, respectively. According to [14], the remapping of luminance and chrominance components of each YC_bC_r color pixel are respectively given by

$$Y_{out}(x,y) = \rho(x,y) Y_{in}(x,y) + 16[1-\rho(x,y)], \text{ and} \tag{8.14}$$

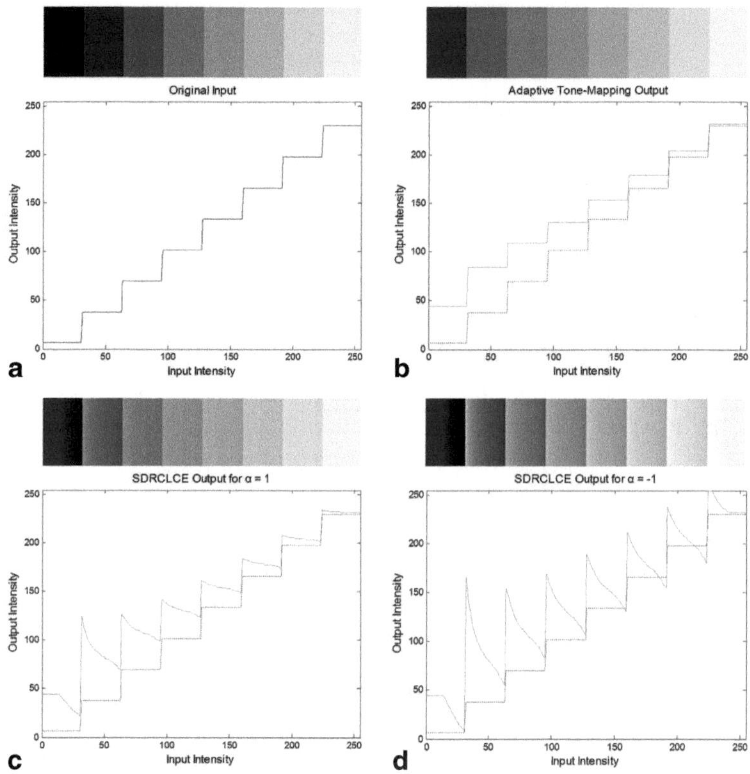

Fig. 8.2 *Top row*. **a** original grayscale image, enhanced result of **b** the adaptive tone-mapping output (8) and the SDRCLCE output (10) with **c** α = 1 and **d** α = − 1. *Bottom row*. the corresponding pixel data of scan lines

$$C^i_{out}(x, y) = \rho(x, y)C^i_{in}(x, y) + 128[1 - \rho(x, y)], \tag{8.15}$$

where $\rho(x, y)$ is the color-mapping ratio defined in (8.7) with $L_{in}(x, y) = Y_{in}(x, y)$, and C^i_j ($i = b, r$, and $j = in, out$) denotes one of the chrominance components. Combining the proposed algorithm with the YC_bC_r linear color restoration formulas (8.14) and (8.15) allows directly enhancing video signals without color space conversion, significantly speeding up the video enhancement process.

8.4 GPU Acceleration of the Adaptive SDRCLCE Algorithm

To achieve real-time performance without hardware acceleration, Ref. [15] reported two lookup table (LUT) acceleration methods, which are also applicable to the proposed algorithm. Figure 8.3 shows block diagrams of these two acceleration methods applied on the proposed algorithm. The first method shown in Fig. 8.3a

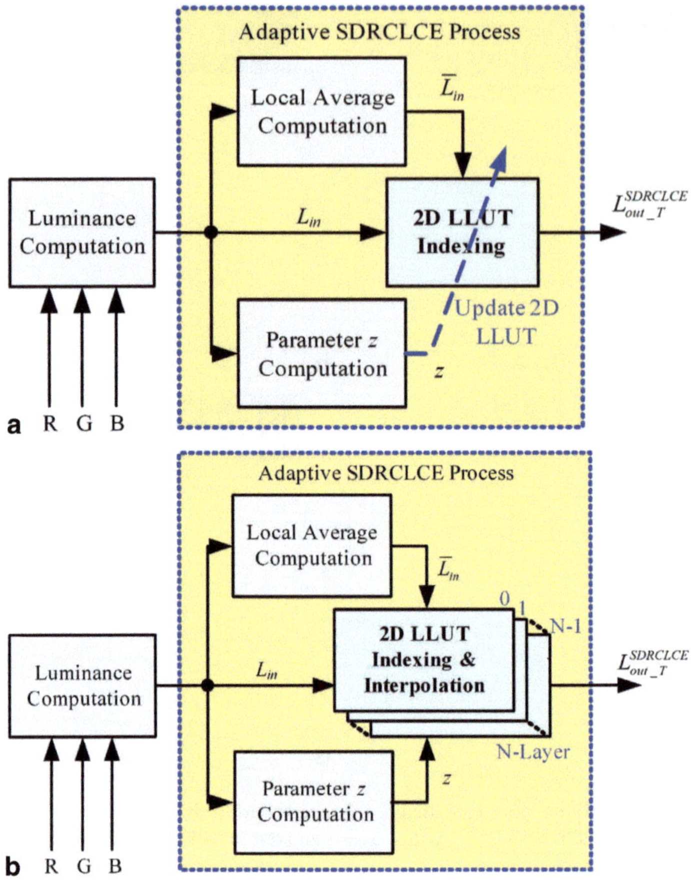

Fig. 8.3 Two LUT acceleration methods of the adaptive SDRCLCE algorithm to achieve real-time performance on the CPU: Accelerated implementation **a** by using a 256-by-256 2D LLUT required updating according to the current value of w, and **b** by using a N-by-256-by-256 3D LLUT combined with a linear interpolation method

uses Eqs. (8.10)–(8.13) to construct and update a 256-by-256 two-dimensional luminance lookup table (2D LLUT), which simplifies the enhancement process to a 2D LLUT indexing operation. On the other hand, the second one shown in Fig. 8.3b employs a N-by-256-by-256 3D LLUT to avoid the LUT update operation, but it requires an extra 2D LUT indexing operation, a linear interpolation operation and costs more memory resources. When implemented on the central processing unit (CPU), we observed that the second method has better computational efficiency than the first one as it does not need floating-point computations, since the linear interpolation operation can be performed via fixed-point approximation. By contrast, the first method can obtain better computational efficiency on the GPU as it needs fewer memory access operations and memory resources than the second one.

Therefore, we chose the first LUT acceleration method [Fig. 8.3a] to implement on the GPU.

In Fig. 8.3a, the proposed adaptive SDRCLCE algorithm consists of four operations: local average computation, parameter z computation, 2D LLUT update, and 2D LLUT indexing. Not all of these operations are suitable to be implemented on the GPU. For an instance, the operation of parameter z computation has to compute the CDF of the input intensity image and determine the intensity level L indicating the darkness of the input image. This operation is apparently a single-threaded sequential process and is unsuitable to be parallelized by multi-threaded programming. Based on this observation, the proposed GPU-accelerated color image enhancement method is thus divided into a CPU/host and a GPU/device process, as shown in Fig. 8.4. In the device process, the GPU handles three parallelizable operations of the proposed adaptive SDRCLCE algorithm, including local average computation, 2D LLUT update and 2D LLUT indexing. On the other hand, the CPU handles the rest operations of the proposed color image enhancement processing, including luminance computation, parameter computation, linear color restoration, and data transfer between the CPU and GPU. In the rest of this section, we focus the discussion on the design of the device processing as it represents a core process of the proposed adaptive SDRCLCE algorithm.

8.4.1 2D LLUT Update

In this work, we use NVIDIA Compute Unified Device Architecture (CUDA) [18] as the GPU development tool due to its high efficiency and programmability. NVIDIA GPUs consist of an array of streaming multiprocessors (SMs), each of which supports a large number of threads. A thread processes a single element of a large data array, such as a pixel of an image or an element of a LUT. All parallel threads executed on a single SM are grouped into a block, and all blocks executed on a CUDA device are grouped into a grid, which corresponds to a CUDA kernel function call in an application. Each block/thread in a grid/block has its own index (ID) number called block/thread ID. Let (b_{idx}, b_{idy}) denote, respectively, the block ID within a grid in x and y dimensions, and (t_{idx}, t_{idy}), respectively, the thread ID within a block in x and y dimensions. Then a global thread ID in x and y dimensions, denoted by (t_x, t_y), can be simply computed by

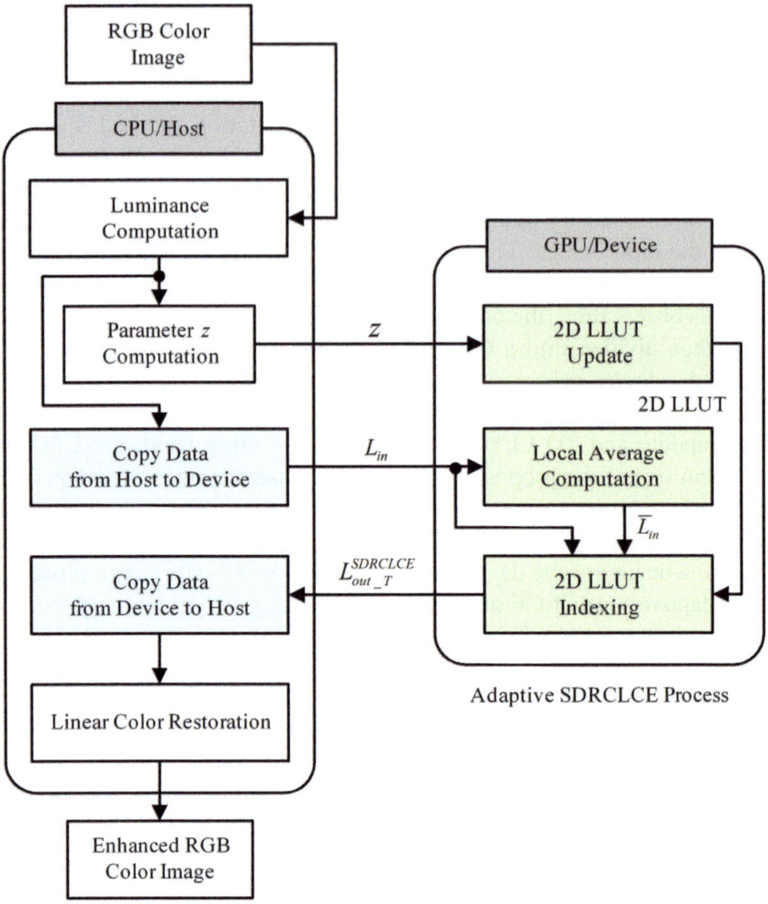

Fig. 8.4 Flowchart of the proposed GPU-accelerated color image enhancement method, in which the GPU handles the parallelizable operations of the proposed adaptive SDRCLCE algorithm

$$t_x = b_{idx} \times b_{dmx} + t_{idx}, \text{ and } t_y = b_{idy} \times b_{dmy} + t_{idy}, \tag{8.16}$$

where (b_{dmx}, b_{dmy}) denote the block size in x and y dimensions, respectively. Based on this computational model, the operation of updating 256-by-256 2D LLUT can be efficiently achieved by a CUDA kernel function call with 256 blocks, each having 256 threads. **Algorithm 1** shows a pseudo code of the proposed CUDA-based 2D LLUT updating kernel function, which is a 1D grid, 1D block, pixel-wise CUDA kernel function associated with four parameters (z, α, S, φ) defined in the previous section. In **Algorithm 1**, each thread performs the proposed algorithm described in Eqs. (8.10)–(8.13) to compute a value of the 2D LLUT at the location $(L, \bar{L})$, where $L \in [0,255]$ and $\bar{L} \in [0,255]$ represent the thread ID and block ID number, respectively. Therefore, the processing time of 2D LLUT update can be significantly reduced by parallelizing with 256 × 256 threads.

Algorithm 1: CUDA-based 2D LLUT updating kernel function

Input: t_{idx}, the thread ID within a block
b_{idx}, the block ID within a grid
z, the brightness level of an image
α, the two-valued parameter
S, the nonzero positive constant
φ, the positive parameter controlling dynamic range compression

Output: $\mathrm{LUT_{2D}}$, the 2D LLUT

$L = t_{idx}$;
$\bar{L} = b_{idx}$;
$t_f = 256/(\bar{L}+1)$;
$F = t_f \times T(1,z,S,\varphi) + (1-t_f)\times\alpha\times T'(1,z,S,\varphi)$;
if $F > 1$ then
    $F = 1$;
else if $F < \varepsilon$ then
    $F = \varepsilon$;
end if
$t_f = L/255$;
$L_{out} = \{T\,(t_f,z,S,\varphi)\times[(L+1)/(\bar{L}+1)] + [1-(L+1)/(\bar{L}+1)] \times \alpha \times T'(t_f,z,S,\varphi)\times(L/255)\}\times(255/F)$;

Return: $\mathrm{LUT_{2D}}(L,\bar{L}) = (L_{out} > 255)\,?\,255 : (L_{out} < 0)\,?\,0 : L_{out}$;

8.4.2 Local Average Computation

To compute the local average luminance of the image, a conventional method is to convolute the luminance image with a spatial low-pass filter such that $\bar{L}_{in} = L_{in} \otimes F_L$, where the operator $\otimes$ denotes the 2D convolution operation, and F_L denotes a spatial low-pass filter kernel function satisfying the condition $\sum_x \sum_y F_L(x, y) = 1$. As mentioned in Sect. 2.2, the kernel function F_L can be selected as a blurring filter or an edge-preserving filter. In this work, we employ a Gaussian blur kernel function given by $F_L(x, y) = Ke^{-(x^2+y^2)/\sigma^2}$, where K is the scalar of the filter to normalize the sum of filter coefficients to 1, and σ is the standard deviation of the Gaussian blur kernel. The filter kernel size is determined by $\lfloor(2\sigma + 3) \times (2\sigma + 3)\rfloor$, where the operator $\lfloor X \rfloor$ means the largest integer smaller than or equal to X. Thus, increasing the value of σ will increment the level of local-contrast enhancement and the processing time of the proposed method. In CUDA, the 2D convolution operation can be simply implemented by using functions in the NVIDIA Performance Primitives (NPP) library [19], which is a collection of GPU-accelerated image, video, and signal processing functions. For example, NPP 6.0 library provides an image filter function, nppiFilterGauss_8u_C1R, which performs 2D Gaussian blur filtering on single-channel 8-bit unsigned images with kernel sizes up to 5×5 ($\sigma = 1$). Therefore, the computation of the 2D Gaussian blur filter can be easily implemented on NVIDIA CUDA devices by using the latest NPP library.

Algorithm 2: CUDA-based 2D LLUT indexing kernel function

Input: t_{idx}, the thread ID within a block in x dimension
t_{idy}, the thread ID within a block in y dimension
b_{idx}, the block ID within a grid in x dimension
b_{idy}, the block ID within a grid in y dimension
b_{dmx}, the block size in x dimension
b_{dmy}, the block size in y dimension
W, the image width
H, the image height
$\overline{L}_{in}$, the local average luminance image
L_{in}, the input luminance image
LUT_{2D}, the 2D LLUT

Output: $L_{out\_T}^{SDRCLCE}$, the SDRCLCE output image

1: $t_x = b_{idx} \times b_{dmx} + t_{idx}$;
2: $t_y = b_{idy} \times b_{dmy} + t_{idy}$;
3: if $t_x < W$ and $t_y < H$ then
4: $\quad L_{out\_T}^{SDRCLCE}(t_x, t_y) = LUT_{2D}(L_{in}(t_x, t_y), \overline{L}_{in}(t_x, t_y))$
5: end if

Return: $L_{out\_T}^{SDRCLCE}(t_x, t_y)$;

Note that, according to [14], the parameter σ controls the level of local-contrast enhancement. That is, increasing the value of this parameter gets more contrast enhancement on the resulting image, but probably reducing runtime performance as the kernel size of the Gaussian filter increases. Therefore, there is a trade-off between the runtime performance and the level of local-contrast enhancement if the value of σ is increased. In this study, we set the value of σ equal to 1 because the kernel size of the current nppiFilterGauss_8u_C1R function is only up to 5 $\times$ 5.

8.4.3 2D LLUT Indexing

As we obtained the local average image, the input luminance image can be enhanced by a simple 2D LLUT indexing operation. Similar to the 2D LLUT update, this operation is also pixel-wise, but is implemented by a 2D grid, 2D block CUDA kernel function. Let W denote the input image width, and H the input image height. Given two positive integers b_{dmx} and b_{dmy} determine the desired 2D block size in x and y dimensions, respectively. Then, the 2D grid size (g_{dmx}, g_{dmy}) used in the kernel function call is set as

$$g_{dmx} = \left\lceil {}^{W}\!/_{b_{dmx}} \right\rceil, \; and \; g_{dmy} = \left\lceil {}^{H}\!/_{b_{dmy}} \right\rceil, \tag{8.17}$$

where the operator $\lceil X \rceil$ means the smallest integer larger than or equal to X. The size of each 2D block within the grid is set as (b_{dmx}, b_{dmy}), so that the grid contains least

Table 8.1 Parameter setting for the adaptive SDRCLCE method used in the experiments

Parameter	Value	Description
σ	1	The sigma value used in a single-scale Gaussian filter
α	−1	The two-valued parameter used in Eqs. (8.11) and (8.12)
S	0.4	The nonzero positive constant used in Eqs. (8.8) and (8.13)
φ	0.25	The positive parameter used in Eqs. (8.8) and (8.13)
ε	1/255	Small positive value used in Eqs. (8.7) and (8.13)

$W \times H$ parallel threads, each of them performs a single LUT indexing operation at location (t_x, t_y) determined by Eq. (8.16) under conditions $b_{idx} \in [0, g_{dmx} - 1]$, $b_{idy} \in [0, g_{dmy} - 1]$, $t_{idx} \in [0, b_{dmx} - 1]$, and $t_{idx} \in [0, b_{dmx} - 1]$. **Algorithm 2** shows pseudo code of the proposed CUDA-based 2D LLUT indexing kernel function. In **Algorithm 2**, each thread requires only memory access operation in global memory, which usually has many times more bandwidth than the system memory. Therefore, this design can greatly speed up the entire LUT indexing process.

8.5 Experimental Results

The performance of the proposed CUDA-based color image enhancement algorithm has been tested on a Windows 7 machine equipped with a 2.9 GHz Intel Core i7-3520M CPU and a NVIDIA NVS 5200M GPU, which has 96 CUDA cores [20]. The proposed GPU acceleration method was implemented on the GPU by using CUDA 6.0 toolkit and NPP 6.0 library. Table 8.1 tabulates the parameter setting of the adaptive SDRCLCE method used in the experiments, which consist of two parts: the first part presents quantitative and visual comparisons between CPU and GPU versions of the adaptive SDRCLCE algorithm; and the second part is computational performance analysis of the proposed GPU acceleration method. Two image quality metrics were adopted in the experiments to quantify the amount of distortion produced by the GPU-accelerated implementations. The first one is PSNR (in decibels) metric defined as

$$PSNR(\mathbf{P}_{CPU}^{RGB}, \mathbf{P}_{GPU}^{RGB}) = 10\log_{10}\left[255^2\left(\frac{1}{UV}\sum_{y=1}^{V}\sum_{x=1}^{U}\left\|\mathbf{P}_{CPU}^{RGB}(x,y) - \mathbf{P}_{GPU}^{RGB}(x,y)\right\|^2\right)\right], \tag{8.18}$$

where U, V are the total column and row numbers of the image. $\mathbf{P}_{CPU}^{RGB}(x, y)$ and $\mathbf{P}_{GPU}^{RGB}(x, y)$ are the color vectors at the (x, y)th position of the color image produced by the CPU and GPU implementations, respectively. The second one is SSIM index proposed in [21], such that

$$SSIM(\mathbf{L}_{CPU}, \mathbf{L}_{GPU}) = \frac{(2\mu_C\mu_G + C_1)(2\sigma_{CG} + C_2)}{(\mu_C^2 + \mu_G^2 + C_1)(\sigma_C^2 + \sigma_G^2 + C_2)}, \tag{8.19}$$

where $\mathbf{L}_{CPU}$ and $\mathbf{L}_{GPU}$ are the luminance images extracted from the color images $\mathbf{P}_{CPU}^{RGB}$ and $\mathbf{P}_{GPU}^{RGB}$, respectively. (μ_C, σ_{C^2}) and (μ_G, σ_{G^2}) are the mean and variance of $\mathbf{L}_{CPU}$ and $\mathbf{L}_{GPU}$, respectively. σ_{CG} is the covariance between $\mathbf{L}_{CPU}$ and $\mathbf{L}_{GPU}$. C_1 and C_2 are the two constants, and their default values are set as $C_1 = (0.01 \times 255)^2$ and $C_2 = (0.03 \times 255)^2$. Note that the SSIM measure is a floating value between -1 and 1, and an SSIM value close to 1 indicates a high similarity between two images.

8.5.1 *Quantitative and Visual Comparisons Between CPU and GPU Implementations*

The LUT-accelerated CPU implementation shown in Fig. 8.3a was employed in the experiments to evaluate the performance of the proposed GPU implementation. To provide a fair comparison, the CPU implementation also uses a single-scale Gaussian filter with kernel sigma $\sigma = 1$ to compute the local average image. For both CPU and GPU implementations, the values $S = 0.4$ and $\varphi = 0.25$ are used as the default setting, and the small positive value ε is set as 1/255. Figure 8.5 shows six test images used to evaluate the performance of the GPU implementation. Figures 8.5b1–b6 and 8.5c1–c6 present the enhanced results of the test images obtained from the CPU and GPU implementations, respectively. A visual comparison shows that the GPU implementation not only produces satisfactory results in range compression and contrast enhancement without inducing color artifacts, but also obtains similar enhancement results compared to the CPU implementation. Table 8.2 records PSNR and SSIM measures between image enhancement results obtained from the CPU and GPU implementations. It is clear, from Table 8.2, that the GPU implementation produces high PSNR and SSIM values compared to the CPU implementation. This implies that both methods produce similar color image enhancement results, and thus the performance of the proposed GPU acceleration method is validated.

8.5.2 *Computational Performance Analysis*

The computational performance of the proposed GPU implementation was evaluated by comparing the execution time with the CPU implementation in processing images with different resolutions M × M, from 256 × 256 to 4096 × 4096. The execution time of each image resolution is a time average of 20 experiments. Figures 8.6a and 8.6b show the execution time of each step in the LUT-accelerated and GPU-accelerated adaptive SDRCLCE process implemented on the CPU and GPU, respectively. It is clear from Fig. 8.6a that the execution time of local average computation and 2D LLUT indexing increases rapidly as the image resolution increases. By contrast, the execution time of 2D LLUT update is independent of the image resolution; however, this step costs the most execution time (about 27.6 ms in average)

Fig. 8.5 Experimental results: **a1–a6** original images, **b1–b6** enhanced images produced by the CPU implementation, and **c1–c6** by the GPU implementation

Table 8.2 Quantitative measure between enhanced images obtained from the CPU and GPU implementations

Results of the CPU implementation	Figure 8.5 b1	Figure 8.5 b2	Figure 8.5 b3	Figure 8.5 b4	Figure 8.5 b5	Figure 8.5 b6
Results of the GPU implementation	Figure 8.5 c1	Figure 8.5 c2	Figure 8.5 c3	Figure 8.5 c4	Figure 8.5 c5	Figure 8.5 c6
PSNR (dB)	37.7595	41.1548	41.5116	38.9940	41.2804	42.3052
SSIM	0.9986	0.9995	0.9986	0.9993	0.9989	0.9988

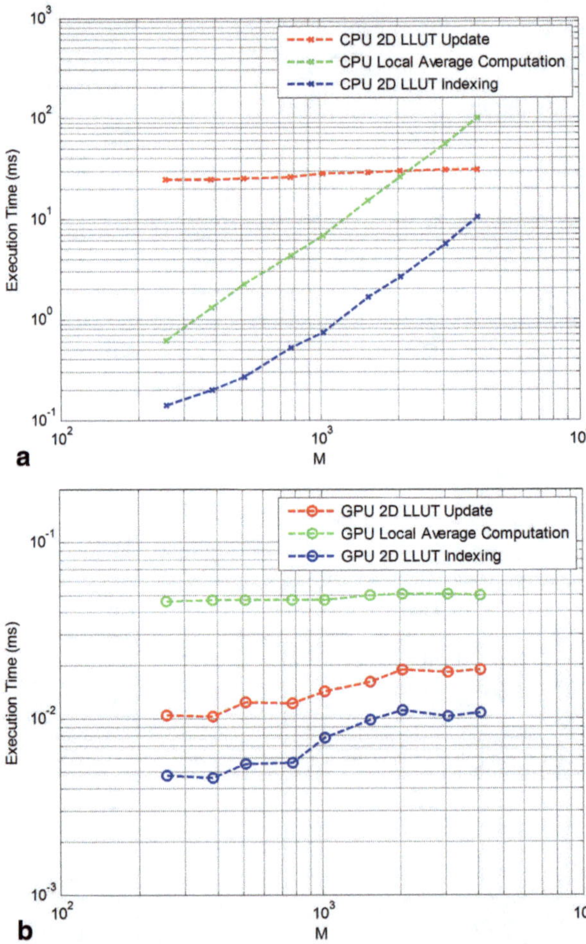

Fig. 8.6 Execution time of each step in the **a** LUT-accelerated and **b** GPU-accelerated adaptive SDRCLCE process with respect to different image resolutions

in processing images with sizes up to 2048 × 2048 pixels. On the other hand, the execution time of 2D LLUT update and 2D LLUT indexing on the GPU just slightly

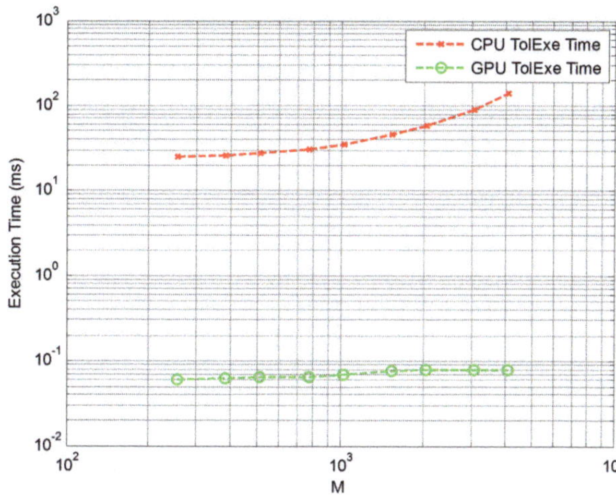

Fig. 8.7 TolExe time comparison of the CPU and GPU implementations

increases about 8 μs and 6 μs, respectively, as image resolution increases from 256 × 256 to 4096 × 4096 [Fig. 8.6b]. Moreover, the execution time of local average computation on the GPU just costs about 48 μs in average independent of the image resolution, significantly improving computational efficiency of the adaptive SDRCLCE process.

We now define a total execution (TolExe) time as the summation of the computational time of each step shown in Fig. 8.6. Figure 8.7 presents a comparison between the CPU and GPU TolExe times. From Fig. 8.7, we observe that the CPU TolExe time increases obviously (from 25.4 ms to 141.8 ms) as image resolution increases. By contrast, the GPU TolExe time just slightly increases from 61.4 μs to 79.5 μs. This experimental result demonstrates a great improvement on the computational efficiency by using the proposed GPU acceleration method in processing high-resolution images up to 4096 × 4096 pixels. However, it is important to notice that the GPU implementation is usually memory bound rather than operationally bound [22]. That is, the performance bottleneck of the GPU implementation is usually dominated by the memory copy (MemCpy) time between the CPU and GPU, as shown in Fig. 8.8. It is clear from Fig. 8.8 that the MemCpy time between the CPU and GPU increases proportionally (from 1.5 ms to 53.6 ms) according to the image resolutions and dominates the actual execution time (TolExe time + MemCpy time) of the GPU implementation.

Figure 8.9 presents the acceleration of the GPU implementation compared to the CPU implementation, which is measured by the following two methods

$$\text{Acceleration of TolExe Time} = \frac{\text{CPU TolExe Time}}{\text{GPU TolExe Time}}, \textit{ and} \tag{8.20}$$

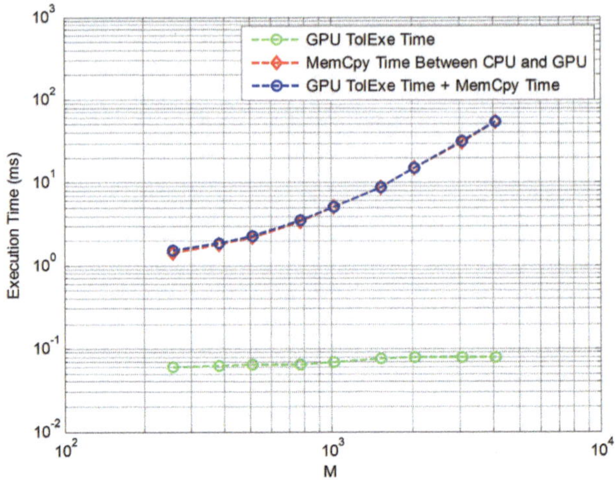

Fig. 8.8 Actual execution time of the GPU implementation, including the cost of memory copy (MemCpy) between the CPU and GPU

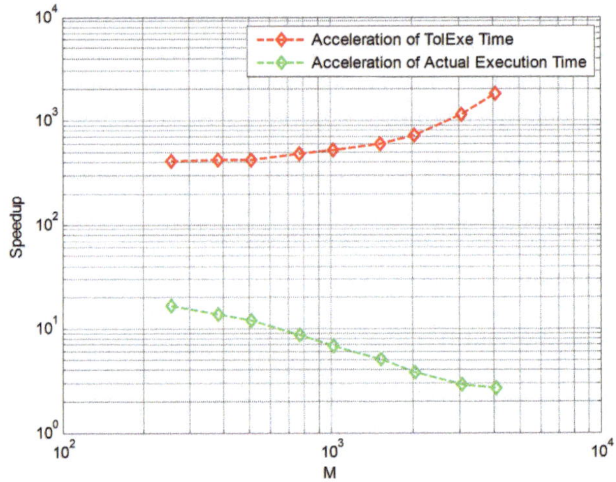

Fig. 8.9 Acceleration of the GPU implementation over the CPU implementation

$$\text{Acceleration of Actual Execution Time} = \frac{\text{CPU TolExe Time}}{\text{GPU TolExe Time} + \text{MemCpy Time}}. \tag{8.21}$$

By observing Fig. 8.9, we notice that the proposed GPU acceleration method can gain at least 414 times acceleration on the TolExe time when compared with the CPU implementation for images with size larger than 256 $\times$ 256 pixels, and it even gains about 1700 times acceleration of TolExe time for the images with 4096 $\times$ 4096 resolution. When considering the cost of memory copy between the

CPU and GPU, the proposed GPU acceleration method still gains about 7 times acceleration for 1 megapixel (1024 $\times$ 124) images and gains about 2.6 times acceleration for 16 megapixel (4096 $\times$ 4096) images. Therefore, these experimental results validate the computational performance of the proposed GPU acceleration method.

8.6 Conclusions

This chapter proposes a GPU-accelerated adaptive SDRCLCE algorithm for real-time color image enhancement. The proposed algorithm consists of an image-dependent nonlinear intensity transfer function, which is able to handle the enhancement of LDR images and produce a satisfactory dynamic range compression result with less color artifacts. By combining the existing SDRCLCE algorithm with the nonlinear intensity transfer function, the proposed method can efficiently improve the visual quality of color images suffered from LDR and poor contrast defects. To improve the real-time performance, a CUDA-based parallel acceleration method is also proposed to accelerate the entire process of the proposed method in processing high-resolution images. Experimental results validate that the proposed GPU-accelerated color image enhancement method not only produces satisfactory color enhancement results, but also achieves a notable improvement on execution time in processing high-resolution color images.

Acknowledgements This work was supported by the National Science Council of Taiwan, ROC under grant NSC 103-2221-E-032-068 and 103-2632-E-032-001-MY3.

References

1. Land, E.: Recent advances in Retinex theory. Vis. Res. **26**(1), 7–21 (1986)
2. Jobson, D., Rahman, Z., Woodell, G.: A multiscale Retinex for bridging the gap between color images and human observation of scenes. IEEE Trans. Image. Process **6**(7), 965–976 (1997)
3. Bertalmío, M., Caselles, V., Provenzi, E., Rizzi, A.: Perceptual color correction through variational techniques. IEEE. Trans. Image. Process. **16**(4), 1058–1072 (2007)
4. Choudhury, A., Medioni, G.: Perceptually motivated automatic color contrast enhancement based on color constancy estimation. EURASIP. J. Image. Video. Process. **2010**(837237), 1–22 (2010)
5. Chen, S.-H., Beghdadi, A.: Natural enhancement of color image. EURASIP. J. Image. Video. Process. **2010**(175203), 1–19 (2010)
6. Ferradans, S., Bertalmío, M., Provenzi, E., Caselles, V.: An analysis of visual adaptation and contrast perception for tone mapping. IEEE. Trans. Pattern. Anal. Mach. Intell. **33**(10), 2002–2012 (2011)
7. Kim, K., Bae, J., Kim, J.: Natural HDR image tone mapping based on retinex. IEEE. Trans. Consum. Electron. **57**(4), 1807–1814 (2011)

8. Saponara, S., Fanucci, L., Marsi, S., Ramponi, G.: Algorithmic and architectural design for real-time and power-efficient Retinex image/video processing. J. Real-Time. Image. Process. (2007). doi:10.1007/s11554-007-0027-z
9. Tao, L., Asari, V. K.: Adaptive and integrated neighborhood-dependent approach for nonlinear enhancement of color images. J. Electron. Imaging. **14**(4) (2005). doi:043006-1-043006-14
10. Tao, L., Tompkins, R., Asari, V. K.: An illuminance-reflectance model for nonlinear enhancement of color images. In: Proceedings of IEEE Computer Society Conference on Computer Vision and Pattern Recognition, pp. 159–166 (2005)
11. Monobe, Y., Yamashita, H., Kurosawa, T., Kotera, H.: Dynamic range compression preserving local image contrast for digital video camera. IEEE. Trans. Consum. Electron. **51**(1), 1–10 (2005)
12. Unaldi, N., Asari, V. K., Rahman, Z.: Fast and robust wavelet-based dynamic range compression with local contrast enhancement. Proceedings of SPIE **6978**, pp. 697805-1-697805-12 (2008)
13. Tsai, C.-Y.: A fast dynamic range compression with local contrast preservation algorithm and its application to real-time video enhancement. IEEE. Trans. Multimed. **14**(4) (Part 2), 1140–1152 (2012)
14. Tsai, C.-Y., Chou, C.-H.: A novel simultaneous dynamic range compression and local contrast enhancement algorithm for digital video cameras. EURASIP J. Image. Video. Process. **2011**(6), 1–19 (2011)
15. Reinhard, E., Kunkel, T., Marion, Y., Brouillat, J., Cozot, R., Bouatouch, K.: Image display algorithms for high and low dynamic range display devices. J. Soc. Inf. Disp. **15**(12), 997–1014 (2007)
16. Marsi, S., Impoco, G., Ukovich, A., Carrato, S., Ramponi, G.: Video enhancement and dynamic range control of HDR sequences for automotive applications. EURASIP J. Adv. Signal. Process. **2007**(080971), 1–9 (2007)
17. Tsai, C.-Y., Huang, C.-H.: An adaptive dynamic range compression with local contrast enhancement algorithm for real-time color image enhancement. J. Real-Time. Image. Process. (2012). doi:10.1007/s11554-012-0299-9
18. NVIDIA's CUDA Toolkit webpage: http://www.nvidia.com/content/cuda/cuda-toolkit.html
19. NVIDIA performance primitives (NPP) webpage: https://developer.nvidia.com/npp
20. NVIDIA's NVS technical specs webpage: http://www.nvidia.com/object/nvs_techspecs.html
21. Wang, Z., Bovik, A. C., Sheikh, H. R., Simoncelli, E. P.: Image quality assessment: From error visibility to structural similarity. IEEE. Trans. Image. Process. **13**(4), 600–612 (2004)
22. Wang, Y.-W., Huang, W.-B.: A CUDA-enabled parallel algorithm for accelerating retinex. J. Real-Time. Image. Process. **9**(3), 407–425 (2014)

Chapter 9
Color Equalization and Retinex

Liqian Wang, Liang Xiao and Zhihui Wei

9.1 Introduction

With the growing popularity of the consumer and professional cameras, images are becoming much easier to obtain. However, because of the influence of the scene conditions and the limitations of photosensitive elements in cameras, many images taken under the condition of dim light or in the high dynamic range scene may have an unsatisfactory visual effect. In order to meet the requirements of display or analysis in some specific applications, the contrast and color of these images need to be enhanced and corrected. Image color correction is a kind of task in image enhancement which is one of the classic topics in image processing. It is an important preprocessing phase of many image analysis tasks such as image segmentation, pattern recognition, and so on. The color corrected image is expected to have more vivid colors or a better visual effect than the original image. There are many techniques which can be used to correct image color. Color equalization and color constancy are two efficient and common techniques. Color equalization adjusts the brightness of colors so that the image histogram is as nearly flat as possible. Color equalization can enhance the contrast in an image and bring out details which were hard to see before. In this chapter, we review the traditional histogram

L. Wang (✉)
School of Telecommunications and Information Engineering, Nanjing University of Posts and Telecommunications, Nanjing, 210003, China
e-mail: alisonwlq@gmail.com

School of Computer Science and Engineering, Nanjing University of Science and Technology, Nanjing, 210094, China

L. Xiao · Z. Wei
Key Lab of Intelligent Perception and Systems for High-Dimensional Information of Ministry of Education, Nanjing, 210094, China
e-mail: xiaoliang@mail.njust.edu.cn

Z. Wei
e-mail: gswei@mail.njust.edu.cn

M. E. Celebi et al. (eds.), *Color Image and Video Enhancement*,
DOI 10.1007/978-3-319-09363-5_9

equalization technique and some recent developments, such as the automatic color equalization [10, 11] and its variational formulation [2].

Color constancy is a feature of the human color perception system which ensures that the color of objects is perceived as constant under varying illumination conditions. As early as 1971, Land formulated the Retinex theory and algorithm [21] to achieve color constancy by computing local ratios of the channel intensity of the image. Retinex is based on the physiological activity of human eye cones and of specialized neurons of the primary visual cortex. Over the years, the Retinex algorithm has been developed in a variety of ways to interpret the human perception system and color constancy. All these Retinex algorithms fall into several categories, such as path-based Retinex [21], recursive Retinex [9], and variational Retinex [18]. In this chapter, we concentrate on some representative algorithms in each category. Furthermore, the relation between automatic color equalization and Retinex is also described.

This chapter is organized as follows: In Sect. 9.2, we review the color equalization methods including histogram equalization, contrast limited adaptive histogram equalization, and automatic color equalization. In Sect. 9.3, we present the Retinex theory for color constancy, and then review and summarize some representative Retinex algorithms. Finally, the discussions are given in Sect. 9.4.

9.2 Color Equalization

9.2.1 Histogram Equalization

The traditional equalization methods are the histogram-based methods. The histogram-based methods enhance image contrast by transforming the values of image intensity. These methods can also be used on color images by respectively applying the same operation on the red, green, and blue channels of a color image in RGB color space, or first converting the image into another color space, for example, lab color space or hue-saturation-value (HSV) color space, and then applying the operation only on the luminance or value channel, keeping the other two channels unchanged. Histogram equalization (HE) [13, 15] is the most traditional histogram-based method. HE, just as its name implies, is to make the histogram of an image equalized, so that the image has high dynamic range and shows more details. We can see from Fig. 9.1 that the histogram of a low contrast image is narrow, while the histogram of the corresponding high contrast image is spread over a broad range of the grayscale. Because it stretches the image histogram to make image intensities better distributed on the histogram, it is very useful when the backgrounds and the foregrounds of the image are both bright or both dark. However, when the image contains noise, HE may increase the contrast of noise, which leads to a bad visual effect.

Given an image I as a discrete function, the pixel intensities range from 0 to $L-1$.Usually L equals to 256 for an 8-bit image. The probability of occurrence of gray level $r_k$is

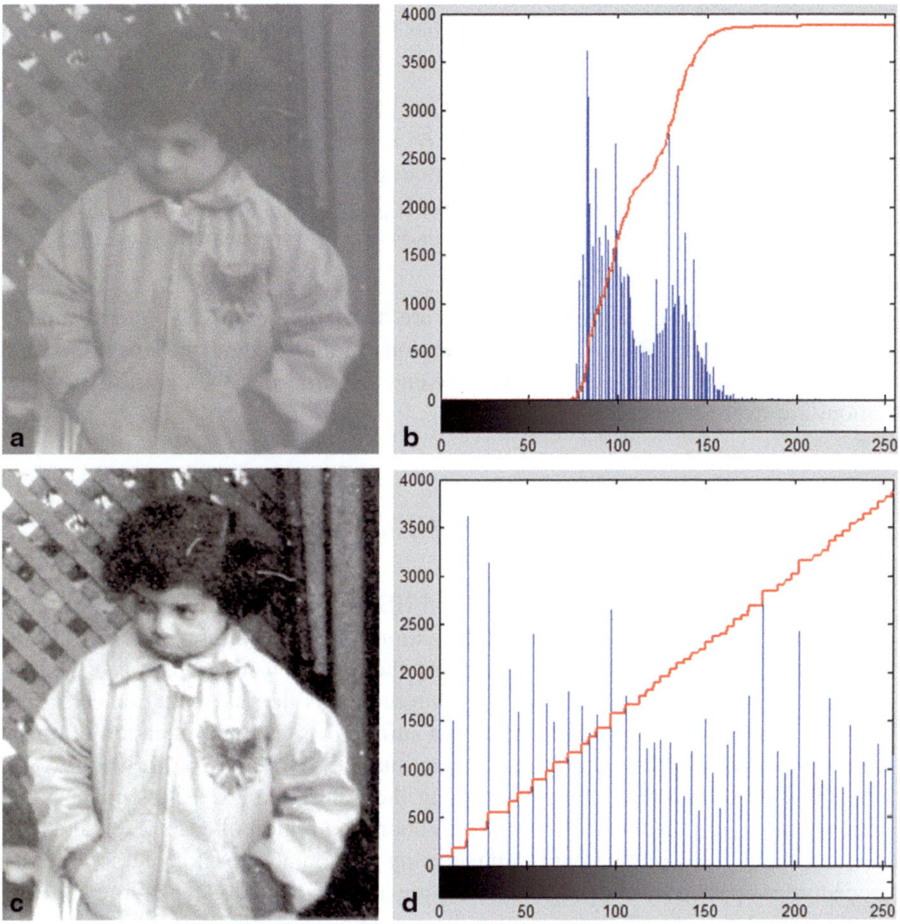

Fig. 9.1 **a** An unequalized image, **b** corresponding histogram and cumulative histogram, **c** the same image after HE, **d** corresponding histogram and cumulative histogram after HE

$$p_r(r_k) = \frac{n_k}{n} \qquad k = 0, 1, \ldots, L-1, \tag{9.1}$$

where n_k is the number of pixels that have gray level $r_k, 0 \leq r_k \leq 1$, and n is the total number of pixels in the image. The cumulative distribution function corresponding to $p_r$is defined as

$$s_k = T(r_k) = \sum_{j=0}^{k} p_r(r_j) = \sum_{j=0}^{k} \frac{n_k}{n} \quad k = 0, 1, \ldots, L-1, \tag{9.2}$$

where $T(r_k)$ is the function transforming every pixel value $r_k$to a level $s_k.T(r_k)$ is a single-valued and monotone increasing function of r_k, and $0 \leq T(r_k) \leq 1$ for $0 \leq r_k \leq 1$. Thus, the resulting image is obtained by transforming the original pixel intensities $r_k$to the new pixel intensities s_k via function $T(r_k)$. The algorithm consists of the following steps:

1. Count $n_k$of the original image, $k = 0,1, \ldots, L - 1$;
2. Compute the original histogram, i.e., $p_r(r_k)$, by Eq. (9.1);
3. Compute the cumulative distribution function by Eq. (9.2);
4. Compute the output gray level $g_k = \text{floor}[(g_{\max} - g_{\min})s_k(r_k) + g_{\min}]/(L - 1)$, $k = 0,1, \ldots, L - 1$, where floor[$\bullet$]rounds down to the nearest integer, $g_{\max} = L - 1$, $g_{\min} = 0$.

Figure 9.1 shows an example of HE. Bins are distributed relatively equally among gray levels after HE. Thus, low contrast regions in the unequalized image become brighter in the output image. However, we can also find that the contrasts of some regions are over-enhanced.

9.2.2 Contrast Limited Adaptive Histogram Equalization

Since the global enhancement of traditional HE may increase the contrast of noise and lose details, adaptive histogram equalization (AHE) [32] improves HE by computing histograms in a local way and using these local histograms to redistribute the intensities of the image. It does bring more details than the traditional HE and enhance local contrast. However, AHE still suffers from the amplification of noise. Contrast limited adaptive histogram equalization (CLAHE) [17] is a variant of AHE, which is able to suppress noise by limiting the contrast of AHE. It enhances the image by applying the HE to the small neighborhood region of each pixel, and then merging the regions into an entire image using interpolation. The contrast limiting procedure is applied by clipping the histogram at a predefined value before computing the cumulative distribution function. Thus, the over-amplification of noise in the image can be avoided. The procedure of CLAHE can be summarized as follows:

1. Partition the image into $M \times N$ nonoverlapping tiles.
2. Clip the histogram of each tile, and redistribute the excess pixels in each bin.

The clip limit, i.e., a threshold above which the histogram is clipped, is computed by

$$L_c = N_{clip} A, \tag{9.3}$$

where N_{clip} is the clip limit coefficient, $A = n_x \times n_y / N_{xy}$ where N_{xy} is the number of bins for the histogram of a tile, $n_x$and n_y are pixel numbers in the tile in x and ydirections, respectively. This clipping makes the number of pixels in each bin no larger than L_c. Let S denote the sum of all clipped pixels, the number of the clipped pixels redistributed in each bin are

$$a_v = S/N_{xy}. \tag{9.4}$$

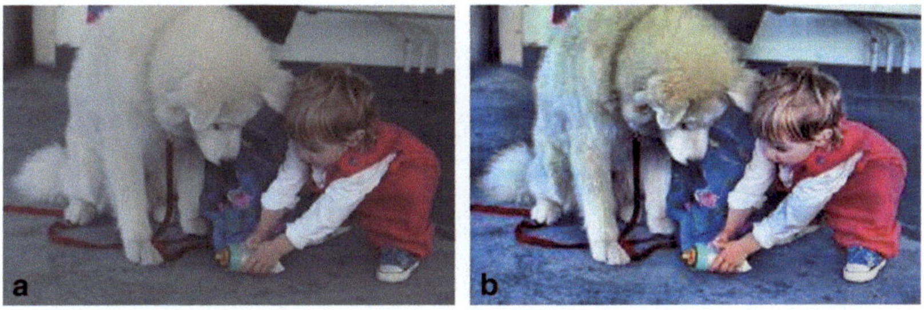

Fig. 9.2 **a** Original image, **b** enhanced image obtained by CLAHE

After the above redistribution, the remaining pixels are redistributed again, one pixel at a time, in looped fashion. Each redistribution loop starts from a low gray level bin to a high gray level bin with a step size. If the number of pixels in the bin is smaller than the clip limit, then one pixel is redistributed to this bin. The loop stops until all the remaining pixels are redistributed to bins. The redistribution step size of the remaining pixels is computed by

$$Step = N_{xy}/n_r, \tag{9.5}$$

where n_r denotes the number of remaining pixels.

3. Equalize the contrast limited histogram of each tile.
4. Combine the neighboring tiles by using bilinear interpolation to eliminate boundary artifacts.

Figure 9.2 shows the result obtained by CLAHE. Local contrast enhancement improves the visibility of the details while avoiding over-enhancement.

9.2.3 Automatic Color Equalization

Automatic color equalization (ACE) [10] is a method for unsupervised global and local color correction by simulating the inner complex behavior of the Human Visual System (HVS), and it was further developed in [2, 39, 40]. The method is inspired by two basic equalization mechanisms, i.e., "Gray world" and "White patch". The "Gray world" refers to the mechanism that HVS performs to adapt to the new luminance condition of the scene. For example, our pupils get wider in the dark to let sufficient light come into the eye, but narrower in bright light to attain acuity. This can also be done in photography by modifying lens aperture and shutter time. The "White patch" refers to the mechanism that HVS uses for normalizing its channel values, maximizing toward a hypothetical white reference area, and achieving color constancy. It is worth mentioning that [39] "Gray world" and "White patch" mechanisms are different from "Gray world" and "White patch" assumptions. The "Gray world" and "White patch" mechanisms refer to the capability of the image

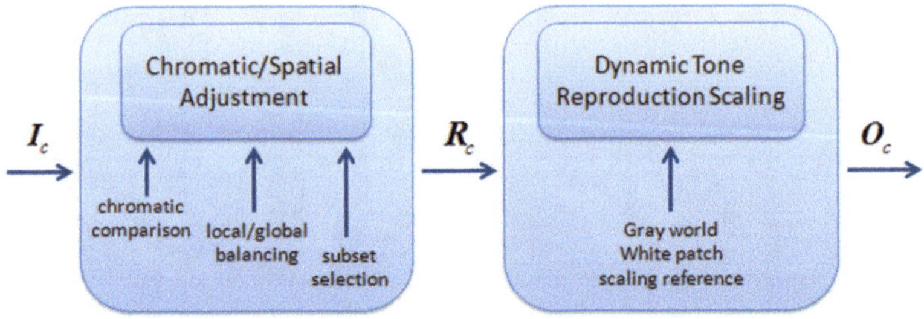

Fig. 9.3 Basic algorithm structure of ACE

correction algorithms to simulate HVS, i.e., to adapt to the lighting condition. The "Gray world" and "White patch" assumptions are hypotheses about the visual naturalness of the scene. ACE merges these two mechanisms into a model that simulates the inner complex behavior of HVS.

There are two phases included in ACE (see Fig. 9.3). The first one is the chromatic spatial adjustment in which a kind of lateral inhibition mechanism is performed and its result is a local–global filter. For each channel c, each pixel x in the output image R is computed as follows:

$$R_c(x) = \sum_{y \in \text{Subset},\, y \neq x} \frac{r((I_c(x) - I_c(y)))}{d(x, y)}, \tag{9.6}$$

where $I_c(x) - I_c(y)$ accounts for the lateral inhibition mechanism that increases contrast and sharpness, $d(\bullet)$ is a distance function which weights the amount of local or global contribution, $r(\bullet)$is the function that accounts for the relative lightness appearance of the pixel. Lateral inhibition mechanism is a concept in neurobiology. It means that an excited neuron is capable of reducing the activity of its neighbors. In visual perception, this mechanism creates contrast and is useful in color discrimination. Generally, in order to compensate for the distance of the pixel from the edge, Eq. (9.6) can be modified with a normalization coefficient as follows

$$R_c(x) = \frac{\sum_{y \in \text{Subset},\, y \neq x} \frac{r((I_c(x) - I_c(y)))}{d(x,y)}}{\sum_{j \in \text{Subset},\, j \neq x} \frac{r_{\max}}{d(x,j)}}, \tag{9.7}$$

where $r_{\max}$ is the maximum value of $r(\bullet)$. Computing the difference between each pixel value and all other pixels of the selected image subset can be viewed as the simulation of the lateral inhibition mechanism, and the function $r(\bullet)$ is used for tuning the difference. The "Gray world" behavior is obtained by setting $r(\bullet)$ to be an odd function, while the "White patch" behavior is obtained by nonlinear enhancements of relative differences between pixels. In [39], the authors tested three kinds of $r(\bullet)$ functions (see Fig. 9.4). As shown in Fig. 9.5, the higher the slope of the saturation function is, the higher the contrast. In the unitary slope case, the saturation function is turned into the linear function, while in the infinite slope case, the saturation function is turned into the signum function.

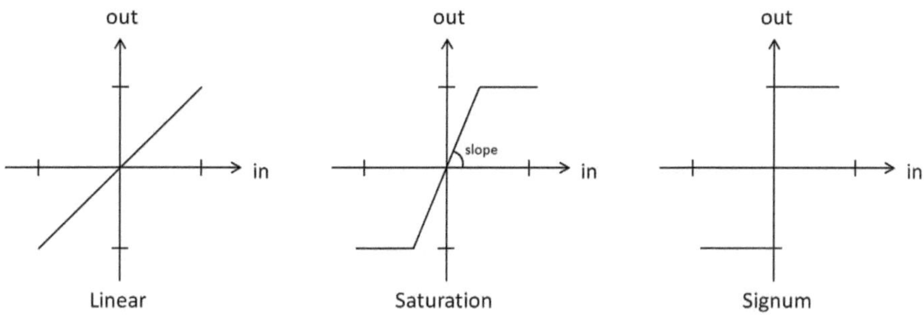

Fig. 9.4 Three tested $r(\bullet)$ functions

The second phase is the dynamic tone reproduction scaling in which the image dynamic is maximized and lightness constancy is performed. In this stage, a linear or nonlinear scaling method can be used for mapping the intermediate pixels matrix R into the final output image O. The authors proposed two linear scaling methods; one is simple linear scaling

$$L_c(x) = \text{round}[s_c(R_c(x) - m_c)], \tag{9.8}$$

where $s_c$is the slope of the segment $[(m_c, 0), (M_c, 255)]$, $i.e. s_c = 255/(M_c - m_c)$, $M_c = \max_x R_c(x)$, $m_c = \min_x R_c(x)$. The advantage of simple linear scaling is that the linear mapping is able to fill the available dynamic range of images without further adjustment. The other one is the "White patch/Gray world" scaling

$$O_c(x) = \text{round}[127.5 + s_c R_c(x)], \tag{9.9}$$

where $s_c = 127.5/M_c$. This scaling takes advantage of the global "Gray world" assumption. Consequently, the dynamic of the final output image is centered on the medium gray. The disadvantage of this scaling is that tones around the very dark values could be lost. Figure 9.6 shows the results obtained by ACE.

9.2.3.1 Variational Framework for Contrast Enhancement

Motivated by the variational formulation of HE [41], Bertalmío et al. [2] noticed that ACE's structure is analogous to the one in the variational formulation of HE. The method in [2] performs color correction by considering the three color channels separately. Here we describe this approach by using the same notation as in [2]. Let I denote a chromatic channel of an input image. Let us assume that the intensity values of I have been normalized to range over the unit real interval [0,1], and let x and y be two arbitrary pixels in I. Bertalmío et al. proposed that the variational framework for contrast enhancement and color correction consists of an average local contrast measure

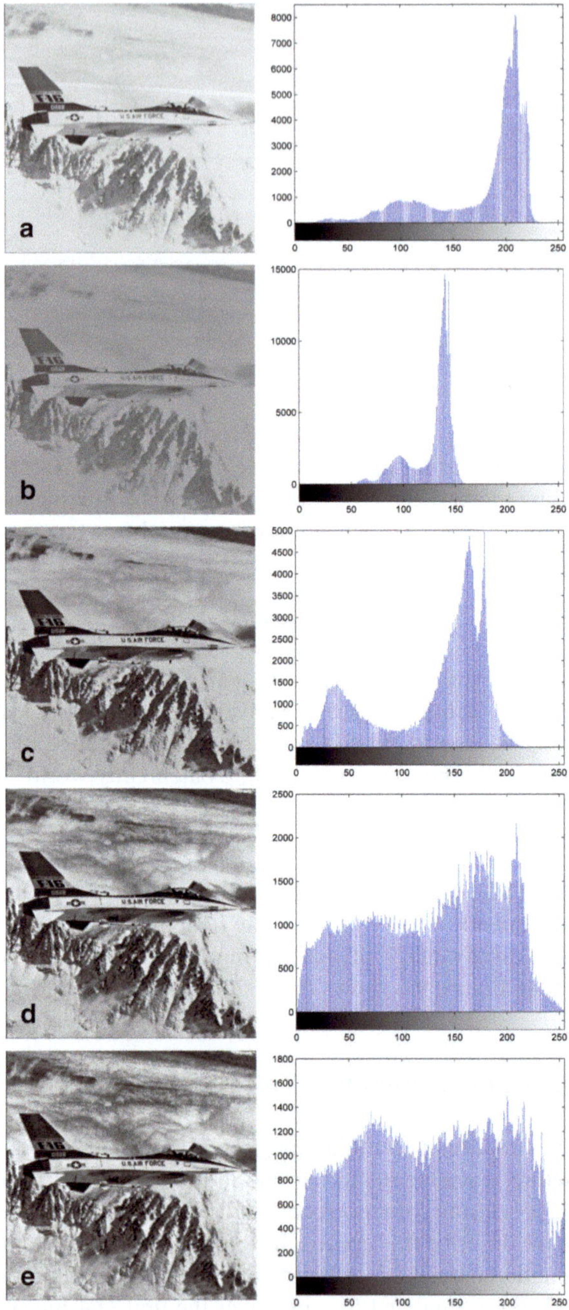

Fig. 9.5 Sample image filtered with different $r(\bullet)$ function

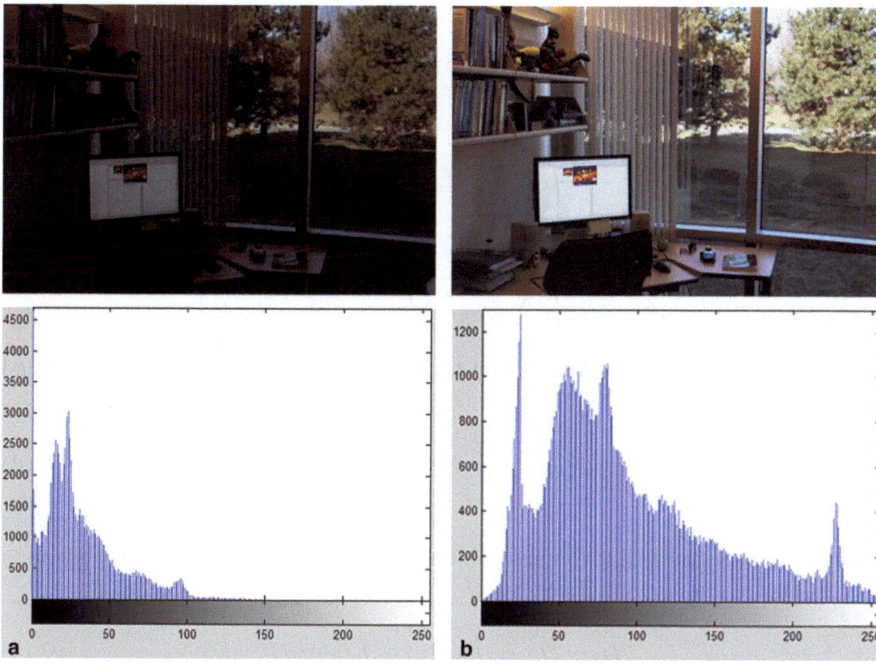

Fig. 9.6 **a** Original image, **b** enhanced results obtained by ACE (*top*: enhanced image; *bottom*: corresponding dynamic range)

$$C_{\omega,J}(I) := \iint_{\mathrm{T}^2} \omega(x, y) J(I(x) - I(y)) dx dy, \tag{9.10}$$

and an average quadratic local dispersion measure

$$D_{\tilde{\omega}}(I) \equiv \iint_{\mathrm{T}^2} \tilde{\omega}(x, y)(I(y) - \bar{I}^{\tilde{\omega}}(x))^2 dx dy, \tag{9.11}$$

where $J:\mathbb{R} \to [0, \infty)$ is a convex even function, T^2 is the spatial domain of image, ω and $\tilde{\omega}$ are two positive symmetric normalized weight functions representing the weight of mutual chromatic influence between the pixels x and y, $\int_{\mathrm{T}} \omega(x, y) dy = \int_{\mathrm{T}} \tilde{\omega}(x, y) dy = 1$, $\bar{I}^{\tilde{\omega}}(x)$ is a local average. The authors in [2] provided three choices for $\bar{I}^{\tilde{\omega}}(x)$. The first choice is the "Gray world," which can be translated into

$$\bar{I}^{\tilde{\omega}}(x) = \frac{1}{2}. \tag{9.12}$$

Consequently, the corresponding dispersion measure can be given as follows

$$D_{\tilde{\omega},GW}(I) \equiv \iint_{\mathrm{T}^2} \tilde{\omega}(x,y)\left(I(y)-\frac{1}{2}\right)^2 dxdy$$
$$= \int_{\mathrm{T}} \left(\int_{\mathrm{T}} \tilde{\omega}(x,y)dx\right)\left(I(y)-\frac{1}{2}\right)^2 dy \tag{9.13}$$
$$= \int_{\mathrm{T}} \left(I(y)-\frac{1}{2}\right)^2 dy.$$

The second choice is the "Attachment to Original Data," which can be translated into

$$\bar{I}^{\tilde{\omega}}(x) = \int_{\mathrm{T}} \tilde{\omega}(x,y)I_0(y)dy, \tag{9.14}$$

where I_0 denotes the original image which we want to enhance. The corresponding dispersion measure is formulated by

$$D_{\tilde{\omega},I_0}(I) = \int_{\mathrm{T}} (I(x)-I_0(x))^2 dx. \tag{9.15}$$

The third choice is a linear combination of "Gray world" and "Attachment to Original Data," i.e.,

$$D_{\tilde{\omega}}(I) = \beta D_{\tilde{\omega},GW}(I) + \gamma D_{\tilde{\omega},I_0}(I), \beta \geq 0, \gamma > 0. \tag{9.16}$$

The final energy functional is defined as

$$E_{\tilde{\omega},\omega}(I) := D_{\tilde{\omega}}(I) - C_{\omega,J}(I). \tag{9.17}$$

Minimizing the functional Eq. (9.17) increases the contrast while controlling the local variance of the resulting image. By assuming that J is differentiable, we have the first variation of $E_{\tilde{\omega},\omega}(I)$, i.e.,

$$\begin{aligned}\delta E_{\tilde{\omega},\omega}(I) &= \delta D_{\tilde{\omega}}(I) - \delta C_{\omega,J}(I)\\ &= 2\beta\left(I(x)-\frac{1}{2}\right) + 2\gamma\left(I(x) - \int_{\mathrm{T}} \tilde{\omega}(x,y)\bar{I}^{\tilde{\omega}}(y)dy\right)\\ &\quad - 2\int_{\mathrm{T}} \omega(x,y)J'(I(y)-I(x))dy.\end{aligned} \tag{9.18}$$

The minimum of $E_{\tilde{\omega},\omega}$ satisfies $\delta E_{\tilde{\omega},\omega}(I) = 0$.

In a discrete framework, Eq. (9.18) can be written as

$$E^d_{\tilde{\omega},\omega}(I) = D^d_{\tilde{\omega}}(I) - C^d_{\omega,J}(I), \tag{9.19}$$

where

$$D^d_{\tilde{\omega}}(I) = \beta D^d_{\tilde{\omega},GW}(I) + \gamma D^d_{\tilde{\omega},I_0}(I), \beta \geq 0, \gamma > 0, \tag{9.20}$$

$$D^d_{\tilde{\omega},I_0}(I) = \sum_x \sum_y \tilde{\omega}(x, y)(I(y) - \bar{I}^{\tilde{\omega}}(x))^2, \tag{9.21}$$

$$\bar{I}^{\tilde{\omega}}(x) = \sum_y \tilde{\omega}(x, y) I_0(y), \tag{9.22}$$

$$D^d_{\tilde{\omega},GW}(I) = \sum_x \left(I(x) - \frac{1}{2}\right)^2, \tag{9.23}$$

and

$$C^d_{\omega,J}(I) = \sum_x \sum_y \omega(x, y) J(I(x) - I(y)). \tag{9.24}$$

Figure 9.7 shows the results obtained by the variational method reported above. We see that the bins of the color histogram of the enhanced image are distributed more equally among gray levels than those of the histogram of the original image.

9.2.3.2 Variational Formulation of ACE

By formulating ACE into a discrete version, we compute the chromatic spatial adjustment by

$$R(x) = \frac{\sum_{y \neq x} \omega(x, y) s_\alpha(I(x) - I(y))}{\sum_{y \neq x} \omega(x, y)}, \tag{9.25}$$

where $\omega(x, y)$ is a weight function, α is a real constant called slope, s_α is the slope function defined by

$$s_\alpha(r) = \begin{cases} -1, & \text{if } -1 \leq r \leq -1/\alpha \\ \alpha r, & \text{if } -1/\alpha < r < 1/\alpha \\ +1, & \text{if } 1/\alpha \leq r \leq 1 \end{cases} \tag{9.26}$$

In Eq. (9.25), $r = I(x) - I(y)$. The normalization factor of Eq. (9.25) can be incorporated into the definition of the weights (for simplicity, we keep the same notation which is similar to the notation in [2])

$$\omega(x, y) = \frac{A}{\|x - y\|}, x \neq y, \tag{9.27}$$

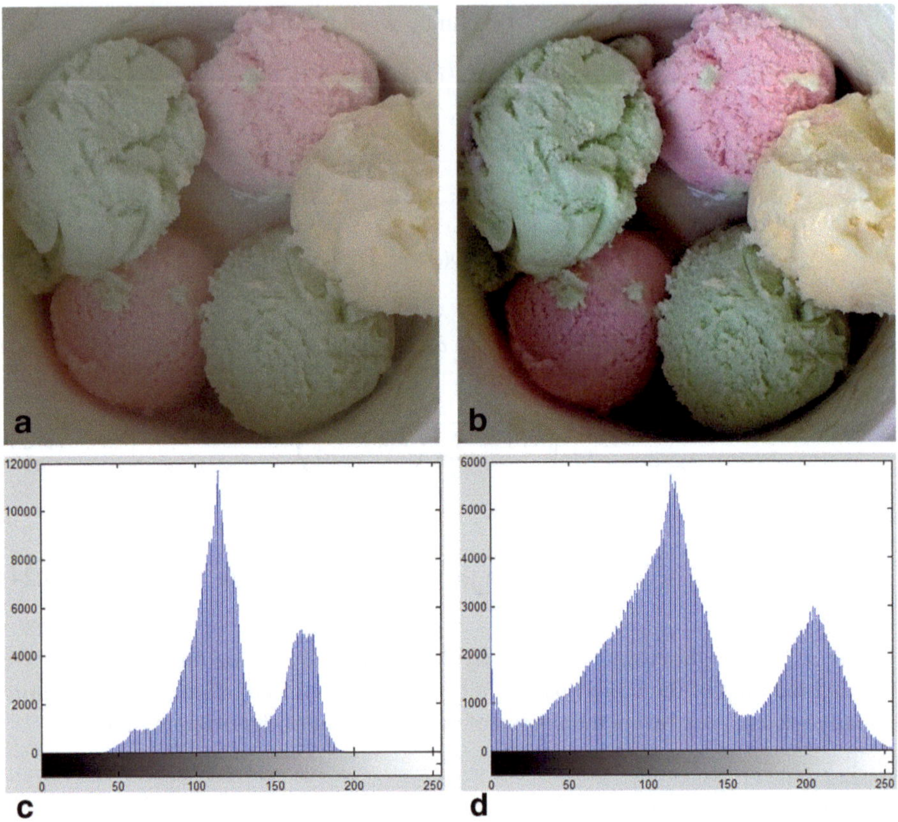

Fig. 9.7 **a** Original image, **b** color enhanced image, **c** original *red* channel histogram, and **d** *red* channel histogram of the color enhanced image

where

$$A(x) = \left(\sum_y \frac{1}{\|x - y\|} \right)^{-1} \tag{9.28}$$

is actually a constant A. Then Eq. (9.25) can be rewritten as

$$R_I(x) := \sum_y \omega(x, y) s_\alpha(I(x) - I(y)), \tag{9.29}$$

and consequently, the variational formulation of ACE can be written as

$$\text{ACE}(I)(x) := L(x) = \frac{1}{2} + \frac{\sum_y \omega(x, y) s_\alpha(I(x) - I(y))}{2M}, \tag{9.30}$$

where $M \equiv \max_x \{R_I(x)\}$.

Fig. 9.8 **a** Original image, **b** enhanced image obtained by variational formulation of ACE

In order to transform $I(x)$ into $L(x)$using the variational techniques presented in Sect. 9.2.3.1, the authors in [2] consider the following functional

$$E(I) = \frac{1}{2}\sum_x \left(I(x) - \frac{1}{2}\right)^2 - \frac{1}{4M}\sum_x \sum_y \omega(x, y) S_\alpha(I(x) - I(y)), \tag{9.31}$$

where S_α is a convex and even function, its derivative is the odd function s_α, *i.e.*, $S'_\alpha = s_\alpha$, and $0 < M \leq 1$ is a constant. The coefficients in front of the terms are used for normalization. If I is a minimum of E, then $\delta E(I) = 0$ and

$$I(x) = \frac{1}{2} + \frac{\sum_y \omega(x, y) s_\alpha(I(x) - I(y))}{2M} = L(x). \tag{9.32}$$

Using a gradient descent strategy, one can minimize the energy $E(I)$by solving

$$\frac{\partial I}{\partial t} = -\delta E(I). \tag{9.33}$$

Then the explicit discretization of Eq. (9.33) can be written as

$$\frac{I^{k+1}(x) - I^k(x)}{\Delta t} = \frac{1}{2} - I^k(x) + \frac{R_{I^k}(x)}{2M}, \tag{9.34}$$

where $\Delta t > 0$, $I^k(x) = I(k\Delta t, x)$, $I^0(x) = I_0(x)$, $R_{I^k}(x) = \sum_y \omega(x, y) s_\alpha(I^k(x) - I^k(y))$, and consequently, the numerical scheme to get to a steady state of ACE can be obtained as follows

$$I^{k+1}(x) = (1 - \Delta t) I^k(x) + \Delta t \left(\frac{1}{2} + \frac{R_{I^k}(x)}{2M}\right). \tag{9.35}$$

The result obtained by variational formulation of ACE is shown in Fig. 9.8.

9.3 Color Constancy

9.3.1 Retinex Theory

Color constancy is one important characteristic of human vision. It ensures that the perceived color remains unchanged under varying illumination conditions. Retinex is one of the most famous algorithms that attempts to explain human color constancy. The Retinex theory [19, 21] is based on the behavior of HVS. The word "Retinex" is a portmanteau formed from two words: retina and cortex. Land and McCann believed that the retina and cortex system may treat a color as a code for a three-part report from the retina, independent of the flux of radiant energy, but correlated with the reflectance of the object.

In order to study the color sensations of HVS, Land and McCann made a series of experiments in which the observers were asked to distinguish different colors on a color board under controllable illuminations. They first illuminated a Mondrian-like colored paper by long-, middle-, and short-wave illuminators, respectively, at a time, and then turned on all three illuminators. The radiation from each chosen area of the colored paper was received and measured by a telescopic photometer. By changing the color of paper from white to other colors, while adjusting the illuminators so that the radiation reflected from the colored paper remained the same, they found that the white paper still appeared white and the other colored papers still appeared their original color. Land and McCann's experiments show that the processing mechanism of HVS is not affected by the wavelength composition of the light coming to the eye. From these experiments, they drew the following conclusions: a color sensation involves the interaction of at least three independent retinal-cortical systems, one for long waves, one for middle waves, and one for short waves; the color sensation is dependent on reflectance rather than illumination or flux.

Thus, Retinex is effective in enhancing images with nonuniform lighting. Many algorithms have been developed based on the Retinex theory. These algorithms are usually categorized [18, 23, 26] as path-based algorithms [19, 21, 34–36], recursive algorithms [8, 9, 24], center/surround algorithms [16, 37, 38], partial differential equation (PDE) based algorithms [5, 14, 26], and variational algorithms [3, 7, 18, 23, 27, 30, 33, 43]. In the following subsections, we introduced some representative algorithms.

In Retinex algorithms, the input image I is assumed to be formed by a product of the illumination L and the reflection R, i.e., $I = L \bullet R$. The aim of the Retinex algorithm is to estimate the illumination L from the original image I. Thus, discarding the estimated illumination from the image can remove the effect of nonuniform lighting and improve the visual effect of the image. For the sake of numerical convenience, the multiplication is usually transformed to addition by a logarithm operation, i.e., $i = l + r$, where $i = \log I, l = \log L, r = \log R$.

Generally, two approaches can be used to apply Retinex algorithms to color images. The traditional approach is to deal with each color channel separately, which is commonly known as "RGB Retinex." Another approach is to map the colors into

HSV color space, then apply the Retinex only to the value channel, and finally map back to the RGB domain. This approach is referred to as "HSV Retinex" [18, 27].

9.3.2 Path-Based Retinex

9.3.2.1 Original Retinex

The original Retinex algorithm proposed by Land and McCann [21] is based on a random path. Consider two different areas on a piece of colored paper under a nonuniform light condition; their luminances are measured by two light detectors. Obviously, the luminances of the two areas become the same when the two detectors are placed closer. Thus, given a certain number of paths $\gamma_1, \ldots, \gamma_N$ (N denotes the number of paths) starting in j_k and ending in a target pixel i in an image I, the lightness value of the end point pixel is obtained by computing the average of products of ratio between the intensity values of two adjacent pixels in the paths as follows [34]

$$L(i) = \frac{1}{N} \sum_{k=1}^{N} \prod_{t_k=1}^{n_k-1} \delta_k(R_{t_k}), \tag{9.36}$$

where $R_{t_k} = I(x_{t_k+1})/I(x_{t_k})$ with $R_0 \equiv 1, t_k = 1, \ldots, n_k$ is the parameter of the kth path γ_k, $\gamma_k(1) = j_k$ and $\gamma_k(n_k) = i, n_k$, denotes the number of pixels travels by γ_k, δ_k is function defined in this way: $\delta_k(R_0) = 1$ and for $t_k = 1, \ldots, n_k - 1$

$$\delta_k(R_{t_k}) = \begin{cases} R_{t_k}, & \text{if } 0 < R_{t_k} \leq 1 - \varepsilon \\ 1, & \text{if } 1 - \varepsilon < R_{t_k} < 1 + \varepsilon \\ R_{t_k}, & \text{if } 1 + \varepsilon \leq R_{t_k} \leq 1 + \varepsilon \\ 1/\prod_{m_k=0}^{t_k-1} \delta_k(R_{m_k}), & \text{if } R_{t_k} > (1+\varepsilon)/\prod_{m_k=0}^{t_k-1} \delta_k(R_{m_k}) \end{cases} \tag{9.37}$$

where $\varepsilon > 0$is a fixed threshold. As the analyses in [34], the first and the third option indicate that δ_k is an identity function which leads to the typical "chain of ratios" of Retinex. When the change of intensity between two subsequent pixels $\gamma_k(t_k) = x_{t_k}$ and $\gamma_k(t_k + 1) = x_{t_k+1}$(for $t_k = 1, \ldots, n_k - 1$) is very small, the second option occurs and indicates that the product of ratios is unchanged. When the fourth option is satisfied, δ_k resets the chain of products to 1 because the local white reference has been reached. This operation determines the "White Patch" behavior of the algorithm. The ratio-threshold-reset mechanism of Retinex is defined by these four options and, in [34], it has been proved that the reset is the most important mechanism. By fixing the threshold $\varepsilon = 0$, it can be proved [34, 35] that Eq. (9.36) can be written as follows

$$L(i) = \frac{1}{N} \sum_{k=1}^{N} \frac{I(i)}{I(x_{H_k})}, \tag{9.38}$$

where x_{H_k} is the pixel with highest intensity traveled by the path $\gamma_k, k = 1, \ldots, N$. We can see that $L(i)$ averages the contributions of paths, and every contribution is characterized by $I(x_{H_k})$which is the highest intensity on the path γ_k. Therefore, Retinex is a pure local "White Patch" algorithm, and the difference between the path-based algorithms is the structure of the path.

9.3.2.2 Random Spray Retinex

Usually, the traditional pathwise Retinex contains redundant paths that have same contributions in Eq. (9.38). In order to reduce the redundancy, Provenzi et al. [35] proposed the random spray Retinex (RSR) in which paths are replaced by 2D pixel sprays. Let $Spray_k(i)$ denote a spray, which plays the role of a path, composed by n_k pixels and centered in i, and (i_x, i_y) denotes the coordinates of i. The polar coordinates of a generic pixel $j \equiv (j_x, j_y)$ belonging to $Spray_k(i)$ are defined by

$$\begin{cases} j_x = i_x + \rho \cos(\theta) \\ j_y = i_y + \rho \sin(\theta) \end{cases} \tag{9.39}$$

where $\rho \in \text{RAND}_{n_k}[0, R]$, $\theta \in \text{RAND}_{n_k}[0,2\pi]$, $\text{RAND}_{n_k}[0, R]$, and $\text{RAND}_{n_k}[0,2\pi]$ are generated by extending a uniform random distribution of n_k values from the real unit interval [0,1] to the intervals [0, R] and [0,2π], respectively, R is a given positive real number presenting the radius of the spray. Then consider a circle C_r of arbitrary radius $r, 0 < r < R$, centered on i, the area of C_r is $A = \pi r^2, r = \sqrt{A/\pi}$, and the mean number of points inside C_r is $n_k r/R = n_k\sqrt{A/\pi}/R$. By computing the derivative of n_k with respect to A, we obtain the variation of the mean areolar density $\delta(r)$as follows

$$\delta(r) = n_k/\left(2R\sqrt{\pi A}\right) = (n_k/2\pi R)/r. \tag{9.40}$$

Figure 9.9 shows an example of a spray with 400 pixels and radius $R = 1$.

We can also apply a function $f:\mathbb{R}^+ \to \mathbb{R}^+$ on the coordinate ρ to change the radial density of the spray pixels around the center i, and consequently the modified spray is defined by

$$\begin{cases} j_x = i_x + f(\rho) \cos(\theta) \\ j_y = i_y + f(\rho) \sin(\theta) \end{cases} \tag{9.41}$$

where $\rho \in \text{RAND}_{n_k}[0, R], \theta \in \text{RAND}_{n_k}[0,2\pi]$. In [35], the authors compared five different radial density functions including the identity function, the log function, the sinh function, the exponential function, and the square function. The experimental results showed that in both subjective quality tests and quantitative tests, the identity function produces the best results, which implies that for the RSR method the most suitable spray to reproduce the behavior of HVS is the naturally localized

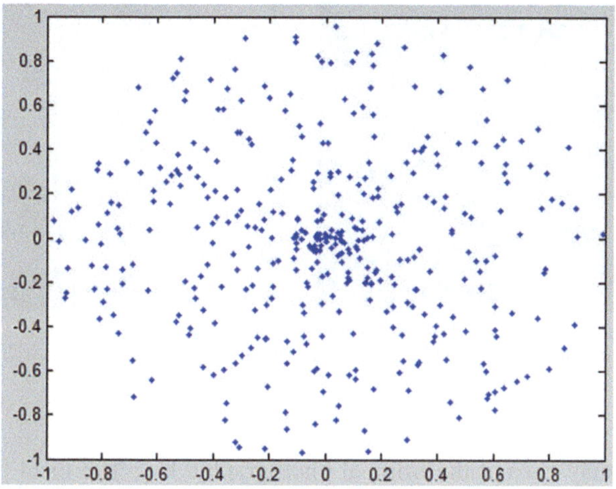

Fig. 9.9 Example of "naturally localized" spray

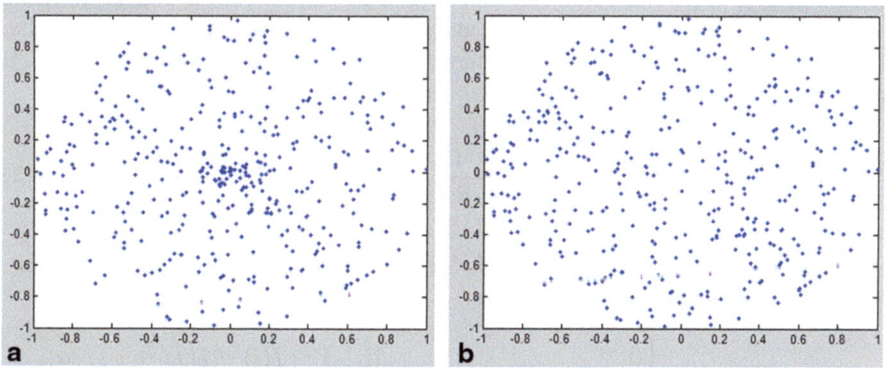

Fig. 9.10 **a** Spray with $f(\rho) = (\log(1+\rho)/\log(2))$, **b** spray with $f(\rho) = \sqrt{\rho}$

spray. Figure 9.10 shows two sprays obtained with 400 pixels, radius $R = 1$, and different functions f. Figure 9.11 shows the enhancement result obtained by the RSR.

9.3.2.3 RACE

In order to take advantages of both "White Patch" and "Gray World" algorithms, Provenzi et al. presented a model named RACE (Retinex + ACE) in [36] by combining RSR, which is a pure "White Patch" algorithm, and ACE, which is an essentially "Gray World" algorithm. RSR tends to preserve the original photographic key of the images, but it cannot recover details in dark areas. ACE is capable of recovering

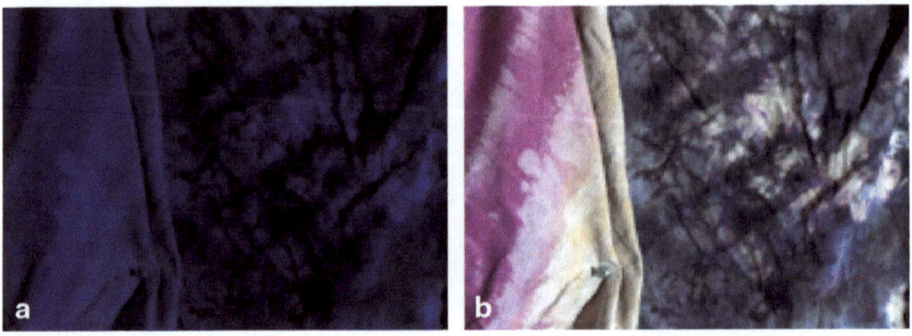

Fig. 9.11 **a** Original image [1], **b** enhanced image (using identity function with 25 sprays and 400 pixels per spray)

details, but it may corrupt the original photographic key. The local fusion of RSR and ACE into RACE shows an improvement of features of both models.

By considering that for both RSR and ACE, the mutual chromatic induction is inversely proportional with the Euclidean distance between pixels, in [36] the authors first establish localized sprays, and then define a new slope function. Finally, by averaging N spray contributions, they define the lightness computed by RACE as

$$L^{\mathrm{RACE}}(i) = \frac{1}{N}\sum_{k=1}^{N}\frac{1}{2}\left(\frac{I(i)}{I(x_{H_k})} + \frac{1}{\tilde{n}_k}\sum_{j\in Spray_k(i)\setminus\{i\}} \tilde{s}_\alpha(I(i)-I(j))\right), \tag{9.42}$$

where $\tilde{n}_k$ is the number of spray pixels (center excluded) that lie inside the image, $\tilde{s}_\alpha$ is the slope function defined as

$$\tilde{s}_\alpha(I(i)-I(j)) = \begin{cases} 0 & \text{if} -1 \le I(i)-I(j) \le -1/2\alpha, \\ 1/2+\alpha(I(i)-I(j)) & \text{if} -1/2\alpha < I(i)-I(j) < 1/2\alpha, \\ 1 & \text{if } 1/2\alpha \le I(i)-I(j) \le 1. \end{cases} \tag{9.43}$$

For every $j \in Spray_k(i)\setminus\{i\}$, when $I(i) = I(x_{H_k})$ and $\tilde{s}_\alpha(I(i)-I(j))$, the ideal white condition is satisfied; when $I(i) = I(x_{H_k})/2$ and $\frac{1}{\tilde{n}_k}\sum_{Spray_k(i)\setminus\{i\}}\tilde{s}_\alpha(I(i)-I(j)) = \frac{1}{2}$, the ideal middle gray condition is satisfied; when $I(i) = 0$ and $\tilde{s}_\alpha(I(i)-I(j)) = 0$, the ideal black condition is satisfied.

Equation (9.42) is an algebraic average between the contribution of RSR and the contribution of ACE. Similarly, a geometric average can be defined as

$$L^{\mathrm{RACE}}(i) = \frac{1}{N}\sum_{k=1}^{N}\sqrt{\frac{I(i)}{I(x_{H_k})}\cdot\frac{1}{\tilde{n}_k}\sum_{j\in Spray_k(i)\setminus\{i\}}\tilde{s}_\alpha(I(i)-I(j))}. \tag{9.44}$$

Equation (9.44) also satisfied the above three ideal conditions. Since the geometric average consumes more time than the algebraic one, and the resulting image produced by the geometric average may have poor details in shadows, the authors in [36] prefer to choose the algebraic average for RACE.

Because the differential operations, e.g., differences, ratios, and derivatives, become trivial on large uniform image zones, algorithms based on local "Gray World" or "White Patch" mechanisms may induce chromatic noise in uniform image areas. In order to avoid this problem, the authors in [36] further present a regularization of the RACE formula by taking a convex linear combination of the original intensity value $I(i)$ and the lightness contribution $L_k^{\mathrm{RACE}}(i)$ associated to a spray $Spray_k(i)$ on the target pixel:

$$\tilde{L}_k^{\mathrm{RACE}}(i) = \beta_k(i) L_k^{\mathrm{RACE}}(i) + (1 - \beta_k(i)) I(i), \tag{9.45}$$

where the coefficient $\beta_k(i) \in [0,1]$ varies with i, and is defined as

$$\beta_k(i) = (2\sigma_k(i))^{\gamma}, \tag{9.46}$$

where $\gamma > 0$ is a real constant, $\sigma_k(i) = \sqrt{\sum_{j \in Spray_k(i)} (I(j) - m_k(i))^2 p_{k,i}(I(j))}$ is the standard deviation associated to $Spray_k(i)$, $p_{k,i}(I(j)) = \mathrm{card}\{l \in Spray_k(i), I(l) = I(j)\}/\tilde{n}_k$, is the occurrence probability of value $I(j)$ inside $Spray_k(i)$, $m_k(i) = \sum_{j \in Spray_k(i)} I(j) p_{k,i}(I(j))$ is the mean intensity value inside $Spray_k(i)$.$\sigma_k(i)$ can be interpreted as a local measure of average contrast around the center i.

To automatically select the parameter γ and distinguish between images with dominate color and color cast, the authors in [36] set $\gamma = \frac{\sigma_{\min}}{\sigma_{\max}}$, where $\sigma_{\min} = \min\{\sigma_c, c \in R, G, B\}$, $\sigma_{\max} = \max\{\sigma_c, c \in R, G, B\}$, σ_c as the global standard deviation in the chromatic channel c.

Finally, a general formula to implement the local and global contrast-based regulator mechanism can be written as

$$L^{\sigma-\mathrm{alg}}(i) = \frac{1}{N} \sum_{k=1}^{N} \left[(2\sigma_k(i))^{\frac{\sigma_{\min}}{\sigma_{\max}}} L_k^{\mathrm{alg}}(i) + \left(1 - (2\sigma_k(i))^{\frac{\sigma_{\min}}{\sigma_{\max}}}\right) I(i) \right], \tag{9.47}$$

where the superscript "alg" denotes the chosen algorithm, i.e., $L_k^{\mathrm{alg}}(i)$ could be chosen as $L_k^{\mathrm{RSR}}(i)$ or $L_k^{\mathrm{RACE}}(i)$ or $L_k^{\mathrm{ACE}}(i)$ or other algorithms, the definition of $\sigma_k(i)$ can vary according to each chosen algorithm.

The results produced by RACE, σ − RSR, σ − RACE, and σ − ACE are shown in Fig. 9.12, respectively. We can see that σ − RACE has an intermediate behavior between σ − RSR and σ − ACE.

9.3.3 Recursive Retinex

Recursive algorithms [8, 9, 24] are computationally more efficient than path-based algorithms. In recursive algorithms, the path computation is replaced by a recursive

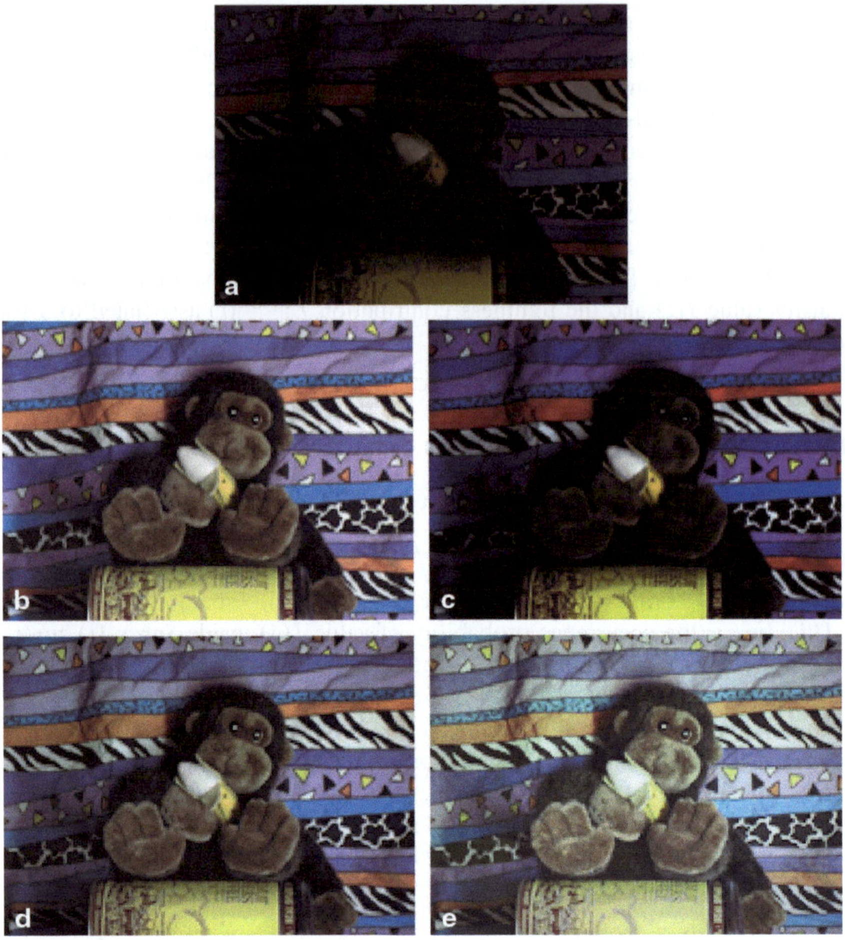

Fig. 9.12 **a** Original image [1], **b** result obtained by RACE, **c** result obtained by σ − RSR, **d** result obtained by σ − RACE, **e** result obtained by σ − ACE (using identity function with 20 sprays and 400 pixels per spray)

matrix calculation. One version of recursive Retinex was Frankle and McCann's Retinex (F-M Retinex) [8]. It first computes long-distance iterations between pixels, and then progressively moves to short-distance interactions. The spacing between pixels being compared decreases at each step in clockwise order. At each step the comparison is implemented using the ratio-product-reset-average operation. The process stops when the spacing decreases to 1 pixel. There are no paths in F-M Retinex.

Another version of recursive Retinex was proposed in [24], which we refer to as McCann99 Retinex as the same as in [9]. In this method, a multiresolution pyramid

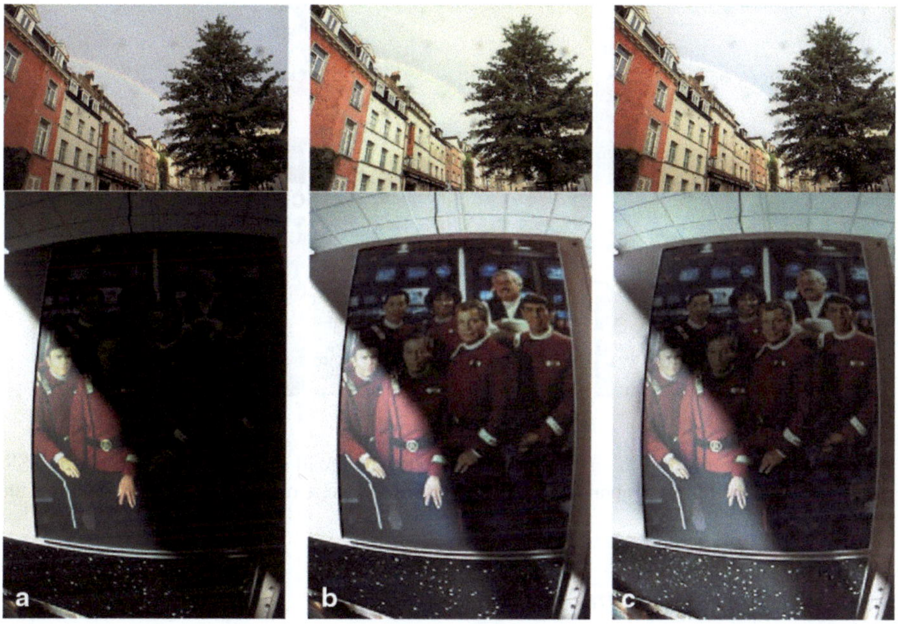

Fig. 9.13 **a** Original images, **b** result obtained by F-M Retinex, **c** results obtained by McCann99 Retinex

is created from the input by averaging image data in log space. The pixel comparisons begin at the most highly averaged or top level of the pyramid. The initial lightness is first computed on the image at a reduced resolution. Then by pixel replication, the lightness values are propagated down to the pyramid's next level as the initial lightness of that level. The pixel replication is an operation to scale up the current level image by a scale factor 2, i.e., the current level image is one fourth of the scaled-up image size. The lightness estimates of the higher resolution level are refined by pixel comparisons. For each pixel, the pixel comparison is implemented by applying ratio-product-reset-average operations [24] on eight immediately neighboring pixels, one pixel a time, in clockwise order. In the ratio step, the neighbor's log luminance is subtracted. In the product step, the subtracted result is added to the scaled-up image. In the reset step, if the lightness is greater than the given maximum, it is reset to the maximum. In the average step, the result obtained by the above steps is averaged with the initial lightness of the current level. Finally, the new lightness estimates are propagated down a level in the pyramid. The process stops when the computation is implemented for the bottom level of the pyramid.

Figure 9.13 shows the results obtained by the F-M Retinex and McCann99 Retinex.

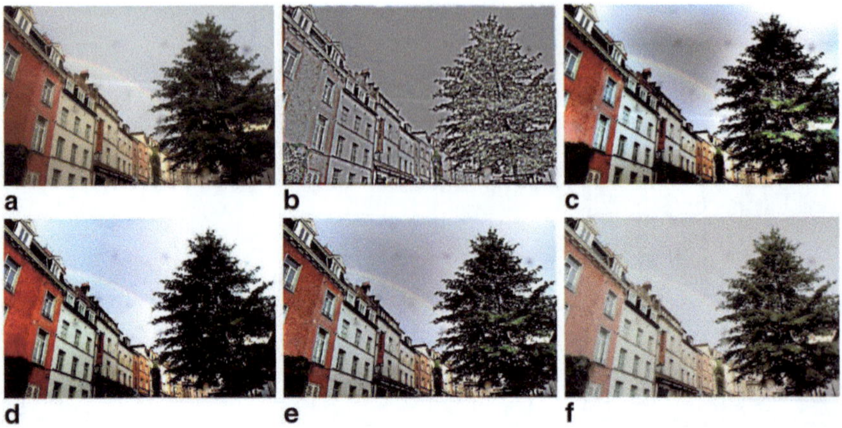

Fig. 9.14 **a** Original image, **b** result obtained by SSR with small scale, **c** result obtained by SSR with medium scale, **d** result obtained by SSR with large scale, **e** result obtained by MSR, **f** result obtained by MSRCR

9.3.4 Center/Surround Retinex

In center/surround algorithms, the illumination is estimated as a convolution of the original image and the surround function. Single scale Retinex (SSR) [16] is the most basic center/surround algorithm. The intensity of the result image is the ratio of the original pixel intensity and weighted average of intensities of the neighbor pixels. In SSR, the surround function is the Gaussian, in which a parameter is used to control the scale of the neighbor. The formula of SSR is given by

$$R_i(x, y) = \log I_i(x, y) - \log [F(x, y) * I_i(x, y)], \tag{9.48}$$

where x and y denote the coordinates of a pixel in image, $R_i(x, y)$ is the Retinex output, $\log I_i(x, y)$ is the image distribution in $i$thspectral band, the symbol $*$ denotes the convolution operator, and $F(x, y)$ is the surround function $F(x, y) = Ke^{-r^2/c^2}$, where c is the Gaussian surround space constant, and Kis also a constant such that $\iint F(x, y)dxdy = 1$. The constant c is used for controlling the scale of $F(x, y)$. From Fig. 9.14 we can see that a small scale results in strong dynamic range compression and halo artifacts near strong edges, while a large scale results in weak dynamic range compression and strong tonal and color rendition.

In order to combine the dynamic range compression and the tonal rendition, SSR is extended to multiscale Retinex (MSR) [16, 37, 38] and further multiscale Retinex with color restoration (MSRCR) [16]. The result of MSR is the weighted sum of the results of SSR with different scales. The ith spectral component of MSR output is mathematically expressed by the following equation

$$R_{MSR_i} = \sum_{n=1}^{N} w_n R_{n_i}, \tag{9.49}$$

where N is the number of scales, R_{n_i} is the ith component of the nth scale, and w_n is the weight associated with the nth scale. R_{n_i} is defined by

$$R_{n_i}(x, y) = \log I_i(x, y) - \log\left[F_n(x, y) * I_i(x, y)\right], \tag{9.50}$$

where $F_n(x, y) = Ke^{-r^2/c_n^2}$, c_n is the constant of the nth scale.

The result of MSRCR is the product of the result of MSR and a color restoration function. Mathematically,

$$R_{\mathrm{MSRCR}_i}(x, y) = C_i(x, y) R_{\mathrm{MSR}_i}(x, y), \tag{9.51}$$

where $C_i(x, y) = f[I'_i(x, y)]$ is the ith band of the color restoration function (CRF) in the chromaticity space. The authors [16] provided the best overall color restoration as follows

$$\begin{aligned} C_i(x, y) &= \beta \log\left[\alpha I'_i(x, y)\right] \\ &= \beta\left\{\log\left[\alpha I_i(x, y)\right] - \log\left[\sum_{i=1}^{S} I_i(x, y)\right]\right\}, \end{aligned} \tag{9.52}$$

where β is a gain constant, and α is used for controlling the strength of the nonlinearity, S is the number of spectral channels.

The results obtained by SSR, MSR, and MSRCR are shown in Fig. 9.14.

9.3.5 PDE-based Retinex

The PDE-based Retinex [26] is based on the assumption that the illumination is spatially smooth and the reflection is piecewise constant. Thus the derivative of the illumination should be close to zero everywhere, and the derivative of the reflection should get high values along the edges and vanish almost everywhere. Then a clipping operation is taken after the application of the Laplacian on image, and the estimated illumination can be obtained by solving a Poisson equation with Neumann boundary condition

$$\begin{cases} -\Delta_x L(x, y) = F(c), & x \neq y, x \in \mathcal{R} \\ L(x, y) = 0, & x = y, x \in \mathcal{R} \\ \dfrac{\partial_d L(x, y)}{\partial n} = 0, & x \in \partial\mathcal{R} \end{cases} \tag{9.53}$$

where $L(x, y)$ is the relative lightness between two pixels x and y, $-\Delta_x L$ denotes the discrete Laplace operator with respect to x, $\mathcal{R}$:$L(x, y)$ denotes the initial discrete lightness image model, and

$$F(x) = f\left(\frac{I(x)}{I(x_{-0})}\right) + f\left(\frac{I(x)}{I(x_{+0})}\right) + f\left(\frac{I(x)}{I(x_{0-})}\right) + f\left(\frac{I(x)}{I(x_{0+})}\right), \tag{9.54}$$

where $x_{-0} = (i-1, j)$, $x_{0-} = (i, j-1)$, $x_{+0} = (i+1, j)$, $x_{0+} = (i, j+1)$, $f(x) = \delta(\log(x))$, δ is defined as

$$\delta(s) = \begin{cases} s, \text{ if } |\text{s}| \geq \text{t} \\ 0, \text{ if } |\text{s}| < \text{t} \end{cases} \tag{9.55}$$

where t is a fixed contrast threshold.

The discrete partial differential Eq. (9.53) can be easily solved by using the discrete Fourier transform. To enforce the Neumann boundary condition, the right and bottom sides of the input image are first padded with each mirror reflection to obtain an image four times larger. The larger image is symmetric with respect to its vertical and horizontal medial axes.

The discrete Fourier transform of a two-dimensional function $f(n, m)$ defined on a $N \times M$ grid is defined for $(k, l) \in \{0, \ldots, M-1\} \times \{0, \ldots, N-1\}$ by

$$\hat{f}(k,l) = \frac{1}{NM} \sum_{n=0}^{N-1} \sum_{m=0}^{M-1} f(n,m) e^{-i2\pi kn/N} e^{-i2\pi lm/M}, \tag{9.56}$$

and the discrete inverse Fourier transform for $(m, n) \in \{0, \ldots, M-1\} \times \{0, \ldots, N-1\}$ by

$$f(n,m) = \sum_{k=0}^{N-1} \sum_{l=0}^{M-1} \hat{f}(k,l) e^{i2\pi kn/N} e^{i2\pi lm/M}, \tag{9.57}$$

The discrete Fourier transform has the following property

$$f(n-n_0, m-m_0) = \sum_{k=0}^{N-1} \sum_{l=0}^{M-1} \hat{g}(k,l) e^{i2\pi kn/N} e^{i2\pi lm/M}, \tag{9.58}$$

where

$$\hat{g}(k,l) = \hat{f}(k,l) e^{-i2\pi kn_0/N} e^{-i2\pi lm_0/M}. \tag{9.59}$$

Applying the discrete Fourier transform to Eq. (9.53) and using the property (9.59) yields

$$\hat{L}(k,l) \left(4 - 2\cos\frac{2\pi k}{N} - 2\cos\frac{2\pi l}{M}\right) = \hat{F}(k,l), \tag{9.60}$$

which entails (for $k, l \neq 0$)

$$\hat{L}(k,l) = \frac{\hat{F}(k,l)}{4 - 2\cos\frac{2\pi k}{N} - 2\cos\frac{2\pi l}{M}}. \tag{9.61}$$

Then using the inverse Fourier transform in Eq. (9.57), we obtain the value of L at each point of the grid. The final lightness can be obtained by normalizing the value of L to the interval [0,255]. Figure 9.15 shows the result obtained by PDE-based Retinex.

Fig. 9.15 **a** Original image, **b** result obtained by PDE-based Retinex [28]

9.3.6 Variational Retinex

9.3.6.1 Kernel-Based Retinex and Contrast Enhancement

Kernel-Based Retinex

In [3] Bertalmío et al. provided a variational investigation of Land and McCann's original Retinex theory [21]. Their model, named Kernel-Based Retinex (KBR), relies on the computation of the expectation value of a suitable random variable weighted with a kernel function. This model shares the same intrinsic properties with the original implementation of Retinex. Given a pixel x in an image I, the kernel function $\omega(x, y)$ is a positive, symmetric, and normalized weight representing the probability density of picking a pixel y in the neighborhood of x. Let $Y_{\omega,x}$ denote the random variable modeling the selection of a pixel in the neighborhood of x according to $\omega(x, y)$. The output of KBR at the pixel x is defined as the conditional expectation of the scaled integrated reflectance with respect to ω, i.e.,

$$\begin{aligned} L_\omega(x) &= \mathrm{E}_{Y_{\omega,x}}\left[\hat{f}\left(\frac{I(x)}{I(Y_{\omega,x})}\right)\right] \\ &= \sum_{I(y)\geq I(x)} \omega(x, y) f\left(\frac{I(x)}{I(y)}\right) + \sum_{I(y)<I(x)} \omega(x, y), \end{aligned} \tag{9.62}$$

where $\hat{f}(\bullet)$ is the scaling function defined by

$$\hat{f}(r) = \begin{cases} f(r) & \text{if } r \in (0,1] \\ 1 & \text{if } r \in [1, +\infty] \end{cases}. \tag{9.63}$$

The function f is a strictly increasing function $f{:}(0,1] \to (0,1]$ such that $f(r) \geq r$ for all $r \in (0,1]$. In [25] and [20], f is suggested to be approximated by Glasser et al.'s [12] power law or logarithmic function. The function $\hat{f}(I(x)/I(Y_{\omega,x}))$ implements both the scaling and the reset mechanism.

In order to analyze the intrinsic characteristics of KBR, the authors in [3] first rewrite Eq. (9.62) as

$$L(x) = \sum_y \omega(x, y) f\left(\frac{I(x)}{I(y)}\right) \text{sign}^+(I(y) - I(x)) + \sum_y \omega(x, y) \text{sign}^-(I(y) - I(x)), \tag{9.64}$$

where

$$\text{sign}^+(\xi) = \begin{cases} 1 & \text{if } \xi > 0, \\ 1/2 & \text{if } \xi = 0, \\ 0 & \text{if } \xi < 0, \end{cases} \quad \text{sign}^-(\xi) = 1 - \text{sign}^+(\xi). \tag{9.65}$$

It has been verified that KBR always increases brightness as the original Retinex implementation, i.e., $L(x) \geq I(x)$. This property implies that both the original Retinex and KBR could not enhance overexposed images. Then by comparing the analytical structure of KBR with that of a variational model of color image enhancement discussed in [30], Bertalmío et al. proposed a two-sided contrast modification to enhance both overexposed and underexposed images by anti-symmetrizing the analytic expression of Eq. (9.64).

In Sect. 9.3.6.1 we briefly review the perceptually inspired variational framework in [30], and then we present two anti-symmetrized KBR versions proposed by Bertalmío et al.

Perceptually Inspired Variational Framework for Color Enhancement

The variational framework proposed in [30] provides a sort of "universal house" where existing models, such as Retinex and ACE, can be rigorously compared and understood. Therefore, here we briefly recall this variational framework, and then we show the anti-symmetrized KBR versions proposed by Bertalmío et al.

Inspired by the basic phenomenology of color perception of HVS, in [30], Palma-Amestoy et al. proposed a perceptual energy functional $E(I)$ composed by a contrast energy term and a dispersion energy term $D_{I_0,1/2}(I), i.e. E(I) = C_\omega(I) + D_{I_0,1/2}(I)$.

The contrast energy term $C_\omega(I) = \lambda \sum_x \sum_y \omega(x, y) c(I(x), I(y)), \lambda > 0$ is a normalization constant, c is an inverse contrast function symmetric in $(I(x), I(y)), i.e., c(I(x), I(y)) = c(I(y), I(x))$. Two basic examples of c are $c(I(x), I(y)) = \min(I(x), I(y)) - \max(I(x), I(y)) = (\min - \max)(I(x), I(y))$ and $c(I(x), I(y)) = \frac{\min(I(x), I(y))}{\max(I(x), I(y))} = \frac{\min}{\max}(I(x), I(y))$. The function c decreases when $\min(I(x), I(y))$ decreases or $\max(I(x), I(y))$ increases. Since we want an overall change in the intensity that does not affect the visual sensation, as HVS does, we require c to be able to disregard the change of intensity and to be homogeneous. Then the minimization of $C_\omega(I)$ leads to a local contrast enhancement for image I.

In [30], the entropic distance is used to model the dispersion energy term, i.e., $D^{\varepsilon}_{\alpha,\beta}(I) = \alpha \sum_x \left[\frac{1}{2}\log\frac{1}{2I(x)} - \left(\frac{1}{2} - I(x)\right)\right] + \beta \sum_x \left[I_0(x)\log\frac{I_0(x)}{I(x)} - (I_0(x) - I(x))\right]$ where $\alpha, \beta > 0$ control the strength of the departure from the middle gray 1/2 and from the original image $I_0(x)$.

To search for the minimum of $E(I)$, a semi-implicit discrete gradient descent strategy is used in [30]. Furthermore, Palma-Amestoy et al. proposed a general procedure to reduce the computational cost of the algorithms by applying a polynomial approximation strategy.

Two Anti-symmetrized KBR Versions

In [3], the inverse contrast functions which are homogenous of degree 0 is written as $G\left(\frac{\min}{\max}(I(x), I(y))\right) \equiv G\left(\frac{\min(I(x),I(y))}{\max(I(x),I(y))}\right)$, where $G(r)$ is a monotone nondecreasing function of r. The corresponding contrast energy functional can be written as

$$C_\omega^{\frac{\min}{\max}}(I) = \frac{1}{2}\sum_x\sum_y \omega(x,y) G\left(\frac{\min}{\max}(I(x), I(y))\right). \tag{9.66}$$

Similarly, the contrast energy functional corresponding to a homogeneous function of degree 1 can be written as

$$C_\omega^{\min-\max}(I) = \frac{1}{2}\sum_x\sum_y \omega(x,y)(\min(I(x), I(y)) - \max(I(x), I(y))). \tag{9.67}$$

The first variation of $C_\omega^{\frac{\min}{\max}}(I)$ and $C_\omega^{\min-\max}(I)$ are respectively

$$\begin{aligned} \delta C_\omega^{\frac{\min}{\max}}(I) = &\sum_y \omega(x,y) g\left(\frac{I(x)}{I(y)}\right)\frac{1}{I(y)}\mathrm{sign}^+(I(y) - I(x)) \\ &- \sum_y \omega(x,y) g\left(\frac{I(y)}{I(x)}\right)\frac{I(y)}{I(x)^2}\mathrm{sign}^-(I(y) - I(x)) \end{aligned} \tag{9.68}$$

and

$$\delta C_\omega^{\min-\max}(I)(x) = \sum_y \omega(x,y)\mathrm{signo}_0(I(y) - I(x)), \tag{9.69}$$

where

$$\mathrm{sign}(z) = \begin{cases} 1 & \text{if } z > 0 \\ [-1,1] & \text{if } z = 0 \\ -1 & \text{if } z < 0 \end{cases} \tag{9.70}$$

and $\text{sign}_0(z)$ is defined as in (9.70), but with the particular choice 0 when $z = 0$.

The variational contrast lightness of homogeneity degree n is defined as

$$L_n(x) = \frac{1}{2} - \frac{1}{2} I(x)^{1-n} \delta C_\omega(I)(x), \forall \text{n} \in \mathbb{Z}, \tag{9.71}$$

where 1/2 is used for assuring that $L_n(x) \in [0,1]$. When $n = 0,1$.

$$L_0(x) = \frac{1}{2} - \frac{1}{2} I(x) \delta C_\omega(I)(x), \tag{9.72}$$

$$L_1(x) = \frac{1}{2} - \frac{1}{2} \delta C_\omega(I)(x). \tag{9.73}$$

Substituting the explicit expressions of $\delta C_\omega^{\frac{\min}{\max}}(I)$ and $\delta C_\omega^{\min - \max}(I)$ in Eqs. (9.72) and (9.73), respectively, we can see that the computation of the lightness can be realized through modulations of positive and negative signs around 1/2, thus the corresponding algorithm is able to enhance both under and overexposed images. It is worth noticing that $L_1(x) = \frac{1}{2} + \frac{1}{2} \sum_y \omega(x, y) \text{sign}_0(I(x) - I(y))$ corresponds to the variational formulation of ACE (see Eq. (9.30) in Sect. 9.2.3.2), which means that ACE can be seen as a particular anti-symmetrization of the KBR model. This analysis reveals novel insights about the Retinex action on contrast and makes a link between Retinex and ACE.

Finally, by adding the entropic dispersion energy term to perform the mechanism of attachment to original data, Bertalmío et al. defined the energy functionals of the two anti-symmetrized KBR versions as

$$E_{\alpha,\beta,\omega}^{F(\frac{\min}{\max})}(I) = D_{\alpha,\beta}^{\varepsilon}(I) + C_\omega^{F(\frac{\min}{\max})}(I) \tag{9.74}$$

and

$$E_{\alpha,\beta,\omega}^{\min - \max}(I) = D_{\alpha,\beta}^{\varepsilon}(I) + C_\omega^{\min - \max}(I). \tag{9.75}$$

To reduce the computational complexity of the algorithms, the strategies of the minimization of $E_{\alpha,\beta,\omega}^{F(\frac{\min}{\max})}(I)$ and $E_{\alpha,\beta,\omega}^{\min - \max}(I)$ are similar to that proposed in [30].

In Fig. 9.16, we show the resulting images of KBR and two anti-symmetrized KBR.

9.3.6.2 Variational Framework for The PDE-based Retinex

Based on the same assumption as the PDE-based Retinex, Kimmel et al. [18] proposed a variational framework for Retinex which unifies Retinex methods using a different interpretation than KBR. Their framework is modeled as follows

$$\begin{aligned} &\text{Minimize: } F(l) = \int_\Omega (|\nabla l|^2 + \alpha(l-s)^2 + \beta|\nabla(l-s)|^2) dxdy \\ &\text{Subject to: } l \geq s, \text{ and } \langle \nabla l, \vec{n} \rangle = 0 \text{ on } \partial\Omega, \end{aligned} \tag{9.76}$$

Fig. 9.16 **a** Original image, **b** result obtained by KBR, **c** result obtained by anti-symmetrized KBR (Eq. (9.74)), **d** result obtained by anti-symmetrized KBR (Eq. (9.75))

where Ω is the support of the image, $\partial\Omega$ is the boundary, and $\overrightarrow{n}$ is the normal to the boundary. α and β are non-negative real parameters. In the functional $F(l)$, the first term forces spatial smoothness on the illumination image. The second penalty term forces a proximity between the illumination l and the original image s. The third term represents a Bayesian penalty expression.

Since the problem defined in penalty functional (9.76) has a quadratic programming form [4, 22], the minimization of the functional can be easily obtained via the Euler–Lagrange equations

$$\begin{aligned} &\forall(x, y) \in \Omega \\ &\frac{\partial F(l)}{\partial l} = 0 = -\Delta l + \alpha(l - s) - \beta\Delta(l - s) \text{ and } l > s \text{ or } l = s \end{aligned} \tag{9.77}$$

Usually, the reflectance image obtained by the Retinex process is an over-enhanced image. Kimmel et al. [18] proposed to add a corrected version of the reconstructed illumination back to the reconstructed reflectance image. First, the reflectance image

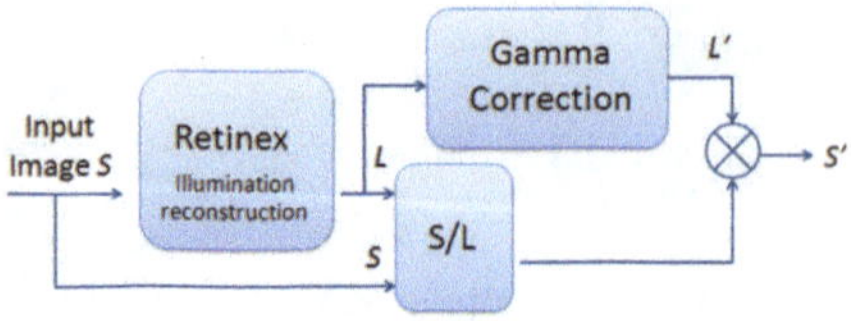

Fig. 9.17 The flow chart of Retinex algorithm

is computed by $R = S/L$. Then the illumination image L is corrected by a Gamma Correction operation with a free parameter γ

$$L' = W \cdot \left[\frac{L}{W} \right]^{\frac{1}{\gamma}}, \tag{9.78}$$

where L'denotes the corrected illumination image, W is the White value (equal to 255 in 8-bit images). Then the output image is obtained by

$$S' = L' \cdot R = \frac{L'}{L} S = W \frac{(L/W)^{1/\gamma}}{L} S = \frac{S}{(L/W)^{1-1/\gamma}}. \tag{9.79}$$

The flowchart of the operation is illustrated in Fig. 9.17.

9.3.6.3 TV Regularized Model for Retinex

Based on this variational Retinex framework, Ma and Osher [23] established a total variational (TV) and nonlocal TV regularized model for Retinex. These two models are respectively written in the following forms

$$r = \arg\min_r \left\{ t \int_\Omega |\nabla r| + \frac{1}{2} \|\nabla (r - s)\|_2^2 \right\}, \tag{9.80}$$

and

$$r = \arg\min_r \left\{ t \int_\Omega \sqrt{d_x^2 + d_y^2} + \frac{1}{2} \|d - \nabla s\|_2^2 \right\} \text{ s.t. } d = \nabla r, \tag{9.81}$$

where t is a parameter for balancing the two terms, $d = (d_x, d_y)$is an auxiliary variable such that $d_x = \nabla_x r$ and $d_y = \nabla_y r$.

In their model, the TV aTV Regularized Model for Retinexnd nonlocal TV penalty terms are used to exploit the prior knowledge of the reflectance image. These two models can be efficiently solved by a Bregman iteration [6, 29]. Figure 9.18 shows the result obtained by the TV regularized model for Retinex.

Considering the relationship between reflectance and the illumination, Ng and Wang [27] improved the TV Bregman iteration model by adding some constraints and a fidelity term in the proposed energy functional. The constraints are based on the following assumptions: first, the illumination function is spatially smooth; second, the reflection function is piecewise constant; third, based on the reflectivity, the constraints $l \geq s$ and $r \geq 0$ are added.

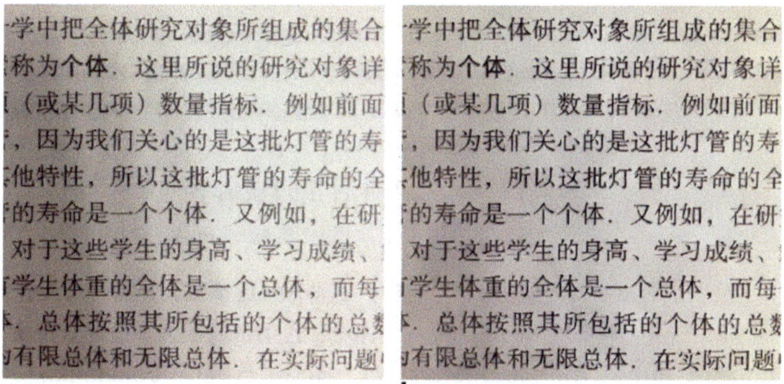

a b

Fig. 9.18 **a** Original image, **b** result obtained by TV regularized model

Fig. 9.19 **a** Original image, **b** enhanced image by the TV model for Retinex

Then the proposed energy functional for Retinex is proposed as follows

$$E(r,l) = \int_\Omega |Dr| + \frac{\alpha}{2}\int_\Omega |\nabla l|^2 dx + \frac{\beta}{2}\int_\Omega (l - r - s)^2 dx + \frac{\mu}{2}\int_\Omega l^2 dx, \quad (9.82)$$

where $r = -\log R > 0, \alpha, \beta$, and $\mu$are positive regularization parameters, the first TV term forces piecewise constant on the reflectance image, the second regularization term forces spatially smooth on the illumination image, the third term is used for the fidelity, and the last term is used only for the theoretical setting. Numerically, the parameter μ can be set to a very small number or even zero. The minimization problem of the energy functional (9.82) can be solved by using an alternating minimization scheme proposed in [27]. Figure 9.19 shows the enhanced image produced by this method.

9.3.6.4 Variational Bayesian Method for Retinex

Recently, Wang et al. [43] proposed a variational Bayesian method for Retinex. To construct a hierarchical Bayesian model, Gibbs distributions are used as prior distributions for the reflection and the illumination, and gamma distributions are used for the model parameters. Let the $N \times 1$vectors $\mathbf{r}$, $\mathbf{l}$, and $\mathbf{s}$ denote vectorized r, l, and s in energy functional (9.82), respectively, based on the same assumptions proposed in [27], the prior models on the reflection image and the illumination image are given by

$$p(\mathbf{r}|\alpha) = c_{\mathbf{r}}\alpha^{N/2} \exp\left[-\alpha F_{\mathrm{TV}}(\mathbf{r})\right], \tag{9.83}$$

$$p(\mathbf{l}|\sigma) = c_{\mathbf{l}}\sigma^{N/2} \exp\left[-\sigma \sum_{i}^{N} (\Delta_i^h(\mathbf{l}))^2 + (\Delta_i^v(\mathbf{l}))^2\right], \tag{9.84}$$

where α and σ are hyperparameters, $c_{\mathbf{r}}$ and $c_{\mathbf{l}}$ are constants, $F_{\mathrm{TV}}(\mathbf{r}) = \sum_i^N \sqrt{(\Delta_i^h(\mathbf{r}))^2 + (\Delta_i^v(\mathbf{r}))^2}$ is the energy function for TV prior, Δ_i^h and Δ_i^v are linear operators corresponding to horizontal and vertical first order differences at pixel i. Let ε denote the estimated error and let us assume that ε has a Gaussian distribution with zero mean and precision β (the inverse of variance). The conditional distribution of $\mathbf{s}$ is given by

$$p(\mathbf{s}|\mathbf{l}, \mathbf{r}, \beta) \propto \beta^{N/2} \exp\left[-\frac{\beta}{2}(\mathbf{l} - \mathbf{s} - \mathbf{r})^2\right]. \tag{9.85}$$

The distribution for the hyperprior is defined by using Gamma distribution

$$p(h) = \Gamma(h|a_h^o, b_h^o) = \frac{h^{a_h^o - 1} \exp\left[-h/b_h^o\right]}{(b_h^o)^{a_h^o}\Gamma(a_h^o)}, \tag{9.86}$$

where $h > 0, h \in \{\alpha, \beta, \sigma\}$ denotes a hyperparameter, $a_h^o > 0$ and $b_h^o > 0$ denote the shape parameter and the scale parameter, respectively. a_h^o and b_h^o are assumed to be known.

Finally, the global distribution is given by

$$p(\mathbf{s}, \mathbf{l}, \mathbf{r}, \alpha, \beta, \sigma) = p(\alpha)p(\beta)p(\sigma)p(\mathbf{r}|\alpha)p(\mathbf{l}|\sigma)p(\mathbf{s}|\mathbf{l}, \mathbf{r}, \beta), \tag{9.87}$$

where $p(\alpha)$, $p(\beta)$ and $p(\sigma)$ are defined in (9.86), $p(\mathbf{r}|\alpha)$, $p(\mathbf{l}|\sigma)$ and $p(\mathbf{s}|\mathbf{l}, \mathbf{r}, \beta)$ are defined in (9.83), (9.84), and (9.85), respectively. The dependencies in the joint probability model are shown in Fig. 9.20 which presents the relationships between variables for Retinex in a directed acyclic graph.

Let $\Theta = \{\mathbf{l}, \mathbf{r}, \alpha, \beta, \sigma\}$ denote the set of all unknown variables, The Bayesian inference is based on the posterior distribution

$$p(\Theta|\mathbf{s}) = p(\mathbf{l}, \mathbf{r}, \alpha, \beta, \sigma|\mathbf{s}) = \frac{p(\mathbf{s}, \mathbf{l}, \mathbf{r}, \alpha, \beta, \sigma)}{p(\mathbf{s})}. \tag{9.88}$$

Because $p(\mathbf{s}) = \int_\Theta \int_{\mathbf{r}} \int_{\mathbf{l}} p(\mathbf{s}, \mathbf{l}, \mathbf{r}, \alpha, \beta, \sigma) d\mathbf{l} d\mathbf{r} d\Theta$ is analytically intractable, it is difficult to calculate the posterior distribution $p(\Theta|\mathbf{s})$. Thus, a variational Bayes approximation approach [42] is used to obtain the approximation $q(\Theta) = q(\mathbf{l})q(\mathbf{r})q(\alpha)q(\beta)q(\sigma)$ of a factorization form for the posterior distribution. Then the estimated variables can be obtained from the expectation of the approximation of the posterior distribution via an alternant iterative scheme. Figure 9.21 shows the result obtained by the variational Bayesian method for Retinex.

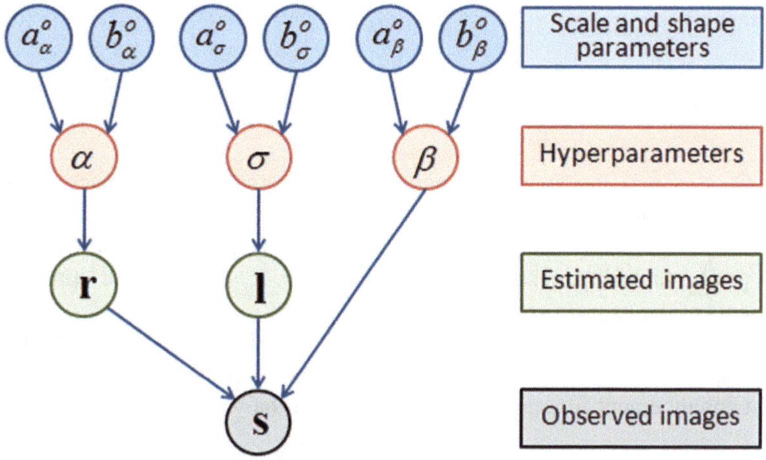

Fig. 9.20 Graphical model describing relationships between variables for Retinex

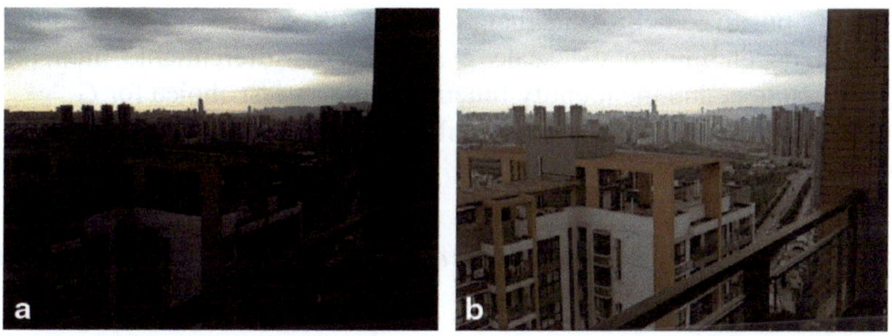

Fig. 9.21 **a** Original image, **b** result obtained by the variational Bayesian method for Retinex

9.3.6.5 Variational Perceptually-Inspired Color Enhancement in Wavelet Domain

Since the KBR technique introduced in [3] reveals a novel insight about the Retinex action on contrast, we would like to introduce a related method which analyzes the perceptually-inspired color enhancement issue based on wavelet theory [33].

Let's first recall the multiscale analysis in contrast measurement introduced by Peli [31]. The contrast at the band of spatial frequencies is defined as

$$C_j^{\text{Peli}}(x, y) = \frac{\psi_j * I(x, y)}{\phi_j * I(x, y)}, \tag{9.89}$$

where ψ_j and ϕ_j are a band-pass, and a low-pass filter of a filter bank at the scale j, $I(x, y)$ is the pixel at coordinate (x, y) in image I. In terms of the wavelet coefficients, the local contrast measure (9.89) can be rewritten as

$$C^{\text{Peli}}_{j,k,l}(x,y) = \frac{d^l_{j,k}}{a^l_{j,k}},\ j \in \{L, \ldots, J\}, k \in \Omega, l \in \{H, V, D\}, \tag{9.90}$$

where $\Omega \subset \mathbb{Z}^2$ is the spatial domain of image area $|\Omega|$, the discrete orthogonal wavelet multiresolution representation between two scales 2^L and 2^J, $L, J \in \mathbb{Z}, L < J$, is given by the horizontal, vertical, and diagonal detail coefficients, i.e., $d^H_{j,k}, d^V_{j,k}$ and $d^D_{j,k}$, respectively, at all scales, $a_{j,k}$denotes the approximation coefficients at the coarser scale.

In order to enhance contrast locally and in an illuminant-independent way as the contrast energy term (9.66) does, Provenzi and Caselles [33] proposed a local contrast enhancement functional at the scale 2^j in the wavelet domain as

$$C_{\omega_j,G,\{a_{j,k}\}}(\{d_{j,k}\}) = \sum_{k\in\Omega} \omega_j G\left(\frac{a_{j,k}}{d_{j,k}}\right), 2^J \geq 2^j \geq 2^{L+1}, \tag{9.91}$$

where ω_j are positive coefficients that permit to differentiate the contrast enhancement action depending on the scale 2^j, $G(r)$ is a differentiable monotonically increasing function, and the identity function is the easiest choice for G.

In order to preserve the original detail coefficients, an entropic distance based conservation term is defined as

$$D_{d^0_{j,k}}(\{d_{j,k}\}) = \sum_{k\in\Omega}\left[d^0_{j,k}\log\frac{d^0_{j,k}}{d_{j,k}} - (d^0_{j,k} - d_{j,k})\right], 2^J \geq 2^j \geq 2^{L+1}, \tag{9.92}$$

Then, combining the contrast enhancement functional (9.91) and the entropic dispersion functional (9.92), Provenzi and Caselles defined the final local contrast enhancement functional as

$$\begin{aligned} E_{\omega_j,G,\{a_{j,k}\},d^0_{j,k}} &= C_{\omega_j,G,\{a_{j,k}\}} + D_{d^0_{j,k}} \\ &= \sum_{k\in\Omega}\left[\omega_j G\left(\frac{a_{j,k}}{d_{j,k}}\right) + d^0_{j,k}\log\frac{d^0_{j,k}}{d_{j,k}} - (d^0_{j,k} - d_{j,k})\right], \end{aligned} \tag{9.93}$$

with $2^J \geq 2^j \geq 2^{L+1}$. The Euler–Lagrange equations for the detail coefficients are

$$\begin{aligned} &\frac{\partial E_{\omega_j,G,\{a_{j,k}\},d^0_{j,k}}}{\partial\{d_{j,k}\}}(d_{j,k}) = 0 \Leftrightarrow \\ &d_{j,k} = d^0_{j,k} + \omega_j G'\left(\frac{a_{j,k}}{d_{j,k}}\right)\frac{a_{j,k}}{d_{j,k}}, \end{aligned} \tag{9.94}$$

where G' denotes the derivative of G.

In [33], the function G is set as the identity function, and the corresponding Euler–Lagrange equations are

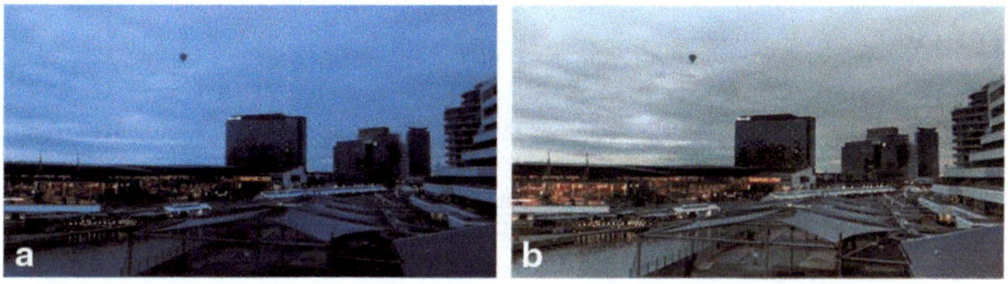

Fig. 9.22 **a** Original image, **b** result obtained by the wavelet-based algorithm

$$d_{j,k} = d^0_{j,k} + \omega_j \frac{a_{j,k}}{d_{j,k}}. \tag{9.95}$$

Newton's method is used in [33] to solve Eq. (9.95), initialized with the original magnitude values $d^0_{j,k}$.

Finally, the general scheme of this algorithm is given as follows [33] :

1. Modify the coarsest approximation coefficients $\{a_{J,k}, k \in \Omega\}$ to implement adaptation to the average gray level in the wavelet domain according to

$$a_{J,k} = \alpha \bar{a}_J + (1-\alpha) a^0_{J,k}, \tag{9.96}$$

where $\{a^0_{J,k}, k \in \Omega\}$ is the original sequence of approximation coefficients at the scale 2^J, and $\alpha \in [0,1]$ is a suitable weight coefficient, $\bar{a}_J = \frac{1}{|\Omega|}\sum_{k\in\Omega} a_{j,k}$ is the average approximation coefficient.

2. Fix $\{a_{J,k}, k \in \Omega\}$ and use them with $\{d^l_{J,k}, k \in \Omega\}, l \in \{H, V, D\}$ to modify the detail coefficients according to Eq. (9.95).
3. Then pass to the scale 2^{J-1} and compute the approximation coefficients by summing the approximation and detail coefficients just computed at the scale 2^J.
4. Fix these approximation coefficients and repeat step 2 at the scale 2^{J-1}.
5. Iterate this scheme until reaching the finest scale.

Figure 9.22 shows the resulting image obtained by this wavelet-based algorithm.

9.4 Discussions

By studying the abovementioned methods, we found that the recent development of both color equalization and Retinex is closely related to the features of HVS. Since HVS is a very complex biological system, the mechanism of human perception of the real world is still not completely understood. Then, how to further understand this mechanism and how to take the advantage of the existent knowledge of HVS to develop a biologically consistent model for image enhancement are the problems to be solved in the future.

References

1. Barnard, H., Martin, L., Funt, B., et al.: A data set of colour research. Color. Res. Appl. **27**(3), 147–151 (2002)
2. Bertalmío, M., Caselles, V., Provenzi, E., et al.: Perceptual color correction through variational techniques. IEEE. Trans. Image. Process. **16**(4), 1058–1072 (2007)
3. Bertalmío, M., Caselles, V., Provenzi, E.: Issues about Retinex theory and contrast enhancement. Int. J. Comput. Vis. **83**(1), 101–119 (2009)
4. Bertsekas, D.P.: Non-Linear Programming. Athena Scientific, Belmont (1995)
5. Blake, A.: Boundary conditions for lightness computation in mondrian world. Comput. Vis. Gr. Image. Process. **32**, 314–327 (1985)
6. Bregman, L.: The relaxation method of finding the common points of convex sets and its application to the solution of problems in convex programming. USSR. Comput. Math. Math. Phys v. **7**, 200–217 (1967)
7. Elad, M., Kimmel, R., Shaked, D., et al.: Reduced complexity Retinex algorithm via the variational approach. J. Vis Commun. Image. Represent. **14**, 369–388 (2003)
8. Frankle, J., McCann, J.: Method and apparatus for lightness imaging. US Patent 4,384,336,17 May (1983)
9. Funt, B., Ciurea, F., McCann, J.J.: Retinex in MATLAB^{TM}. J. Electron. Imaging. **13**, 48–57 (2004)
10. Gatta, C., Rizzi, A., Marini, D.: ACE: An automatic color equalization algorithm. Proceedings of the First European Conference on Color in Graphics Image and Vision (CGIV02) (2002)
11. Getreuer, P. (2012) Automatic color enhancement (ACE) and its fast implementation, Image Processing On Line, 2, pp. 266–277. http://dx.doi.org/10.5201/ipol.2012.g-ace
12. Glasser, L., McKinney, A., Reilly, C., et al.: Cube-root color coordinate system. J. Opt. Soc. Am. **48**, 736–740 (1958)
13. Gonzalez, R.C., Woods, R.E.: Digital image processing, 2nd edn. Prentice-Hall, Englewood Cliffs (2002)
14. Horn, B.K.P.: Determining lightness from an image. Comput. Gr. Image. Process. **3**, 277–299 (1974)
15. Hummel, R.A.: Image enhancement by histogram transformation. Comput Gr. Image. Process. **6**(2), 184–195 (1977)
16. Jobson, D.J., Rahman, Z., Woodell, G.A.: Properties and performance of the center/surround Retinex. IEEE. Trans. Image. Process. **6**(3), 451–462 (1997)
17. Karel, Z.: Contrast limited adaptive histogram equalization. In: Heckbert, P.S. (eds.) Graphic Gems IV, pp. 474–485. Cambridge, Academic Press Professional (1994)
18. Kimmel, R., Elad, M., Shaked, D., et al.: A variational framework for Retinex. Int. J. Comput. Vis. **52**(1), 7–23 (2003)
19. Land, E.H.: The Retinex theory of color vision. Sci. Am. **237**(6), 108–128 (1977)
20. Land, E.: An alternative technique for the computation of the designator in the Retinex theory of color vision. Proc Natl Acad Sci U. S. A. 83, 3078–3080 (1986)
21. Land, E.H., McCann, J.J.: Lightness and the Retinex theory. J. Opt. Soc. Am. **61**(1), 1–11 (1971)
22. Luenberger, D.G.: Linear and Non-linear programming, 2nd edn. Addison-Wesley, Menlo-Park (1987)
23. Ma, W., Osher, S.: A TV Bregman iterative model of Retinex theory. UCLA CAM Report 10–13, UCLA, Los Angeles, CA (2010)
24. McCann, J.J.: Lessons learned from Mondrians applied to real images and color gamuts. In: Proc IS & T/SID 7th Color Imag. Conf., 1–8 (1999)
25. McCann, J.J., McKee, S., Taylor, T.: Quantitative studies in Retinex theory: a comparison between theoretical predictions and observer responses to the 'color mondrian' experiments. J. Vis. Res. **16**, 445–458 (1976)
26. Morel, J.M., Petro, A.B., Sbert, C.: A PDE formalization of Retinex theory. IEEE. Trans. Image. Process. **19**(11), 2825–2837 (2010)

27. Ng, M.K., Wang, W.: A total variation model for Retinex. SIAM. J. Imag. Sci. **4**(1), 345–365 (2011)
28. Nicolas, L., Ana, B.P., Catalina, S., Jean-Michel, M.: Retinex Poisson equation: a model for color perception, Image Processing On Line, 1. http://dx.doi.org/10.5201/ipol.2011.lmps_rpe. (2011)
29. Osher, S., Burger, M., Goldfarb, D., et al.: An iterative regularization method for total variation based image restoration. SIAM. Multiscale. Model. Simul. **4**, 460–489 (2005)
30. Palma-Amestoy, R., Provenzi, E., Bertalmío, M., et al.: A perceptually inspired variational framework for color enhancement. IEEE. Trans. Pattern. Anal. Mach. Intell. **31**(3), 458–474 (2009)
31. Peli, E.: Contrast in complex images. J. Opt. Soc. Am. **7**(10), 2032–2040 (1990)
32. Pizer, S.M., Amburn, E.P., Austin J.D., et al.: Adaptive histogram equalization and its variations. Comput. Vis. Gr. Image. Process. **39**, 355–368 (1987)
33. Provenzi, E., Caselles, V.: A wavelet perspective on variational perceptually-inspired color enhancement. Int. J. Comput. Vis. **106**(2), 153–171 (2014)
34. Provenzi, E., Carli, L.D., Rizzi, A., et al.: Mathematical definition and analysis of the retinex algorithm. J. Opt. Soc. Am. **22**(12), 2613–2621 (2005)
35. Provenzi, E., Fierro, M., Rizzi, A., et al.: Random spray Retinex: A new Retinex implementation to investigate the local properties of the model. IEEE. Trans. Image. Process. **16**(1), 162–171 (2007)
36. Provenzi, E., Gatta, C., Fierro, M., et al.: A spatially variant white-patch and gray-world method for color image enhancement driven by local contrast. IEEE. Trans. Pattern. Anal. Mach. Intell. **30**(10), 1750–1770 (2008)
37. Rahman, Z., Jobson, D.J., Woodell, G.A.: Multiscale Retinex for color image enhancement. Proc IEEE ICIP. 1003–1006, (1996)
38. Rahman, Z., Jobson, D.J., Woodell, G.A.: Retinex processing for automatic image enhancement. J. Electron. Imaging. **13**, 100–110 (2004)
39. Rizzi, A., Gatta, C., Marini, D.: A new algorithm for unsupervised global and local color correction. Pattern. Recognit. Lett. **24**, 1663–1677 (2003)
40. Rizzi, A., Gatta, C., Marini, D.: From Retinex to automatic color equalization: Issues in developing a new algorithm for unsupervised color equalization. J. Electron. Imaging. **13**(1), 75–84 (2004)
41. Sapiro, G., Caselles, V.: Histogram modification via differential equations. J. Differ. Equ. **135**(2), 238–266 (1997)
42. Šmídl, V., Quinn, A.: The variational Bayes method in signal processing. Springer Verlag, New York (2005)
43. Wang, L., Xiao, L., Liu, H., et al.: Variational Bayesian Method for Retinex. IEEE. Trans. Image. Process. **23**(8), 3381–3396 (2014)

Chapter 10
Color Correction for Stereo and Multi-view Coding

Sid Ahmed Fezza and Mohamed-Chaker Larabi

10.1 Introduction

Humans perceive the real world in 3D and this ability allows to have a depth sensation. For many years, researchers in industry and academia have been working on providing realistic feelings of immersion to viewers. The main followed direction is to fool the brain by providing a different view to each eye in order to generate a positive or negative disparity that will be interpreted as depth. To this end, a new set of 3D video applications have been introduced to the user through various consumer devices. Among them, the two famous applications may be cited, 3D television (3DTV) [62] and free viewpoint video (FVV) [57]. 3DTV offers to viewers a more realistic multimedia experience by providing an additional sensation of depth for the perception of the scene, while FVV allows the user to watch the scene freely by the possibility to change his viewpoint and viewing direction in an interactive way.

Multi-view video (MVV) and depth-image-based rendering (DIBR) are considered as the key technologies to ensure such content delivery. In MVV, N views are acquired with a set of synchronized cameras capturing the same scene from different viewpoints. When only two views ($N = 2$) are used, we talk about stereoscopic video, which is well-known and the most widespread 3D video format. Besides 3D sensation, in order to achieve the interactive selection of viewpoint and viewing direction, as in FVV, and provide wide range multi-view autostereoscopic displays [60], a large number of views (28–52) are required. This large number of views can be effectively supported with only transmitting a limited subset of views to the receiver. The remaining views are generated as virtual views by the DIBR process at the decoder side [10, 53]. Such advanced 3D video applications are supported, thanks to the multi-view video plus depth (MVD) format [27, 54]. MVD

S. A. Fezza (✉)
University of Oran 2, Oran, Algeria
e-mail: sidahmed.fezza@gmail.com

M.-C. Larabi
XLIM Institute, SIC Department, University of Poitiers, Poitiers, France
e-mail: chaker.larabi@univ-poitiers.fr

M. E. Celebi et al. (eds.), *Color Image and Video Enhancement*,
DOI 10.1007/978-3-319-09363-5_10

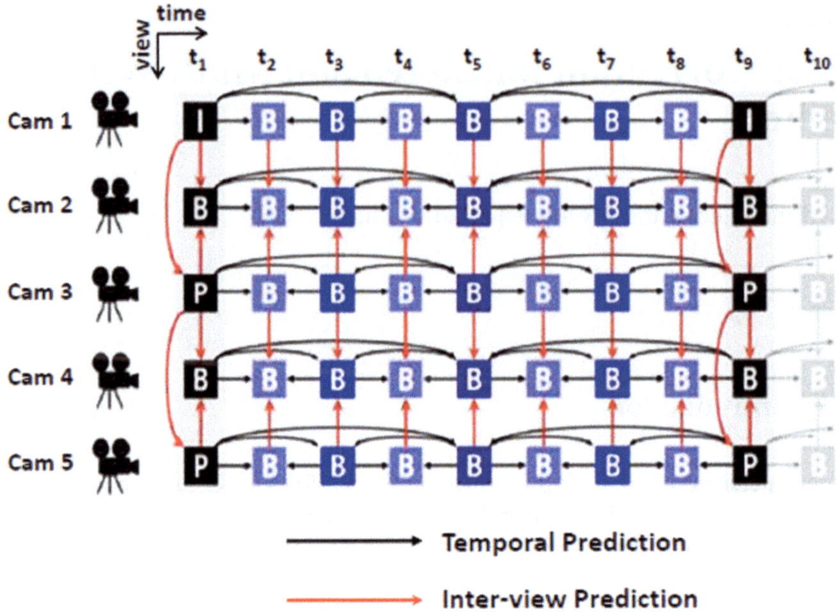

Fig. 10.1 Typical MVC prediction structure

consists of multiple color videos (referred to as texture data) captured from different viewing angles, with a depth value associated with each texture pixel (referred to as depth data). Depth data provide the geometry information and allow rendering intermediate views, in between existing camera positions.

In sight of the huge amount of required data by 3D video applications, which depends linearly on the number of cameras [40], operation of storing or delivering these views to the end-user appears as a challenging issue. Consequently, the design of efficient compression methods is of paramount importance in ensuring the feasibility and utility of such applications. An obvious solution is to independently encode each camera view with conventional coding techniques, such as H.264/AVC (advanced video coding) [20]. This approach is known as the simulcast coding approach. However, as all cameras in the multi-view system are capturing the same scene from slightly different perspectives, they are highly correlated [28]. By taking advantage of this inter-view redundancy existing among the views, compression efficiency can be significantly increased.

According to this vision, the H.264/AVC standard has been amended to support multi-view video coding (MVC) [4, 20, 63]. MVC combines temporal and inter-view predictions, where frames are not only predicted from their temporal neighbors but also from corresponding ones in adjacent views [39]. An example of MVC prediction structure adopted in the standard is depicted in Fig. 10.1. In this prediction structure, in addition to conventional intra and motion-compensated prediction (MCP), MVC exploits also disparity-compensated prediction (DCP) between frames from the same time step but from different camera views.

Fig. 10.2 An example of the color inconsistency between views

Compared to H.264/AVC simulcast coding approach, where views are separately coded, using MCP and DCP at the same time improves coding efficiency. It is reported in [59] that MVC can save an additional 20 % of the overall bit rate. Also, in some specific cases, bit-rate savings of up to 50 % can be obtained [39].

Since MVV is captured from different viewpoints with multiple cameras, significant illumination and color inconsistencies can be observed between views. There are several factors that contribute to these color discrepancies between views. For example, camera orientation, focal length, and manufacturing may be inconsistent, even between cameras of the same model with same parameters setting. Also, nonuniform camera parameters can cause color mismatches. Another important factor is linked to defects in camera calibration. Moreover, even if all cameras are perfectly calibrated, it is often difficult to capture an object under perfectly constant lighting conditions. The reflectance properties over the object surface such as scene geometry, angles between cameras and objects, and light incidence, must also be considered. All these factors negatively impact the color consistency, leading to a possible difference of appearance of the same object on various views, as illustrated by Fig. 10.2.

These color inconsistencies negatively affect the MVV-processing chain. They reduce the correlation existing between views, and result in a significant reduction in compression efficiency; consequently, causing a performance reduction in MVC that exploits the inter-view statistical dependencies. Furthermore, when rendering new virtual views and switching between the different views, observer may perceive color variations across views. This will appear unnatural and will affect his quality of experience (QoE) [46, 47]. In addition, if these color differences exceed a certain threshold, fusion difficulty (i.e., binocular rivalry) and binocular luster can occur inducing visual fatigue and discomfort [6, 14, 16, 31, 48, 66]. Therefore, color calibration or color correction methods for MVV are necessary to enhance the performance of 3D systems [65].

To cope with such important problems, various approaches were proposed to compensate for the illumination and color changes between views. To achieve this, some of them proposed to fine-tuning shooting conditions, such as camera arrangements and parameters, and other methods proposed to customize the codecs in order to take into account the color mismatches. In a completely different way, other approaches performed color correction by preprocessing the captured views to make them color consistent and harmonized.

The goal of this chapter is to review the most important strategies for color correction applied to stereo and/or multi-view images/video. We decompose the existing methods in three categories depending on their objectives and behaviors. The first category tackles the method applied on the capturing stage aiming at significantly reducing color discrepancies between the calibration of the used cameras. The second category is devoted to methods embedded in the core coding, and proposing strategies to apply illumination/color correction. Finally the third category can be seen as a preprocessing applied to captured scene before being fed to coding steps.

A section is dedicated to the experimental comparison of the most common techniques in the literature. The comparison is made by using visual inspection allowing the reader to build his own judgment on the usefulness of the tested methods. In addition, color and quality metrics are applied to provide a quantitative evaluation of color fidelity and coding efficiency. This chapter ends with a conclusion summarizing the different techniques described and opening the floor to the remaining issues to be solved.

10.2 Color Calibration Methods for Multi-view Images/Video

As mentioned previously, several methods have been proposed in the literature to resolve the problem of color inconsistency in MVV. This section is dedicated to their description with a classification into three categories: (1) calibration of multi-camera setup, (2) color compensation in the core coding process called embedded illumination/color compensation, and (3) correction applied as a preprocessing stage before compression.

10.2.1 Calibration of Multi-camera Setup

Over the last few decades, the use of the multi-camera system has grown exponentially. Different applications resort to such acquisition systems, for example 3DTV, FVV, and surveillance, to cite a few examples. However, inherent problems with this type of acquisition are geometric errors and inconsistent color responses. Many methods have been developed to geometrically calibrate the camera arrays [26,36,43,69], but much less attention is being paid to color calibration of cameras, relying only on manual adjustment.

To calibrate color responses of cameras, Nanda et al. [41] proposed a method using scene statistics. In this method, any camera can be chosen as the reference. The latter is color calibrated by adjusting brightness and white balance. The brightness adjustment is performed by acquiring an absolutely black scene at zero exposure, and then by making the mean intensity of the acquired image reach a desired black value. In turn, the white balance is calibrated by adjusting the two parameters, specifically red and blue gain values, to make the amount of green, blue, and red equal in the scene. Once the reference camera has been color calibrated, to ensure color uniformity between cameras, the camera settings of the neighboring cameras are adjusted to make the colors of the overlapping regions across cameras similar. However, this method produces varying results depending on the part of the scene acquired by the camera.

Another color calibration way is to use a Macbeth color checker. For example, in [21,22], the authors used color charts to improve color consistency. Firstly, under static lighting conditions an acquisition of a Macbeth color checker is made with the whole set of cameras. Then, color patches of the color checker chart are automatically detected in the captured images of all cameras. Next, the gain and offset settings of each camera response are iteratively adjusted on each channel to fit a line through the RGB values recorded from the six gray patches of the color chart. Following this step, some errors in the gain and offset settings, nonlinearity in the sensors, and color distortions remain. To address that, the authors proposed a three-stage post processing. The first one is to correct the residual errors in gain and offset by computing them once again in a floating-point domain, allowing more precision and accuracy. The next stage addresses the nonlinearity in the sensor. A look-up table that re-maps the response curve of each channel in each camera to the target color is calculated. Finally the third stage is to correct global color errors by computing a 3×3 RGB to RGB color transform that minimizes the error between each recorded color values from the patches and the average values of the patches recorded by the whole set of cameras. Inspired by this work, Ilie et al. [17] proposed a color calibration method for a multi-camera system. This method is implemented in two stages. Similar to [21], once all the cameras have taken an image of the Macbeth color checker, one of these acquired images is considered as the reference image. The first stage consists of adjusting each camera's parameter settings by minimizing color differences between the reference image and images of the color chart acquired by all cameras. This minimization is iteratively performed using a modified Powell's method. Starting from the obtained initial cameras' parameters, the same optimization process for all cameras is repeated, but by comparing with a new reference image calculated as the average of all cameras images computed in the previous step. In the second step of the method, a post-processing refinement is performed. The refinement is applied to enhance the cameras' parameters obtained from the first stage. However, this method requires a constant lighting condition, which cannot be guaranteed in a real practical system. Also, they are focused only on planar acquisition systems.

Li et al. [35] formulated the multi-camera color calibration problem as an overdetermined linear system, in which each camera collaborates with other cameras on color calibration. The authors begin by establishing global correspondences between all views. Based on these global correspondences, the camera sensor response can be approximated by a linear model with a multiplicative gain and an additive offset. The next step is to find proper gain and offset values so that the color values of the global correspondences will be the same in all camera images, and consequently a consistent color response of the cameras will be obtained. In addition, in order to ensure proper dynamic range, the authors incorporated a dynamic range shaping process into the described linear system. Slavkovikj in [51] first detects and captures a white target to generate calibration data. This is done by capturing a white sphere of diffuse reflectance placed in the scene and by using the cameras' response model described in [52]. Thanks to this step, the gain parameters of the color channels are instantiated for each of the cameras. Then, the author acquired images of a scene containing colored objects. To take into account the nonuniformity of the cameras' sensors, the acquired images are corrected as a matrix containing per-pixel scaling factors. Following that, the cameras' images are white balanced, and the gain and exposure are adjusted according to one reference camera. After applying this initial correction, correspondence points are calculated between adjacent camera pair images. The latter are used to form corresponding color pairs and to derive color correction matrices by performing nonlinear multiple regression. Finally, the correction is performed by using the obtained color correction matrices.

Zhong et al. [68] proposed a method to control each camera in multi-camera systems with the aim to enhance video quality. They proposed to address both inter-camera color inconsistency and object location difference problems. To achieve this objective, each camera is individually and automatically controlled by adjusting its pan angle and focal length. They formulate the problem as a nonlinear optimization problem, and then find the optimal camera control of pan angle and focal length. The authors first estimate initial values for the camera control, and then perform an iterative optimization by the bivariate Newton–Raphson method to get the optimal ones. Unlike the previous described methods, this one has been evaluated for coding purposes. To test the performance of their method, the authors have taken three sequences with and without their camera control method. The obtained MVV sequences have been compressed using MVC. The achieved coding gains range from 0.5 to 1.9 dB.

Despite the improved color consistency among camera views, this class of methods do not come with performance guarantees, because of the impossibility to control the reflectance properties over the object's surface, which is dependent on the capturing angle of each camera, lighting conditions, and so on. As a result, color camera responses are not well balanced. Consequently, several color correction approaches based on image-processing techniques were proposed to compensate the illumination and color changes between views. This is the object of the next sections.

10.2.2 Embedded Illumination/Color Compensation

When illumination/color changes occur between adjacent views, disparity estimation exploited by DCP in the MVC cannot be accurately performed. Therefore, the residual energy will be higher and, consequently, the coding efficiency will decrease. One way to address this problem is to tackle it during the coding process, and specifically, as part of the inter-view prediction process on a block level. To achieve that, several approaches used a modified block matching measure.

For instance, block-based illumination compensation (IC) technique has been introduced in [29, 30]. Based on an idea outlined by Kamikura et al. in [25], the authors proposed a modified matching metric that incorporates an IC model. They modeled illumination mismatches as an additive offset term and a multiplicative scale factor. Before matching, the candidate reference block and the current block to be matched are decomposed into a mean and a mean-removed signal. Next, the IC parameters, namely the scale and the offset, are applied to the reference block. Thus, for each candidate block, optimal IC parameters are computed. This process is repeated until finding the best matching block in the reference frame.

In order to transmit IC parameters to the decoder, the authors predicted IC values from neighboring blocks, and encoded them using the context adaptive binary arithmetic coder (CABAC) [38]. Finally, the decision on whether or not to use the IC model for each block is based on rate–distortion (R–D) criteria. According to simulation results presented in [30], a coding improvement up to 0.8 dB is obtained with this approach.

In the same vein, an adaptive local illumination change compensation method has been presented in [15, 32]. The compensation is performed at the macroblock (MB) level. The authors assume that the discrepancy caused by the illumination change is equal to the discrepancy of the DC component between the current and the reference blocks. Accordingly, the difference of DC values of the current coded MB and the reference one is subtracted during the motion and disparity estimation/compensation. This difference, referred to as the illumination change value (DVIC), which is in fact an offset factor is used to cancel out the illumination change between MBs. As the method described above, DIVC needs to be encoded and signaled to the decoder for each MB in the bitstream. Each DIVC value is predicted from the neighboring MBs of the current MB, and entropy-encoded using CABAC. Also, for each MB, the R–D control is used to decide whether or not the IC process is needed. It is reported by the authors that gains of up to 0.6 dB can be achieved. This approach has been adopted by the joint video team (JVT) standardization committee in the joint multi-view video model (JMVM) software [61], which is the MVC reference software based on H.264/AVC, but not in the standard itself [63].

In [42], the authors proposed to extend the H.264/AVC weighted prediction [2] for illumination compensation in MVC. They proposed to calculate the offset and weight parameters of the weighed prediction for each MB. Firstly, the slice weight value is determined. Thereafter, the best MB partition mode and motion vectors are set based on the R–D optimization using the normalized sum of absolute differences

(NSAD). In the next step, they built the predicted MB with the motion compensation. Then, weight and offset parameters are calculated by minimizing a distance between the current MB and the predicted one. Consequently, the slice header has slice weight values and each MB header has weight and offset to be mandatorily sent to the decoder.

Sohn et al. [55] proposed an inter-view balanced disparity estimation (IBDE) method. As a first step, the authors calculate the global balancing parameters for a whole frame with mean and standard deviation. Using these initial parameters and their probabilities, they derive the local balancing parameters (scale and offset) for each MB. This is achieved by using the 2D Lloyd-Max algorithm. The reported coding gain is about 0.1$\sim$0.3 dB. Li et al. [34] proposed an illumination compensation algorithm based on what they call a template, that is, the reconstructed neighboring region (left and above) of the current block. Their idea lies in the fact that the illumination of the current block has a strong correlation with its adjacent template. Similarly, the reference block and its corresponding template also have this correlation. Therefore, they derived the illumination change between the current block and the reference one from the change between their templates. The key advantage of this approach is that the local illumination change is calculated by the template and compensated without transmitting additional bits.

Another method proposed by Yamamoto et al. [67] applies color compensation in the decoded picture buffer (DPB). They performed color correction of each frame being used for prediction, that is, decoded pictures in DPB. Firstly, each frame is converted from YUV to the RGB color space. After that, they calculate lookup tables of each color channel from RGB. The latter are created first by finding correspondence points between pairs of views. Subsequently, the found correspondence points are used to create corresponding colors. However, to be robust to possible noise in the frames, the authors produced some Gaussian-filtered frames from input frames and detected corresponding colors from them. Once the corresponding colors are defined, they formulated the color-mapping relationships as an energy-optimization problem and resolve it by dynamic programming to obtain lookup tables. Color correction is applied on both encoder and decoder sides, which implies sending lookup tables in the codestream or as side information.

The previous methods are integrated into the H.264/AVC for evaluation purposes, and the coding gains were notable. However, compensating the illumination/color changes during the coding process has some disadvantages. It involves a new syntax and requires design changes affecting low levels of the encoding and decoding processes. In addition, since the compensation is forced to be carried out in the encoder and decoder, the complexity is increased. Also, an overhead is needed to send the IC parameters to the decoder. All the above described methods, except for [67], only deal with variations in brightness and do not compensate for color variation, because they are only applied on the luminance component. Finally, these methods do not perform color correction of the input sequence, they only enhance coding performance with the aim of R–D optimization. Thus the problem of inconsistencies in luminance and color still persists after decoding, which will unfortunately introduce errors on DIBR for the creation of new virtual views.

10.2.3 Preprocessed Color Correction

As an alternative to embedded illumination/color compensation, appropriate preprocessing consisting in color correction methods can be applied in an independent manner before feeding the view sequences to the encoder. In essence, the color correction operation on MVV can be seen as a process transferring color palette of a source view (or reference view) to a target view. The source view is chosen by the user from the set of views, and all the remaining views are corrected to acquire the color palette of that reference. Following this idea, several tentative approaches have been proposed [3, 5, 7, 9, 11, 23, 24, 33, 44, 45, 49, 58]. The key advantage of such approaches is that no changes are necessary to either the encoder/decoder or the bitstream syntax. In the following, we try to present a comprehensive overview of these methods with an emphasis on those that are applied to MVC.

A straightforward solution for correcting color in MVV as a preprocessing step is the histogram matching (HM) method. HM has already demonstrated its efficiency to perform image color correction [13]. Fecker et al. [9] proposed an extension of the HM method to MVV. In their method, all camera views of a multi-view sequence are corrected toward a common reference view in the center of the camera setup. To achieve this, the authors calculate lookup tables using cumulative histograms of the view being corrected and the reference. They adapt the cumulative histogram of each view to match that of the reference. To maintain the temporal correlation, and avoid the introduction of color variation between successive frames within a single view, the authors proposed a time-constant mapping function. They compute the histogram for the whole sequence; it is summed up over all frames of the sequence. So, the same correction is used for every temporal frame in the sequence. However, for the cases when the camera setup is not fixed and in real-time applications, this solution is not sound. The obtained gain in coding efficiency is up to 0.7 dB for luminance and 1.9 dB for chrominance. Similarly, Chen et al. [3] proposed an approach based on HM, where two parameters, namely scaling and offset, are derived optimally for each YUV channel based on the histograms of the chosen reference view and the view being corrected. Both approaches have common drawbacks linked to HM. Since the MVV sequences do not show exactly the same content, especially in a sparse camera system (i.e., widely separated views), frames from different viewpoints contain nonoverlapping regions (occluded regions). Therefore, applying HM to the entire frame produces inaccurate and inconsistent results.

In order to avoid the occlusion problem, several recent approaches used correspondences between views. For instance, Lee et al. [33] proposed to segment the image according to depth, and perform HM for each sub-image. The image segmentation is performed by using an expectation maximization clustering method based on a Gaussian mixture model (GMM). However, this method depends heavily on the quality of depth/disparity information, for which unfortunately the performance of disparity estimation methods are still error prone. Also, disparity estimation requires knowledge about the range of disparities for all views, which is a difficult task as it depends on the scene and camera geometry, and will have a significant effect on the performance of color correction. In [5] a histogram-offset-based color correction method is proposed. As a first step, the authors identify maximum matching regions between the reference view and the target one. To achieve that, disparity estimation is performed on the rank-transformed domain; thus allowing to handle occlusion regions. Then, the histograms of the reference view and the target view are calculated for each YUV channel only within the identified matching regions. According to an iterative threshold, the calculated histograms are divided into high and low occurrences bin classes, and only the histogram bins having high occurrences are calibrated. Finally, the generated histogram offset is exploited to correct the target view. However, this approach considers only the global color discrepancies between views, since it computes only a rough color mapping between views, which does not produce a locally refined result.

Another approach is presented by Pitié et al. [44, 45], where they consider the colors of an image as an N-dimensional probability density function (PDF), and repeatedly project these N-dimensional distribution into a series of random 1-dimensional marginal PDFs. In order to map the PDF of the reference image, the color distribution of the target image is iteratively manipulated until matching that of the reference. To avoid the increase in graininess over the corrected image, the authors proposed in [45] a post-processing algorithm to preserve its gradient field. This method provides reasonable performance, but if the reference and the target images have different color statistics in the nonoverlapping regions, it can produce poor results. Also, this method has been tested on still images only, and the temporal coherence is an important aspect to be considered in MVV.

Recently in [11], Fezza et al. proposed a preprocessing method using a customized HM algorithm. To deal with the occlusion problem, the authors take into account only common regions across views when calculating the histogram, thus providing a more precise color correction. To effectively define this set of common areas, they use the scale invariant feature transform (SIFT) [37], followed by the random sample consensus (RANSAC) [12] to remove matching outliers. Thereafter, the cumulative histogram of each view is calculated based only on corresponding areas already detected. In order to keep the temporal correlation, the HM method is performed on a sliding window, where each color-mapping function is defined using a group of pictures, and this may also ensure an adaptation to time-varying acquisition system, camera moving capture, and real-time broadcasting. Another aspect tackled in this method corresponds to the reference view selection. Indeed, this step ignored by many works may highly influence the correction results and thus the coding efficiency. Most methods choose the central view of the camera arrangement as the reference view. While this choice may appear natural and consistent, as this view shares most parts of content with the other views, the central view may be distorted, leading to the impairment of the whole sequences of MVV. In addition, in some cases, there is no explicit central view, for example, with an even number of views. Furthermore, when using camera arrangements, such as dome type, circular, and array, it is not obvious to choose a central view. To select an appropriate reference view, the authors proposed to perform it in a fully automatic manner, by defining a robust criterion relying on view statistics and structural content. It is reported in [11] that this method increases coding efficiency with gains of up to 1.1 and 2.2 dB for the luminance and chrominance components, respectively.

Another method is presented in [7], where the authors proposed to find the average color of the set of views and modify all views to match this average. Hence, in this approach the average color is considered as the color reference, and was calculated by finding matching points between all views using block-based disparity estimation. The authors expressed each of the corrected YUV channels as a weighted linear sum of the three original ones in addition to an offset. Then, they performed a least-squares regression on the matching points to find optimal weighting parameters leading to a good fit between the average YUV values and those of the view being corrected. With this method, the compression efficiency is increased by up to 1 dB in luminance component, and up to 2.1 dB in chrominance component. Unfortunately, choosing the average color as the color reference cannot guarantee a good quality, because, if some views are distorted, this will strongly influence the average color. In addition, it is difficult to compute the average, especially when we have large distances between the different views. Another limitation lies in the fact that only 10 successive frames are used from the sequence in order to calculate these correction parameters. The extracted set of parameters is used to correct every frame in a given view, under the assumption that correction parameters are time-constant during the whole sequence. However, this assumption does not always hold, especially when the acquiring system of MVV is not fixed. In addition, assigning a single disparity vector for 8×8 pixels blocks, as described in their approach, can be the cause of a number of bad matches. Finally, disparity estimation and least-squares regression can lead to a significant increase of computational cost, which cannot be neglected, even though the process is applied offline. A less complex preprocessing method was proposed in [49]. In order to achieve fast color correction, instead of performing correction in a frame-by-frame basis, the authors divide the set of frames of each view into keyframes and non-keyframes, and adopt different color correction strategies for each of them. The keyframes are corrected directly by the same color correction method proposed by [7]. While for non-keyframes, the color corrected values are derived from previous frames by using a temporal variation model. Basically, the latter consists of two steps, time-invariant detection and temporal variation modeling. The purpose of the time-invariant detection step is to extract the background regions that are unchanged over time. These regions are used for extracting stable temporal variation information from a single view. Once these regions are defined, the color discrepancy relationship between adjacent frames is derived under the assumption that color distributions in background regions are Gaussians. The interval between keyframes is fixed and equals the size of group of picture (GOP) in MVC. This approach has two main advantages. Firstly, by exploiting the temporal variation model, computational complexity is significantly decreased. Secondly, as the color corrected values are derived from previous frames, color variations between successive frames can be avoided.

Jung et al. [23] proposed to improve the color consistency by modeling the characteristics of the cameras and estimating their coefficients. To consider occlusion regions, first, the authors extract correspondences between pairs of views. For this step, they adopted the robust SIFT algorithm. After that, based on the calculated correspondences, the authors estimated the coefficients (gain, gamma, and offset) of the camera characteristic curve via an outlier-removed nonlinear regression based on the Levenberg–Marquardt algorithm. With these estimated coefficients, lookup tables are generated and followed by color correction. The coding improvement made by this method is up to 0.9 db and 0.8 dB for luminance and chrominance components, respectively. The main disadvantage of this approach is linked to the fact that it may be used only when the setup is fixed, because the authors consider fixed coefficients for each camera. The same authors proposed a color correction algorithm using adaptive weights of correspondences [24]. They exploited two types of correspondences: global and local. Global correspondences are common features on all views. While, local correspondences are matched features for a certain stereo pair. The speed up robust feature (SURF) [1] method has been used for the correspondence purpose, and RANSAC algorithm for outliers removed in the correspondences. In order to increase correspondences, an additional matching on Gaussian-filtered images was applied. The authors consider that the change between the reference and target views is nonlinear for the Y component and, linear for U and V components. Accordingly, to compensate for relative color differences between views, they used gain-offset-gamma (GOG) and linear models for luminance and chrominance components, respectively. The coefficients of these linear and nonlinear models were estimated from correspondences. In addition, an adaptive weighting factor is included into these two models to weight correspondences according to from where they are extracted.

Tehrani et al. [58] proposed a similar method to the one developed by Yamamoto et al. [67], with some important differences. Instead of performing the correction in the core coding process, it is achieved as a preprocessing step. Also, there is no unique reference view, for which all target views are corrected. The algorithm can start from any camera view from the set of views and continues by shifting the pair sequentially on the multi-camera system for several iterations, until the average color change between all views becomes small enough. This allows its use for any camera configuration, from sparse to dense multi-camera systems. Another difference lies in the increase of the correction accuracy by appending corresponding intensities from several time instances and several blurred images of the pair, suppressing the outliers, and the nonlinear weighting of model.

The main advantage of these approaches is that the encoder and decoder remain unchanged. Also, since these methods can be applied in an independent manner, they can be used for general purposes, such as coding and DIBR applications [8].

10.3 Simulations

This section is devoted to the presentation of the experimental evaluations of the commonly used color correction methods drawn from the aforementioned categories. More precisely, four methods have been tested. Since most of the proposed methods in the literature belong to the third category, that is, preprocessed color correction, we selected the most representative three approaches namely HM method [9], Fezza et al. [11], and Pitié et al. [44]. The IC method [15] has been chosen as a representative of the second category. Unfortunately, the first category has not been included in the evaluation process. The main reason lies in the hardware character of this family of approaches with controlled environments, making them very hard to reiterate.

The evaluation of the four methods was carried out in three experiments. The first one consists of a visual inspection of the color consistency between views. The second experiment evaluates the coding performance, and finally, the third explores the effect of the tested methods on DIBR results.

For the first experiment, we used two standard MVV test data, *Flamenco2* (five views) and *Race1* (eight views) [56]; both sequences have a resolution of 640×480. In order to allow a visual color comparison, we show examples of frames before and after correction. Due to space limitation, only a few selected frames are shown. As depicted in Figs. 10.3 and 10.4, one frame is taken from each view at the same time step, as well as three views (view 1, view 3, and view 5) among all views are presented.

First, one can notice in both figures, before correction (original views), that there are apparent color differences between the different views. It is thus clear that color consistency among these original views is poor. Consequently, to resolve this color mismatch, the tested color correction methods try to correct the views to make them as similar as possible to a reference chosen, in this case, in the center (view 3).

Based on the results depicted in Figs. 10.3 and 10.4, it is noticeable that method of Fezza et al. [11] provides good and stable results for both sequences. Resulting frames of this method are visually close to the reference image. For Pitié et al. method [44], it provides an acceptable result for *Race1* (Fig. 10.4), and there is almost no visual difference between this method and the one in [11]. However, the achieved results by HM [9] are far from being satisfactory. For *Flamenco2* (Fig. 10.3), it can be observed that Pitié et al. [44] and HM [9] introduce impairments (noise), especially in the background, leading to poor visual result for this sequence. This is usually due to occluded regions but also to the fact that *Flamenco2* contains different local color discrepancies, making global correction such as applied by HM inappropriate. For a better analysis of the visual results of color correction, a zoom on some sensitive parts from *Flamenco2* and *Race1* sequences is shown in Fig. 10.5.

The IC method [15] has not been considered for this first assessment. As mentioned in Sect. 10.2.2, this method does not perform color correction of the input sequence. It is only intended for coding performance. Thus, it provides sequences containing the same color variations, in addition to some distortions due to the coding process.

In order to assess the effect of color correction methods on coding performance, we used three methods (HM method [9], Fezza et al. [11], and Pitié et al. [44]) as a preprocessing before feeding the sequences to the MVC encoder (JMVM version 8.0) [61]. However, for the IC method [15], it is already implemented and integrated in the JMVM reference software [32,61]. Consequently, to ensure a fair comparison

Fig. 10.3 Visual effect of color correction methods on the *Flamenco2* test sequence (*left*: view 1, *middle*: view 3, and *right*: view 5). Comparison between **a** without correction, **b** HM method [9], **c** Fezza et al. [11], and **d** Pitié et al. [44]

with the three preprocessing methods, the IC is disabled when they are used. Experiments were performed using typical settings for MVC (refer to [56] for details regarding the test conditions).

Figure 10.6 shows the PSNR versus bit rate averaged over all views of the test sequence for both luma and chroma. As it can be observed from the plots, the coding performance comparison clearly indicates the benefits of using color correction methods. The quality improvement is reported for the luminance as well as the chrominance components, in comparison with the multi-view encoder that uses original sequences without any correction (denoted in the figure as *no correction*). The previous conclusion does not hold for Pitié et al. method [44], which shows

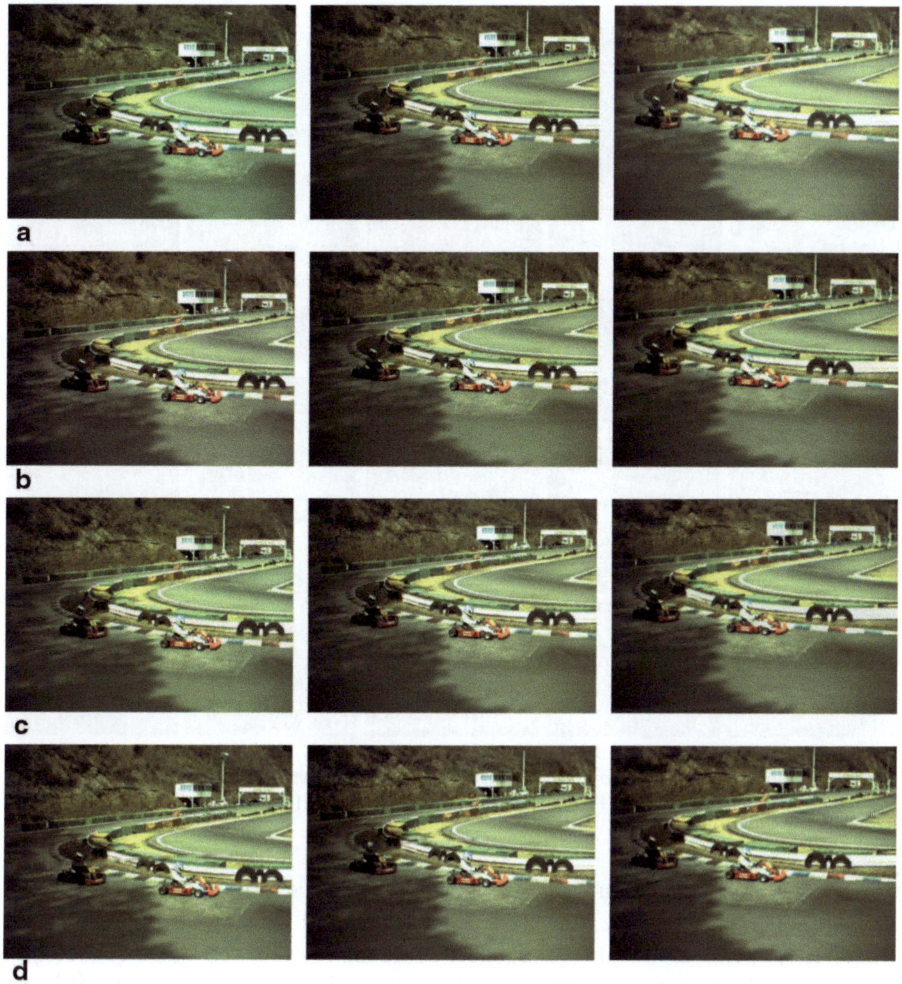

Fig. 10.4 Visual effect of color correction methods on the *Race1* test sequence (*left*: view 1, *middle*: view 3, and *right*: view 5). Comparison between **a** without correction, **b** HM method [9], **c** Fezza et al. [11], and **d** Pitié et al. [44]

worse results than the case without correction for *Flamenco2*. These remarks are consistent with those of the visual inspection. Therefore, the coding results demonstrate that the achieved color calibration helps in reducing color variations across views, and consequently, enhances the inter-view prediction, leading to a significant reduction of the bit rate for a fixed level of PSNR (quality).

In order to show the effect of color correction methods on rendering performance, they have been applied on *Newspaper* sequence (five views) [19]. The latter is in the multi-view video plus depth (MVD) format, with a resolution of 1024×768.

Fig. 10.5 Enlarged parts of *Flamenco2* (*top*) and *Race1* (*bottom*) sequences. **a** Uncorrected. **b** Reference. **c** HM method [9]. **d** Fezza et al. [11]. **e** Pitié et al. [44]

For this part of the simulation, the following test scenario has been adopted in line with [19]. Hence, three views (view 1, view 3, and view 5) have been selected from *Newspaper*. Also, to allow the reader to see the effect of color discrepancies on rendering performance, we manually distorted two views (view 1 and view 5) by an increase/decrease in brightness and color deflection, while the view 3 is kept as is. Then, a correction of the distorted views (view 1 and view 5) is performed separately using the three tested preprocessing methods. During this stage, the center view (view 3) is assumed to be the reference view. Finally, once the correction is done, two virtual views (view 2 and view 4) are generated before and after correction. The generation of these views has been made by the DIBR technique based on the view synthesis reference software (VSRS version 3.5) [18], between the reference view and the distorted (resp. corrected) ones.

First, Fig. 10.7 shows the color correction result of view 1 and view 5. This result confirms the previous conclusion that Fezza et al. [11] and Pitié et al. [44] methods outperform HM [9] in terms of color consistency. The rendering results (view 2 and view 4) for *Newspaper* are depicted in Fig. 10.8. One can notice from this figure that

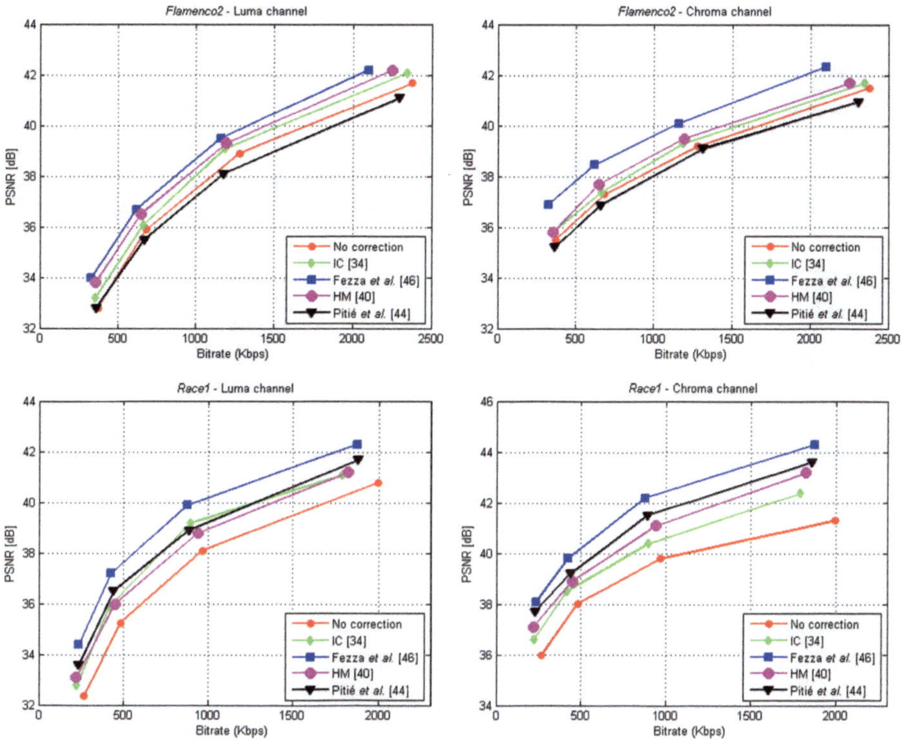

Fig. 10.6 Rate distortion curves for *Flamenco2* and *Race1* test sequences. The chroma PSNR is the average PSNR of the U and V channels

when the distorted views (uncorrected) are used, virtual views are obtained containing visible distortions. For instance, a ghosting effect can be seen in addition to some darker/lighter regions in the foreground and background of the scene, potentially creating an annoyance for the viewer. The distortions can be explained by the fact that the color inconsistency between views induces important errors in the rendered views. However, once the correction is performed, significant improvements are obtained in the rendering quality. For a better analysis of the rendering results, some sensitive parts from *Newspaper* have been enlarged and depicted in Fig. 10.9.

In addition, to objectively measure the color consistency across views, the well-known CIEDE2000 color difference [50], defined and standardized by the CIE (International Commission on Illumination), has been used. The latter is based on the CIELAB color space, which has been designed to represent colors in a perceptual manner. The detailed definition and implementation of the CIEDE2000 equation can be found in studies by Sharma et al. 2005 and Westland et al. 2012 [50, 64]. Figure. 10.10 shows the color difference comparison, before and after correction using the three tested preprocessing methods, between the reference view (view 3) and all remaining views (real and rendered). The curves highlight that the

Fig. 10.7 Visual effect of color correction methods on the *Newspaper* test sequence (*left*: view 1, *middle*: view 3, and *right*: view 5). Comparison between **a** without correction, **b** HM method [9], **c** Fezza et al. [11], and **d** Pitié et al. [44]

color difference generated by the Fezza et al. [11] method is significantly low in comparison with those of the others.

10.4 Conclusion and Future Trends

In this chapter, a review of the used color correction techniques for MVV has been presented, with a specific focus on the improvement of coding performance. The

Fig. 10.8 Visual effect on rendering performance for the *Newspaper* test sequence (*left*: view 2, *middle*: view 3, and *right*: view 4). Comparison between **a** without correction, **b** HM method [9], **c** Fezza et al. [11], and **d** Pitié et al. [44]

description of these methods has been made with a decomposition into three different families, according to their operating procedures, and the stage of the performed correction. The described methods have been evaluated in terms of color consistency, coding performance, and rendering quality. The choice of the method depends strongly on the target application and the available resources. However, relying on the performed experimental evaluation and the reported results of the literature, the preprocessing color correction methods appear to offer the most flexibility and the best performance.

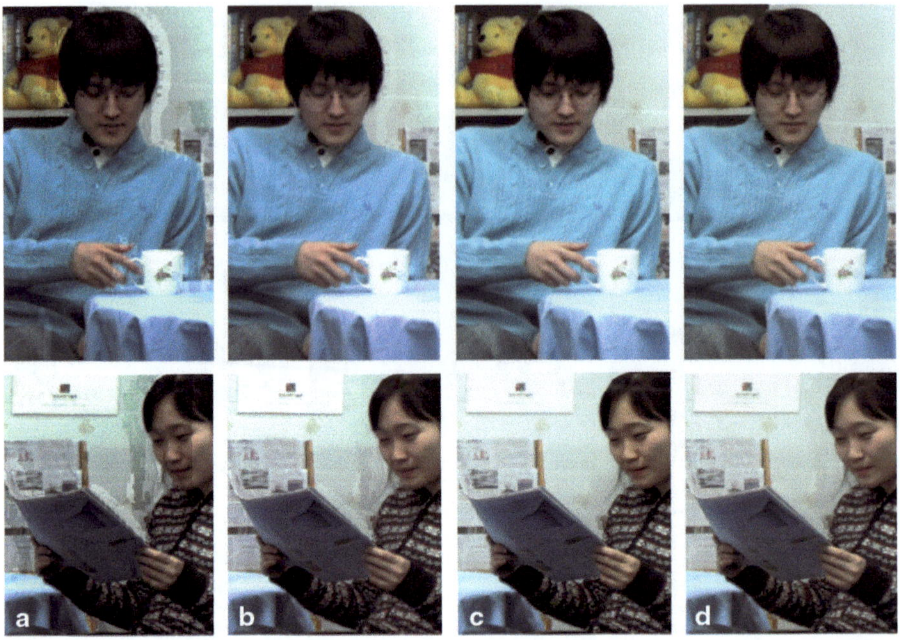

Fig. 10.9 Enlarged parts of the *Newspaper* sequence. **a** Uncorrected. **b** HM method [9]. **c** Fezza et al. [11]. **d** Pitié et al. [44]

There are many open research problems in color correction of MVV. For instance, maintaining the temporal coherence between successive frames within a single view needs to be improved. The only two proposed approaches [11,49] tackling this problem are based on defining group of picture and perform the correction group by group. However, on both approaches the size of group is fixed. Future studies should focus on proposing more advanced strategies allowing to tune adaptively and automatically the size of each group. Also, the reduction of computational complexity remains an important issue, especially, with the emergency of new real-time 3D applications. One solution to address this problem is to parallelize the color correction process by using a graphics processing pipeline or a graphics processing unit (GPU) implementation. Finally, the impact of color variation is not fully investigated, the relation between color inconsistencies and binocular rivalry should be explored. As the latter can induce visual fatigue, the maximum tolerance of color difference between views needs to be quantified.

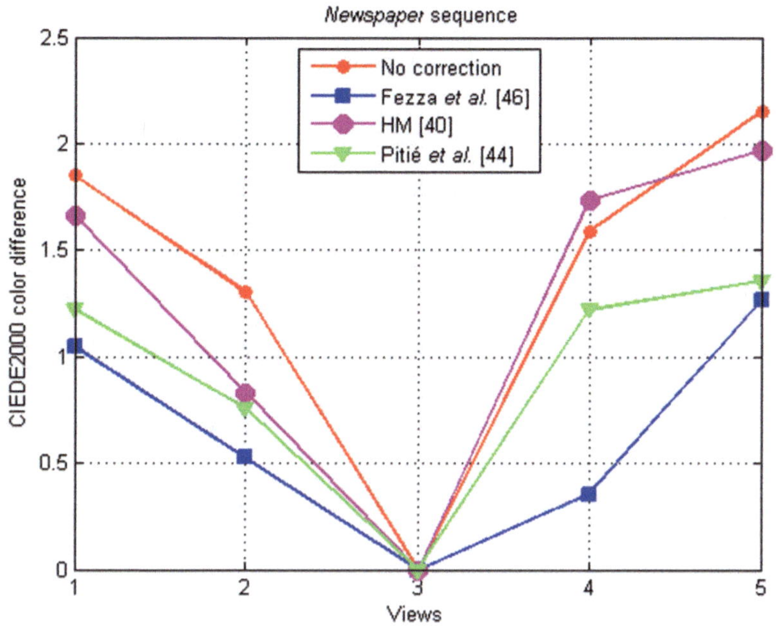

Fig. 10.10 Objective color difference comparison for the *Newspaper* test sequence. CIEDE2000 color difference computed between the reference (view 3) and all remaining views (lower values denote better correction)

References

1. Bay, H., Ess, A., Tuytelaars, T., Gool, L.V.: Surf: speeded up robust features. Comput. Vis. Image Underst. **110**(3), 346–359 (2008)
2. Boyce, J. M.: Weighted prediction in the H.264/MPEG AVC video coding standard. In: Proceedings of IEEE International Symposium on Circuits and Systems (ISCAS), Vancouver, Canada, 789–792 (2004)
3. Chen, Y., Cai, C., Liu, J.: YUV correction for multi-view video compression. In: Proceedings of the International Conference Pattern Recognition (ICPR), Hong Kong, 734–737 (2006)
4. Chen, Y., Wang, Y.-K., Ugur, K., Hannuksela, M. M., Lainema, J., Gabbouj, M.: 3D video services with the emerging MVC standard. EURASIP J. Adv. Signal Process. Article ID 786015 (2009)
5. Chen, Y., Ma, K. K., Cai, C.: Histogram-offset-based color correction for multi-view video coding. In: Proceedings of the 17th IEEE International Conference on Image Processing (ICIP), 977–980 (2010)
6. Chen, J., Zhou, J., Sun, J., Bovik, A. C.: Binocular mismatch induced by luminance discrepancies on stereoscopic images. In: IEEE International Conference on Multimedia and Expo (ICME 2014), 1–6 (2014)
7. Doutre, C., Nasiopoulos, P.: Color correction preprocessing for multiview video coding. IEEE Trans. Circuits Syst. Video Technol. **19**(9), 1400–1406 (2009)
8. Faridul, H. S., Pouli, T., Chamaret, C., Stauder, J., Tremeau, A., Reinhard, E.: A survey of color mapping and its applications. Eurographics State of the Art Report, 43–67 (2014)

9. Fecker, U., Barkowsky, M., Kaup, A.: Histogram-based pre-filtering for luminance and chrominance compensation of multi-view video. IEEE Trans. Circuits Syst. Video Technol. **18**(9), 1258–1267 (2008)
10. Fehn, C.: Depth-image-based rendering (DIBR), compression and transmission for a new approach on 3D-TV. In: Proceedings of the SPIE Conference on Stereoscopic Displays and Virtual Reality Systems XI, San Jose, CA, USA, 93–104 (2004)
11. Fezza, S. A., Larabi, M.-C., Faraoun, K. M.: Feature-based color correction of multiview video for coding and rendering enhancement. IEEE Trans. Circuits Syst. Video Technol. **24**(9), 1486–1498 (2014)
12. Fischler, M.A., Bolles, R.C.: Random sample consensus: a paradigm for model fitting with applications to image analysis and automated cartography. Commun. ACM. **24**(6), 381–395 (1981)
13. Gonzalez, R.C., Woods, R.E.: Digital Image Processing. 3rd ed. Prentice-Hall, Englewood Cliffs (2007)
14. Hong, W., Zhao, Y., Yu, L., Zhu, C.: Detection model of luster effect in binocular rivalry. In Proceedings 19th International Conference on Digital Signal Processing (DSP 2014), 866–869 (2014)
15. Hur, J. H., Cho, S., Lee, Y. L.: Adaptive local illumination change compensation method for H.264-based multiview video coding. IEEE Trans. Circuits Syst. Video Technol. **17**(11), 1496–1505 (2007)
16. Iatsun, I., Larabi, M.-C., Fernandez, C.: Study of visual fatigue/discomfort generated by S3D video using eye-tracking data. In: Proceedings Stereoscopic Displays Appl. XXIV, Burlingame, CA, USA (2013)
17. Ilie, A., Welch, G.: Ensuring color consistency across multiple cameras. In: Proceedings of International Conference on Computer Vision (ICCV), Washington, DC, USA, 1268–1275 (2005)
18. ISO/IEC JTC1/SC29/WG11: Report on Experimental Framework for 3D Video Coding. Doc. N11631, Guangzhou, China (2010)
19. ISO/IEC JTC1/SC29/WG11: Call for Proposals on 3D Video Coding Technology. Doc. N12036, Geneva, Switzerland (2011)
20. ITU-T Recommendation H.264: Advanced Video Coding for Generic Audiovisual Services, ITU-T Rec. H.264 (2013)
21. Joshi, N. S.: Color calibration for arrays of inexpensive image sensors. Masters thesis, Stanford University CA, USA (2004)
22. Joshi, N. S., Wilburn, B., Vaish, V., Levoy, M. Horowitz, M.: Automatic Color Calibration for Large Camera Arrays. Technical Report CS2005-0821, UCSD CSE (2005)
23. Jung, J. I., Ho, Y. S.: Color correction algorithm based on camera characteristics for multi-view video coding. Signal Image Video Process. **8**(5),1–12 (2014)
24. Jung, J.I., Ho, Y.S.: Geometric and colorimetric error compensation for multi-view images. J. Vis. Commun. Image Represent. **25**(4), 698–708 (2014)
25. Kamikura, K., Watanabe, H., Jozawa, H., Kotera, H., Ichinose, S.: Global brightness-variation compensation for video coding. IEEE Trans. Circuits Syst. Video Technol. **8**(8), 988–1000 (1998)
26. Kang, Y. S., Ho, Y. S.: An efficient image rectification method for parallel multi-camera arrangement. IEEE Trans. Consumer Electron. **57**(3), 1041–1048 (2011)
27. Kauff, P., Atzpadin, N., Fehn, C., Müller, M., Schreer, O., Smolic, A., Tanger, R.: Depth map creation and image based rendering for advanced 3DTV services providing interoperability and scalability. Signal Process. Image Commun. **22**(2), 217–234 (2007)
28. Kaup, A., Fecker, U.: Analysis of multireference block matching for multiview video coding. In: Proceedings 7th Workshop Digital Broadcasting, Erlangen, Germany, 33–39 (2006)
29. Kim, J.-H., Lai, P.-L., Ortega, A., Su, Y., Yin, P., Gomila, C.: Results of CE2 on multi-view video coding. In Joint Video Team (JVT), Doc. JVT-T117. Klagenfurt, Austria (2006)
30. Kim, J.-H., Lai, P.-L., Lopez, J., Ortega, A., Su, Y., Yin, P., Gomila, C.: New coding tools for illumination and focus mismatch compensation in multiview video coding. IEEE Trans. Circuits Syst. Video Technol. **17**(11), 1519–1535 (2007)

31. Lambooij, M.T., IJsselsteijn, W.A., Heynderickx, I.: Visual discomfort in stereoscopic displays: a review. In: Proceedings of Stereoscopic Displays and Virtual Reality Systems XIV, San Jose, CA, USA (2007)
32. Lee, Y.L., Hur, J.H., Lee, Y.K., Cho, S.H., Kwon, H.J., Hur, N.H., Kim, J.W.: Results of CE2 on multi-view video coding. Joint Video Team (JVT) Doc. JVT-T110. Klagenfurt, Austria (2006)
33. Lee, D. S., Ko, M. S., Seo, Y. H., Kim, D. W., Yoo, J.: Illumination compensation for multiview video based on layered histogram matching with depth information. Opt. Commun. **286**, 74–84 (2013)
34. Li, X., Jiang, L., Ma, S., Zhao, D., Gao, W.: Template based illumination compensation algorithm for multiview video coding. In: Proceedings Conference on Visual Communications and Image Processing (VCIP), Huangshan, China (2010)
35. Li, K., Dai, Q., Xu, W.: Collaborative color calibration for multi-camera systems. Signal Process. Image Commun. **26**(1), 48–60 (2011)
36. Lou, J.G., Cai, H., Li, J.: A real-time interactive multi-view video system. In Proceedings of the 13th Annual ACM International Conference on Multimedia, Singapore, 161–170 (2005)
37. Lowe, D.G.: Distinctive image features from scale-invariant keypoints. Int. J. Comput. Vis. **60**(2), 91–110 (2004)
38. Marpe, D., Schwarz, H., Wiegand, T.: Context-based adaptive binary arithmetic coding in the H.264/AVC video compression standard. IEEE Trans. Circuits Syst. Video Technol. **13**(7), 620–636 (2003)
39. Merkle, P., Smolic, A., Müller, K., Wiegand, T.: Efficient prediction structures for multiview video coding. IEEE Trans. Circuits Syst. Video Technol. **17**(11), 1461–1473 (2007)
40. Müller, K., Merkle, P., Wiegand, T.: 3-D video representation using depth maps. Proc. IEEE. **99**(4), 643–656 (2011)
41. Nanda, H., Cutler, R.: Practical calibrations for a real-time digital omnidirectional camera. In Technical Sketches, Computer Vision and Pattern Recognition (2001)
42. Park, S., Sim, D.: Core experiment results on CE2 for multi-view video coding. Joint Video Team (JVT) Doc. JVT-T113, Klagenfurt, Austria (2006)
43. Pedersini, F., Sarti, A., Tubaro, S.: Accurate and simple geometric calibration of multi-camera systems. Signal Process. **77**(3), 309–334 (1999)
44. Pitié, F., Kokaram, A. C., Dahyot, R.: N-dimensional probability density function transfer and its application to color transfer. In: Proceedings of the IEEE International Conference on Computer Vision (ICCV), Beijing, China, 1434–1439 (2005)
45. Pitié, F., Kokaram, A. C., Dahyot, R.: Automated colour grading using colour distribution transfer. Comput. Vis. Image Underst. **107**(1), 123–137 (2007)
46. Pölönen, M., Hakala, J., Bilcu, R., Järvenpää, T., Häkkinen, J., Salmimaa, M.: Color asymmetry in 3D imaging: influence on the viewing experience. 3D Res. **3**(3), 1–10 (2012)
47. Reiter, U., Brunnström, K., De Moor, K., Larabi, M.-C., Pereira, M., Pinheiro, A., You, J., Zgank,A.: Factors influencing quality of experience. In: Möller, S., Raake, A. (eds.) Quality of Experience: Advanced Concepts, Applications, and Methods, pp. 55–72. Springer Switzerland (2014)
48. Salmimaa, M., Hakala, J., Pölönen, M., Järvenpää, T., Bilcu, R., Häkkinen, J.: Luminance asymmetry in stereoscopic Content: binocular rivalry or luster. In: Proceedings SID Symposium Digest of Technical Papers, 801–804 (2014)
49. Shao, F., Jiang, G.Y., Yu, M., Ho, Y.S.: Fast color correction for multi-view video by modeling spatio-temporal variation. J. Vis. Commun. Image Represent. **21**(5), 392–403 (2010)
50. Sharma, G., Wu, W., Dalal, E.N.: The CIEDE2000 color-difference formula: implementation notes, supplementary test data, and mathematical observations. Color Res. Appl. **30**(1), 21–30 (2005)
51. Slavkovikj, V.: Color calibration of a multi-camera array. Masters thesis, Gjøvik university college, Norway (2011)
52. Slavkovikj, V., Hardeberg, J. Y., Eichhorn, A.: Characterizing the response of charge-couple device digital color cameras. In: Proceedings SPIE Sensors, Cameras, and Systems for Industrial and Scientific Applications XIII, Burlingame, California, USA (2012)

53. Smolic, A.: 3D video and free viewpoint video—from capture to display. Pattern Recognit. **44**(9), 1958–1968 (2011)
54. Smolic, A., Müller, K., Merkle, P., Atzpadin, N., Fehn, C., Müller, M., Schreer, O., Tanger, R., Kauff, P., Wiegand, T., Megyesi, Z.: Multi-view video plus depth (MVD) format for advanced 3D video systems. Joint Video Team (JVT) of ISO/IEC MPEG & ITU-T VCEG, JVT-W100, San Jose, CA, USA (2007)
55. Sohn, K., Kim, Y., Seo, J., Yoon, J., Kim, G.: Results on CE2 using IBDE for multi-view video coding. Joint Video Team (JVT) Doc. JVT-T114, Klagenfurt, Austria (2006)
56. Su, Y., Vetro, A., Smolic, A.: Common test conditions for multiview video coding. Joint Video Team (JVT) Doc. JVT-U211, Hangzhou, China (2006)
57. Tanimoto, M.: FTV: free-viewpoint television. Signal Process. Image Commun. **27**(7), 555–570 (2012)
58. Tehrani, M.P., Ishikawa, A., Sakazawa, S., Koike, A.: Iterative colour correction of multicamera systems using corresponding feature points. J. Vis. Commun. Image Represent. **21**(5), 377–391 (2010)
59. Tian, D., Pandit, P., Yin, P., Gomila, C.: Study of MVC coding tools, Joint Video Team (JVT) Doc, JVT-Y044, Shenzhen, China (2007)
60. Vetro, A., Yea, S., Smolic, A.: Towards a 3D video format for auto-stereoscopic displays. In Proceedings of the SPIE Conference on Applications of Digital Image Processing XXXI, San Diego, CA, USA (2008)
61. Vetro, A., Pandit, P., Kimata, H., Smolic, A., Wang, Y. K.: Joint multiview video model (JMVM) 8.0. Joint Video Team (JVT), Doc. JVT-AA207, Geneva, Switzerland (2008)
62. Vetro, A., Tourapis, A. M., Müller, K., Chen, T.: 3D-TV content storage and transmission. IEEE Trans. Broadcast. **57**(2), 384–394 (2011)
63. Vetro, A., Wiegand, T., Sullivan, G.J.: Overview of the stereo and multiview video coding extensions of the H.264/AVC standard. Proc. IEEE. **99**(4), 626–642 (2011)
64. Westland, S., Ripamonti, C., Cheung, V.: Computational Colour Science Using MATLAB. 2nd ed. Wiley Chichester, West Sussex, England (2012)
65. Winkler, S., Min, D.: Stereo/multiview picture quality: overview and recent advances. Signal Process. Image Commun. **28**(10), 1358–1373 (2013)
66. Woods, A., Docherty, T., Koch, R.: Image distortions in stereoscopic video systems. In: Proceedings Stereoscopic Displays and Applications IV, San Jose, CA, USA (1993)
67. Yamamoto, K., Kitahara, M., Kimata, H., Yendo, T., Fujii, T., Tanimoto, M., Shimizu, S., Kamikura, K., Yashima, Y.: Multiview video coding using view interpolation and color correction. IEEE Trans. Circuits Syst. Video Technol. **17**(11), 1436–1449 (2007)
68. Zhong, J., Kleijn, B., Hu, X.: Camera control in multi-camera systems for video quality enhancement. IEEE Sens. J. **14**(9), 2955–2966 (2014)
69. Zilly, F., Müller, M., Eisert, P., Kauff, P.: The stereoscopic analyzer—an image-based assistance tool for stereo shooting and 3D production. In: Proceeding 17th IEEE International Conference on Image Processing (ICIP), Hong Kong, 4029–4032 (2010)

Chapter 11
Enhancement of Image Content for Observers with Colour Vision Deficiencies

Neda Milić, Dragoljub Novaković and Branko Milosavljević

11.1 Introduction

The normal colour vision, also called trichromacy, depends on the light absorption by three types of colour photoreceptors within retina, commonly known as L, M and S cones, because of their highest sensitivity to long-wave, middle-wave and short-wave bands of the spectrum, respectively [30]. The cones' spectral sensitivity functions are shown in Fig. 11.1. The combination of three cone responses is processed and interpreted as a particular colour sensation in the vision centre [36].

Colour vision deficiency (CVD) represents an inability to perceive the differences between some colour combinations that can be distinguished in the case of normal colour vision and affects the significant percentage of the human population—approximately 8 % of men and 0.4 % of women according to statistics for Caucasian population [18].

Based on severity, three main categories of CVDs can be distinguished [31]: *monochromacy* is the condition of total blindness for colours where two or all three cone types are absent; *dichromacy* is the condition of partial colour blindness which occurs when one cone type is missing (the L type in the case of protanopia, the M type in deuteranopia and the S type in tritanopia); and *anomalous trichromacy* is the condition of moderate colour blindness where all photoreceptors are present, but one cone type has shifted sensitivity (the L type in the case of protanomaly, the M type in deuteranomaly and the S type in tritanomaly). Sensitivity function of the protanomalous L cones is shifted closer to the sensitivity function of normal M type (Fig. 11.1b), whereas sensitivities of the anomalous cone types, in the cases of deuteranomaly (Fig. 11.1c) and tritanomaly (Fig. 11.1d), are shifted towards longer wavelengths.

The incidences of the CVD types among the Caucasian population are shown in Table 11.1. The malfunction or the absence of L and M types causes over 99 % of

N. Milić (✉) · D. Novaković · B. Milosavljević
Faculty of Technical Sciences, University of Novi Sad,
Trg Dositeja Obradovića 6, Novi Sad, Serbia
e-mail: milicn@uns.ac.rs

M. E. Celebi et al. (eds.), *Color Image and Video Enhancement*,
DOI 10.1007/978-3-319-09363-5_11

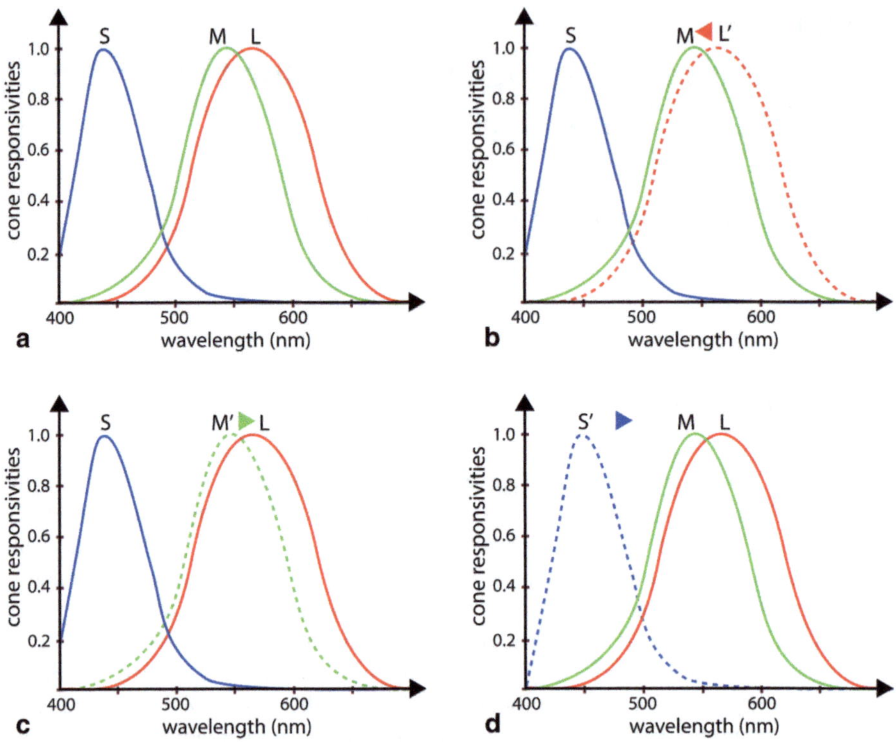

Fig. 11.1 The spectral sensitivity functions. **a** Normal trichromacy. **b** Protanomaly (anomalous L cone type). **c** Deuteranomaly (anomalous M cone type). **d** Tritanomaly (anomalous S cone type)

all CVD cases with a common name, red-green blindness [31]. One quarter of them are dichromats, while three quarters are anomalous trichromats.

While prevalence of CVDs among women is similar for different ethnic groups, Asian (4.2 %) and African (2.6 %) males are significantly less affected than Caucasian [18]. Considering the overall number of colour-deficient observers, the colour information accessibility in the digital content becomes an important issue.

11.1.1 Normal Colour Vision

If the spectral power distribution of light falling on the retina is noted as a function of wavelength E(λ), the cone responses can be modelled as a vector [30]:

$$[L, M, S] = \int [l(\lambda),\ m(\lambda),\ s(\lambda)]\, E(\lambda)\, d\lambda, \tag{11.1}$$

where l(λ), m(λ) and s(λ) represent the *spectral sensitivity functions*. Once the light has been absorbed, all the information about its spectral composition is lost [23].

Table 11.1 Categorization and incidence of CVD types among the Caucasian population [18]

CVD category	Number of cone types	CVD type	Colour discrimination	Incidence (%) men	Incidence (%) women
Monochromacy	0	Monochromacy without cones (achromatopsia)	Total colour blindness and low vision acuity	0.003	0.00001
	1	Monochromacy with cones	Total colour blindness		
Dichromacy	2	Protanopia (L cones absent)	Severely reduced colour discrimination	1.01	0.02
		Deuteranopia (M cones absent)		1.27	0.01
		Tritanopia (S cones absent)		0.002	0.001
Anomalous trichromacy	3	Protanomaly (L cones abnormal)	Continuous range from slightly to severely reduced colour discrimination	1.08	0.03
		Deuteranomaly (M cones abnormal)		4.63	0.36
		Tritanomaly (S cones abnormal)		0.0001	0.0001

Hence, any stimulus can be characterized as a point in 3D space, spanned by quantum catches of cone types [30]. Stimuli with different spectral compositions that plot at the same point in the space will appear alike [23].

This *trichromatic theory* only approximates colour vision mechanism, as it cannot explain some colour appearance phenomena, such as the opponent nature of afterimages[1] or simultaneous contrast[2] [8]. According to Hering's *opponent-colour theory*, the appearance of colour depends on three opponent channels: one achromatic, *white-black* and two chromatic, *red-green* and *yellow-blue* [8]. The colour vision system is best modelled with the *stage theory* which combines the trichromatic theory as the first retinal stage and an opponent-colour theory as the second stage [36]. Figure 11.2 illustrates the signal encoding in the stage theory.

The LMS colour space is not an adequate solution for quantifying colour in colour reproduction industry because it is difficult to measure cone responses directly, and they vary from individual to individual. The first colorimetry standard defines the *CIE 1931 standard observer* with the set of three colour-matching functions. Using standardized $x(\lambda)$, $y(\lambda)$ and $z(\lambda)$ colour-matching functions and the

[1] Afterimage is a phenomenon when the long exposition to one colour stimulus results in the appearance of its opponent colour on a white background after cancelling the stimulus.

[2] Simultaneous contrast is a phenomenon when stimuli shift in colour appearance with the change of background colour.

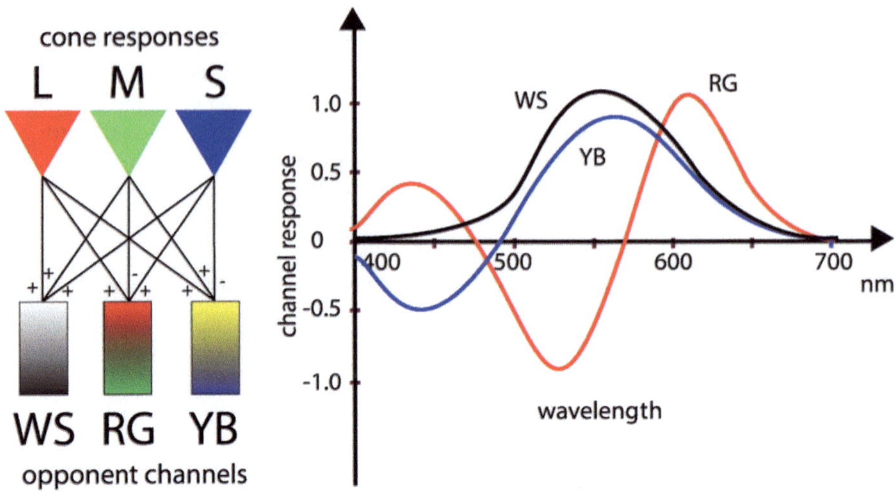

Fig. 11.2 Schema of the encoding of cone response into opponent channels and spectral response functions of opponent channels in the case of an average normal trichromat—one achromatic channel, *white-black* (*WS*) and two chromatic channels, *red-green* (*RG*) and *yellow-blue* (*YB*)

coefficient of normalization k, any colour can be uniquely identified with its *XYZ tristimulus values* [36]:

$$[X, Y, Z] = k \int [x(\lambda), y(\lambda), z(\lambda)]E(\lambda)d\lambda. \tag{11.2}$$

The cone responses (LMS) can be obtained from tristimulus values (XYZ) by linear transformation with a 3 × 3 matrix M_{XYZ_LMS}:[3]

$$[L, M, S]^T = M_{XYZ\_LMS} [X, Y, Z]^T. \tag{11.3}$$

If tristimulus vector gets normalized to lie in the unit plane, a plot of colours on that plane is called a CIE 1931 xy chromaticity diagram and it represents all chromaticities that people with normal colour vision can perceive. However, equal distances in this diagram do not correspond to equal perceptual differences [17]. The CIE u'v' diagram, a linear transformation of the CIE 1931 xy chromaticity diagram, is defined as an attempt to define an encoding with uniformity in the perceptibility of colour differences [25].

[3] There are many similar transformation matrices available in scientific literature, but none of them has been defined as a standard. Currently, the CIE technical committee TC1-36 is working on this recommendation for the CIE 1931 20 and the CIE 1964 100 standard observers.

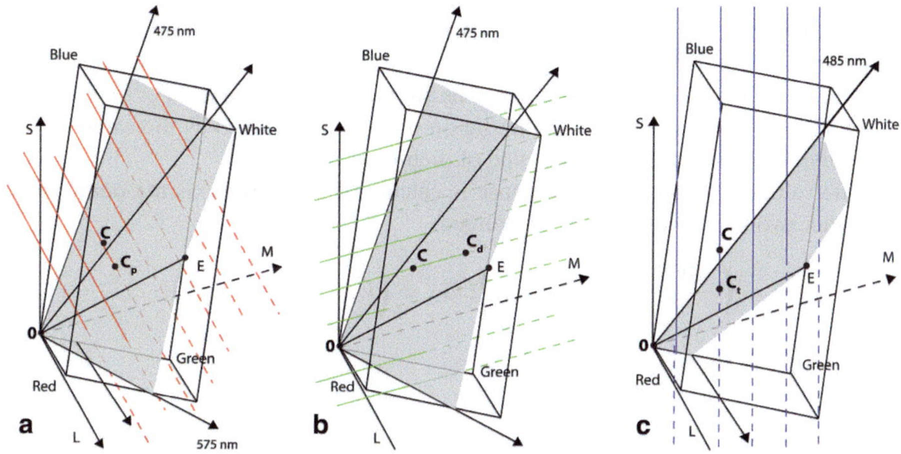

Fig. 11.3 Projection of colour stimulus to dichromatic colour gamut (*grey* half planes) in LMS space. The dichromatic version for a given colour stimulus C is found with projection to a particular half plane in direction parallel to the missing cone response. **a** Protanopia (CC_p ∥ L axis). **b** Deuteranopia (CC_d ∥ M axis). **c** Tritanopia (CC_t ∥ S axis)

11.2 Simulation of Colour Vision Deficiencies

Compared with trichromatic vision, CVD entails reduction of colour gamut manifested with a loss of hue discrimination. Deeper comprehension of CVD starts with the report of unilateral dichromats, which experimentally determines invariant hues appearing the same to dichromats and to normal observers [4]: achromatic colours for all CVDs, the stimuli of 475 nm (a blue hue) and 575 nm (a yellow hue) for protanopia and deuteranopia and the stimuli of 485 nm (a blue-green hue) and 660 nm (a red hue) for tritanopia.

11.2.1 Simulation of Dichromatic CVDs

There are many proposed methods for simulating dichromatic deficiencies based on the assumptions about invariant hues, but the method defined by Brettel et al. and Vienot et al. [4], [33] is widely accepted and used. The method explains the mechanism of mapping trichromatic into dichromatic colour gamut using LMS space in which the orthogonal axes represent cone responses.

Figure 11.3 presents, for each type of dichromacy, a reduced colour gamut—two half planes divided by the achromatic axis and anchored on points specifying the invariant monochromatic stimuli [4].

Figure 11.3 also demonstrates mapping of colour stimulus to the dichromatic colour gamut. As dichromats lack one cone type, they confuse colours that differ only in the intensity of the missing cone response [32]. Hence, the dichromatic version for a given colour stimulus C is found with projection to a particular half plane in direction parallel to the missing cone response: CC_p || L axis, CC_d || M axis and CC_t || with the S axis. These directions are called *confusion lines* or *dichromatic isochromatic lines,* as the points that lie on them are perceived as the same colour by individuals with the particular type of dichromacy.

The value of missing cone response is calculated from two existing ones, while the other two responses remain unchanged:

$$C_P = \left(-\frac{b\,M + c\,S}{a},\ M,\ S\right);\ C_D = \left(L,\ -\frac{a\,L + c\,S}{b},\ S\right); \tag{11.4}$$

$$C_T = \left(L,\ M,\ -\frac{a\,L + b\,M}{c}\right),$$

where a, b and c coefficients are obtained from LMS values of stimulus E (dichromatic white point) and the nearest invariant monochromatic stimulus A: $a = M_E S_A - S_E M_A$, $b = S_E L_A - L_E S_A$, $c = L_E M_A - M_E L_A$.

Vienot et al. [33] also presented simplified model of dichromacy simulation with following linear transformations:

$$L_p = 2.023\,M - 2.526\,S,\ M_D = 0.4942\,L + 1.248\,S,\ S_T = -0.0122\,L + 0.0720\,M. \tag{11.5}$$

Figure 11.4 illustrates the perception of dichromats simulated using this approach.

When confusion lines are converted from LMS colour space (Fig. 11.3) to the CIE u'v' chromaticity diagram (Fig. 11.5), they are no longer parallel—they converge to specific confusion point. Excluding lightness value, dichromatic chromaticity gamut is represented by a line connecting invariant monochromatic stimuli on the edge of the locus. The dichromatic version for a given chromaticity is found as the intersection of the confusion line and the gamut line.

As the human eye has limited accuracy, there is a region on a chromaticity diagram, that is, a MacAdam ellipse, which encloses all colours that are indistinguishable at the same level of lightness [17]. While on the CIE xy chromaticity diagram, the size and orientation of the discrimination ellipses varied widely depending on the position of chosen colour, in the CIE u'v' diagram ellipses of normal colour distinction become nearly circular [30].

Figure 11.6 shows an example of discrimination ellipses for normal (with axes r_1^T and r_2^T) and for dichromatic colour vision (with axes r_1^D and r_2^D).

Based on the experiments of Mollon and Regan [23], every confusion line in CIE u'v' chromaticity diagram has its corresponding threshold ellipse oriented along a confusion line with large axis ratio ($r_1^D << r_2^D$), which represents the area of indistinguishable chromaticities (Fig. 11.6).

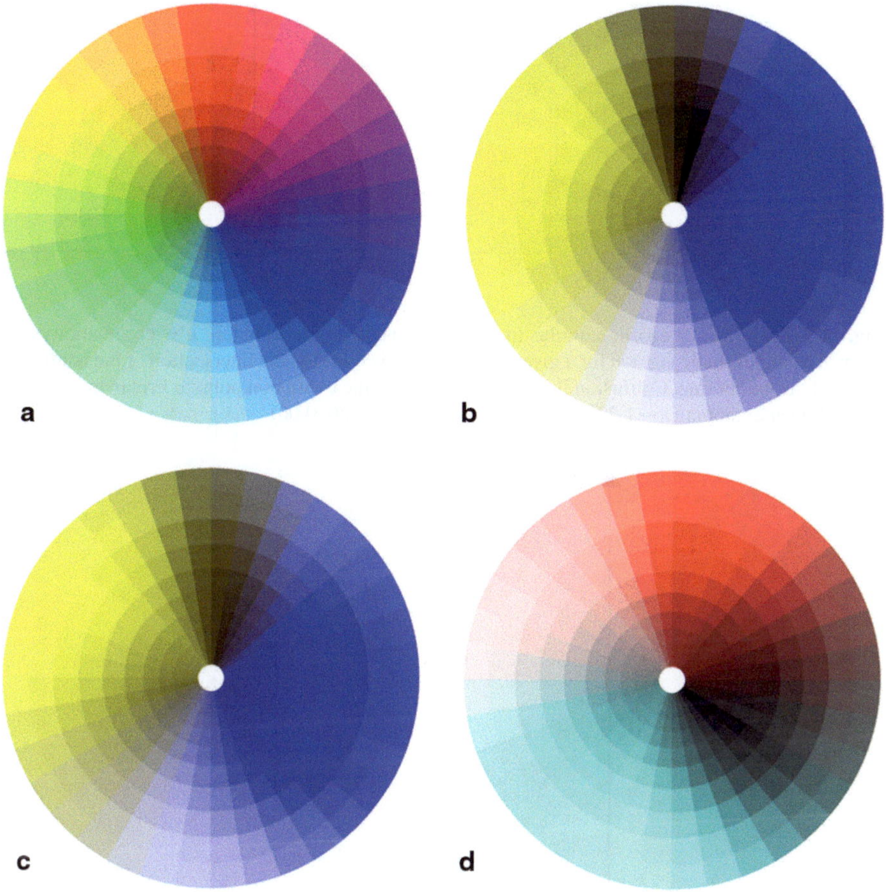

Fig. 11.4 Simulation of dichromatic perception (Brettel et al.'s approach): **a** original hue circle and the perception of **b** protanopes, **c** deuteranopes and **d** tritanopes

11.2.2 Simulation of Anomalous Trichromatic CVDs

Unlike dichromats, anomalous trichromats have three distinctive cone responses. However, there is no universal mapping from normal to anomalous trichromatic colour gamut, as the term anomalous trichromacy covers a large variety of the degree of colour deficiency from almost normal colour vision to weak colour discrimination, such as that exhibited by the corresponding type of dichromacy. Hence, their discrimination ellipses lie between these extremes ($r_1^T < r_1^{AT} < r_1^D$ and $r_2^T < r_2^{AT} < r_2^D$). The r_2^{AT} axis should be taken to indicate the quantitative measure of the subject's colour deficiency.

The recent progress in the simulation of anomalous trichromatic CVDs is done in the work of Machado et al. [19]. Their method simulates anomalous trichromacy

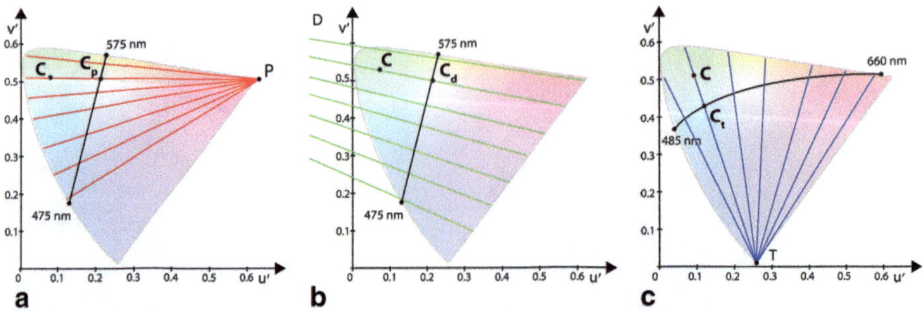

Fig. 11.5 Projection of colour stimulus to dichromatic colour gamut in CIE u'v' chromaticity diagram. The dichromatic version for a given chromaticity is found as the intersection of the confusion line and the gamut line. Confusion lines converge to specific confusion point: **a** Protanopia P(0.68, 0.50), **b** Deuteranopia D(− 1.22, − 0.78), **c** Tritanopia T(0.26, 0.00)

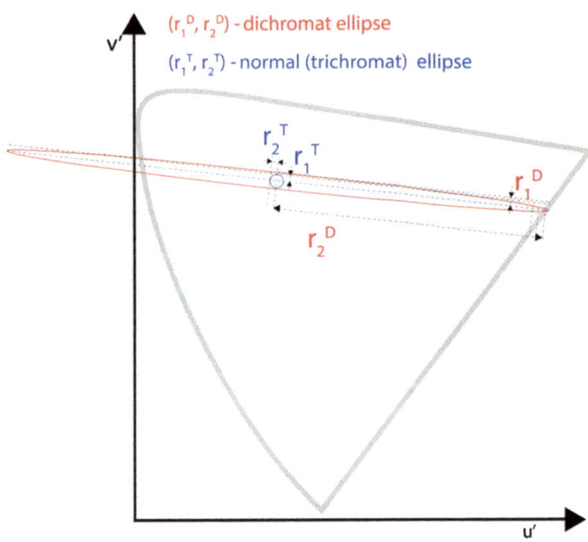

Fig. 11.6 An example of discrimination ellipses for normal and dichromatic colour vision

by shifting the spectral sensitivity function of the anomalous cone type according to the level of deficiency as follows:

$$L_P(\lambda) = l(\lambda + \Delta\lambda_L);\ M_d(\lambda) = m(\lambda + \Delta\lambda_M);\ S_T(\lambda) = s(\lambda + \Delta\lambda_S), \qquad (11.6)$$

where $l(\lambda)$, $m(\lambda)$ and $s(\lambda)$ are the cone spectral sensitivity functions for an average trichromat and $\Delta\lambda_L$, $\Delta\lambda_M$ and $\Delta\lambda_S$ represent the amount of shift applied to the L, M and S anomalous cones, respectively. According to Machado's simulation model, based on the stage theory of the colour vision, simulated RGB values are obtained by a single matrix multiplication ϕ_{CVD}:

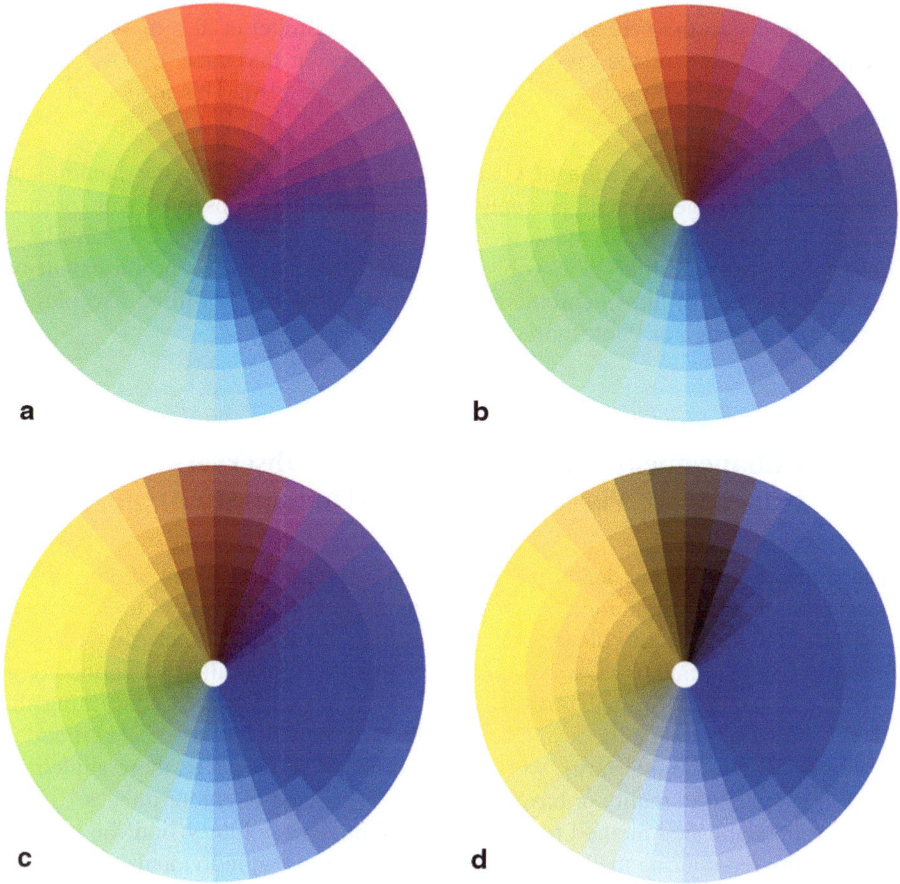

Fig. 11.7 Simulation of protanomalous perception (Machado et al.'s approach): **a** original hue circle and the perception of protanomalous with the shift in spectral sensitivity of anomalous L cones of: **b** 4 nm, **c** 8 nm and **d** 20 nm

$$\begin{bmatrix} R_{\mathrm{CVD}} \\ G_{\mathrm{CVD}} \\ B_{\mathrm{CVD}} \end{bmatrix} = \phi_{\mathrm{CVD}} \begin{bmatrix} R \\ G \\ B \end{bmatrix} = \Gamma_{\mathrm{normal}}^{-1} \Gamma_{\mathrm{CVD}} \begin{bmatrix} R \\ G \\ B \end{bmatrix}, \tag{11.7}$$

where Γ represents the class of transformation matrices that map the RGB values to the opponent channels:

$$\begin{bmatrix} WS \\ YB \\ RG \end{bmatrix} = \Gamma \begin{bmatrix} R \\ G \\ B \end{bmatrix} = \begin{bmatrix} WS_R & WS_G & WS_B \\ YB_R & YB_G & YB_B \\ RG_R & RG_G & RG_B \end{bmatrix} \begin{bmatrix} R \\ G \\ B \end{bmatrix}. \tag{11.8}$$

Figure 11.7 simulates different severities of protanomalous vision using this approach. A severe case of protanomaly with a spectral shift of approximately 20 nm is very similar to the perception of a protanope by Brettel et al.'s method, as the sensitivity function of the anomalous cone type is, in that case, almost entirely overlapping with the function of normal M cone type [18]. This also applies for deuteranomaly.

Simulation Tools Simulation of CVD can be used for better understanding the difference in image content perception comparing to an observer with normal vision. Nowadays, there are many software solutions for simulating CVD available online like Visolve (http://www.ryobi-sol.co.jp/visolve/en/) and a Chrome extension (Chrome Daltonize, https://chrome.google.com/extensions) or offline like plug-in for Adobe InDesign or Photoshop software.

11.3 Enhancement of Image Content for Observers with Colour Vision Deficiencies (Daltonization)

The simulation of colour deficients' perception illustrate the amount of their gamut compression. The simulation tools enable graphic and web designers to check how their work will look to colour-deficient persons and identify confusing colours. Based on these simulations, some guidelines on web accessibility are defined (e.g. *Techniques for Web Content Accessibility*) which solve problems concerning the choice of colour scheme for design (especially for text with coloured background in documents, software interfaces, web presentations, etc.). However, these guidelines are rarely applied in reality or have limited use in the case of colour images.

A well-established method for simulating anomalous colour vision can be used for defining recolouring method that enhances image so that individuals with colour vision deficiencies can distinguish more details or, in other words, see the image content much similar to an average observer. This colour adaptation, commonly called daltonization,[4] is a complex and still actual problem of mapping image gamut to a reduced one. It represents a specific type of colour quantization process, as it reduces the number of image colours in accordance with deficient hue discrimination while trying to maintain the visual appearance of the original image. The aim of quantization is to find the optimal subset of representative colours from CVD users' gamut and map the image into this subset [5]. Table 11.2 contains a categorization of daltonization methods.

The colour-barrier-free design for particular target population can be achieved by colour universal design or colour customization. In the case of colour universal design, or "design for all", the applied colour schemes should be "safe" for anybody. A single design conveys the information accurately both to observers with normal and impaired vision, for example, a public metro map. Colour customization, on the

[4] Colour vision deficiency is also known as daltonism due to John Dalton, the first researcher who made notes about CVDs and was colour-blind himself.

Table 11.2 The categorization of daltonization methods

Criteria	Categories	Key features of image processing
Content dependency	1) Content independent	One pixel's colour is always remapped into same the colour regardless of the image content
		Simple and suitable for real-time adaptation
	2) Content dependent	The same colour is in two different images remapped to different colours based on image content
		Suitable for complex content (e.g. natural images)
Processing area	1) Total	A complete image is modified
	2) Partial	Only a certain zone (image segment) or a particular hue range (e.g. red hues) is modified
		Suitable for interactive (event driven) and situation-specific adaptation
Target group	1) Universal	Image (design) should be adapted for all—both normal visions and all types and severities of CVDs
		Suitable for public printed design (e.g. metro map)
	2) Customized	Image (design) should be adapted for only one type of CVD or specific CVD user
		Suitable for digital content (e.g. web design, images, computer games etc.)
Processing control	1) User assisted	The result is highly dependent on the chosen parameters
	2) Automated	The result is obtained without interaction with the user
		Suitable for real-time adaptation
Enhancement type	1) Recolouring	Image enhancement is done by recolouring
	2) Texturing (pattering)	Suitable for natural images, textual elements
		Image enhancement is done by adding texture and hatching
		Suitable for maps, graphics and infographics
	3) Labelling	Image enhancement is done by adding colour label of selected image area
		Suitable for maps/infographics and colour recognition issues

other hand, provides the most suitable content for a certain type of CVD or even for a specific individual. This concept can be achieved in the digital environment where one image content can be rendered differently on request.

Besides recolouring, some daltonization tools suggest the use of texture hatching with variable angles and patterns [29] and adding textual labels of selected colours [28] as useful features to compensate the lack of colour information.

Daltonization methods can modify a complete image, only a certain image segment or a particular hue range while the processing control can be automated or assisted by user's choice of parameters.

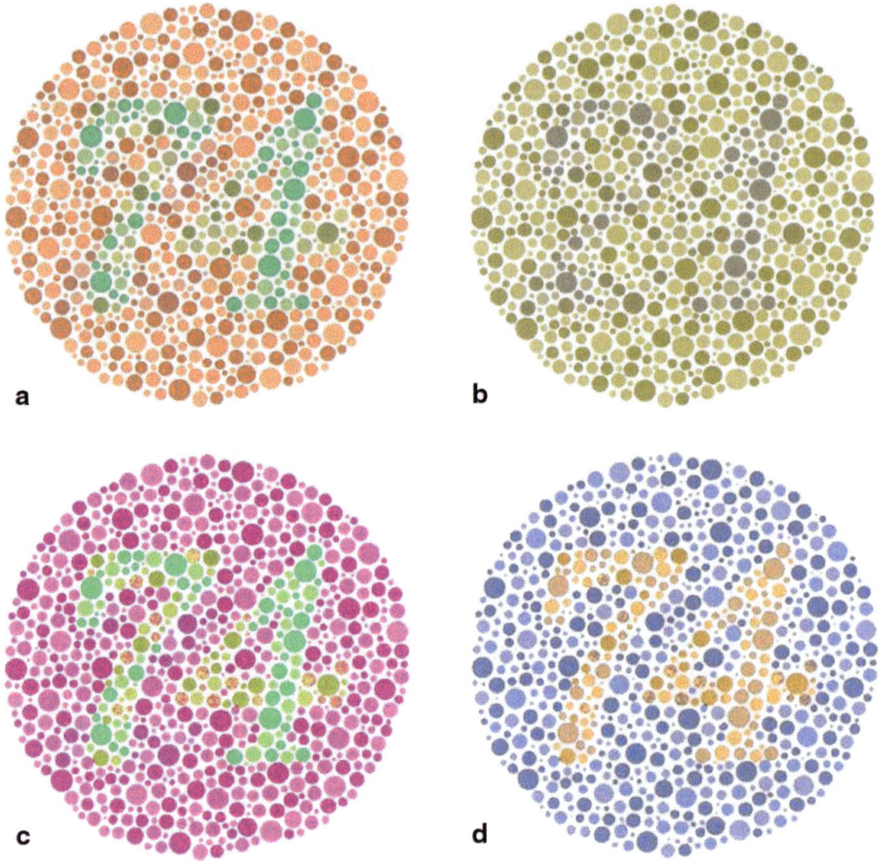

Fig. 11.8 Daltonization example. **a** An Ishihara plate. **b** Deuteranopic simulation. **c** Daltonized image. **d** Deuteranopic simulation of daltonized image

Frequently used examples for demonstrating efficiency of daltonization methods are Ishihara plates (Fig. 11.8) or natural images with red-green confusing combination (Fig. 11.9).

11.3.1 Content-Independent Methods

The content-independent methods do not assure colour differentiation by dichromats because they use a global pixel-based processing that does not take into account the image content and spatial distribution of confusing colours in images.The entire image is treated in a uniform manner and the resulting value of a processed pixel is a function of its original value, regardless of its location within the image. This way, it often happens that recolouring solves the initial problem, but

Fig. 11.9 Daltonization example. **a** A natural image with *red-green* confusing combination. **b** Daltonized image. **c** Protanopic simulation. **d** Protanopic simulation of daltonized image. **e** Protanopic range of chromaticities in inital image. **f** Protanopic range of chromaticities in daltonized image

creates a new confusing pair between the new (remapped) colour and the colour of another neighbouring area. Even worse, colours that are clearly distinctive can be remapped into indistinguishable combinations.

Vischeck Daltonize One of the first daltonization algorithms is Vischeck Daltonize filter, available online (http://www.vischeck.com). It adapts images for protanopes and deuteranopes with pixel-based processing using three parameters to stretch the histogram of red-green channel, project it into the luminance and yellow-blue channels. As adaptation parameters are manually chosen by the user, based on "trial and error" method, this method is inconvenient and cumbersome.

Chrome Daltonize Today, the frequently used solution for the colour accessibility in the web environment is Chrome Daltonization extension based on the work of Fidaner et al. [9]. The concept of this pixel-based method is using an error matrix obtained by subtracting simulated dichromatic values from the original RGB values. This error matrix reveals the colours that cannot be perceived correctly by a CVD user. The image processing steps are:

1. Transformation of normalised RGB to LMS values

$$\begin{bmatrix} L \\ M \\ S \end{bmatrix} = M_{\text{lms}\to\text{rgb}} \begin{bmatrix} R \\ G \\ B \end{bmatrix} = \begin{bmatrix} 17.8824 & 43.5161 & 4.1193 \\ 3.4557 & 27.1554 & 3.8671 \\ 0.02996 & 0.18431 & 1.467 \end{bmatrix} \begin{bmatrix} R \\ G \\ B \end{bmatrix}, \tag{11.9}$$

2. Calculation of dichromatic LMS values

$$\begin{bmatrix} L_{\text{dic}} \\ M_{\text{dic}} \\ S_{\text{dic}} \end{bmatrix} = M_{\text{dic}} \begin{bmatrix} L \\ M \\ S \end{bmatrix}, \text{ e.g. protanopia } \begin{bmatrix} L_p \\ M_p \\ S_p \end{bmatrix} = \begin{bmatrix} 0 & 2.02344 & -2.52581 \\ 0 & 1 & 0 \\ 0 & 0 & 1 \end{bmatrix} \begin{bmatrix} L \\ M \\ S \end{bmatrix}, \tag{11.10}$$

3. Transformation of dichromatic LMS values back to RGB values

$$\begin{bmatrix} R_{\text{dic}} \\ G_{\text{dic}} \\ B_{\text{dic}} \end{bmatrix} = M^{-1}_{\text{lms}\to\text{rgb}} \begin{bmatrix} L_{\text{dic}} \\ M_{\text{dic}} \\ S_{\text{dic}} \end{bmatrix} = \begin{bmatrix} 0.0809 & -0.1305 & 0.1167 \\ -0.0102 & 0.0540 & -0.1136 \\ -0.0003 & -0.0041 & 0.6935 \end{bmatrix} \begin{bmatrix} L_{\text{dic}} \\ M_{\text{dic}} \\ S_{\text{dic}} \end{bmatrix}, \tag{11.11}$$

4. Adding error matrix multiplied by daltonization matrix M_{dal} to original RGB values

$$\begin{bmatrix} R_{\text{dal}} \\ G_{\text{dal}} \\ B_{\text{dal}} \end{bmatrix} = \begin{bmatrix} R \\ G \\ B \end{bmatrix} + \text{M}_{\text{dal}} \begin{bmatrix} R - R_{\text{dic}} \\ \text{G} - G_{\text{dic}} \\ \text{B} - B_{\text{dic}} \end{bmatrix}, \text{ e.g. protanopia } \text{M}_{\text{dal}} = \begin{bmatrix} 0 & 0 & 0 \\ 0.7 & 1 & 0 \\ 0.7 & 0 & 1 \end{bmatrix}. \tag{11.12}$$

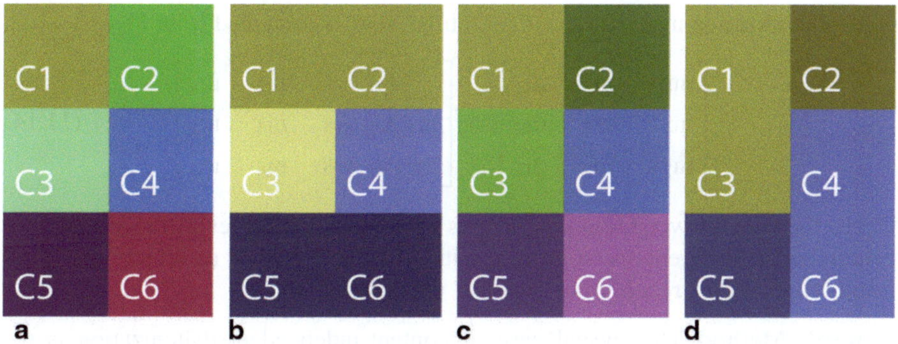

Fig. 11.10 Recolouring problem with content-independent Chrome Daltonization. **a** Initial colour schema. **b** Protanopic version of initial schema. **c** Daltonized schema. **d** Protanopic version of daltonized colours

It can be noticed from weighting coefficients in M_{dal} (Eq. 11.12) that recolouring for protanopia means shifting red values, dominant in the protanope's error matrix, towards the blue side of the spectrum.

Figure 11.10 demonstrates a typical problem with content-independent methods on the example obtained with Chrome Daltonize algorithm. The RGB values of image segments are: C1(155,155,35), C2(73,165,35), C3(100,227,113), C4(90,112,187), C5(76,36,91) and C6(159,25,90).

Figure 11.10b shows that the original schema has two problematic colour pairs, (C1, C2) and (C5, C6), for individuals with protanopia. The image adaptation conveyed with Chrome Daltonize solves initial problematic combinations but creates two new ones—between C1 and C3 and between C4 and C6 segments (Figure 11.10d).

For these reasons, Anagnostopoulos et al. [1] proposed the upgrading of the method with an automatic iteration module, which eliminates the possibility of matching between the original and the daltonized colours. The additional algorithm steps include colour clustering and checking for indistinguishable combinations after the first daltonization. Possible indistinguishably coloured image cluster would be isolated by logical masking from the rest of the image and daltonized again with changed weighting coefficients in matrix M_{dal}.

The coefficients of initial daltonization matrix $M_{dal,0}$ are defined as:

$$M_{dal,0} = \begin{bmatrix} m_1 & m_2 & m_3 \\ m_4 & m_5 & m_6 \\ m_7 & m_8 & m_9 \end{bmatrix} = \begin{bmatrix} -1 & 0 & 0 \\ 1 & 1 & 0 \\ 1 & 0 & 1 \end{bmatrix}, \quad (11.13)$$

while daltonization matrix in the tth iteration, $M_{dal,t}$ is obtained from $M_{dal,t-1}$:

$$M_{dal,t} = \begin{bmatrix} m_1 & m_2 & m_3 \\ m_{4,t} & m_5 & m_6 \\ m_{7,t} & m_8 & m_9 \end{bmatrix} = \begin{bmatrix} m_1 & m_2 & m_3 \\ m_{4,t-1} - s & m_5 & m_6 \\ m_{7,t-1} + s & m_8 & m_9 \end{bmatrix}, \quad (11.14)$$

where the sum of weighting coefficients is always 3 in order to redistribute the colours to the different spectrum side without exaggerated or extremely unnatural results (keeping the recolouring "energy" unchanged).

Kotera's Method The special case of content-independent daltonization is the most current spectral-based method by Kotera [14]. Although this method uses projection of spectral data to dichromatic cone response space to recover the lost spectra for dichromats, the nature of recolouring is content independent—one pixel's colour is always remapped into the same colour. The fundamental spectral data are obtained from the RGB values by a pseudo-inverse projection and the lost spectra ΔC_{DIC} are calculated as the difference between fundamental spectra visible to the trichromats (C^*) and to dichromats (C^*_{DIC}). The daltonization has two aims: to maximize the spectral visibility for dichromats and to minimize the difference in perception of the daltonized image between normal and dichromatic vision. The image processing steps are:

1. Calculation of fundamental spectral curve of colour stimulus (C^*) from sRGB values by a pseudo-inverse projection (P^{inv}):

$$C^* = P^{inv}\,[L, M, S]^T = P^{inv}\,M_{lms\rightarrow rgb}[R, G, B]^T, \quad (11.15)$$

where pseudo-inverse projection from LMS cone response space to n-dimensional spectral space is based on the matrix R-theory:

$$P^{inv} = A_{lms}\,({A_{lms}}^T A_{lms})^{-1}, \quad (11.16)$$

where A_{lms} is a vector with cones' spectral sensitivity functions as components ($A_{lms} = [l(\lambda), m(\lambda), s(\lambda)]$).

2. Extraction of visible (C^*_{DIC}) and invisible spectra ($\Delta C_{DIC} = C^* - C^*_{DIC}$) for dichromats

$$C^*_{DIC} = R_{DIC} C^*. \quad (11.17)$$

Dichromatic projector operator R_{DIC} is calculated using existing cone responses:

$$R_{DIC} = A_{DIC}\,({A_{DIC}}^T A_{DIC})^{-1} {A_{DIC}}^T, \quad (11.18)$$

where, for protanopia $A_{DIC} = [m(\lambda), s(\lambda)]$, for deuteranopia $A_{DIC} = [l(\lambda), s(\lambda)]$ and for tritanopia $A_{DIC} = \Delta[l(\lambda), m(\lambda)]$.

Fig. 11.11 Recolouring problem with content-independent Kotera's algorithm. **a** The initial image. **b** Protanopic version of the initial image. **c** Daltonized image. **d** Protanopic version of daltonized image

3. Optimal image daltonization using lost spectra ΔC_{DIC}

$$\Delta C_{DIC} = C^* - C^*_{DIC}. \tag{11.19}$$

Daltonization is done by shifting lost spectral distribution into the visible wavelength region by wavelength interval λ_{shift} in the following way:

if $\lambda_{max} \geq \lambda \geq \lambda_{shift} + \lambda_{min}$ then $\Delta C_{shift}(\lambda - \lambda_{shift}) = \Delta C_{DIC}(\lambda)$ and
if $\lambda_{shift} + \lambda_{min} > \lambda \geq \lambda_{min}$ then $\Delta C_{shift}(\lambda_{max} + \lambda_{min} + \lambda - \lambda_{shift}) = \Delta C_{DIC}(\lambda)$.

Figure 11.11 demonstrates the problem with content-independent methods where daltonization results in reduced chromatic diversity compared to the original image.

Kotera's method is useful for creating colour universal design, as it minimizes the difference between trichromatic and dichromatic perceptions of the adapted image.

11.3.2 Content-Dependent Methods

The content-dependent category includes histogram-based, neighbourhood-based, region-based and other methods, where pixel colour value after daltonization depends on initial image gamut, histograms or location of the pixel in the image. These methods are more complex and computationally demanding, but give more successful differentiation of coloured image elements.

However, there are problems with the preservation of naturalness and colour identification, as users can now distinguish between all the coloured elements, but the colours are often remapped to completely different ones. That is why this type of recolouring often includes objective function that defines the best remapping in the sense of balancing colour contrast and naturalness.

The problem of naturalness conservation should be emphasized in the cases of real-life images and in the cases where observers have moderate anomalous trichromacy with hue discrimination close to normal vision. These methods usually aim to adapt optimally the image for dichromatic CVD types, although the number of anomalous trichromats is three times larger. Minimization of the visual gap between normal trichromatic and dichromatic vision results in a reduction of image colour gamut for normal vision observers. That is justified in the case of a colour universal design like public maps, but represents unnecessary constraint for average observers and anomalous trichromats in the case of digital content.

The content-dependent methods also have a problem with time coherence, needed for remapping interactive or video content, as the same colour in two different images is often remapped to entirely different colours.

Methods from this category calculate the optimal colours using genetic algorithm [13], mass-spring optimization [15], affine transformation [26], neural networks [16], etc.

Self-Organizing Daltonization Ma et al. [16] define a self-organizing map (SOM), an artificial neural network model, for nonlinear colour mapping of initial image gamut to reduced dichromatic gamut. The colour gamut of the initial image is mapped to the new one maintaining the neighbouring relations between colours. However, transformed images often look very unnatural for both normal and colour-deficient observers. Moreover, the method cannot be considered for the practical use in real-time applications because of its computational complexity. Figure 11.12 presents a daltonization result obtained using SOM processing.

The image processing steps are:

1. Sampling reduced colour set from the original image

The sampling of the original image colours into reduced colour set is done because neighbouring pixels have similar colours, especially in the uniform regions. The sampling reduces the computational cost of following processing steps, where the number of samples is determined by the image size.

2. Mapping sampled colours to SOM

A feature vector with RGB values of sampled pixel represents the input of the SOM network. Each input vector is connected to all the neurones of the SOM network so that the dimension of the weight matrix is 3-by-n, where n is the number of neurones.

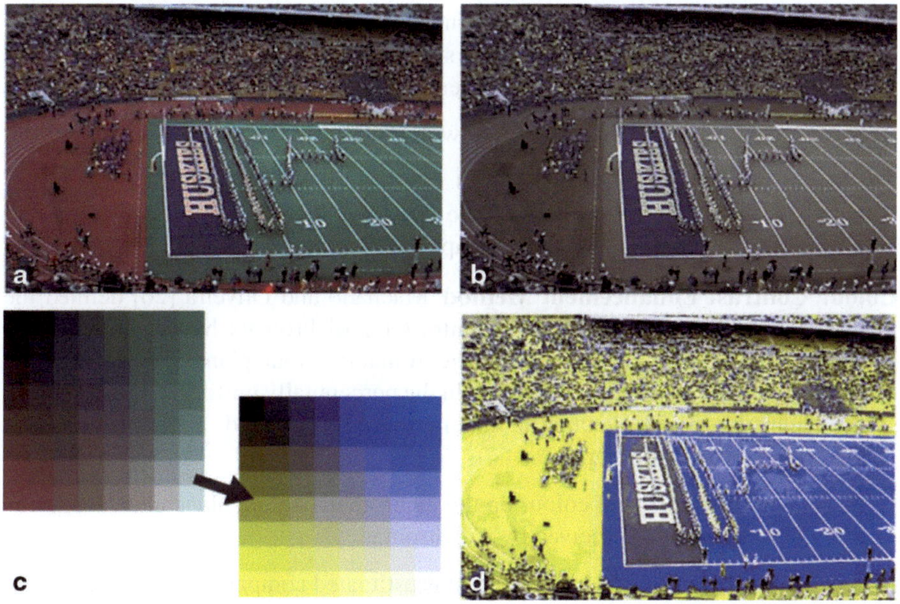

Fig. 11.12 Daltonization result using SOM processing [16]. **a** Original image. **b** Protanopic simulation of original image. **c** The "original colours" of the codebook vectors and their version in new space. **d** Daltonized image perceived in the same manner by normal trichromats, protanopes and protanomals

3. Self-organizing process

The SOM uses a learning algorithm to cluster input feature vectors, so that each input vector is paired with its "winner"—the neurone whose weight is most similar to the input vector:

$$\|\mathrm{x} - \mathrm{m_c}\| = \min\{\|\mathrm{x} - \mathrm{m_i}\|\}, \tag{11.20}$$

where $\|\bullet\|$ represents the Euclidian norm, x is the input vector, $\mathrm{m_i}$ is the weight vector of the ith neurone and $\mathrm{m_c}$ is the "winner". The weights of both the "winner" and its neighbours are adjusted iteratively to approach the input vector:

$$m_i(\mathrm{t}+1) = \begin{cases} m_i(\mathrm{t}) + \alpha(\mathrm{t})[\mathrm{x}(\mathrm{t}) - m_i(\mathrm{t})], & i \in N_c(\mathrm{t}) \\ m_i(\mathrm{t}), & \text{else} \end{cases}, \tag{11.21}$$

where $\alpha(\mathrm{t})$ is the learning rate (between 0 and 1) and $\mathrm{N_c}(\mathrm{t})$ is the neighbourhood of the "winner". The resulting weights are called "codebook vectors".

4. Mapping codebook colour space to a new (dichromatic) colour space

A new colour space has colours of two opponent channels in the corners (white and black in one diagonal, blue and yellow in the other) and a smooth interpolation in between. The size and resolution of new space are the same as for the codebook colour space. The codebook colour space is rotated to make its corner colours closely related to the corresponding corner colours of the new space.

5. Mapping pixels of the original image with most similar codebook vectors

The feature vector of each pixel (its RGB values) in the original image is compared with all the codebook vectors to find the most similar one, whose corresponding colour in the new space is set as the remapped colour for the pixel.

Colour Contrast Enhancement Method Machado and Oliveira [20] defined the method that aims to recover colour contrast for dichromats by orthographically projecting the original colours onto a dichromatic colour plane aligned with the direction that maximizes loss of contrast in the perceptually uniform CIE Lab colour space. Figure 11.13 gives geometric illustration of this method.

The image processing steps are:

1. Projection of the original colour on the dichromat's gamut plane and calculation of local contrast loss perceived by a dichromat

For each pixel p_i, the loss of local contrast is estimated comparing to a single neighbouring pixel p_j. The relative loss of contrast between an initial pair of colours (c_i,c_j) and dichromatic version (c_i',c_j') is given by:

$$l_{(C_i,C_j)} = \frac{\|c_i - c_j\| - \|c_i' - c_j'\|}{\|c_i - c_j\|}, \tag{11.22}$$

where $\|.\|$ represents the vector length operator and $w_i = l_{(ci,cj)}\ \vartheta_{ij}$ is a vector of contrast loss along the direction of individual contrast loss $\vartheta_{ij} = c_i - c_j$ associated to pixel p_i.

2. Estimation of the direction v_{ab} that maximizes the contrast loss

Direction v_{ab} that maximizes the local contrast loss is calculated in a–b chromaticity plane (CIE Lab space) as the main eigenvector of the matrix M^TM, where M is a matrix whose rows contain chromaticity coordinates (a, b) of the contrast loss vectors for all image pixels:

$$M = \begin{bmatrix} w_1^a & w_1^b \\ w_2^a & w_2^b \\ \cdots & \cdots \\ w_n^a & w_n^b \end{bmatrix}. \tag{11.23}$$

The elements of the ith row of M are the projections of w_i on the a–b plane.

The precise estimation of v_{ab} would require evaluation of all pixels p_j in a neighbourhood N_i around p_i rather than a single neighbouring pair (p_i, p_j). However, this implies unnecessary redundancy as the neighbouring pixels usually have similar

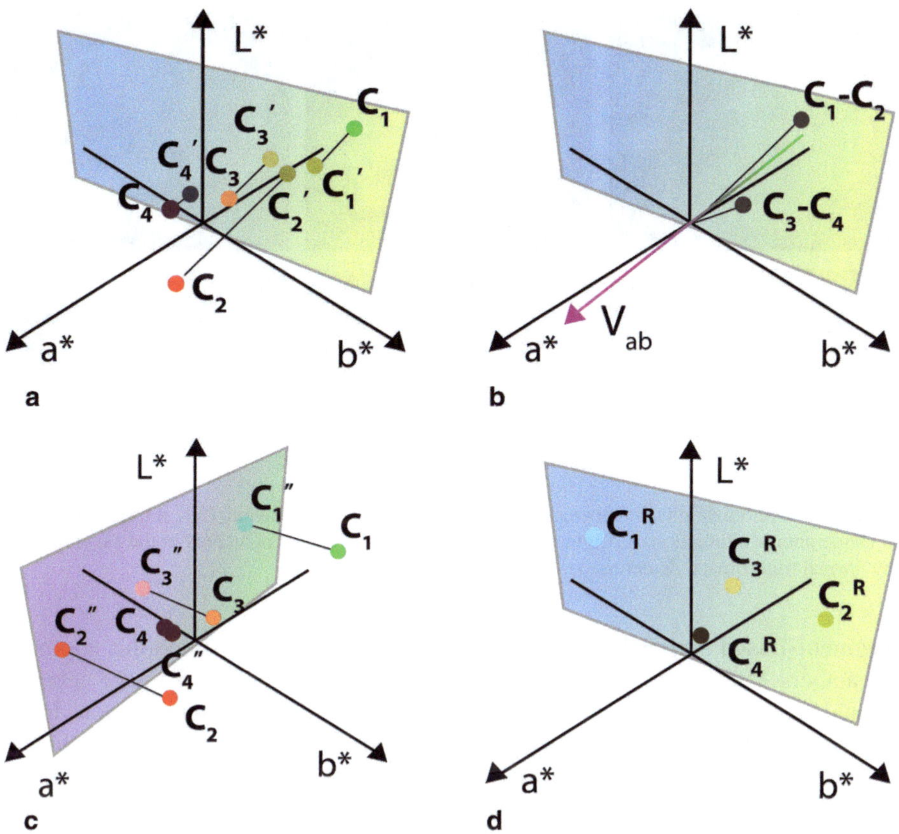

Fig. 11.13 Geometric illustration of image processing steps. **a** Calculating dichromatic versions (c_1' to c_4') of colours c_1 to c_4. **b** Finding direction v_{ab} that maximizes the contrast loss. **c** Projection of the original colours on the plane defined by v_{ab} and the lightness axis. **d** Rotation of the projected colours around L axis until they align with the dichromat's gamut plane

values. Since directions v_{ab} based on a single neighbour and the entire neighbourhood would be sufficiently close to each other, sampling of one pixel pair from the neighbourhood is justified and reduces the computational cost.

3. Orthographic projection of the original colours on the plane defined by v_{ab} and the lightness (L) axis, which spreads original colours compared to their initial dichromatic projections
4. Rotation of the projected colours around L axis until they align with the dichromat's plane and recolouring image with new colour values

Figure 11.14 demonstrates daltonization results using this approach.

Figure 11.14 demonstrates that the method enhances the local colour contrast and overall image quality for dichromats, but it is not suitable for anomalous trichromats because of the exaggerated image gamut reduction.

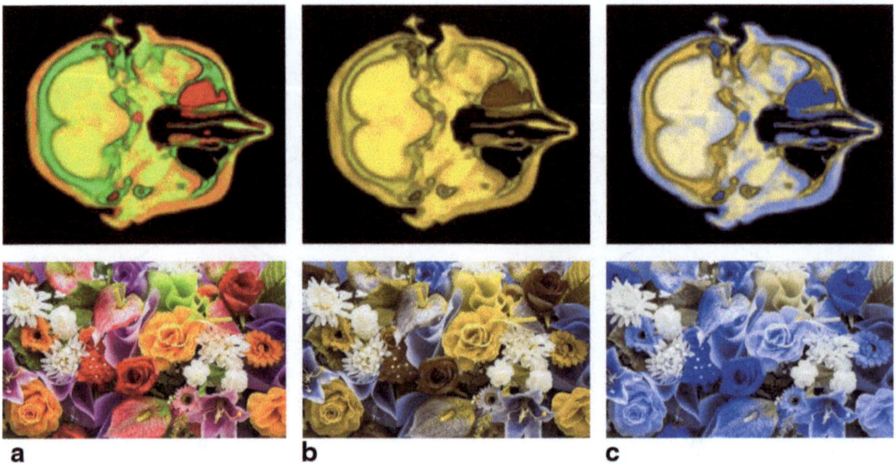

Fig. 11.14 Daltonization result using the method by Machado and Oliveira [20]. **a** Original images. **b** Deuteranopic simulation of original images. **c** Daltonized images perceived in the same manner by normal trichromats, deuteranopes and deuteranomals

Segment-Based Naturalness-Preserving Method The idea behind this approach by authors Milić et al. [22] is that the colour centre of only one image segment should be placed on one confusion line. If the colours of two or more regions lie on the same confusion line, then they should be remapped in direction perpendicular to confusion line taking into account the overall content—the colour distribution of other segments. The method conserves the image naturalness by restricting, for each region centre, an area of admissible remapping. This achieves recolouring balance, where the colours are made sufficiently distinguishable from each other, but they do not deviate too much from the original image colours.

The proposed recolouring concept is applicable for all types of dichromacy and anomalous trichromacy with adjustment for their set of confusion lines. The image processing steps are (see also Fig. 11.15):

1. Image segmentation based on chromaticity and defining centres as representative colours of each segment

The clustering into regions is based on the Euclidean distance metric in the Lu'v' space between chromaticity values excluding the variations in lightness. K-means clustering segments groups of pixels based on their location in u'v' plane such that pixels within each cluster are as close to each other as possible, and as far as possible from pixels in other clusters [6]. A representative colour centre is then selected for each segment by averaging all colours in it.

2. Remapping of colour centres in the direction perpendicular to confusion line inside corresponding admissible area

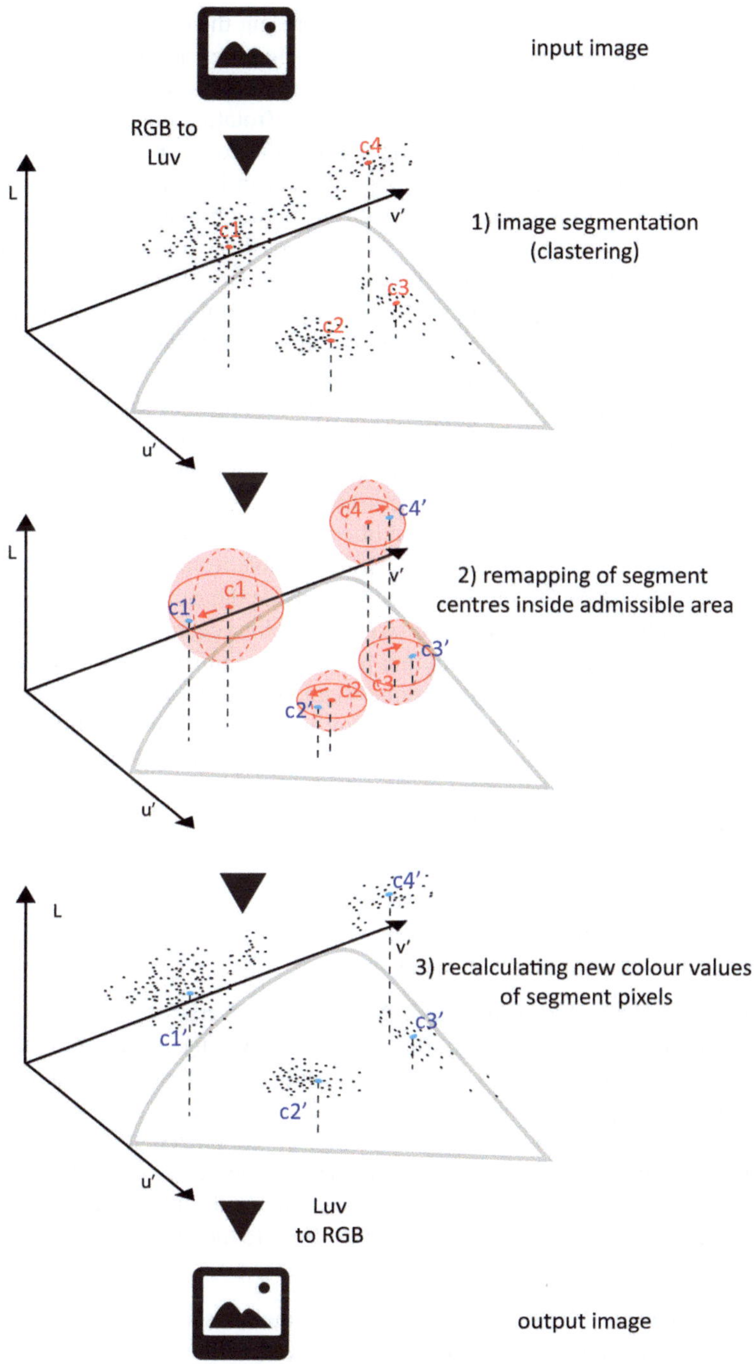

Fig. 11.15 Graphic presentation of the daltonization steps

The colour points $c_0, c_1, \ldots, c_i, \ldots, c_n, c_{n+1}$ represent the colour centres of image regions. For each colour point, there is a corresponding admissible area and an ellipse oriented along the confusion line, where it is allowed for a point to be remapped. The aim is to redistribute these colours (rotate around confusion point) in a way that the lines from the confusion point to the colour points are as far from each other as possible. The set of colour positions can be compared to a model in which points represent the positions of electrons or small magnets and due to their magnetic charge they want to be pushed as far as possible from each other, but their movement is restricted by their individual intervals. This model tends to seek the equilibrium state. For the purpose of finding equilibrium state, the first colour is fixed at the leftmost position and the last colour is set at the rightmost position, while other colour points are iteratively moved. The resulting positions of colour centres are content dependent: they depend on the number of segments, the initial positions of segment centres and the defined admissible area.

Restricting the area admissible for remapping colour centre and maintaining the original angle sequence of the colour centres after remapping conserve the image naturalness. This ensures that the change of colour is not exaggerated.

A request for the colour centres to be distinguishable from each other for a particular CVD type is that each pair of centres lies outside the contour of corresponding discrimination ellipse. However, when too many colour centres are concentrated in the same area of the chromaticity diagram for the initial image or when there are too many image regions in general, this request can be hard to achieve. Enlarging the admissible area can provide satisfying recolouring solution for the situations when more colour centres are placed near the same confusion line, although it reduces image naturalness. When there are too many image regions then adjusting the admissible area cannot make a difference, and it is convenient for the lightness value to be included in remapping. Redistributing colour centres in the direction of lightness axis optimizes the daltonization.

3. Replacing the original colour centres with the new ones and recalculating the colour values of other pixels in segments.

Difference in colour values (Δu', Δv', ΔL) between every recoloured pixel in the segment and the remapped centre of the segment is the same as before remapping.

Figure 11.16 demonstrates daltonization result in example of a natural image with confusing red-green combination.

The protanope and protanomalous versions of daltonized image (Fig. 11.16) show that the regions are made distinguishable from each other for the CVD users. Additional advantage is that the images remained quite natural to the CVD observer—colours in an image remained quite close to the ones that CVD observers are used to.

Daltonization Tools Many content-independent methods are, because of their simplicity and computational efficiency, implemented in the web environment and modify HTML pages on pixel basis. Chrome Daltonization extension performs daltonization of <image> elements in HTML documents through support of

Fig. 11.16 The daltonization results for protanopia and protanomaly for a natural image with *red-green* confusing combination. **a** Original image. **b** Protanomalous versions of original image. **c** Image segmentation. **d** Remapping of segment centres. **e** Daltonized image. **f** Protanomalous versions of daltonized image

Javascript. Iaccarino et al. [12] created a simple colour conversion filter CBFS (Color Blind Filter Service) for increasing contrast. Ruminski et al. [27] offers a web proxy system and a mobile application for rendering web pages by modifying only certain range of red and green hues. Their tools also have colour labelling option for a selected image pixel. Eyepilot software available online provides combination of daltonization techniques—hatching, zoned recolouring and colour labelling. The Colour Blind Aid iPhone app [2] enables partial daltonization of selected image area.

Viewing Conditions Despite the relevance of viewing environment in which colour sample is observed, little has been published on how it affects people with common colour vision deficiencies. The analysis of images simulated under different viewing conditions revealed that decreasing colour temperature of the illuminant leads to even more reduced colour gamut and, thus, to enhanced number of hardly distinguishable colour combinations (see Fig. 11.17), while changing viewing surround from average to dark leads to better image contrast [21]. These conclusions can be used for defining compensating adaptation for certain viewing environment in the cases of protanopia/maly and deuteranopia/maly. This compensation would include enhancing the colour temperature of the display's white point and reducing the luminance of the viewing surround.

Compensating Advantage of Colour-Deficient Population Although colour blindness is classified as a disability, a few studies revealed that red-green colour-blind people are better at seeing through certain colour camouflages indicating an evolutionary reason for the high prevalence of CVDs. In a study published by Morgan et al. [24], the participants were asked to locate a flashing pattern (7-by-7 "target area" of horizontally oriented rectangles) on a background (30-by-30 grid of vertically oriented rectangles). In the case when all the rectangles were of the same colour, both trichromatic and dichromatic participants had no difficulties with spotting the target area. However, in the case when the rectangles were randomly coloured red or green, observers with the normal colour vision did poorly, while the red-green dichromats located the target as easy as before without colour to distract them and interfere with the detection of the texture. The same results were obtained when rectangles of pattern and background were replaced with the letters "A" and "B". These findings are supported by some military reports that discovered the advantage of red-green colour-deficient soldiers in picking out camouflage from background foliage. A possible explanation for this superiority in detecting texture and lightness variations can be the cross-modal organization[5] in the brain of colour-deficient observers similar to visually-impaired individuals who compensate for their lack of sight with enhanced auditory or other abilities.

Besides dichromats' advantageous detection of texture, anomalous trichromats have a colour advantage in detecting differences between stimuli that represent metameric pair[6] for normal trichromats under a particular illuminant. There are certain colour combinations that anomalous trichromacy can distinguish in given viewing conditions but people with normal colour vision cannot and vice versa [3].

[5] The cross-modal organization is a phenomenon describing the brain's capability of reorganizing itself when it is deprived of input in one sensory modality in order to enhance other senses.

[6] The metameric colours are stimuli with different spectral compositions that appear the same when viewed under one set of viewing conditions but distinct under another.

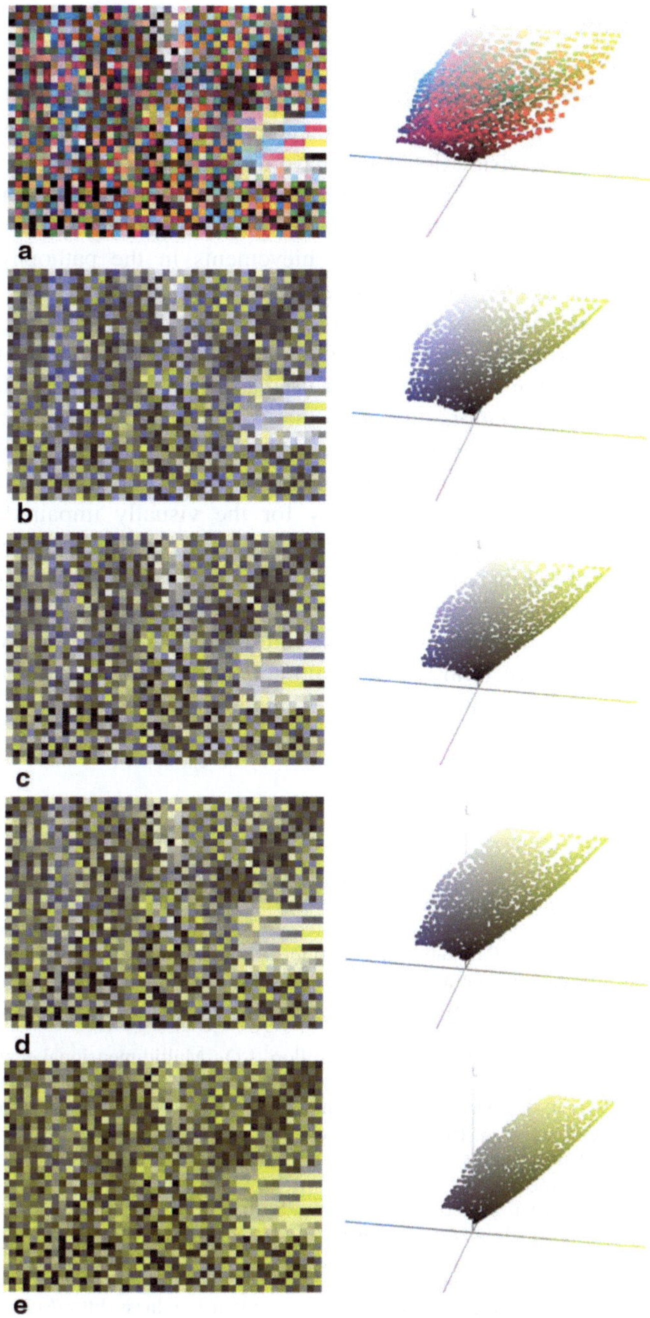

Fig. 11.17 Simulation of protanopic image perception under different viewing conditions. **a** Initial image—ECI2002 random layout chart (white point D65) and its colour gamut. Simulated protanopic version and correspondent colour gamut (gamut plane) in the case of illuminant. **b** D65-6500K. **c** D50-5000K. **d** F11-4000K. **e** A-2856K

11.4 Summary

In contemporary life, information in colour surrounds us in both printed and digital media. While these colour images are taken for granted by a majority of viewers, people with colour vision deficiency have difficulties in discriminating certain colour combinations that are easily distinguishable to people with normal vision and, consequently, have difficulties in perceiving the image features.

Despite significant discoveries and achievements in the pathophysiology of colour vision deficiencies, "cure" for them does not exist. No image processing can give CVD individuals back the colours that they do not perceive correctly in the first place. Our contribution can be taking into account the limitation of their visual system in every situation when colours are used as primary means of information.

Well-established simulation tools show that colour vision deficiencies affect the entire spectrum and they are much more complex than the stereotypical inability to tell red from green. Daltonization methods use these simulation techniques to optimize the perceptual image quality for the visually impaired observers. The enhancement of image content is achieved with better colour combinations from their reduced colour gamut rather than by magical appearance of the new "unseen-so-far" colours. The synergy of simulation and daltonization provides "colour-barrier-free" environment. Possible future research should analyze existing daltonization methods, recommend optimal method(s) for every category of CVDs and define procedure and criteria for their evaluation.

References

1. Anagnostopoulos, C.N., Tsekouras, G., Anagnostopoulos, I., Kalloniatis, C.: Intelligent modification for the daltonization process of digitized paintings. Paper presented at the 5th international conference on computer vision systems, University of Bielefeld, Bielefeld, 2007
2. Asada, K., Kobayasi, M., Furukawa, S.: A portable assistance tool for color-deficient individuals with real-time color changing function. Paper presented at the AIC 2011 midterm meeting of the International Colour Association (AIC), International Colour Association, Zurich, 7–10 June 2011
3. Bosten, J.M., Robinson, J.D., Jordan, G., Mollon, J.D.: Multidimensional scaling reveals a color dimension unique to "color-deficient" observers. Curr. Biol. **15**(3), R950–R952 (2005)
4. Brettel, H., Vienot, F., Mollon, J.: Computerized simulation of colour appearance for dichromats. J. Op. Soc. Am. **14**, 2647–2655 (1997)
5. Celebi, M.E.: Improving the performance of K-means for color quantization. Image Vis. Comput. **29**(4), 260–271 (2011)
6. Celebi, M.E.: Partitional Clustering Algorithms. Springer (2015)
7. Chrome, D.: Chrome Daltonize extension. https://chrome.google.com/webstore/detail/chrome-daltonize/efeladnkafmoofnbagdbfaieabmejfcf Accessed 20 June 2014 (2011)
8. Fairchild, M.: Colour Appearance Models. Wiley, Chichester (2005)
9. Fidaner, P., Poliang, L., Ozguven, N.: Analysis of colour blindness. http://scien.stanford.edu/pages/labsite/2005/psych221/projects/05/ofidaner/colourblindness_project.htm (2005). Accessed 20 June 2005
10. Huang, J., Tseng, Y.C., Wu, S.I., Wang, S.J.: Information preserving colour transformation for protanopia and deuteranopia. IEEE Signal Process Lett. **14**, 711–714 (2007)

11. Huang, J., Wu, S., Chen, S.: Enhancing colour representation for the colour vision impaired. Paper presented at the ECCV international workshop on computer vision applications for the visually impaired in conjunction with European conference on computer vision, Marseille, 2008
12. Iaccarino, G., Malandrino, D., Percio, M.D., Scarano, V.: Efficient edge-services for colourblind users. Paper presented at the 15th WWW'06. 22–26 May 2006, Edinburgh, 2006
13. Ichikawa, M., Tanaka, K., Kondo, S., Hiroshima, K., Ichikawa, K., Tanabe, S., Fukami, K.: Web-page colour modification for barrier-free colour vision with genetic algorithm. LNCS. **2724**, 2134–2146 (2003)
14. Kotera, H.: Optimal daltonization by spectral shift for dichromatic vision. Proceedings of paper presented at the 20th colour and imaging conference, 2012
15. Kuhn, G.R., Oliveira, M.M., Fernandes, L.A.F.: An efficient naturalness-preserving image-recolouring method for dichromats. IEEE Trans. Vis. Comput. Graph. **14**, 1747–1754 (2008)
16. Ma, Y., Gu, X., Wang, Y.: Colour discrimination enhancement for dichromats using self-organizing colour transformation. Inf. Sci. **179**, 830–843 (2009)
17. MacAdam, D.: Visual sensitivities to colour differences in daylight. J. Opt. Soc. Am. **32**, 247–274 (1942)
18. Machado, G.M.: A model for simulation of color vision deficiency and a color contrast enhancement technique for dichromats. Dissertation, Universidade Federal do Rio Grande do Sul (2010)
19. Machado, G.M., Oliveira, M.M., Fernandes, L.A.F.: A physiologically-based model for simulation of colour vision deficiency. IEEE Trans. Vis. Comput. Graph. **15**, 1291–1298 (2009)
20. Machado, G.M., Oliveira, M.M.: Real-time temporal-coherent colour contrast enhancement for dichromats. Comput. Graph. Forum. **29**, 933–942 (2010)
21. Milić, N., Novaković, D., Zeljković, Ž.: Development of plug-in for optimizing colours of graphic and web designs for persons with dichromatic vision deficiencies. Paper presented at the AIC 2011 midterm meeting of the International Colour Association (AIC), International Colour Association, Zurich, 7–10 June 2011
22. Milić, N., Hoffmann, M., Tómács, T., Novaković, D., Milosavljević, B.: A content-dependent naturalness-preserving daltonization method for dichromatic and anomalous trichromatic colour vision deficiencies. J. Imaging Sci. Technol. **59**(1), in press (2015)
23. Mollon, J., Regan, B.C.:Cambridge Colour Test: Cambridge Research Systems. Cambridge (2000)
24. Morgan, M.J., Adam, A., Mollon, J.D.: Dichromats detect colour-camouflaged objects that are not detected by dichromats. Proc. Bio. Sci. **248**(1323), 291–295 (1992)
25. Ohta, N., Robertson, R.: Colourimetry: Fundamentals and Applications. Wiley, Chichester (2005)
26. Rasche, K., Geist, R., Westall, J.: Detail preserving reproduction of colour images for monochromats and dichromats. IEEE Comput. Graph. Appl. **25**, 22–30 (2005)
27. Ruminski, J., Wtorek, J., Ruminska, J., Kaczmarek, M., Bujnowski, A., Kocejko, T., Polinski, A.: Color transformation methods for dichromats. Paper presented at the 3rd International Conference on Human System Interactions (HSI), IEEE, Rzeszow, 13–15 May 2010.
28. Ruminski, J., Bajorek, M., Ruminska, J., Wtorek, J., Bujnowski, A.: Computerized color processing for dichromats. Hum. Comput. Syst. Interact. **98**, 453–470 (2012)
29. Sajadi, B., Majumder, A., Oliveira, M., Schneider, R.G., Raskar, R.: Using patterns to encode color information for dichromats. Trans. Vis. Comput. Graph. **17,** 1–13 (2011)
30. Sharma, G.: Digital Colour Imaging. CRC Press, London (2003)
31. Sharpe, T.L.: Color Vision: From Genes to Perception. Cambridge University Press, Cambridge (1999)
32. Vienot, F., Brettel, H., Ott, L., Ben M'Barek, A., Mollon, J.: What do colour-blind people see? Nature. **376**, 127–128 (1995)
33. Vienot, F., Brettel, H., Mollon, J.: Digital video colourmaps for checking the legibility of displays by dichromats. Colour Res Appl. **24**, 243–252 (1999)
34. Vischeck.Vischeck filter. http://www.vischeck.com. Accessed 20 June 2014 (2009)
35. Visolve. http://www.ryobi-sol.co.jp/visolve/en/. Accessed 20 June 2014 (2013)
36. Wyszecki, G., Stiles, W.S.: Color Science: Concepts and Methods, Quantitative Data and Formula. Wiley, Chichester (2000)

Chapter 12
Overview of Grayscale Image Colorization Techniques

Adam Popowicz and Bogdan Smolka

12.1 Introduction

Colorization is the process of introducing color to grayscale digital images. In these images, each pixel has a scalar value representing its intensity. However, the pixels of color images contain more complex, three-dimensional information. Depending on a color model, the pixel attributes correspond to a three-value color representation. For the purpose of grayscale image colorization, we strongly advise the use of color representation, which allows easy luminance extraction. This makes the colorization simpler— we only need to estimate two chrominance channels, while the luminance of the colorized image remains unchanged.

Despite plenty of color models, we cannot directly transform a one-dimensional grayscale value to a three-dimensional color space. We only get the luminance of the colorized image, while the two remaining color channels are still unknown. Therefore, the colorization may be performed in many different ways, achieving various outcomes. Additionally, it is difficult to objectively assess the final colorization results. We can compare the colorized version using standard image quality measures only if we have the color version of the image. The colorization of grayscale images without a reference has to be evaluated mainly subjectively.

In general, there are two main types of colorization frameworks. The first one is the semiautomatic approach. In such a range of methods, a user has to indicate the color hints in form of scribbles within the grayscale image. The scribbles enable the algorithms deciding which colors have to be put in corresponding parts of the image. The colorization procedure in the semiautomatic approach may be visualized as color spilling over the image areas starting from the indicated scribbles.

The automatic approach employs a color image that resembles the colorized grayscale picture. The algorithms rely on the similarities between the luminance

B. Smolka (✉) · A. Popowicz
Silesian University of Technology, Gliwice, Poland
e-mail: Bogdan.Smolka@polsl.pl

A. Popowicz
e-mail: adam.popowicz@polsl.pl

M. E. Celebi et al. (eds.), *Color Image and Video Enhancement*,
DOI 10.1007/978-3-319-09363-5_12

channels of both images, and the matching procedure compares the features and textures of an analyzed reference grayscale image. The algorithms require no user effort and it is impossible to modify or enhance their final results manually. The user may only change the parameters, without interfering in particular areas of the image. Hence, this type of algorithm may be disadvantageous in some applications.

At the beginning of this chapter, we review the methods presented in literature, including both the semiautomatic and automatic algorithms. Then, we concentrate on semiautomatic colorization, based on the so-called distance maps. We show four methods in detail. We also provide examples of the automatic approach, explaining its pros and cons. Finally, we describe the methodology of assessing colorization results with various quality measures. At the end of the chapter, we compare the colorization quality of the four semiautomatic methods, using images of different complexity.

12.2 Literature Overview

The literature provides many colorization approaches. The authors of [36] and [6] developed pioneer work in this field. Their idea is based on so-called color transfer, which enables a user to introduce colors to grayscale images with minimum effort. Such a method uses the similarities between a reference color image and a grayscale one. The analysis involves the textures in the swatches, indicated either by a user or automatically by the algorithm. This work also introduces the idea of manipulating only the chrominance channel, without changing the luminance level. It is to be noted that the concept of color transfer between images, employed in numerous colorization algorithms, was adopted from the article [27], where the authors present the procedure of color manipulation, so that the image may take on another image's look and feel.

Zhao et al. [40] provide a similar example of the automatic colorization approach. They initially convert the standard red-green-blue (RGB) into decorrelated $l\alpha\beta$ [29] space for both the reference and the target images. Then they calculate mean and standard deviation of luminance in a moving square window. Finally, by comparing two features previously obtained in each window, they transfer the colors from the parts of the reference image into the grayscale one.

In [9], we find further development of color transfer-based colorization. In the algorithm, the authors employ a robust supervised classification scheme that analyzes low-level features of the image swatches. After the segmentation, each pixel is assigned a color using a neighborhood matching. The authors also perform spatial filtering to improve final colorization outcomes.

The utilization of the color transfer concept [36] is also present in article [19]. Lipowezky shows how automatic colorization may be implemented for a very specific application: space images of the Earth. The advantage of using monochrome image sensors, mainly much higher spatial resolution and sensitivity, makes them suitable for satellite missions. Since the original concept [36] is not able to transfer color from images of different resolution, the author proposes Bayesian texture classification utilizing texture prototypes which enables automatic colorization of satellite images even without color reference images.

The idea of automatic transfer of selective colors from a reference image to a grayscale one is presented by Xiang et al. in [37]. The expectation-maximization (EM) algorithm was enhanced to model color distribution using the Gaussian mixture model (GMM). The images are segmented into an optimal number of regions before matching procedure. The authors also present an original metric of similarity between image regions which combines colorfulness and structural similarity.

An algorithm utilized for cartoon movies colorization was presented by Sýkora et al. in [33]. Their method is dedicated for cartoons where images contain two planar layers, stable background and dynamic foreground, which are colorized independently. While the reference frame is colorized manually by a human operator, the next frames are first segmented by a robust outline detector and then, the segments are compared and matched with the reference frames by so-called patch-based structural similarity.

The article [41] presents another automatic colorization method. Contrary to most of the methods, this one tries to use pattern continuity and spatial consistency in grayscale image, instead of considering the luminance properties. The segmentation is initially performed and then, the Gabor wavelets are applied to find the vector of image features. Finally, the colorization is performed by a modified classification algorithm.

The paper [35] presents the next example of fully automatic colorization method for infrared video sequences. The reference images are obtained from the flash-assisted photos during the registration of the infrared video. The authors transfer the colors by luminance matching in 3×3 local windows between infrared and the reference image, and then employ the technique of searching for the so-called key frame to find the moments when the scene changes significantly.

Several automatic colorization algorithms employing machine learning methods are widely discussed in [1]. All presented algorithms require an initial learning stage utilizing an auxiliary set of exemplary color source images. The authors make use of state-of-the-art machine learning methods, mainly the support vector machine (SVM). They try to predict the color exploiting the knowledge derived from various sources of information included within the image.

The most popular colorization algorithms of grayscale images or movies involve the semiautomatic colorization approach. Several such methods have already been published. In [17], the minimization technique is applied for the purpose of colorization. Levin et al. assume that neighboring pixels with similar intensities should also have similar colors. When inserting the scribbles, the algorithm uses standard optimization methods to achieve the aforementioned assumption while preserving required colors within the scribbles. Such algorithm were also used for video colorization [26], as well as for biomedical digital image visualization [32].

Sapiro, inspired by [17], in his article [31] proposes the use of the optimization formula which leads to realistic colorization results. The optimal colorization solution is obtained from the partial differential equation, which is derived from the assumption, that the difference between luminance and chrominance gradients should be minimal. The advantage of the algorithm is its simplicity and efficiency, as the differential equation is of Poisson type and can be solved efficiently by any Poisson solver.

The article [8] presents a method which uses seed pixels propagation. The local Markov property is utilized to properly transfer scribbled seed pixels. The colors are propagated from a seed pixel to its neighbors by minimizing the color difference of adjacent pixels. The use of the local Markov property for colorization may be also found in another article by the same author [7].

In [38], Yatziv and Sapiro use the Dijkstra algorithm [3] to find the cost of the shortest path between two image pixels. They calculate the cost by integrating the squared difference of intensities between pixels on the path. The resulting chrominance in a grayscale pixel is calculated as a weighted average of scribbled colors, where the shortest path analysis provides the weights. The authors also suggest reducing the number of colors during the final blending, hence their algorithm does not require high-computational burden.

Another example of employing the Dijkstra algorithm is presented in article series [11–13]. Kawulok and Smolka show that we can calculate the cost of a digital path by using different approaches. While for the plain parts of the image we prefer to choose paths with the smallest intensity gradients, for the complex textures, we select a path with high-luminance variations. Either a user or the algorithm may decide which one to apply. By proper selection of cost function, the authors achieved high-quality colorization outcomes for images containing both the plain and complex parts.

A somewhat similar approach was adopted for manga colorization. Yingge Qu et al. in [25] presented a semiautomatic colorization which is based on pattern continuity. In manga drawing, hatching and screening techniques are widely used to show different effects like structures, shading or reflectance, thus it is required that the regions of the same textures should also have similar colors. The authors utilize the Gabor wavelet transform to obtain the structure features. While most of the methods are universal, the usefulness of this technique is mainly limited to a very specific application.

The next derived approach is presented in article [14]. The color blending is performed using the procedure of Random Walk with Restarts. A virtual particle starts its walk from a scribbled pixel and the probability of reaching a given pixel is calculated. As the obstacle in form of edges in the image is encountered, it stops the particle and the chance of reaching further pixels is reduced. In this way, the authors obtain an interesting measure of a distance between pixels in the image. Finally, utilizing this kind of distance, the idea of color blending presented in [38] leads to satisfactory results. The advantage of this method, in comparison with the classic concept [38], is the very high sensitivity to boundaries between regions, even if there are some small inlets, which result in color flooding in the original method [38]. It is to be noticed that the random walks technique is also utilized by authors of [20]. The initial coarse color propagation and farther image segmentation for accurate colorization are both performed employing the aforementioned algorithm.

Lagodzinski and Smolka in [16] present the morphological distance transformation as a possible way to obtain the distance between a seed and any other image pixel. The authors use the double-scan algorithm, which is able to obtain a very good estimation of Euclidean distance between the image points. We may modify the distance, so that it makes the combination of intensity differences and topographic distance. The final color of a pixel is also calculated as a weighted average of all scribbled colors; thus, the larger the distance from a given scribble, the smaller the influence of its color on the final result.

Zhang et al. [39] also present the distance transformations. They perform fuzzy clustering of pixels, where they choose the initial centers of the clusters as scribbled pixels. The fuzzy membership is assigned using the aforementioned distance transformation approach.

In [10], a somewhat different design based on the image segmentation, is presented. The so-called "rainwater simulation technique" is a segmentation method of properly clustering the parts of the image into regions with the same colors. After initial over-segmentation, the authors continue merging the clusters, until the number of clusters equals the number of indicated scribbles.

Luan et al. in [21] try to deal with complex natural images where both, smooth regions and those with complicated textures, are present. They describe the colorization system, where the user indicates exemplary regions, which should be colorized with a given color. The algorithm not only floods the colors to neighboring pixels of similar intensity but also finds remote pixels of similar textures. In the method, authors incorporate two similarity measures based on the intensity continuity and on the textural features.

The original concept of automatic scribble generation is presented in [4]. The authors suggest to reduce operator interaction only to select proper colors for automatically inserted, reliable, and precise scribbles. First, the algorithm performs over-segmentation using graph-based image segmentation [5], and then the scribbles are placed within the regions of especially high-information density. This is achieved by defined spatial distribution entropy and is based on a number of different intensity levels present in the given image patch.

Konushin and Vezhnevets in [15] present the idea of semiautomatic colorization which is concentrated on the computational burden. For most of the algorithms, colorization enhancement by an operator is a time-consuming procedure, since the whole colorization has to be recomputed. In the proposed algorithm, authors utilize the idea of coupled map lattices which evolutionary nature allows for fast recolorization. Their algorithm can be exploited, e.g., for changing colors of some regions within color images.

The authors of [18] suggest to use of graph topologies and edge weights to obtain the gradients of intensities in grayscale images. In contrast to previous methods based, for example, on distance transformations, the proposed idea utilizes the concept of an extended pixel neighborhood which may include more than 8 pixels (9×9 pixels patches are considered in presented examples). This leads to a nonlocal colorization design and improves the outcomes in some cases.

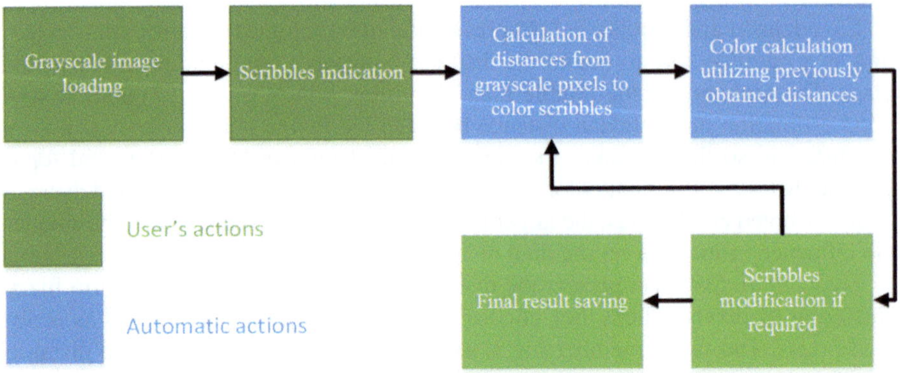

Fig. 12.1 Schema of semiautomatic colorization steps

The use of an image matting procedure for complex images is adopted for colorization purpose by Chen et al. in [2]. The algorithm requires user interaction in the form of an exemplary background and foreground pixels indication. By solving maximum a posteriori problems, the rest of the pixels (unknown) are also classified into either the background or the foreground. The concept of Welsh [36] is utilized in the last part of this algorithm.

12.3 Basics of Semi-Automatic Grayscale Image Colorization

The framework for nearly all semiautomatic algorithms of image colorization is usually similar. Figure 12.1 presents the schema of such algorithms. The procedures include several actions performed by both a user and a computer system.

At the beginning, a user should introduce the color scribbles for the most crucial parts of the image. This initial procedure may be performed with the use of the paint-like interface, as in Fig. 12.2. In some cases, the standard color palettes may be insufficient, thus we recommend retrieving colors from other images. It may be especially useful when colorizing portraits, due to the unique color of human skin. Some other options, like zooming, undo, or eraser, should be also considered when developing colorization software.

Although the next part of the colorization differs from method to method, the general idea remains the same. For each pixel not included in the inserted scribbles, the algorithms seek the pseudo-distance between each of the given scribbles or colors. Such distances represent the membership of the a pixel to the scribble, as well as enable a further decision-making process while selecting proper colors or their mixture. The distances are usually calculated by the luminance differences encountered on the digital path between a scribble and a pixel. Sometimes, an additional penalty is added as the topographic distance increases. Subsections of this

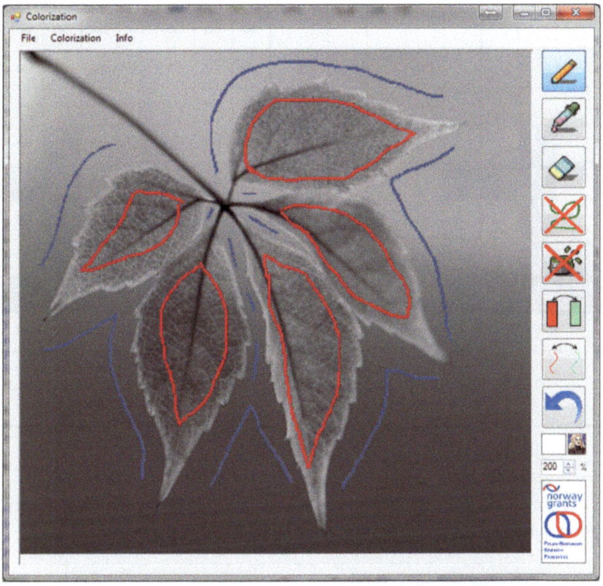

Fig. 12.2 Example of a colorization software layout. Grayscale image with indicated *color scribbles* and a paint-like toolbox

chapter dedicated to individual algorithms provide further details of implementing a distance map calculation.

The final selection of the color for a given pixel is performed using different approaches. One of them is to calculate the weighted average of each indicated color, where the distances are used as weights. When working with YCbCr color space, we calculate the average of Cb and Cr channels independently. Additionally, we may calculate the color as the mixture of only two or three closest colors to reduce the computation time. Additionally, we can avoid the leakage of colors from distant scribbles by modifying the weights, so the smallest distances are promoted.

We can analyze the outcomes of colorization to change the colors or the positions of previously indicated scribbles. Here, the computation time is very important. If the algorithm is computationally efficient, we can quickly apply the user's" feedback. Moreover, it is possible to perform colorization and to present the results after each scribble indication. In this real-time fashion, the colorization process is the most efficient and user-friendly.

12.4 Distance Transformation Based Colorization

Distance transformation is one of the ways to measure the distance between a scribble and a pixel [28]. Although several methods based on this approach have already been published, in this section we would like to concentrate on one of them [16].

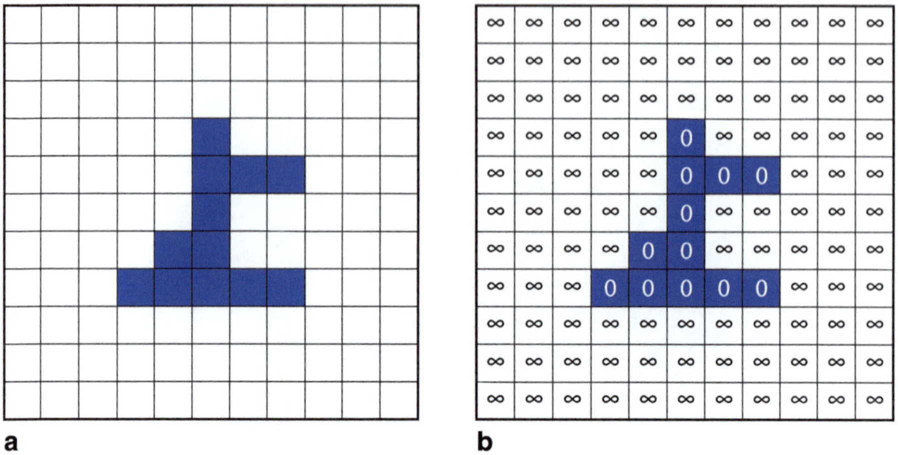

Fig. 12.3 Initial stage of exemplary distance transformation. **a** Scribble position within image matrix. **b** Initial values of the array of distances

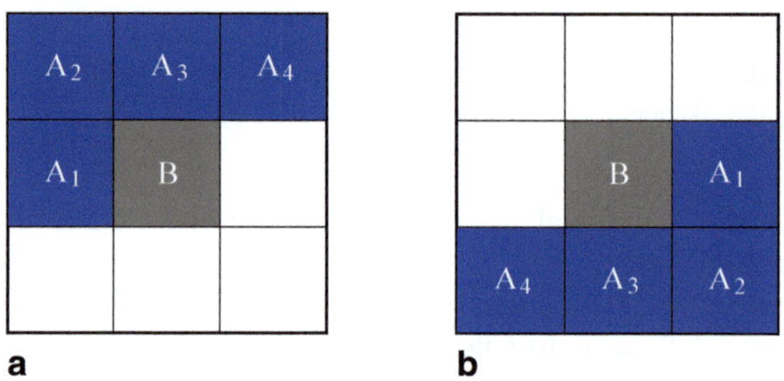

Fig. 12.4 Local 2 × 3 window in double-scan algorithm. **a** Window for the first scan. **b** Window for the second scan

The classic concept of such transformation relies on the so-called double-scan algorithm, which provides a very good estimation of Euclidean distance. In the array of distances, we may present the scribbles as zeros while other pixels are set to infinity (see Fig. 12.3). It is a starting point for this transformation algorithm.

On such an initial array, a double-scan algorithm is performed [28]. It consists of a few operations in the local sliding window. In the first scan, the window shifts from left to right and top to bottom. In the second scan—from bottom right to bottom left, toward the top. Figure 1.5 shows simple 2 × 3 local window utilized for the first and the second scan.

In the simplest version of the algorithm, we perform the following operation of minimum finding:

$$B \leftarrow \min(A_i + 1, B),\ i = 1, 2, 3, 4. \tag{12.1}$$

Figure 12.5 presents the results of double-scan operations, performed on the exemplary scribble indicated in Fig. 12.3.

∞	∞	∞	∞	∞	∞	∞	∞	∞	∞	∞
∞	∞	∞	∞	∞	∞	∞	∞	∞	∞	∞
∞	∞	∞	∞	∞	∞	∞	∞	∞	∞	∞
∞	∞	∞	∞	∞	0	1	2	3	4	5
∞	∞	∞	∞	1	0	0	0	1	2	3
∞	∞	∞	2	1	0	1	1	1	2	3
∞	∞	3	2	0	0	1	2	2	2	3
∞	4	3	0	0	0	0	0	1	2	3
5	4	1	1	1	1	1	1	1	2	3
5	2	2	2	2	2	2	2	2	2	3
3	3	3	3	3	3	3	3	3	3	3

a

5	4	3	3	3	3	3	3	3	4	4
5	4	3	2	2	2	2	2	3	3	4
5	4	3	2	1	1	1	2	2	2	3
5	4	3	2	1	0	1	1	1	2	3
5	4	3	2	1	0	0	0	1	2	3
4	3	2	1	1	0	1	1	1	2	3
3	2	1	1	0	0	1	1	1	2	3
3	2	1	0	0	0	0	0	1	2	3
3	2	1	1	1	1	1	1	1	2	3
3	2	2	2	2	2	2	2	2	2	3
3	3	3	3	3	3	3	3	3	3	3

b

Fig. 12.5 Results of distance transformation after first and second scan. **a** First scan results. **b** Second scan results

The basic version of a double scan (see above) provides information about the distance between the scribble and each of the image pixels. To adapt this algorithm to colorization purposes, the authors [16] suggest to use the luminance difference between the pixels in a local window. This proposition aims to enlarge the distance when the intensity difference, encountered between subsequent pixels, is significant. We may employ this concept by modifying Eq. 12.1 in the following manner:

$$B \leftarrow \min(A_i + \delta + \Delta_i, B), \quad \Delta_i = |I(A_i) - I(B)|^{\lambda}, \; i = 1, 2, 3, 4, \qquad (12.2)$$

where: $I(A_i)$ and $I(B)$ are the grayscale values at the corresponding pixels A_i and B, whereas λ and δ are constant coefficients allowing to balance the metric and the luminance distance.

Within the smooth areas of the image, the distance increases slowly and proportionally to the Euclidean distance, while within the sharp edges, it rises very quickly. Thus, only the areas of similar intensity are given high weights in the final color blending part. An example of distance maps for two scribbles is depicted in Fig. 12.6.

To obtain the color C_j of a pixel at position j, the authors of [16] inspired by the chrominance blending concept [38], perform the weighted averaging that should be calculated independently for both of the chrominance channels, as the luminance should be preserved:

$$C_j = \frac{\sum_{i=1}^{N} C_i \, \exp(-d(i, j)/\sigma)}{\sum_{i=1}^{N} \exp(-d(i, j)/\sigma)}, \qquad (12.3)$$

where C_i is i-th seed color, N is the number of scribbles, $d(i, j)$ is the distance between the i-th scribble and a pixel at j-th position on the image domain, σ is the parameter of the color blending extent.

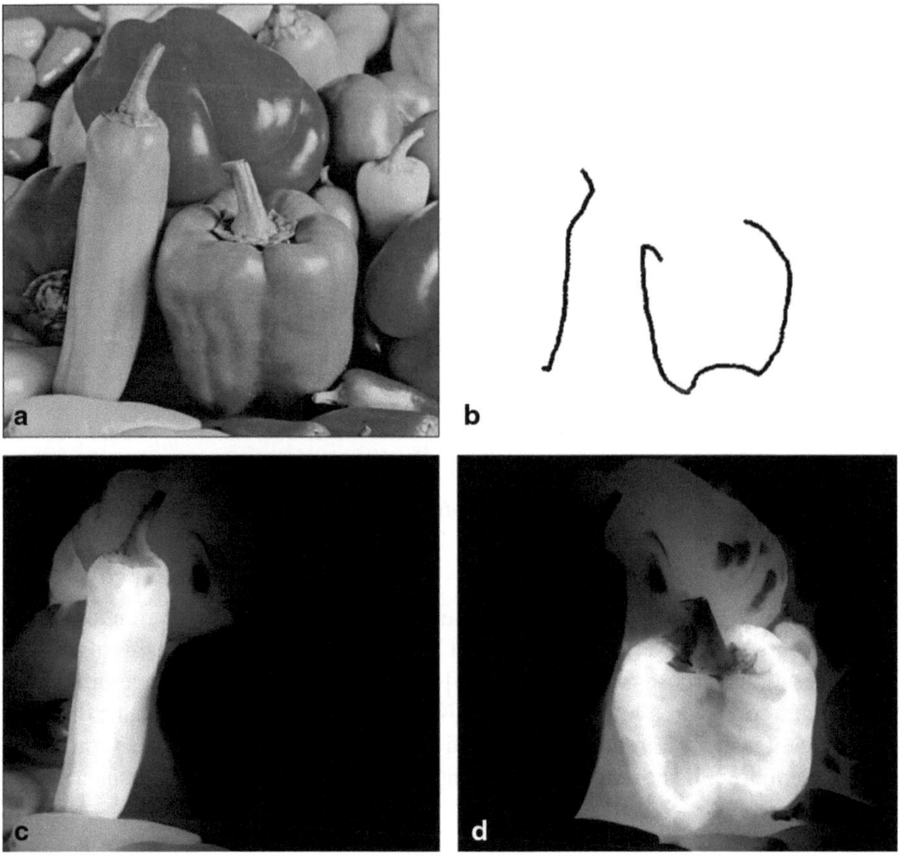

Fig. 12.6 Exemplary distance maps for two colorized images (the darker the image pixels, the greater the distance to the scribble). **a** Grayscale image. **b** Scribbles. **c** Distance map for the left scribble. **d** Distance map for the right scribble

12.5 Chrominance Blending Colorization

In the article [38], we may find another way to obtain the distance map representing the level of pixel membership to a given scribble. The authors suggest to use the Dijkstra algorithm [3] to properly estimate the path cost between the pixel and the scribble.

The original concept of the Dijkstra algorithm is applied to solve the shortest path problem in the graph, where each connection between the nodes has an individual transition cost. We can effectively implement this classic algorithm for distance map calculation. To do so, we treat each pixel in the image as a node. The cost of movement from a pixel to its neighbor is proportional to the absolute value of the difference between the luminance of the pixels. This way, each image pixel eventually has the value of the smallest cumulative luminance gradient encountered on the path originating in the scribbled feature.

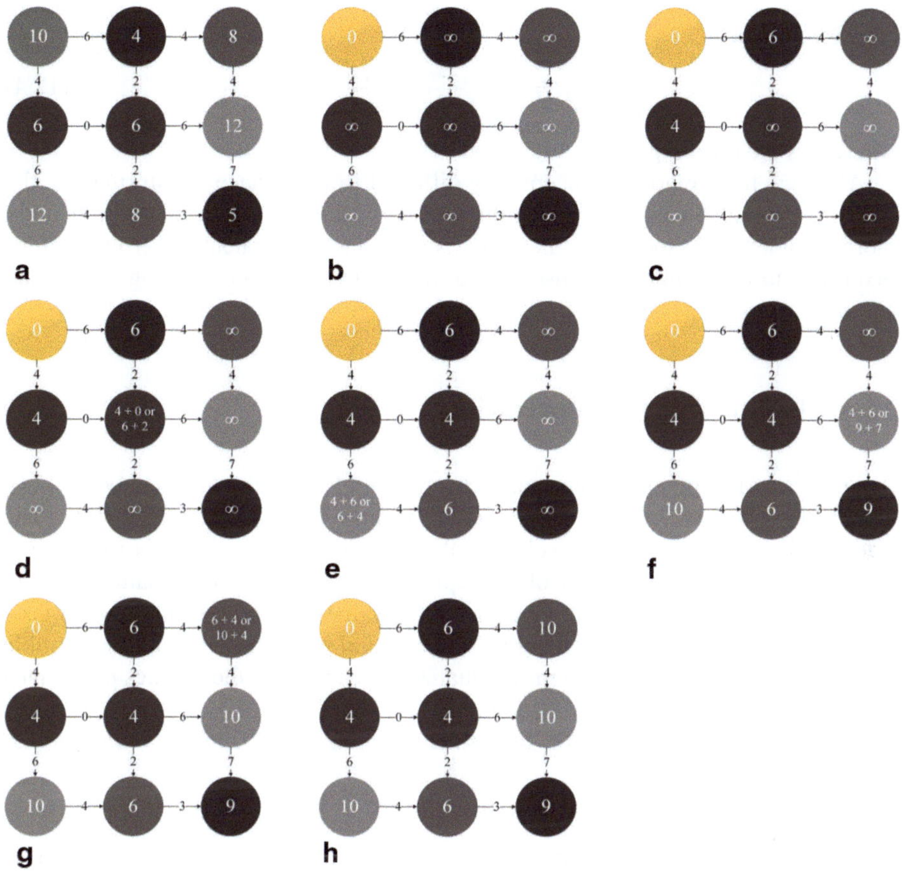

Fig. 12.7 Subsequent stages of Dijkstra algorithm on digital image. **a** Initial image with given intensities and transition costs. **b** Initialization of distance values, the *yellow* node is the seed pixel. **c–h** The stages of shortest path calculation

The Dijkstra algorithm starts from the seed node and analyzes the neighbors. Each neighboring pixel pair is linked by a corresponding path cost, which is proportional to the intensity difference (Fig. 12.7a). Initially all the nodes, except for the seed node, are given infinite distance (Fig. 12.7b). Then, the distance is calculated for all neighbors of the seed pixel; however, only the one with the shortest distance is assigned. If there is more than one available link, the shortest distance is always selected (e.g., center pixel in Fig. 12.7d). The procedure recurs as long as any unassigned pixels are left.

In [38], to obtain modified weights, the authors propose to raise the distance to a negative power, instead of calculating the exponents in weighted averaging. The value of the power (b) indicates how smooth the color blending is in the final colorized image:

$$C_j = \frac{\sum_{i=1}^{N} C_i \, d(i,j)^{-b}}{\sum_{i=1}^{N} d(i,j)^{-b}}, \tag{12.4}$$

where b is the blending level, (suggested range: $1 < b < 6$).

Although the method is conceptually different from distance transformation, the resulting distance maps are very similar (see e.g., 12.6). However, it is significantly more complicated, especially when the distance has to be calculated from a multi-pixel structure (scribble) because of the large number of neighboring pixels and sorting operations.

12.6 Isoline Image Colorization

Another way to obtain distance maps from scribbled pixels is presented in [24]. The *Isoline Colorization* uses the concept of isolines in the geographical maps. The interpretation of the intensity level is similar to the elevation or other property visualized on the maps. While the isolines represent the locations of the same value, in this colorization algorithm, the pixels have the values that correspond to the maximum intensity difference encountered on the digital path, starting from the scribble and ending at a given pixel. We may visualize the concept of the isolines in distance maps, as in Fig. 12.8.

The corresponding definition of distance calculated in *Isoline Colorization* is given as:

$$d(i,j) = \min_{\Phi} \left(\max |I_m - I_n|, \; m \in M_i, \; n \in N_j \right), \tag{12.5}$$

where $d(i,j)$ is the distance between i-th scribble and a pixel at position j, Φ is a set of all digital paths between pixel j and i-th scribble, I_m and I_n are pixel intensities at respective positions m and n, M_i is a set of all pixels included in i-th scribble, N_j is a set of all pixels on digital path linking pixels at positions m and j, ($N_j \in \Phi$).

The following one-dimensional example helps to understand the idea of isoline distance calculation. For a simple one-dimensional array, isoline distance is calculated for each cell, as presented in Fig 12.9. We need to note that although the last pixel has the luminance of 9, and the intensity difference from the seed pixel with intensity 10 is only 1, there is a larger intensity difference on the path (the pixel number 6 has isoline distance of 3) and, of course, the distance cannot decrease.

The calculations for two-dimensional image array are much more complicated due to plenty of possible paths between the pixels. To mitigate this, the authors suggest an approach similar to the double-scan algorithm. The idea involves the running square windows, identical to the ones depicted in Fig. 12.4. In contrast to *Distance Transformation-based Colorization*, the algorithm additionally transfers the intensity of the scribbled pixel. Thus in each step, while sliding the local window, we get information about the intensity of the so-called seed pixel located at the beginning of a current digital path. In this way, the authors of [24] achieve a very good estimation of previously defined isoline distance (Eq. 12.5). The distances are utilized in the final color blending as presented in Eq. 12.4.

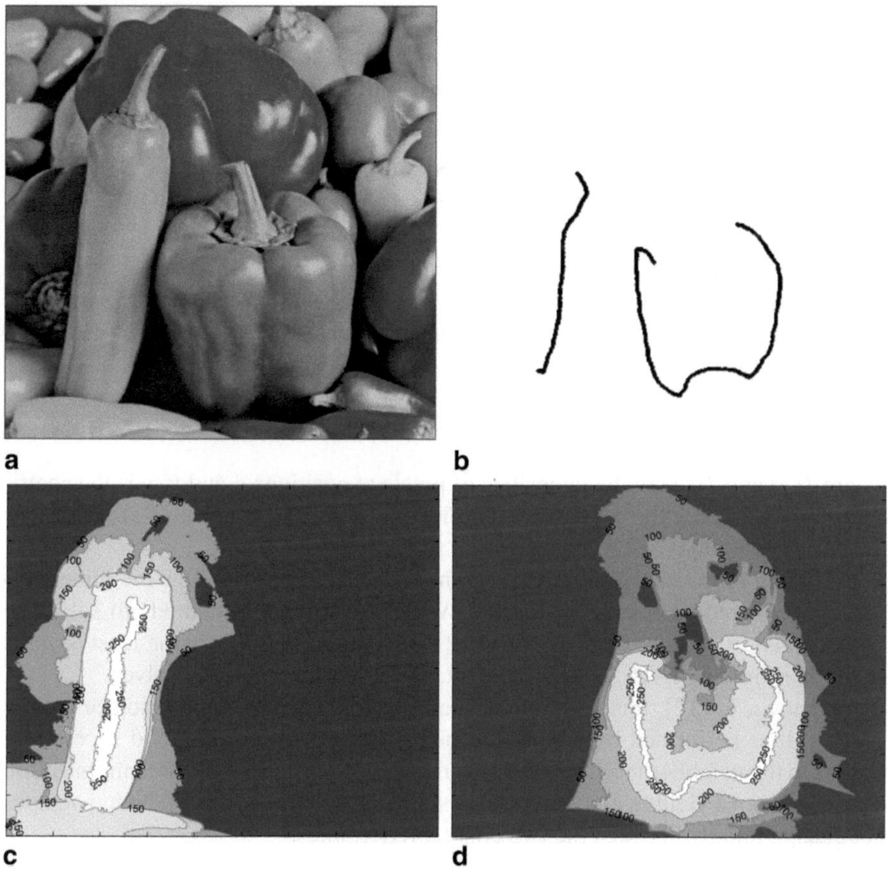

Fig. 12.8 Visualization of isoline concept for distance maps. **a** Original grayscale image. **b** Indicated scribbles. **c** Isoline representation of distance map for the first scribble. **d** Isoline representation of distance map for the second scribble

Intensity	10	9	10	12	8	7	9
Isoline distance	0	1	1	2	2	3	3
Pixel number	1	2	3	4	5	6	7

Fig. 12.9 Isoline distance calculations. Scribbled seed value is filled with *gray*

12.7 Colorization Using Optimization

The methods described above include semiautomatic image colorization using distance maps for each scribble indicated by a user. There is another way to achieve high-quality colorization results without the distance maps concept [17]. It is based on the optimization approach, assuming the constraints that the neighboring pixels with similar luminance should have the same color. The authors work in YUV color space and suggest the following formula for minimization:

$$J(U) = \sum_j \left(U(j) - \sum_{k \in \mathcal{N}(j)} w(j,k) \cdot U(k)) \right)^2 , \tag{12.6}$$

where $U(j)$ is the U chrominance component of a pixel at position j, $\mathcal{N}(j)$ is the set of pixels adjacent to the pixel at position j, $w(j,k)$ is the measure of the intensity similarity between pixels at respective position j and k and may be defined using the Gaussian function:

$$w(j,k) = \exp\left(- \left[I(j) - I(k) \right]^2 / 2\eta^2 \right), \tag{12.7}$$

where $I(j)$ and $I(k)$ are the intensities of pixels at positions j and k, η is the coefficient responsible for the degree of color blending (the lower the value—the stronger the color separation).

To obtain the second chrominance component, we have to follow the same minimization procedure for the V channel. We can use several methods to solve the linear equation system, defined in Eq. 12.6. The authors [17] use the least squares solver built in the MATLAB environment, as well as the multigrid solver.

As shown in [17], this concept can be also used for the colorization of simple movies. It is performed by adding the pixels from the preceding and succeeding frames to the neighborhood $\mathcal{N}(j)$ of a current pixel. In such a case, the minimization described above yields satisfying results, since the color of pixels in the movie does not change rapidly between the two consecutive frames.

12.8 Examples of Semi-Automatic Image Colorization

Semiautomatic colorization has several important applications such as: old black and white photography, biomedical imaging, night vision colorization, or astronomical image colorization. Here, in Fig. 12.10 we present some sample results obtained with the four aforementioned algorithms: *Distance Transformation-based Colorization* [16], *Chrominance Blending Colorization* [38], *Isoline-based Image Colorization* [24], and *Colorization Using Optimization* [17]. Although the methods differ one from the other, we have to note that all the algorithms provide very realistic and high-quality color images.

Additionally, we may adapt the semiautomatic colorization to the image segmentation purposes. If a human operator indicates the initial scribbles, it is possible to flood the area of interest by pointing out some important features. It is important in medical imaging, where we can highlight the given areas and calculate significant coefficients such as the level of some pathological changes. As an example, we show the colorization of two ultrasound images of joint inflammation. Figures 12.11a and

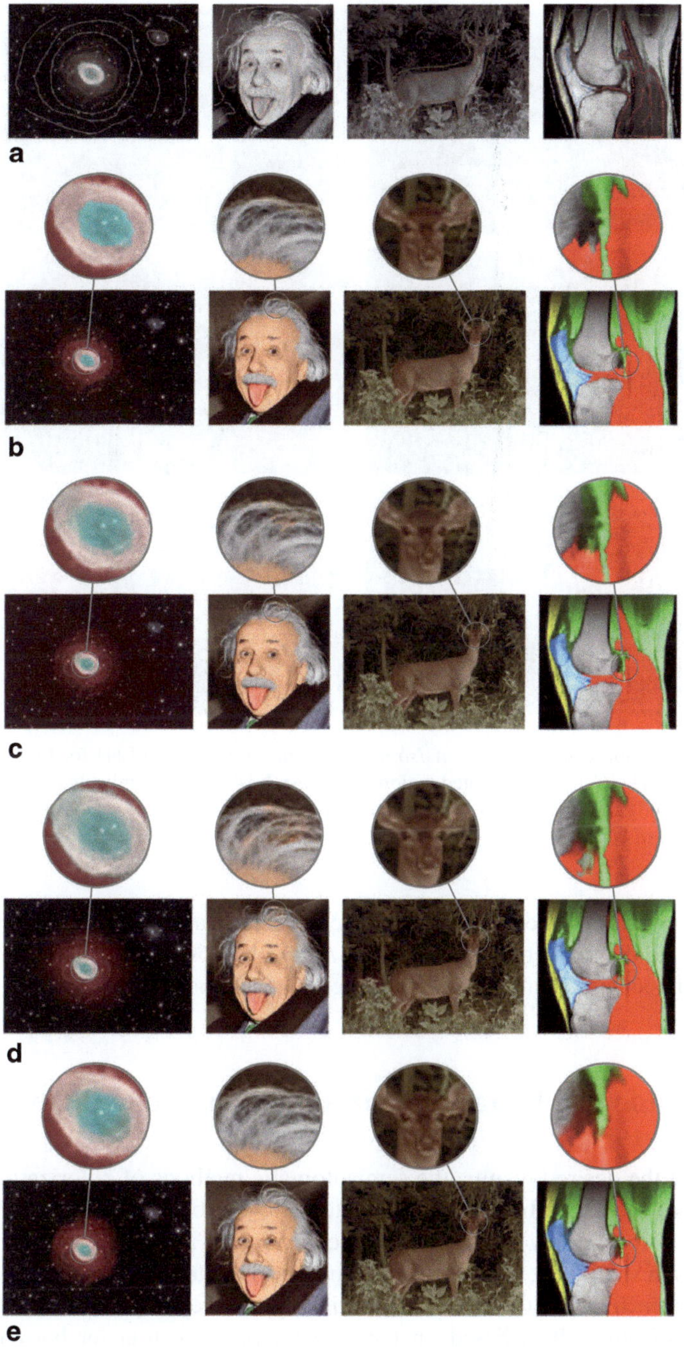

Fig. 12.10 Exemplary results of semiautomatic image colorization. **a** Original grayscale images with indicated scribbles. **b** Outcomes of distance transformation based colorization [16]. **c** Outcomes of chrominance blending colorization [38]. **d** Outcomes of isoline based image colorization [24]. **e** Outcomes of colorization using optimization [17]

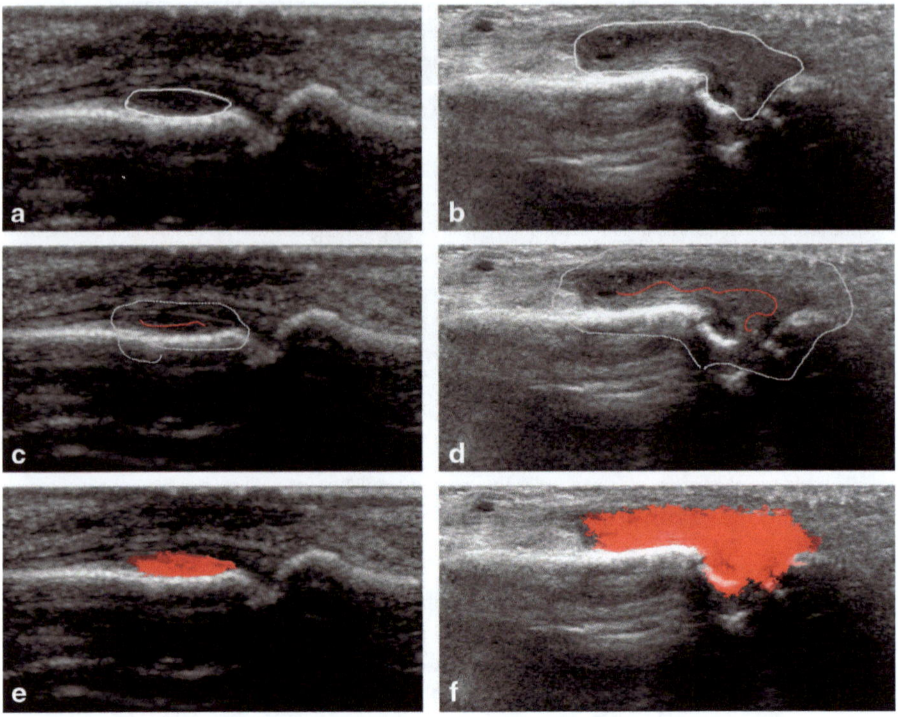

Fig. 12.11 Examples of utilization of *Isoline Colorization* algorithm [24] for highlighting joint inflammation. **a** USG 1 with indicated region of interest. **b** USG 2 with indicated region of interest. **c** Scribbles configuration for USG 1. **d** Scribbles configuration for USG 2. **e** Colorization of USG 1. **f** Colorization of USG 2

b depict the areas delineated by the physician. The images below show the inserted scribbles and the final colorization result. As can be observed, the highlighted areas have similar shapes as the regions indicated by the expert.

12.9 Automatic Image Colorization Using Color Transfer

To achieve the desired results, the semiautomatic methods of colorization requires the effort of the users. In contrast, the automatic approach does not need any hints about the colors in any part of the image. However, it requires a color reference image which should be, to some extent, similar to the target grayscale image.

The *Color Transfer Colorization* [36] is a typical representative of a fully automatic colorization. It is based on the concept of color transfer between images described in pioneer work [27]. The authors of [27] suggested to utilize $l\alpha\beta$ decorrelated space [29]. They applied simple statistics transformations in each separate, decorrelated color channel of the target image, so that the source and the target had a similar look. The method was extended further for the purpose of colorization in [36]. In this work, the authors analyzed luminance similarities between the source and the target grayscale image, and then they transferred chrominance values α and β from selected regions of the source image.

The basics of the mentioned colorization concept [36] are, in general, much simpler than those of the semiautomatic methods. To achieve better colorization effects, the standard color space of analyzed images is converted into $l\alpha\beta$ decorrelated space [29]. We may perform the corresponding conversion from RGB space by the following matrix operations:

$$\begin{bmatrix} L \\ M \\ S \end{bmatrix} = \begin{bmatrix} 0.3811 & 0.5783 & 0.0402 \\ 0.1967 & 0.7244 & 0.0782 \\ 0.0241 & 0.1288 & 0.844 \end{bmatrix} \begin{bmatrix} R \\ G \\ B \end{bmatrix}, \tag{12.8}$$

$$\begin{bmatrix} l \\ \alpha \\ \beta \end{bmatrix} = \begin{bmatrix} 1/\sqrt{3} & 0 & 0 \\ 0 & 1/\sqrt{6} & 0 \\ 0 & 0 & 1/\sqrt{2} \end{bmatrix} \begin{bmatrix} \log(L) \\ \log(M) \\ \log(S) \end{bmatrix}. \tag{12.9}$$

To convert the image back to RGB space, the following transformation has to be applied:

$$\begin{bmatrix} L' \\ M' \\ S' \end{bmatrix} = \begin{bmatrix} 1 & 1 & 1 \\ 1 & 1 & -1 \\ 1 & -2 & 0 \end{bmatrix} \begin{bmatrix} \sqrt{3}/3 & 0 & 0 \\ 0 & \sqrt{6}/6 & 0 \\ 0 & 0 & \sqrt{2}/2 \end{bmatrix} \begin{bmatrix} l \\ \alpha \\ \beta \end{bmatrix}, \tag{12.10}$$

$$\begin{bmatrix} R \\ G \\ B \end{bmatrix} = \begin{bmatrix} 4.4679 & -3.5873 & 0.1193 \\ -1.2186 & 2.3809 & -0.1624 \\ 0.0497 & -0.2439 & 1.2045 \end{bmatrix} \begin{bmatrix} 10^{L'} \\ 10^{M'} \\ 10^{S'} \end{bmatrix}. \tag{12.11}$$

As the source and the target image may have different luminance ranges (e.g., the histogram may be shifted or extended), the authors suggest that we initially remap the target image luminance channel:

$$l'_T = \frac{\sigma_S}{\sigma_T}(l_T - \mu_T) + \mu_S, \tag{12.12}$$

where l_T and l'_T are the luminance of the target image pixel before and after the remapping, μ_T and μ_S are the mean luminance of target (T) or source (S) image; σ_T, σ_S are the standard deviations of respective images.

In the next step, the reference image is divided into 5×5 pixel parts, which consist of various objects and textures. The selection of the parts is randomized and their amount is much lower than the total number of such segments, thus it significantly reduces the computational burden. The authors of [36] claim that about 200 random samples are enough to cover most of the details within analyzed images. For each such sample, both the mean and the standard deviation of luminance are calculated. Moreover, the color of the central pixel is stored, so that the 5×5 pixel part is saved in a form of a three-component feature: color, the mean luminance, and the standard deviation of luminance in a segment.

In the next part, before transferring the α and β color components, the algorithm tries to match the samples from the target image with the ones from the source. While the authors of original work [36] utilized the measure of similarity based on the sum of luminance and standard deviation difference, more flexible formula may be found in a subsequent article based on color transfer [40]:

$$D = k \left|1 - \frac{\mu_T}{\mu_S}\right| + \left|1 - \frac{\sigma_T}{\sigma_S}\right|, \tag{12.13}$$

where D is a measure of similarity between the parts of the images, k is the parameter responsible for balancing the impacts of luminance deviation σ, and the mean value μ on the resulting measure of similarity.

Figure 12.12 presents several astronomical, biomedical, and multimedia examples of automatic colorization. We need to note that, although it does not require human interaction, the automatic colorizing provides worse results than semi-automatic colorizing. The parts of the colorized image containing undesirable color usually cannot be corrected or enhanced manually. Hence, in [36], the authors additionally propose color transfers between user-defined swatches. First, a human operator indicates the areas, both in the reference and the target image, then the pixel matching occurs only between the given swatches. With these additional interactions, we may achieve some enhancement of final colorization outcomes.

The automatic colorization usually works better for images containing clearly distinguishable features. Sometimes, like in the biomedical image in the middle of Fig. 12.12, automatic colorization can be useful, because it would be extremely hard to manually scribble all the objects. However, often it is difficult to find matching reference color images, thus the applications of such a range of methods are usually limited.

12.10 Colorization Quality Assessment

The evaluation of colorization results is important, especially when the colorization algorithm can be controlled with some parameters. The quality measures in such cases should indicate the range of values of the parameters which leads to the optimal colorization performance. Hence, the role of proper colorization results evaluation is an important step in the development of novel colorization algorithms. Ideally, the measures should be consistent with human perception, which is very sensitive to any kind of disturbances within the color image.

Peak signal-to-noise ratio (PSNR) is one of the most popular quality measures of comparing the enhanced images with their original version. We may implement this fidelity measure to assess the quality of both the color and the grayscale images. For the colorization purpose, the PSNR of color image is derived from mean square error (MSE) as follows:

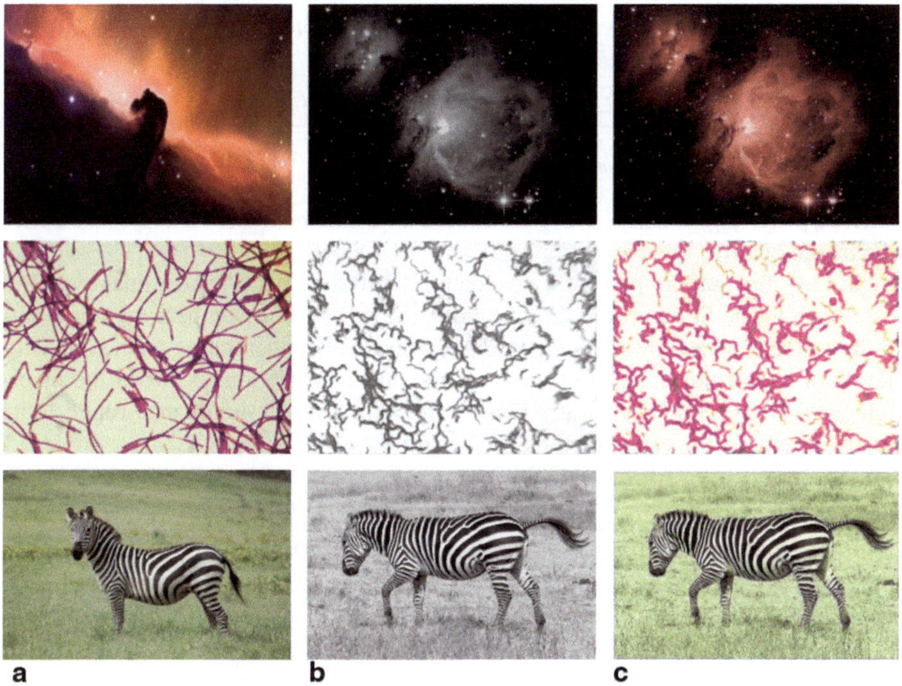

Fig. 12.12 Colorized images obtained by automatic algorithm [36]. **a** Color source image. **b** Target grayscale image. **c** Colorized image

$$\mathrm{MSE} = \frac{1}{MNQ}\sum_{q=1}^{Q}\sum_{n=1}^{N}\sum_{m=1}^{M}\left[O_q(n,m) - I_q(n,m)\right]^2, \tag{12.14}$$

$$\mathrm{PSNR} = 10\log_{10}\left(\frac{255^2}{\mathrm{MSE}}\right), \tag{12.15}$$

where: Q is the number of color channels (3 for RGB space), M and N are the image width and height, O is the original reference image and I is the analyzed image, 8-bit color channel resolution is assumed.

This classic quality measure is extremely sensitive to the offsets in color or luminance channels. It is possible that, although the visual impression of the image does not change, the PSNR value may be much lower, indicating the poor quality. To compensate for such effects, we should consider initial remapping of luminance to match the intensity range of the reference image, as in Eq. 12.12.

Normalized color difference (NCD) is another way to compare the images. This metric is focused on the distances between the colors in a given color space. It is especially useful when employing the CIE L*a*b* color space, since it considers the color difference in a similar way the human does. Taking into account the CIE

L*a*b* color space, we use the following equation for NCD calculation [22,23,30]:

$$\mathrm{NCD} = \frac{\sum_{m=1}^{M}\sum_{n=1}^{N}\sqrt{\sum_{q=1}^{Q}\left[O_q(m,n) - I_q(m,n)\right]^2}}{\sum_{m=1}^{M}\sum_{n=1}^{N}\sqrt{\sum_{q=1}^{Q}\left[O_q(m,n)\right]^2}}, \tag{12.16}$$

where $q_1 = \mathrm{L}^*$, $q_2 = \mathrm{a}^*$, $q_3 = \mathrm{b}^*$. We should note that the lower the NCD, the better the image quality.

In [34], the authors present a conceptually different approach to image quality assessment. The Structural Similarity index (SSIM) aims to compare the images as if they were analyzed by a human observer. Instead of concentrating on the intensity levels of each color channel, it analyzes the cross correlations between the pixel' intensity. High resistance to any biasing or scaling of the luminance range is its main advantage. We calculate the SSIM as follows:

$$\mathrm{SSIM} = \sum_{m=1}^{M}\sum_{n=1}^{N}\frac{[2\mu_O(m,n)\mu_I(m,n) + c_1]\,[2\sigma_{OI}(m,n) + c_2]}{\left[\mu_O(m,n)^2 + \mu_I(m,n)^2 + c_1\right]\left[\sigma_O(m,n)^2 + \sigma_I(m,n)^2 + c_2\right]}, \tag{12.17}$$

where:

$$\mu_O(m,n) = Z\sum_{k,l=-w}^{w} O(m+k, n+l), \quad \mu_I(m,n) = Z\sum_{k,l=-w}^{w} I(m+k, n+l), \tag{12.18}$$

$$\sigma_O(m,n) = Z\sum_{k,l=-w}^{w}\left(O(m+k, n+l) - \mu_O(m,n)\right)^2, \tag{12.19}$$

$$\sigma_I(m,n) = Z\sum_{k,l=-w}^{w}\left(I(m+k, n+l) - \mu_I(m,n)\right)^2, \tag{12.20}$$

$$\begin{aligned}\sigma_{OI}(m,n) = Z\sum_{k,l=-w}^{w} &\left(I(m+k, n+l) - \mu_I(m,n)\right)\cdot \\ &\cdot\left(O(m+k, n+l) - \mu_O(m,n)\right),\end{aligned} \tag{12.21}$$

$$Z = \frac{1}{(2w+1)^2}, \tag{12.22}$$

where $O(m,n)$ is the intensity of grayscale reference pixel at m, n position, $I(m,n)$ is the luminance of the pixel of target image, $2w + 1$ is the size of the window in which the calculation of statistical features is performed.

Since the SSIM is defined only for grayscale images, it can be adapted for color image colorization. We may accomplish that by calculating the SSIM for every single color channel independently, then by calculating the mean. When taking into consideration decorrelated color spaces, where the channels are orthogonal, we may

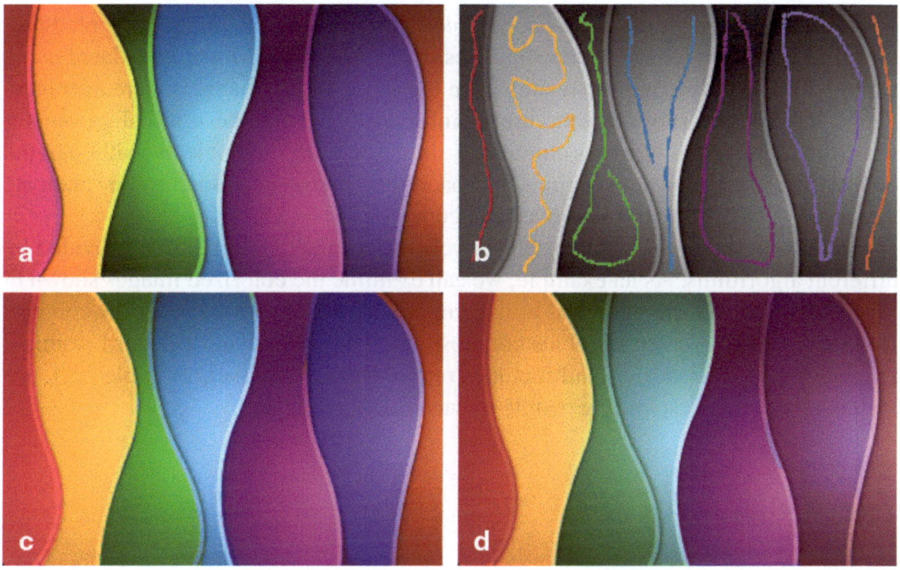

Fig. 12.13 Impact of color blending parameter on colorization outcomes. **a** Original color image. **b** Scribbled grayscale image. **c** Colorized image, $\sigma = 5$. **d** Colorized image, $\sigma = 50$

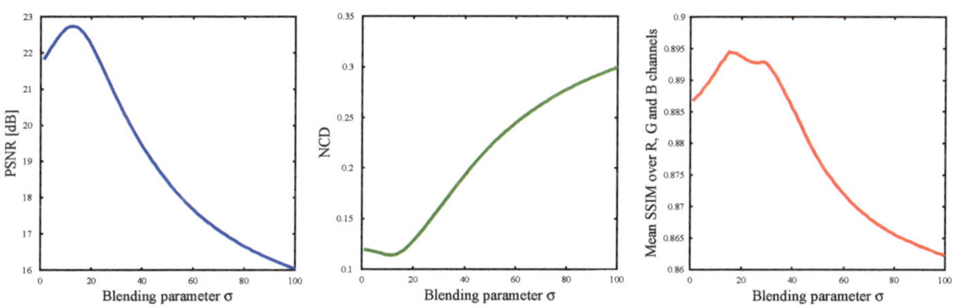

Fig. 12.14 Dependence of structural similarity index (*SSIM*), normalized color difference (*NCD*), and peak signal to noise ratio (*PSNR*) quality measures of colorization outcomes on blending parameter σ

assess the quality with the root of the sum of squared SSIM results obtained for each channel.

We present the analysis of the impact of blending parameter σ in distance transformation on colorization fidelity, as an example of the aforementioned assessment of colorization. The blending parameter σ in Eq. 12.3 is responsible for color mixing at the boundary of the image regions. While this parameter increases, the border between two or more colors blurs. In contrast, when the value of σ is small, the edges are more steep. Figure 12.13 presents the results of colorization with different values of blending parameter σ.

If a reference color image is available, it is possible to calculate the values of quality measures as a function of the blending parameter. The corresponding results of the *Distance Transformation-based Colorization* and of the image presented previously are shown as quality dependencies in Fig 12.14. SSIM was calculated using a 5×5 local window and the results from the R, G, and B channels were averaged. Such analysis can provide very useful information about the optimal parameter selection for a range of images containing similar features and objects. Although the quality measures represent various approaches to image assessment, they indicate that the best σ parameter for a given image is $\sigma = 13 \sim 15$.

It is also possible to compare subjectively the results of colorization provided by different algorithms. As an example, we can examine the colorized images shown in Fig. 12.15. The first example was utilized in previous tests; however, for the comparison, it contains simpler scribbles composition. The second and the third example is much more complicated and includes real objects. Moreover, in Table 12.1, we summarize the corresponding results of each quality measure.

12.11 Summary

Colorization—the process of adding colors to grayscale images—is a significant aspect of modern image processing. Since colorized images are more pleasing to the human eye, the colorization algorithms may be widely applied in such disciplines as medicine, astronomy, or education. However, science is not the only beneficiary—general consumer electronics may also employ colorization in many multimedia applications.

In this chapter, we presented a review of grayscale image colorization methods. An extended literature analysis has shown the variety of approaches leading to high-quality colorization outcomes. We concentrated on semiautomatic algorithms, as they seem to be more applicable. We presented four such state-of-the-art methods. They apply the idea of user-indicated color scribbles within a grayscale image. Additionally, we showed an exemplary automatic approach. It may be useful for some specific images, where the number of required scribbles would be very high. Finally, we provided a methodology of assessing colorization results, based on several well-known quality measures.

Within the chapter, we also presented some exemplary colorization results. They proved that the modern colorization algorithms are able to meet most of the user's' requirements. Although the methods presented in the chapter differ, all of them are able to provide high-quality outcomes.

The development of colorization methods is important and remains an open issue. Not only do the novel colorization methods aim to improve quality but their goal is also to minimize the computational burden. Thus, the algorithms could be run in real-time way. With the increasing power of modern computers, colorization tools aim at allowing the user to observe results immediately, making colorization much easier and more comfortable. The efficiency is also crucial, considering the increasing resolution of images. Since the quality of colorization with modern methods achieved a satisfactorily high level, in the future, the algorithms should definitely concentrate on the aspects of computational efficiency.

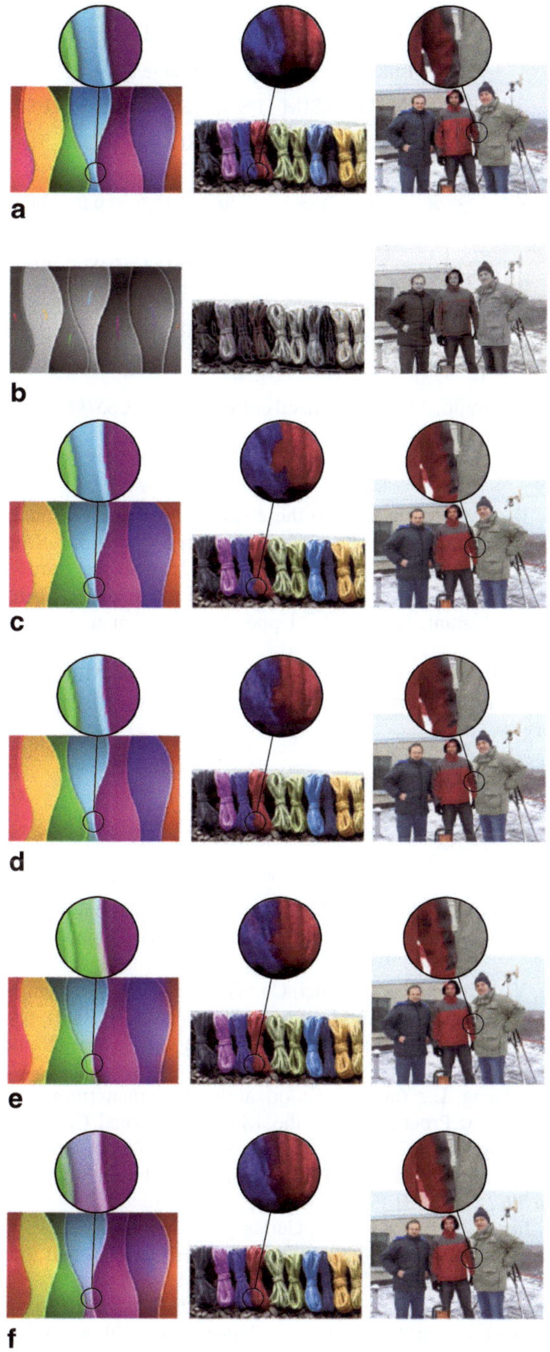

Fig. 12.15 A comparison of colorization quality achieved by different methods. **a** Original color images. **b** Scribbled grayscale images. **c** Distance transformation based colorization [16]. **d** Chrominance blending colorization [38]. **e** Isoline image colorization [24]. **f** Colorization using optimization [17]

Table 12.1 Results of colorization quality in images shown in Fig. 12.15

Method	Image 1			Image 2			Image 3		
	PSNR	NCD	SSIM	PSNR	NCD	SSIM	PSNR	NCD	SSIM
Distance transformation-based colorization [16]	23.19	*0.115*	0.880	25.05	0.116	0.933	31.07	0.032	0.9781
Chrominance blending colorization [38]	*23.34*	0.116	0.880	25.86	*0.107*	0.940	31.14	0.032	0.9787
Isoline image colorization [24]	22.59	0.139	0.870	25.25	0.114	0.936	30.77	0.033	0.9779
Colorization using optimization [17]	19.63	0.181	*0.888*	*25.97*	0.108	*0.944*	*31.40*	*0.031*	*0.9802*

PSNR peak signal to noise ratio, *NCD* normalized color difference, *SSIM* structural similarity index

Acknowledgments The research leading to these results has received funding from the Polish-Norwegian Research Programme operated by the National Center for Research and Development under the Norwegian Financial Mechanism 2009–2014 in the frame of Project Contract No. Pol-Nor/204256/16/2013. This work was performed using the infrastructure supported by POIG.02.03.01-24-099/13 grant: GeCONiI - Upper Silesian Center for Computational Science and Engineering.

Ultrasound images depicted in Fig. 12.11, were created for MEDUSA project at Section for Rheumatology; Department for Neurology, Rheumatology and Physical Medicine, Central Hospital, Førde, Norway.

References

1. Bezrukov, I., Charpiat, G., Hofmann, M.: Computational photography methods and applications. Machine learning methods for automatic image colorization. Boca Raton: CRC Press (2010)
2. Chen, T., Wang, Y., Schillings, V., Meinel, Ch.: Grayscale image matting and colorization. In: Proceedings of Asian Conference on Computer Vision (ACCV 2004), pp. 1164–1169 (2004)
3. Dijkstra, E.W.: A note on two problems in connexion with graphs. Numer. Math. **1**(1), 269–271 (1959)
4. Ding, X., Xu, Y., Deng, L., Yang, X. Colorization using quaternion algebra with automatic scribble generation. In: Proceedings of the 18th International Conference on Advances in Multimedia Modeling, MMM'12, pp. 103–114. Springer-Verlag, Berlin (2012)
5. Felzenszwalb, P.F., Huttenlocher, D.P.: Efficient graph-based image segmentation. Int. J. Comput. Vision **59**(2):167–181 (2004)
6. Hertzmann, A., Jacobs, C.E., Oliver, N., Curless, B., Salesin, D.H.: Image analogies. In: Proceedings of the 28th Annual Conference on Computer Graphics and Interactive Techniques, SIGGRAPH '01, pp. 327–340. ACM, New York (2001)
7. Horiuchi, T.: Estimation of color for gray-level image by probabilistic relaxation. In: Proceedings of the 16th International Conference on Pattern Recognition, 2002., vol. 3, pp. 867–870 (2002)
8. Horiuchi, T., Hirano, S.: Colorization algorithm for grayscale image by propagating seed pixels. In: Proceedings of International Conference on Image Processing, 2003. ICIP 2003., vol. 1, pp. I–457–60 (Sept 2003)

9. Irony, R., Cohen-Or, D., Lischinski, D.: Colorization by example. In: Proceedings of the Sixteenth Eurographics Conference on Rendering Techniques, EGSR '05, pp. 201–210. Eurographics Association, Aire-la-Ville (2005)
10. Jacob, V.G., Gupta, S.: Colorization of grayscale images and videos using a semiautomatic approach. In: Proceedings of the 16th IEEE International Conference on Image Processing, ICIP'09, pp. 1653–1656. IEEE Press, Piscataway (2009)
11. Kawulok, M., Kawulok, J., Smolka, B.: Discriminative textural features for image and video colorization. IEICE Trans. Inf. Syst. **95**, 1722–1730 (2012)
12. Kawulok, M., Smolka, B.: Competitive image colorization. In: 17th IEEE International Conference on Image Processing (ICIP 2010), pp. 405–408 (Sept 2010)
13. Kawulok, M., Smolka, B.: Texture-adaptive image colorization framework. EURASIP J. Adv. Sig. Proc. **2011**, 99 (2011)
14. Kim, T.H., Lee, K.M., Lee, S.U.: Edge-preserving colorization using data-driven random walks with restart. In: 16th IEEE International Conference on Image Processing (ICIP 2009), pp. 1661–1664 (Nov 2009)
15. Konushin, V., Vezhnevets, V.: Interactive image colorization and recoloring based on coupled map lattices. In: GraphiCon, pp. 231–234 (2006)
16. Lagodzinski, P., Smolka, B.: Application of the extended distance transformation in digital image colorization. Multimed. Tool Appl. **69**(1), 111–137 (2014)
17. Levin, A., Lischinski, D., Weiss, Y.: Colorization using optimization. ACM Trans. Graph. **23**(3), 689–694 (Aug 2004)
18. Lezoray, O., Ta, V.T., Elmoataz, A.: Nonlocal graph regularization for image colorization. In: 19th International Conference on Pattern Recognition (ICPR 2008), pp. 1–4 (Dec 2008)
19. Lipowezky, U.: Grayscale aerial and space image colorization using texture classification. Pattern Recognit. Lett. **27**(4), 275–286 (2006)
20. Liu, X., Liu, J., Feng, Z.: Colorization using segmentation with random walk. In: Jiang, X., Petkov, N. (eds.) Computer analysis of images and patterns, volume 5702 of Lecture Notes in Computer Science, pp. 468–475. Springer, Berlin (2009)
21. Luan, Q., Wen, F., Cohen-Or, D., Liang, L., Xu, Y.Q., Shum, H.Y.: Natural image colorization. In: Proceedings of the 18th Eurographics Conference on Rendering Techniques, EGSR'07, pp. 309–320, Aire-la-Ville, Switzerland, Switzerland (2007)
22. Lukac, R., Smolka, B., Martin, K., Plataniotis, K. N., Venetsanopoulos, A.N.: Vector filtering for color imaging. Sign. Process. Mag., IEEE, **22**(1), 74–86 (Jan 2005)
23. Plataniotis, K.N., Venetsanopoulos, A.N.: Color Image Processing and Applications. Springer-Verlag, Berlin (2000)
24. Popowicz, A., Smolka, B.: Isoline based image colorization. In: Proceedings of the 2014 UKSim-AMSS 16th International Conference on Computer Modelling and Simulation, UKSIM '14, pp. 280–285. IEEE Computer Society, Washington, DC (2014)
25. Qu, Y., Wong, T.T., Heng, P.A.: Manga colorization. ACM Trans. Graph. **25**(3), 1214–1220 (July 2006)
26. Rao Veeravasarapu, V.S., Sivaswamy, J.: Fast and fully automated video colorization. In: International Conference on Signal Processing and Communications (SPCOM 2012), pp. 1–5 (July 2012)
27. Reinhard, E., Adhikhmin, M., Gooch, B., Shirley, P.: Color transfer between images. Comput. Gr. Appl., IEEE **21**(5), 34–41 (Sep 2001)
28. Rosenfeld, A., Pfaltz, J.L.: Distance functions on digital pictures. Pattern Recognit. **1**(1), 33–61 (1968)
29. Ruderman, D.L., Cronin, T.W., Chiao, C.C.: Statistics of cone responses to natural images: Implications for visual coding. J. Opt. Soc. Am. A **15**, 2036–2045 (1998)
30. Russo, F.: Performance evaluation of noise reduction filters for color images through normalized color difference (NCD) decomposition. ISRN Machine Vision (2014)
31. Sapiro, G.: Inpainting the colors. In: IEEE International Conference on Image Processing, (ICIP 2005), vol. 2, pp. II–698–701 (Sept 2005)

32. Shah, A., Gandhi, M., Shah, K.: Medical image colorization using optimization technique. Int. J. Sci. Res. Publ. **3**(3), 1–6 (2013)
33. Sýkora, D., Buriánek, J., Žára, J.: Unsupervised colorization of black-and-white cartoons. In: Proceedings of the 3rd International Symposium on Non-photorealistic Animation and Rendering, NPAR '04, pp. 121–127. ACM, New York (2004)
34. Wang, Z.B., Sheikh, H.R., Simoncelli, E.P.: Image quality assessment: from error visibility to structural similarity. IEEE Trans. Imag. Process. **13**(4), 600–612 (April 2004)
35. Wei-Qi, Y., Kankanhalli, M.S.: Colorizing infrared home videos. In: Proceedings of International Conference on Multimedia and Expo, (ICME 2003), vol. 1, pp. I–97–100 (July 2003)
36. Welsh, T., Ashikhmin, M., Mueller, K.: Transferring color to greyscale images. ACM Trans. Graph. **21**(3), 277–280 (July 2002)
37. Xiang, Y., Zou, B., Li, H.: Selective color transfer with multi-source images. Pattern Recognit. Lett. **30**(7), 682–689 (2009)
38. Yatziv, L., Sapiro, G.: Fast image and video colorization using chrominance blending. IEEE Trans. Imag. Process. **15**(5), 1120–1129 (May 2006)
39. Zhang, Z., Cui, H., Lu, H., Chen, R., Yan, Y.: A colorization method based on fuzzy clustering and distance transformation. In: 2nd International Congress on Image and Signal Processing 2009 (CISP 2009), pp. 1–5 (Oct 2009)
40. Zhao, Y., Wang, L., Jin, W. , Shi, S.: Colorizing biomedical images based on color transfer. In: International Conference on Complex Medical Engineering (CME 2007), pp. 820–823 (May 2007)
41. Zhen, Z., Yan, G., Lizhuang, M.: An automatic image and video colorization algorithm based on pattern continuity. In: International Conference on Audio, Language and Image Processing (ICALIP 2012), pp. 531–536 (July 2012)

Chapter 13
Computationally Efficient Data and Application Driven Color Transforms for the Compression and Enhancement of Images and Video

Massimo Minervini, Cristian Rusu and Sotirios A. Tsaftaris

13.1 Introduction

The red-green-blue (RGB) color model is currently the standard for acquisition and display of color digital images. However, this representation is not efficient for coding, due to high correlation between color bands of natural images [51].

To reduce spectral redundancy, image and video compression algorithms operate on luminance/chrominance representations of the color information, achieved through linear transformations of the RGB color space [47, 50]. Each color band is coded independently, therein deploying a variety of techniques to address spatial and, for video, also temporal correlation. A family of such color models is the YC_bC_r, defined by the International Telecommunication Union [30] and adopted by many coding standards [55]. However, due to high variability in source image characteristics, a fixed transform may easily result in suboptimal performance, thus motivating the adoption in some contexts of a data-dependent one.

Among the linear transformations, the energy compaction and decorrelation properties of the Karhunen–Loève transform (KLT) [32] make it desirable for color image compression [19, 37, 51, 52, 64]. The KLT has been adopted in many coding schemes (e.g., for dimensionality reduction [1, 9]), and it was shown to be superior to other approaches in a variety of contexts, both for color [8, 23, 35] and hyperspectral [13] imagery, and has formed the basis for new fixed transforms [24, 36, 42]. However, the computational complexity of calculating the color covariance matrix limits its applicability in real-time video enhancement applications (e.g., denoising [67], contrast and color [20, 21] enhancement, color to gray scale conversion [11]),

M. Minervini (✉) · C. Rusu · S. A. Tsaftaris
IMT Institute for Advanced Studies, Piazza San Ponziano 6, Lucca, Italy
e-mail: massimo.minervini@imtlucca.it

C. Rusu
e-mail: cristian.rusu@imtlucca.it

S. A. Tsaftaris
e-mail: stsaft@gmail.com

M. E. Celebi et al. (eds.), *Color Image and Video Enhancement*,
DOI 10.1007/978-3-319-09363-5_13

and sensing environments with low computational power (e.g., surveillance cameras or visual sensor networks operating under low lighting conditions).

A variety of approaches have been proposed to circumvent this bottleneck. Kountchev et al. rely on covariance matrix approximations [38], whereas [11] uses numerical methods to estimate eigenvalues and eigenvectors of the covariance matrix. Subsampling strategies to limit complexity by reducing the amount of input data are discussed in [14, 48]. Du et al. adopt a learning approach based on neural networks to estimate projection directions [14]. Porikli et al. [49] propose an algorithm based on integral images for fast computation of the covariance matrix. Others focus on schemes that favor parallel implementations [4, 66] of the KLT, or implementations optimized for graphics processing units [4, 34, 41]. In this work, we propose a new data-dependent color transform, the approximate KLT (aKLT), rooted in the orthogonal Procrustes problem, that preserves energy compaction and performs similar to the KLT, but is less computationally complex.

Although KLT and aKLT are designed to match the statistical properties of the image data, they are agnostic to the semantics of the scene (e.g., distinction between foreground and background regions). In present days, more often than not, image data are analyzed by computer vision algorithms (e.g., surveillance applications [33], industrial inspection [2], or medical image analysis [6]) and their transmission over channels necessitates their compression, to reduce bandwidth costs. It was shown recently that considering the application and designing data codecs appropriately not to maximize fidelity type criteria (e.g., mean squared error [61]) or psycho-visual criteria (e.g., structural similarity [62]), but considering how would an analysis algorithm (e.g., a classifier) perform on compressed data, is beneficial from a bit rate perspective [57]. This notion was explored in [12] and [57] with respect to quantization, however, as of now the design of color transforms optimized particularly for application accuracy has not been considered yet. In this work we consider such application to be a classification task, motivated by its broad range of real-world applications. Thus, we propose for the first time to obtain a color transform using supervision (e.g., previously labeled image data), aiming at the preservation of the image features relevant to the application. We formulate our methodology as a supervised learning problem, envisioning two alternative approaches to find a solution, relying either on the Foley–Sammon transform (FST) [17] or on metric learning methods [39]. From an application-aware image compression perspective, it is desirable to achieve: (a) classification accuracy, pursued by enhancing separability of the transformed data, and (b) compression performance, achieved with decorrelation and energy compaction. These two requirements appear conflicting and designing a color transform that optimally accommodates both remains an open challenge. Therefore, we propose to adopt an optimization approach to obtain application-dependent color transforms that while aiming to retain energy compaction properties, also try to maximize separability of the transformed data.

This chapter builds upon our previous work [46], where we first investigate our proposed color transforms and present preliminary results. The rest of the chapter is organized as follows. Sect. 13.2 details our methodology to learn color transforms from the data. Sect. 13.3 demonstrates the proposed approaches on a variety of

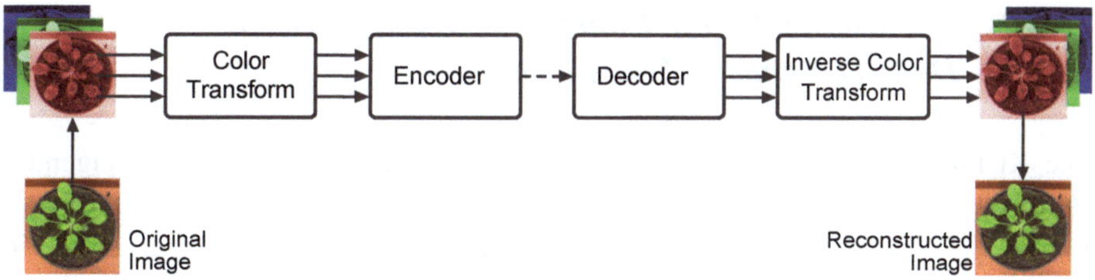

Fig. 13.1 Schematic of the typical encoding/decoding process of a color image

different image datasets, using the JPEG 2000 standard to compress test images. Finally, Sect. 13.4 offers concluding remarks.

13.2 Methodology

We represent an RGB image as a $3 \times n$ matrix $\mathrm{X} = \left(\mathbf{r}, \mathbf{g}, \mathbf{b}\right)^{\mathrm{T}}$, where $\mathbf{r}$, $\mathbf{g}$, and $\mathbf{b}$ are the linearized color components, and n is the number of pixels. Prior to lossy coding, X is projected into a new color space by $\mathrm{T} \in \mathbb{R}^{3\times 3}$. Each pixel value $\mathbf{x}_i = \left(r_i, g_i, b_i\right)^{\mathrm{T}}$ in X is transformed by the linear relation $\mathbf{y}_i = \mathrm{T}\mathbf{x}_i$. Upon reconstruction, the color transform is inverted, obtaining the approximation $\widetilde{\mathbf{x}}_i = \mathrm{T}^{-1}\widetilde{\mathbf{y}}_i$ in the RGB domain. To ensure that the output range of $\mathbf{y}_i$ is the same as $\mathbf{x}_i$ (e.g., 0–255, for 8-bit unsigned integer representation), we scale the directions (rows) of T with respect to the ℓ_1 norm [36].

We operate in the general framework depicted in Fig. 13.1, in which color transformation is decoupled from encoding/decoding operations. Therefore, any compression scheme can be adopted (in the experiments we use the JPEG 2000 standard [56]) and the resulting bit stream will be standard compliant.

In the following, we address the problem of obtaining data-driven color space transformations[1], that change according to the image to be encoded or the application the images will be used for. In Sect. 13.2.1, based on a heuristic, we derive a new low-complexity transform (aKLT) that adapts to the content using statistical information from the image being processed. In Sect. 13.2.2, we propose a novel approach to obtain transforms that adapt according to the application (here a pixel-level classifier for foreground–background segmentation), relying on supervised learning methods and labeled training data. Finally, in Sect. 13.2.3, we combine the unsupervised and the supervised transforms using an optimization approach.

[1] Approaches to efficiently compute the color space conversion for a known transform are discussed, for example, in [7, 65].

13.2.1 The aKLT: A Low-Complexity Unsupervised Data-Dependent Transform

The KLT produces an orthogonal transformation, K, obtained from the eigendecomposition of the color covariance matrix $\Sigma = \frac{1}{n}\sum_{i=1}^{n}(\mathbf{x}_i - \boldsymbol{\mu})(\mathbf{x}_i - \boldsymbol{\mu})^{\mathsf{T}}$, where $\boldsymbol{\mu} = \frac{1}{n}\sum_{i=1}^{n}\mathbf{x}_i$ is the mean color vector [51]. The eigenvectors of Σ, sorted in decreasing order of magnitude of the corresponding eigenvalues, define the directions of K. The KLT achieves complete statistical decorrelation of the color signals and energy compaction in the first channel, thus favoring efficient representation and subsampling of the other two channels [40, 52].

However, estimation of Σ can be computationally demanding in memory and computing power, particularly when images are large, and its application in resource-constrained sensing devices can be problematic. Thus, we seek to find a transform that is close to the KLT but less computationally complex to obtain.

Let $\overline{\mathrm{X}} \in \mathbb{R}^{3\times n}$ be the matrix obtained by normalizing each column (pixel) of X with respect to the ℓ_2 norm. We seek an orthogonal transform $\Omega \in \mathbb{R}^{3\times 3}$ that maps $\overline{\mathrm{X}}$ to a given reference matrix $\mathrm{W} \in \mathbb{R}^{3\times n}$, and formulate it as:

$$\begin{aligned} \underset{\Omega}{\text{minimize}} \quad & \|\mathrm{W} - \Omega\overline{\mathrm{X}}\|_F \\ \text{subject to} \quad & \Omega^{\mathsf{T}}\Omega = \mathrm{I}\,, \end{aligned} \tag{13.1}$$

where $\|\cdot\|_F$ denotes the Frobenius norm, and I is the identity matrix. Let $\mathrm{Z} = \mathrm{W}\overline{\mathrm{X}}^{\mathsf{T}}$, and $\mathrm{Z} = \mathrm{USV}^{\mathsf{T}}$ be the singular value decomposition (SVD) of Z. The optimization problem of Eq. (13.1), known as the orthogonal Procrustes problem, admits closed-form solution UV^{T} [54]. In order to concentrate energy in the first direction, we impose structure to W:

$$\mathrm{W} = \begin{pmatrix} 1 \ldots 1 \\ 0 \ldots 0 \\ 0 \ldots 0 \end{pmatrix}.$$

Notably, this leads to a simplified form of Z:

$$\mathrm{Z} = \begin{pmatrix} \sum_{i=1}^{n} r_i & \sum_{i=1}^{n} g_i & \sum_{i=1}^{n} b_i \\ 0 & 0 & 0 \\ 0 & 0 & 0 \end{pmatrix},$$

with only a single direction, $\mathbf{a}_1 = \mathbf{z}_1^{\mathsf{T}}/\|\mathbf{z}_1\|_2$, that corresponds to the principal direction, thus making the SVD computation unnecessary. We adopt the vector $\mathbf{a}_1$ as an approximation of the principal direction of the KLT.

To obtain the full transform, we proceed by constructing the 3×3 matrix $\mathrm{A} = \big(\mathbf{a}_1, \mathbf{a}_2, \mathbf{a}_3\big)$, where $\mathbf{a}_2$ and $\mathbf{a}_3$ are initialized with random elements, e.g., uniformly distributed on the interval [0, 1] (the effect of randomness on performance is

Table 13.1 Comparison of KLT approaches as a function of input size n, where n denotes the number of image pixels

KLT, Power method [11], ACKLT [38]	Penna et al. [48]	IPCA[a] [14]	Porikli et al. [49]	aKLT
$15n$	$\rho 15n$	$15n$	$15n$	$12n$

[a] Complexity reported by the authors of [14] only takes into account multiplications
aKLT approximate Karhunen–Loève transform

explored in Sect. 13.3). Subsequently, we use QR factorization to decompose A into the product $A = QR$, where $Q \in \mathbb{R}^{3\times 3}$ has orthogonal columns and $R \in \mathbb{R}^{3\times 3}$ is upper triangular. The aKLT transformation matrix, $\widetilde{K} = Q^T$, shares relevant properties with the regular KLT: (a) orthogonality and (b) energy compaction capabilities. Although there is no guarantee on sorting and relative amount of energy of second and third channel, this is not of concern from a compression standpoint (e.g., chroma subsampling strategies would downsample the lower-energy components using the same scheme).

Complexity comparison between KLT approaches

The computation of the KLT requires $15n$ floating point operations in total, where n is the number of pixels, and is dominated by computation of mean color vector $\boldsymbol{\mu}$ ($3n$ operations) and covariance matrix Σ (due to symmetry, $12n$ operations are necessary to compute its 6 distinct entries). Notice that while most authors center the original data on the mean prior to calculating Σ (a step that would require additional $3n$ operations, since it is performed on all image pixels), the covariance matrix can also be defined as $\Sigma = \frac{1}{n}\left(\sum_{i=1}^{n} \mathbf{x}_i \mathbf{x}_i^T\right) - \boldsymbol{\mu}\boldsymbol{\mu}^T$ [3]. Thus, mean subtraction can be performed on the small 3×3 autocorrelation matrix, with fixed computational cost [16]. We also ignore cost of subsequent eigenvalue decomposition of Σ to obtain K, since this step does not depend on n.

Approaches that speed up the eigendecomposition of Σ (e.g., power method [11], or adaptive color Karhunen-Loève transform (ACKLT) [38]) provide negligible benefit in this context (particularly as image resolution increases). As shown in Table 13.1, the iterative principal component analysis (IPCA) [14], based on neural networks, achieves an approximation of the principal direction using $15n$ multiplications (additions are not reported by the authors), whereas [48] necessitates to keep a fraction $\rho = 0.8$ of the data to match the aKLT. Porikli et al. [49] propose a method based on integral images for fast construction of feature covariance matrices of all possible rectangular regions in an image. However, when used in the full image, this approach does not provide any benefit, computing the color covariance matrix using also $15n$ operations.

On the other hand, our proposed aKLT estimates the statistical properties of the source image and computes the color transform with lower computational complexity. It requires $9n$ operations to normalize the input data, that is, $3n$ to square the pixel values, $2n$ to add the components of each pixel, n to compute the square root for each pixel, and $3n$ to divide each channel by the so-obtained ℓ_2 norm (notice that

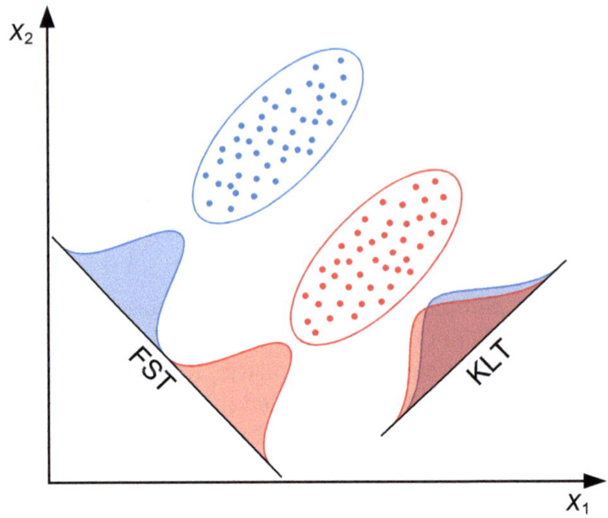

Fig. 13.2 The Karhunen–Loève transform (*KLT*) seeks directions of maximum variance in the projected data. On the other hand, the Foley–Sammon transform (*FST*) as other discriminative methods seeks directions that maximize class separability in a lower-dimensional subspace [59]

mean subtraction is not necessary for the aKLT). Furthermore, $3n$ operations are required to calculate $\mathbf{z}_1$, and a small fixed cost (27 operations) for the QR decomposition of A, resulting in $12n$ operations in total to obtain $\widetilde{\mathrm{K}}$, that is, 20 % reduction in complexity compared to the regular KLT. Computational complexity of the aKLT can be further reduced if combined with the subsampling strategies proposed in [48].

13.2.2 A Supervised Approach to an Application-Dependent Color Transform Using Labeled Pixels

It is known that projecting to principal components is not always optimal from a pattern recognition perspective: clusters of points belonging to semantically different objects in the scene may overlap now in the projected color space [59] (cf. Fig. 13.2). Introducing distortions due to lossy compression may affect this separability further. With an application-aware compression setting in mind, we seek to identify a transform that maintains (a) class separation as well as (b) decorrelation and energy compaction properties.

We assume that the computation of the supervised color transform will occur in an offline fashion and we will use a training set (pixels partitioned in two classes), thus it is supervised. Compression of newly acquired images at the sensor occurs as before, with the transform now known. The calculation of a new transform is necessary only if the scene conditions change (depending on the process being observed) and if new training data are available.

Let C_1 and C_2 be disjoint sets of pixel values (i.e., $C_1 \cap C_2 = \emptyset$) representative of distinct pattern classes (e.g., foreground and background). We seek an orthogonal transform $\mathrm{D} \in \mathbb{R}^{3\times 3}$ that projects data points belonging to distinct classes, $\mathbf{x}_1 \in C_1$ and $\mathbf{x}_2 \in C_2$, in a domain where they are maximally separated according to a measure of separability $\mathcal{C}$:

$$\underset{\mathrm{D}}{\text{maximize}} \quad \mathcal{C}\{\mathrm{D}\mathbf{x}_1, \mathrm{D}\mathbf{x}_2\} \tag{13.2a}$$

$$\text{subject to} \quad \mathrm{D}^{\mathsf{T}}\mathrm{D} = \mathrm{I}\,, \tag{13.2b}$$

$$\|\mathrm{D}\Sigma\mathrm{D}^{\mathsf{T}} - \Lambda\|_F \le \varepsilon\,, \tag{13.2c}$$

where I is the identity matrix, Σ is the color covariance matrix, Λ is a diagonal matrix whose elements are the eigenvalues of Σ, and $\varepsilon \ge 0$. The objective function of Eq. (13.2a) accounts for the class separation property of D, while the constraints of Eq. (13.2b) and (13.2c) allow for decorrelation and energy compaction. The parameter ε determines the trade-off between class separation on the one hand and decorrelation and energy compaction on the other hand. Therefore, solving the complete problem of Eq. (13.2) would lead to an orthogonal transform with the full set of the desired properties. On the other hand, imposing the orthogonality constraint (non-convex in nature) renders the optimization problem of Eq. (13.2) non-convex, possibly admitting multiple local optima. Non-convex problems are generally difficult to solve (i.e., finding the global optimum), and require to resort to global optimization methods, which are however computationally intensive. In the remainder of this section, we relax the problem by ignoring Eq. (13.2c), but we revisit the complete problem in the next section.

In the following paragraphs, we discuss two supervised learning approaches to obtain a color transform with class separation capabilities: (a) the Foley–Sammon transform, based on the linear discriminant analysis, and (b) metric learning approaches.

Foley–Sammon transform

An effective measure of class separability is the Fisher's criterion [15,59], defined as the difference between the means of the class data points, normalized by a measure of the within-class dispersion. This notion is formally expressed in the form of the generalized Rayleigh quotient:

$$J(\mathbf{d}) = \frac{\mathbf{d}^{\mathsf{T}}\mathrm{S}_b\mathbf{d}}{\mathbf{d}^{\mathsf{T}}\mathrm{S}_w\mathbf{d}}\,, \tag{13.3}$$

where $\mathbf{d} \in \mathbb{R}^3$ and $\mathrm{S}_b, \mathrm{S}_w \in \mathbb{R}^{3\times 3}$ are, respectively, between-class scatter matrix and within-class scatter matrix:

$$\mathrm{S}_b = \sum_{i=1}^{2} (\mathbf{m}_i - \boldsymbol{\mu})(\mathbf{m}_i - \boldsymbol{\mu})^{\mathsf{T}}\,,$$

$$\mathrm{S}_w = \sum_{i=1}^{2} \sum_{\mathbf{x}\in C_i} (\mathbf{x} - \mathbf{m}_i)(\mathbf{x} - \mathbf{m}_i)^{\mathsf{T}}\,,$$

where $\boldsymbol{\mu} = \sum_{i=1}^{2} P_i \mathbf{m}_i$ is the mean sample vector, and $\mathbf{m}_i \in \mathbb{R}^3$ and $P_i \in \mathbb{R}$ are, respectively, mean and a priori probability of class i.

A closed-form solution to finding orthogonal discriminant vectors that maximize the Fisher's criterion, can be obtained adopting the Foley–Sammon transform (FST) [17]. The first direction, $\mathbf{d}_1$, termed Fisher's discriminant vector (or Fisher's linear discriminant [15]), corresponds to the projection direction that yields maximum between-class scatter and minimum within-class scatter, that is, $\mathbf{d}_1 = \arg\max_{\mathbf{d}} J(\mathbf{d})$, and is obtained as the eigenvector associated with the largest eigenvalue α_1 satisfying $\mathrm{S}_b\mathbf{d}_1 = \alpha_1 \mathrm{S}_w \mathbf{d}_1$ [59]. The following directions to complete the three-dimensional transform are found recursively, by maximizing the Fisher's criterion with an orthogonality constraint. If $\mathrm{D} = \left(\mathbf{d}_1, \ldots, \mathbf{d}_r\right)^{\mathsf{T}}$ is the set of previously obtained directions, $\mathbf{d}_{r+1}$ corresponds to the eigenvector associated with the largest eigenvalue α_{r+1} satisfying $\mathrm{M}\mathrm{S}_b\mathbf{d}_{r+1} = \alpha_{r+1}\mathrm{S}_w\mathbf{d}_{r+1}$, where $\mathrm{M} = \mathrm{I} - \mathrm{D}^{\mathsf{T}}(\mathrm{D}\mathrm{S}_w^{-1}\mathrm{D}^{\mathsf{T}})^{-1}\mathrm{D}\mathrm{S}_w^{-1}$ [31], and I is the identity matrix. For three-dimensional RGB data, the final color transformation matrix is defined by $\mathrm{D} = \left(\mathbf{d}_1, \mathbf{d}_2, \mathbf{d}_3\right)^{\mathsf{T}}$.

In this chapter we consider only a two-class classification problem (i.e., foreground vs. background), however, the FST formulation can be easily generalized to an arbitrary number of pattern classes [18]. According to the class distribution of our test image datasets (cf. Sect. 13.3), we use the standard FST formulation to find a linear separation between foreground and background. Kernel formulations of the FST [68] could be adopted to accommodate nonlinearly separable classes.

Metric learning approaches

Metric learning methods (see [39] for a comprehensive survey) seek to estimate from supervised information a Mahalanobis distance function over data points, $\mathcal{D}_{\mathrm{A}}(\mathbf{x}_i, \mathbf{x}_j) = (\mathbf{x}_i - \mathbf{x}_j)^{\mathsf{T}}\mathrm{A}(\mathbf{x}_i - \mathbf{x}_j)$, parametrized by a positive semidefinite matrix $\mathrm{A} = \mathrm{L}^{\mathsf{T}}\mathrm{L}$. Computing the distance in the input space is equivalent to applying a linear transformation L of the input space, such that data points with small distance according to $\mathcal{D}_{\mathrm{A}}$ are close (in a Euclidean sense) in the projected space, that is, the matrix L minimizes $\mathcal{D}_{\mathrm{A}}(\mathbf{x}_i, \mathbf{x}_j) = \|\mathrm{L}\mathbf{x}_i - \mathrm{L}\mathbf{x}_j\|_2^2$, which is another view of the objective function in Eq. (13.2a). While the FST matrix D is orthogonal, the L transformation matrix obtained by a metric learning approach is positive semidefinite, and in general not orthogonal. Observe that orthogonality is beneficial for the numerical stability of the color transformation, that is, errors introduced by compression and decompression operations are not magnified when forward and reverse color space conversions are computed. Therefore, with respect to the framework defined by Eq. (13.2), metric learning approaches optimize solely for class separability.

Recent metric learning methods include relevant component analysis (RCA) [5], large margin nearest neighbors (LMNN) [63], and information theoretic metric learning (ITML) [10], which can all be used to find L.

13.2.3 Combining Unsupervised and Supervised Approaches

Our approach for finding the supervised transform D in Sect. 13.2.2 relaxed the constraint of optimal decorrelation and energy compaction of Eq. (13.2), finding one that only optimizes for separation. In the previous section we also obtained orthogonality when using the FST, however, this does not guarantee energy compaction, which is achieved by the KLT (or can be approximated by the aKLT). Therefore, we consider now a different approach, removing the orthogonality constraint to obtain a convex relaxation of the problem of Eq. (13.2).

We seek to find a new transform $\mathrm{D}' \in \mathbb{R}^{3\times 3}$ that is close to D whilst trying to satisfy Eq. (13.2c), or equivalently, since we know that the KLT (or the aKLT) optimizes Eq. (13.2c), we can pose the following unconstrained optimization problem:

$$\underset{\mathrm{D}'}{\text{minimize}} \; \|\mathrm{D}' - \mathrm{D}\|_F + \lambda \|\mathrm{D}' - \widetilde{\mathrm{K}}\|_F \,, \tag{13.4}$$

thus, finding a transform that is between D (application-aware, obtained offline using labeled data) and the aKLT (obtained at the sensor and computed based on the unseen image), where the trade-off is controlled by the value of the regularization parameter λ (playing here a role similar to ε in Eq. (13.2)). In the same fashion, the L transform obtained with metric learning methods could be used in Eq. (13.4) instead of D. Although D and $\widetilde{\mathrm{K}}$ in Eq. (13.4) are orthogonal, in general D' will not be orthogonal. Approaches for finding the nearest orthonormal matrix to D' can be adopted, for example, relying on the polar decomposition [27], or the square root matrix [28] of D'.

While this approach adapts the supervised transform to unseen data on the sensor and is expected to gain decorrelating and compacting capabilities, from a computational perspective may be less attractive. In this setting, with the FST (or the RCA) known, the encoder is required to compute the (a)KLT and then solve Eq. (13.4) to obtain the final color transform.

The approaches presented in Sects. 13.2.1 and 13.2.2 admit closed-form solutions, whereas D' is found relying on iterative optimization procedures computing the solution path along the λ parameter. On the other hand, Eq. (13.4) involves only color transforms (i.e., small 3×3 matrices), rather than the original image pixels.

Fig. 13.3 (*Left*) Example images of Arabidopsis plants from different experiments [53], and (*right*) corresponding ground truth segmentations delineated manually

13.3 Results and Discussion

13.3.1 Experimental Settings

The proposed methodology is evaluated on color image data from a variety of application domains. We demonstrate the unsupervised approach on standard test images, including natural, aerial, and retinal [58] images (Fig. 13.4). We showcase the supervised transform using images of different size (up to 18 megapixel) downloaded from the Internet[2], including horses, balloons, and fish (Fig. 13.6). The approaches are also evaluated on a dataset of 20 images (width $\times$ height: 3108×2324 pixels) from a time-lapse sequence of Arabidopsis plant subjects (Fig. 13.3a), arising from plant phenotyping experiments [53]. We use images from this application since they are usually large and due to design requirements they may need to be communicated via the Internet to centralized locations for processing [44]. Thus, any bit rate savings possible are desirable.

We include in the comparison plain RGB (i.e., no color transform) and YC_bC_r (ITU-R BT.601) [30]. KLT and aKLT are computed for each image. We also adopt the RCA [5], a metric learning approach to find a supervised transform L that aims to preserve variability in the data relevant to the classification task at hand. For brevity and clarity of presentation we do not include other popular metric learning approaches, such as LMNN [63] and ITML [10], because they perform similar to the RCA in our image compression context, while being more computationally

[2] http://www.flickr.com/.

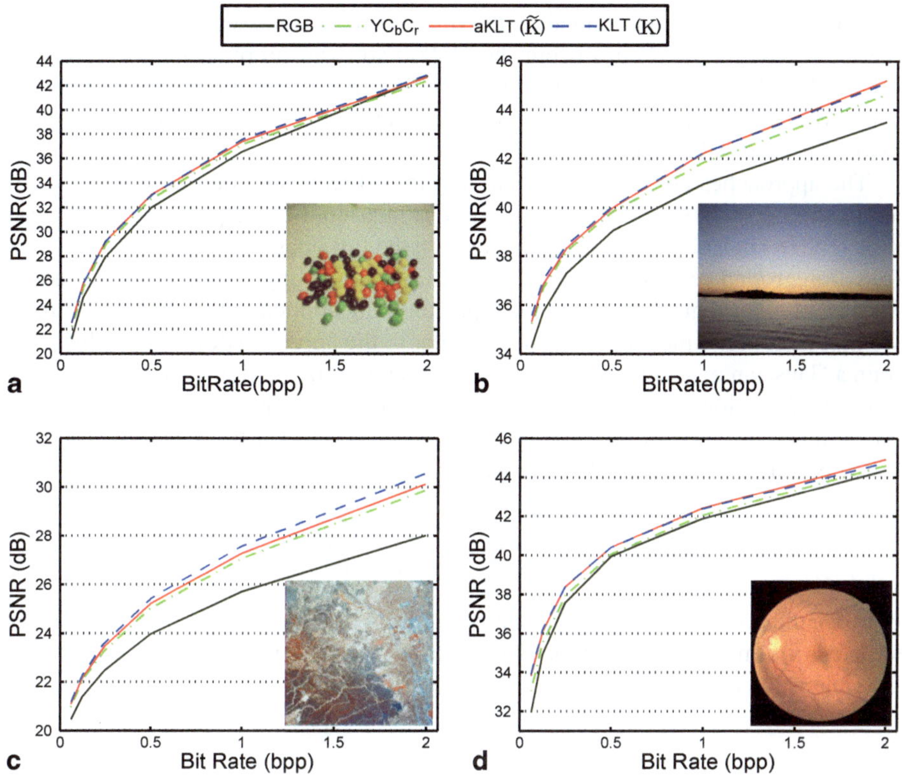

Fig. 13.4 Reconstruction accuracy of standard test images, using fixed and data-dependent color transforms (proposed transform is shown with *solid red curve*). **a** Jelly beans (width × height: 256×256 pixels). **b** Landscape of Bretagne (width × height: 2592×1944 pixels). **c** Aerial photograph of Woodland Hills, Ca. (width × height: 512×512 pixels). **d** Human retina [58] (width × height: 565×584 pixels). For the aKLT, average results are shown, obtained using 100 different initializations. (see Sect. 13.2.1)

demanding (they rely on iterative optimization procedures). The supervised transforms (FST, RCA) are estimated on manually labeled training image data (excluded from testing). On the plant dataset, the supervised transforms (D and L) are estimated from the first image of the time-lapse sequence using pixel label information obtained manually. The so-obtained D and L are then applied to all subsequent images of the same sequence and also to a test image of Arabidopsis plants with different scene conditions (Fig. 13.3c). While the other transforms included in the comparison are either fixed (RGB, YC_bC_r) or present closed-form solutions (KLT, aKLT, FST, RCA), to solve the optimization problem of Eq. (13.4) we use CVX[3], a package for specifying and solving convex programs [22].

[3] http://cvxr.com/cvx.

After color space transformation, the images are compressed at various bit rates (between 0.0625 and 2 bpp) using the JJ2000 software implementation[4], version 5.1, of the JPEG 2000 coding standard [56]. We implement the proposed methods using Matlab R2011b, and conduct all experiments on a machine with Intel Core 2 Duo CPU E8200 2.66 GHz and 4 GB memory.

The approaches are evaluated according to: (a) reconstruction accuracy and (b) application error. Reconstruction accuracy is measured using peak signal-to-noise ratio (PSNR) in RGB image domain, either in the full image or in regions of interest (e.g., foreground regions as in Fig. 13.3). To estimate application error, we adopt the task of plant segmentation for plant phenotyping applications [44,45], therefore, we first build a rudimentary classifier. Similar to the approach described in [45], we train a Gaussian mixture model, $\mathcal{M}$, on color features ($a*$ and $b*$ components of the CIE $L*a*b*$ color space [29]), using labeled foreground (plant) data from the first uncompressed image of the time-lapse sequence (excluded from testing). At each tested bit rate, we calculate the average application error:

$$E_{\mathcal{M}} = \frac{\sum_{i=1}^{n} (\mathcal{M}(\widetilde{\mathbf{x}}_i) - \mathcal{M}(\mathbf{x}_i))^2}{\sum_{i=1}^{n} (\mathcal{M}(\mathbf{x}_i))^2}, \tag{13.5}$$

between the posterior probabilities estimated by $\mathcal{M}$ on the n original, $\mathbf{x}_i$, and reconstructed, $\widetilde{\mathbf{x}}_i$, image pixels. Application error is expressed in percentage, where best possible value of $E_{\mathcal{M}}$ is 0%.

13.3.2 Results

In this section, we present rate-distortion performance of the proposed approaches. We first compare them in terms of overall reconstruction accuracy. Next, we demonstrate the supervised approach in an application-aware context.

Reconstruction accuracy

On the benchmark images of Fig. 13.4, all of the decorrelating transforms provide considerable PSNR improvement with respect to the plain RGB color space, with the data-dependent transforms (KLT, aKLT) outperforming the fixed YC_bC_r. Notably, our proposed low-complexity aKLT, $\widetilde{K}$, exhibits performance very close to the regular KLT, or in some cases slightly superior (cf. red line in Figs. 13.4b and 13.4d, higher bit rates).

[4] http://code.google.com/p/jj2000/.

Table 13.2 Reconstruction accuracy comparison for the plant dataset [53]. (cf. Fig. 13.3a)

	Average PSNR (dB)					
Bit Rate (bpp)	RGB	YC_bC_r	KLT	aKLT	FST	RCA
0.0625	26.75	27.07	27.28	27.23	26.81	25.68
0.125	27.86	28.31	28.44	28.39	27.92	26.26
0.25	29.09	29.53	29.58	29.55	29.13	27.21
0.5	30.53	30.78	30.81	30.90	30.49	28.05
1.0	32.39	32.39	32.28	32.46	32.07	29.23
2.0	34.86	34.68	34.48	34.81	34.46	30.42

RGB red-green-blue, *FST* Foley-Sammon transform, *RCA* relevant component analysis, *PSNR* peak signal-to-noise ratio, *aKLT* approximate Karhunen–Loève transform

Table 13.2 reports image fidelity results for the Arabidopsis plant dataset[5]. At low bit rates (< 1 bpp), decorrelating transforms (YC_bC_r, KLT, aKLT) achieve better performance than RGB (0.25 to 0.6 dB improvement in PSNR). Performance of the aKLT is always superior to the YC_bC_r, and for bit rates greater than 0.5 bpp it surpasses the KLT. As also found by others in some cases [43], at higher bit rates the RGB representation may result in higher performance, due to noise amplification effects of the other transformations and reduced quantization (see solid green line in Fig. 13.4a, in the range of bit rates close to 2 bpp). The supervised FST, D, shows PSNR performance comparable to RGB, with slight improvement only at low bit rates. On the other hand, the supervised RCA, L, performs worse than baseline RGB, probably due to the lack of orthogonality (Gershikov et al. [19] observe a dependence of PSNR performance on the condition number of the color transformation matrix).

Figure 13.5 offers a visual comparison between the components of the color spaces. The RGB channels appear highly redundant (particularly the first two, i.e., red and green), total signal energy is spread across all channels, and the distributions of intensity values span the entire 0–255 range. In the YC_bC_r, the distributions of second and third channel cover a smaller range of values, however signal energy is again dispersed over all three channels. On the other hand, KLT and aKLT present highly similar output, with most of the signal energy (66–70 %) compacted in the first channel, and narrow and peaked distributions in second and third channel, containing a relatively low amount of information. On the other hand, the supervised FST concentrates more energy (64 %) in the second channel, as the first one (i.e., projection on Fisher's discriminant vector) is purposely designed to exhibit good discrimination capabilities of the plant objects. Such features render the KLT, aKLT,

[5] Observe that, in general, major bit rate savings are attained by compression schemes with the combined use of several coding tools. Thus, seemingly small differences in PSNR observed here (i.e., in the order of a fraction of dB) are accounted for by the fact that only the effect of color transformation is tested.

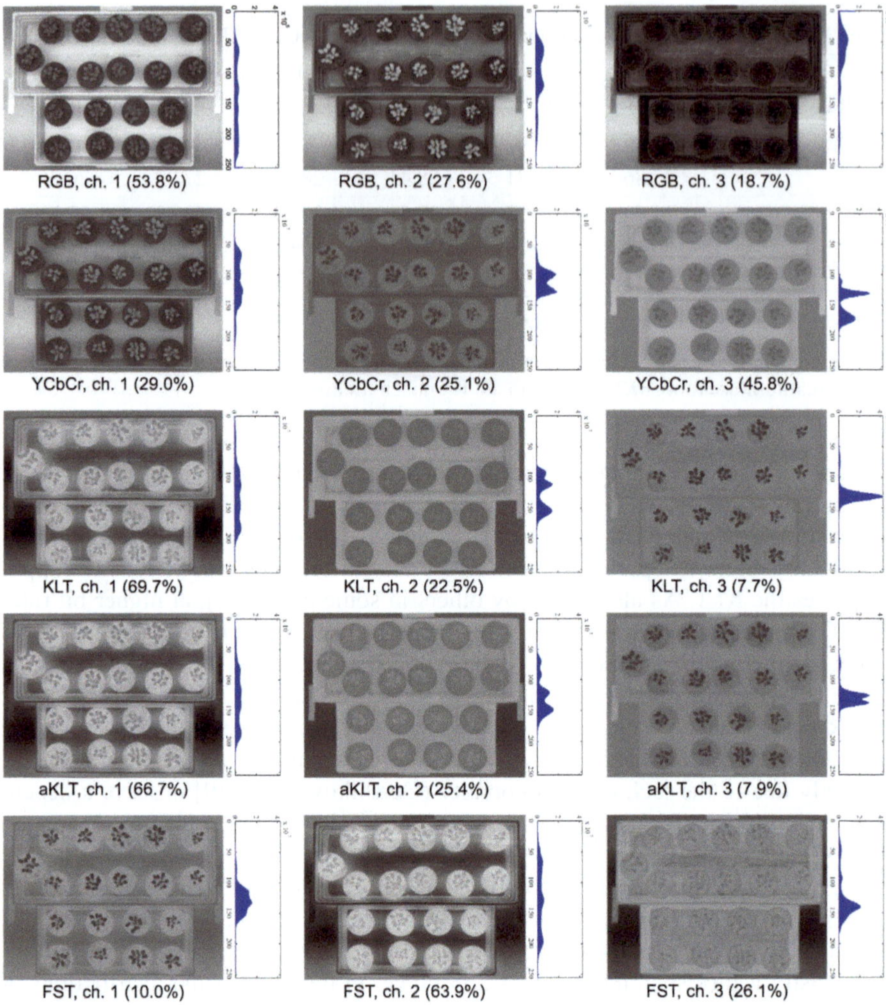

Fig. 13.5 Projection of the image of Fig 13.3a in a variety of color spaces. Next to each channel is shown the corresponding histogram of intensity values, and in parentheses the percentage of signal energy contained in that component

and FST ideal for the coding of color images, because the channels accounting for less energy can be effectively subsampled.

Unsupervised transform

To assess the sensitivity of the aKLT to the random initialization of the vectors $\mathbf{a}_2$ and $\mathbf{a}_3$ in the matrix A (cf. Sect. 13.2.1), we compute 100 different realizations of $\widetilde{\mathrm{K}}$ for each of the test images in Fig. 13.4. As shown in Table 13.3, the aKLT behaves

Table 13.3 Mean and standard deviation of reconstruction accuracy performance for the images of Fig. 13.4, using the aKLT and 100 different initializations

	Average PSNR (dB)			
Bit Rate (bpp)	Jelly Beans	Bretagne	Aerial	Retina
0.0625	22.52 ± 0.08	35.32 ± 0.06	21.09 ± 0.03	33.82 ± 0.08
0.125	25.71 ± 0.06	36.85 ± 0.05	22.16 ± 0.07	36.10 ± 0.07
0.25	29.22 ± 0.11	38.33 ± 0.04	23.48 ± 0.09	38.39 ± 0.06
0.5	33.05 ± 0.06	39.98 ± 0.04	25.20 ± 0.10	40.40 ± 0.05
1.0	37.41 ± 0.04	42.25 ± 0.02	27.27 ± 0.12	42.43 ± 0.05
2.0	42.71 ± 0.06	45.20 ± 0.02	30.12 ± 0.17	44.92 ± 0.06

PSNR peak signal-to-noise ratio, *aKLT* approximate Karhunen–Loève transform

Table 13.4 Average interchannel linear correlation of the test images of Fig. 13.4. For the aKLT, average results are shown, obtained using 100 different initializations (see Sect. 13.2.1)

	Correlation		
Transform	ch. 1–2	ch. 1–3	ch. 2–3
RGB	0.84	0.71	0.91
YC_bC_r	-0.39	0.11	-0.71
aKLT	0.04	0.09	0.11
KLT	0.00	0.00	0.00

RGB red-green-blue, *aKLT* approximate Karhunen–Loève transform

consistently, and variations in PSNR performance due to different initial values are on average approximately only 0.2 %.

Furthermore, the aKLT exhibits good decorrelating capabilities. As shown in Table 13.4, in the RGB domain, the channels of the test images of Figure 13.4 present on average strong linear correlation. Interchannel linear correlation is only moderately reduced by the YC_bC_r, whereas the aKLT is able to achieve the almost complete decorrelation obtained by the optimal KLT.

Supervised transforms and application-aware compression

Figure 13.6 provides several visual examples of the supervised transform on a variety of different images, showing its ability to identify the objects of interest in the test images, even when major changes occur in the scene with respect to the training data (e.g., compare background appearance of the images in Figs. 13.6u and 13.6v). This approach is chiefly based on color information, therefore after learning the transform D on the image of a black horse (Figs. 13.6a and 13.6g), only the black stripes of the zebra in Fig. 13.6f result in a high response, whereas the white stripes are regarded as background (cf. Fig. 13.6l). On the other hand, the transform D estimated from training data in Figs. 13.6m and 13.6q, is able to selectively identify only the red balloons in the image of Fig. 13.6p.

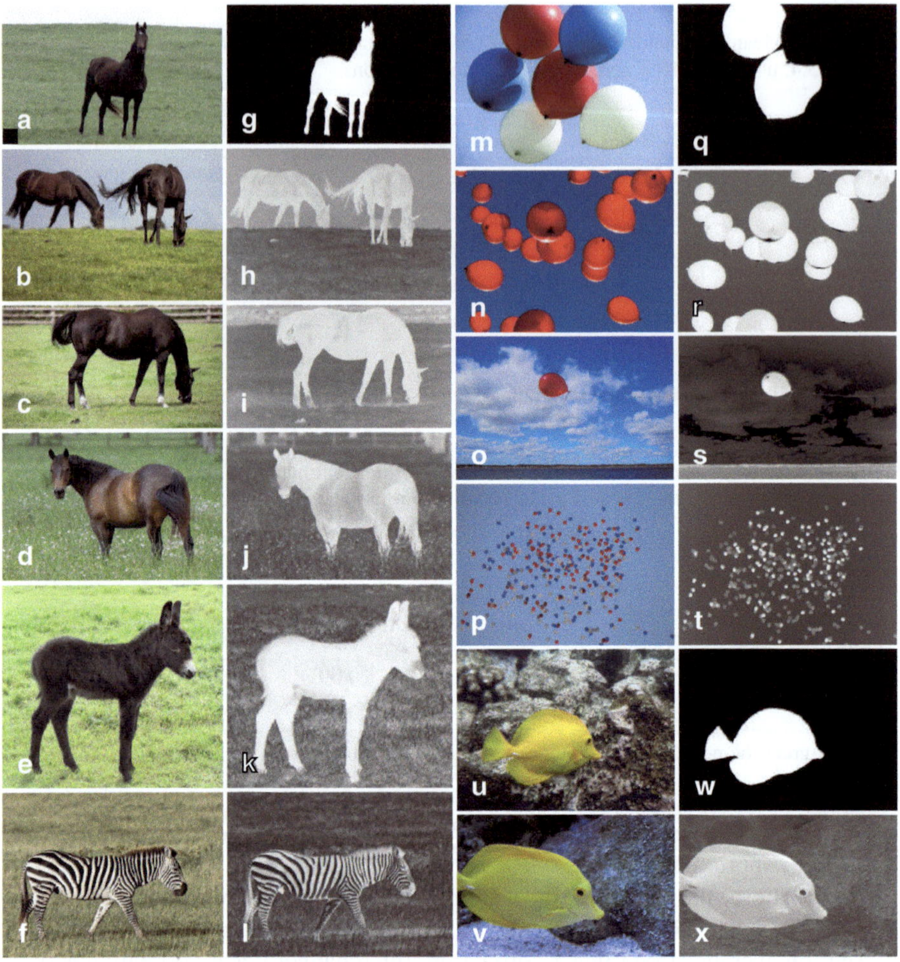

Fig. 13.6 Demonstration of the supervised transform, using images of: **a**–**f** horses, **m**–**p** balloons, and **u**–**v** fish. For each category, a single FST was obtained, using for training, respectively, images in (**a**), (**m**), and (**u**), and corresponding ground truth segmentations (i.e. (**g**), (**q**), and (**w**), respectively). Images in (**h**–**l**), (**r**–**t**), and (**x**) visualize the projections of the test images on the first component of the FST

Figure 13.7 compares the approaches from an application standpoint. Color transformation alone provides up to 1.26 dB improvement in PSNR of the foreground (plant) regions relative to RGB, with the FST now obtaining competitive performance. The supervised transforms do not show remarkable improvements with respect to the other approaches, probably due to lacking decorrelation capabilities for these images, causing losses in bit rate performance.

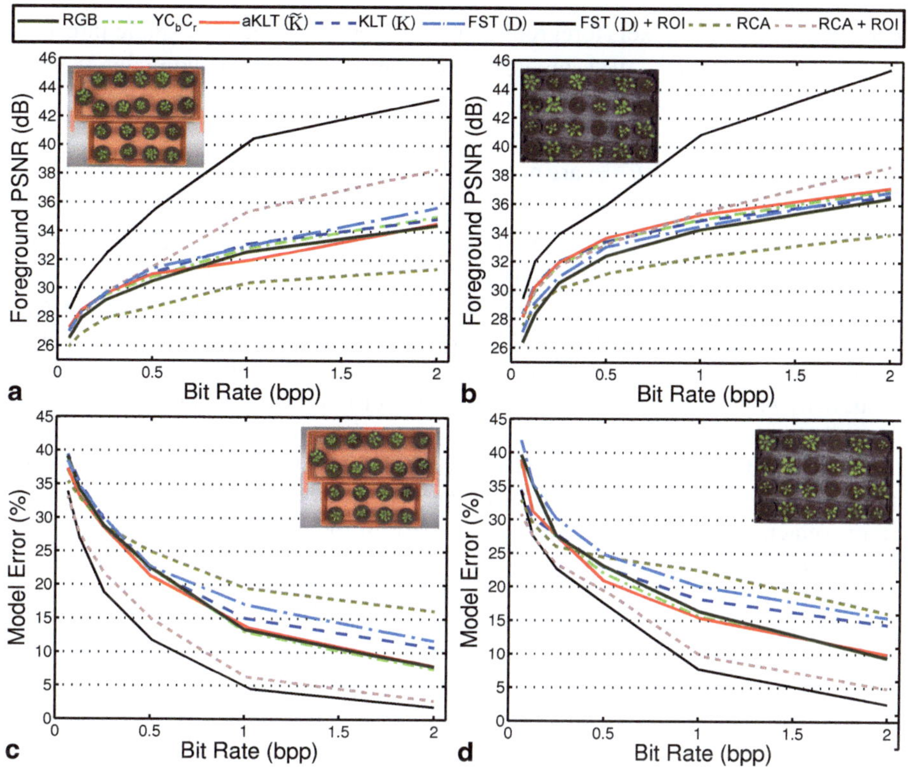

Fig. 13.7 R-D performance using the proposed transforms (in *solid curves*) in comparison to others (*dashed curves*) on the plant image data of Fig. 13.3 using application-aware metrics: **a**–**b** reconstruction accuracy of the objects of interest, and **c**–**d** model error $E_{\mathcal{M}}$ of Eq. (13.5). Results in (**a**) and (**c**) are averaged over 19 test images

Supervised transform for ROI detection

The separation property of the supervised color transform can be further exploited in applications in which the objects of interest can be discriminated by color features (e.g., plant objects in our dataset can be separated from the background based on color information). Therefore, we envision the use of the supervised color transform to obtain from the transformed image a region of interest (ROI) estimate, that can be used in an encoder with ROI coding capability (e.g., the JPEG 2000 standard [56]). With respect to other approaches obtaining the ROI information from a detection module external to the encoder [44], we propose for the first time to combine color transformation and ROI estimation in a single framework, identifying potential ROI masks solely on the basis of the class separation capabilities of the supervised transform, thus reducing computational overhead at the encoder.

When using the FST approach, the first channel of the FST domain, $y_i^{(1)} = \mathbf{d}_1^\top \mathbf{x}_i$, corresponds to the projection on Fisher's discriminant vector (cf. Fig. 13.5, bottom row). In an unseen image, to obtain an ROI estimate, $\Gamma(\mathrm{D}, \theta^*) \in \{0, 1\}^n$, we decide the class of a pixel (foreground or background) based on a single threshold θ^* on the values of $y_i^{(1)}$. We estimate θ^* from our training set, maximizing the dice similarity coefficient (DSC):

$$\theta^* = \arg\max_{\theta} \frac{2 \cdot |\Gamma_{\mathrm{GT}} \cap \Gamma(\mathrm{D}, \theta)|}{|\Gamma_{\mathrm{GT}}| + |\Gamma(\mathrm{D}, \theta)|}, \tag{13.6}$$

between the ground truth of pixels, Γ_{GT}, and the classification, $\Gamma(\mathrm{D}, \theta)$, obtained using D and threshold θ on the training data. Supervised transform D and threshold θ^* are generally assumed to be obtained offline, therefore we estimate θ^* using a parameter sweeping strategy. On the other hand, if an application requires that θ^* be obtained at the sensor, statistical assumptions on the distribution of the data (e.g., Gaussian) would lead to closed-form solutions for finding the optimal θ^* efficiently [59]. When using the RCA approach, ROI estimation proceeds analogously.

When used in a spatial decorrelation context to estimate an ROI, combined with the ROI coding feature of JPEG 2000, the FST + ROI approach obtains a major improvement at all bit rates: 2–8.8 dB increase in foreground PSNR, and 13–77 % reduction in application error (cf. black solid line in Figs. 13.7a and 13.7c). When using the same FST on a test image of Arabidopsis plants acquired under significantly different scene conditions (Fig. 13.3c), the FST + ROI approach proves robust, obtaining again best performance (cf. Figs. 13.7b and 13.7d). On the contrary, although the RCA approach is capable of detecting the regions of interest in an image in both testing scenarios, when projecting the images in the so-obtained color space, the new intensity values are altered in a way that the benefits of the application-aware transform are diminished (or surpassed) by numerical errors introduced by the combination of forward and reverse color transformation and compression (cf. yellow dashed line in Fig. 13.7).

A visual comparison of reconstructed images after compression with JPEG 2000 and all color transforms adopted in this work is shown in Fig. 13.8. The RGB image appears oversmoothed, whereas the decorrelating transforms ($\mathrm{YC_bC_r}$, aKLT, and KLT) exhibit higher image fidelity and appear increasingly richer in details (cf. Figs. 13.8b, 13.8c, 13.8d, and 13.8e). The supervised FST alone already provides good reconstruction accuracy, however, the FST + ROI outperforms all other methods (cf. Figs. 13.8g and 13.8h). The artifacts introduced by the RCA are evident in Fig. 13.8i, and even when coupled with ROI coding the approach produces a noisy image (Fig. 13.8j).

The results envision different use cases for the proposed approaches. The aKLT is general purpose and can be efficiently calculated on a per image basis to target reconstruction accuracy. On the other hand, the supervised approach is best suited for application-aware compression or enhancement scenarios, and since it does require supervision (which can be costly to obtain at the sensor) is assumed to be computed offline. The regularized versions of Eq. (13.4) are highly dependent on the free parameter λ and their performance is found to lie within the bounds of the

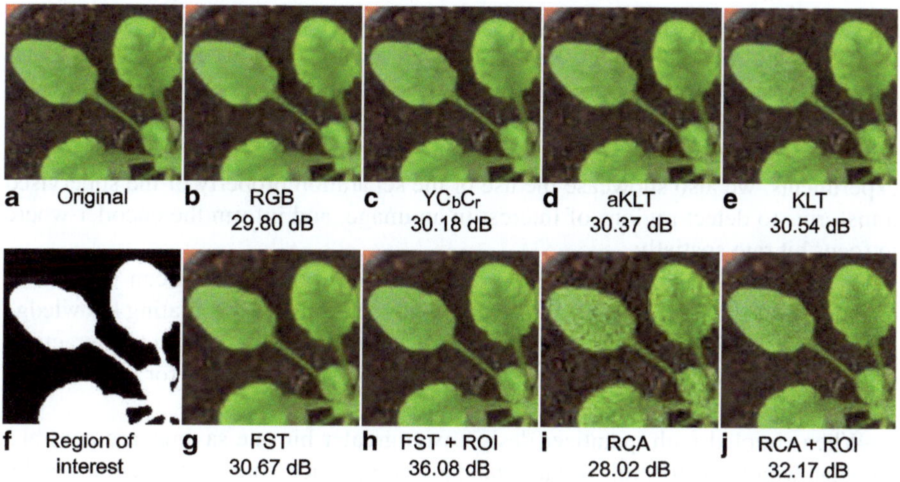

Fig. 13.8 **a** Detail of the image in Fig. 13.3a. **b–e**, **g–j** Reconstructed images after compression at 0.5 bpp using the JPEG 2000 standard and several color space transformations. Foreground PSNR between (**a**) and each of the reconstructed versions, calculated for the plant region indicated in (**f**), is also reported

other two. When varying the value of λ, the new transform D' exhibits behavior very close to either the supervised or the unsupervised transform, respectively. Therefore, it is best to exploit the classification abilities of the supervised FST to focus bits in appropriate places in the image, which is considerably less computationally demanding.

13.4 Conclusions

We address the problem of designing image-adaptive color space transformations for coding and enhancement applications. In recognition of the superior data-dependent KLT with respect to fixed transforms such as the YC_bC_r (as also confirmed by our experimental results), we derive a low-complexity approximation, the aKLT, capable of comparable performance. Our proposed aKLT achieves lower computational complexity than other KLT approaches in the literature, which is expected to result in proportionally reduced computation time, when devising optimized implementations. This will ease adoption on resource-constrained devices or in time-critical applications.

We also consider an application-aware compression setting, in which prior knowledge is available on the objects of interest in the scene. We formulate a novel approach to design color transforms with class separation capabilities, using supervised learning methods. Inspired by the linear discriminant analysis, we measure class separability using the Fisher's discrimination criterion, and adopt the Foley–Sammon transform to obtain an orthogonal application-aware color transform. We also adopt metric learning approaches, however they focus only on class separation (renouncing also orthogonality) and are found to result in lower performance in a compression context. The proposed unsupervised and supervised approaches, for which closed-form solutions are presented, address different requirements, therefore we also consider optimization strategies to combine the two approaches. In the experiments, we also showcase the use of the separation property of the supervised transforms to detect regions of interest in an image, and inform the encoder where to focus bit rate spatially.

In an enhancement context, the proposed supervised approach can be used to enhance the contrast of objects of interest in the scene, also incorporating knowledge of the application and the expert system, and facilitate human–computer interaction, or for automatic content-aware cropping or resizing of large images for visualization on small displays [60].

When coupled with quantizer design even greater bit rate savings are possible, but that would in general violate standard compliance. Increased image resolution or video applications are expected to emphasize the benefits of the proposed approaches. While we adopt the JPEG 2000 standard, our methodology is general and can be adapted to other coding schemes. Reversible integer approximations of the proposed transforms can also be obtained [25, 26], for lossless or progressive lossy-to-lossless compression of color images.

Acknowledgments This work was partially supported by a Marie Curie Action: "Reintegration Grant" (grant number 256534) of the EU's Seventh Framework Programme (FP7).

References

1. Abadpour, A., Kasaei, S.: Color PCA eigenimages and their application to compression and watermarking. Image Vision Comput. **26**(7), 878–890 (2008)
2. Agin, G.J.: Computer vision systems for industrial inspection and assembly. Computer **13**(5), 11–20 (1980)
3. Anderson, T.W.: An Introduction to Multivariate Statistical Analysis, 3rd edn. Wiley, New York (2003)
4. Andrecut, M.: Parallel GPU implementation of iterative PCA algorithms. J. Comput. Biol. **16**(11), 1593–1599 (2009)
5. Bar-Hillel, A., Hertz, T., Shental, N., Weinshall, D.: Learning a Mahalanobis metric from equivalence constraints. J. Mach. Learn. Res. **6**, 937–965 (2005)
6. Celebi, M.E., Schaefer, G. (eds.): Color Medical Image Analysis, Lecture Notes in Computational Vision and Biomechanics, vol. 6. Springer, Berlin (2013)
7. Celebi, M.E., Kingravi, H.A., Celiker, F.: Fast colour space transformations using minimax approximations. IET Image Process. **4**(2), 70–80 (2010)
8. Chen, Y., Hao, P., Dang, A.: Optimal transform in perceptually uniform color space and its application in image coding. In: International Conference on Image Analysis and Recognition (ICIAR), pp. 269–276 (2004)
9. Clausen, C., Wechsler, H.: Color image compression using PCA and backpropagation learning. Patt. Recognit. **33**(9), 1555–1560 (2000)
10. Davis, J.V., Kulis, B., Jain, P., Sra, S., Dhillon, I.S.: Information-theoretic metric learning. In: International Conference on Machine Learning (ICML), pp. 209–216. ACM (2007)

11. Dikbas, S., Arici, T., Altunbasak, Y.: Chrominance edge preserving grayscale transformation with approximate first principal component for color edge detection. In: International Conference on Image Processing (ICIP), vol. 2, pp. 261–264 (2007)
12. Dogahe, B.M., Murthi, M.N.: Quantization for classification accuracy in high-rate quantizers. In: Digital Signal Processing Workshop, pp. 277–282 (2011)
13. Du, Q., Fowler, J.E.: Hyperspectral image compression using JPEG2000 and principal component analysis. IEEE Geosci. Remote Sens. Lett. **4**(2), 201–205 (2007)
14. Du, Q., Fowler, J.E.: Low-complexity principal component analysis for hyperspectral image compression. Int. J. High Perform. Comput. Appl. **22**(4), 438–448 (2008)
15. Fisher, R.A.: The use of multiple measurements in taxonomic problems. Annal. Eugen. **7**(2), 179–188 (1936)
16. Fleury, M., Downton, A.C., Clark, A.F.: Karhünen-Loève transform: An exercise in simple image-processing parallel pipelines. In: Euro-Par '97 Parallel Processing, Lecture Notes in Computer Science, vol. 1300, pp. 815–819. Springer, Berlin (1997)
17. Foley, D.H., Sammon, J.W.: An optimal set of discriminant vectors. IEEE Transac. Comput. **C-24**(3), 281–289 (1975)
18. Fukunaga, K.: Introduction to Statistical Pattern Recognition, 2nd edn. Academic, San Diego (1990)
19. Gershikov, E., Porat, M.: Does decorrelation really improve color image compression? In: International Conference on Systems Theory and Scientific Computation (ISTASC), pp. 306–309 (2005)
20. Gillespie, A.R., Kahle, A.B., Walker, R.E.: Color enhancement of highly correlated images. I. Decorrelation and HSI contrast stretches. Remote Sens. Environ. **20**(3), 209–235 (1986)
21. Gillespie, A.R., Kahle, A.B., Walker, R.E.: Color enhancement of highly correlated images. II. Channel ratio and "chromaticity" transformation techniques. Remote Sens. Environ. **22**(3), 343–365 (1987)
22. Grant, M.C., Boyd, S.P.: Graph implementations for nonsmooth convex programs. In: Recent Advances in Learning and Control, Lecture Notes in Control and Information Sciences, vol. 371, pp. 95–110. Springer, London (2008)
23. Han, S.E., Tao, B., Cooper, T., Tastl, I.: Comparison between different color transformations for the JPEG 2000. In: PICS 2000, pp. 259–263 (2000)
24. Hao, P., Shi, Q.: Comparative study of color transforms for image coding and derivation of integer reversible color transform. In: International Conference on Pattern Recognition (ICPR), vol. 3, pp. 224–227 (2000)
25. Hao, P., Shi, Q.: Matrix factorizations for reversible integer mapping. IEEE Transac. Signal Process. **49**(10), 2314–2324 (2001)
26. Hao, P., Shi, Q.: Reversible integer KLT for progressive-to-lossless compression of multiple component images. In: International Conference on Image Processing (ICIP), vol. 1, pp. 633–636 (2003)
27. Higham, N.J.: Functions of Matrices: Theory and Computation. SIAM, Philadelphia (2008)
28. Horn, B.K.P., Hilden, H.M., Negahdaripour, S.: Closed-form solution of absolute orientation using orthonormal matrices. J. Opt. Soc. Am. A **5**(7), 1127–1135 (1988)
29. International Commission on Illumination (CIE): CIE Colorimetry - Part 4: 1976 L*a*b* Colour Space. Joint ISO/CIE Standard ISO 11664-4:2008(E)/CIE S 014-4/E:2007, CIE Publication (2007)
30. International Telecommunication Union: Studio encoding parameters of digital television for standard 4:3 and wide screen 16:9 aspect ratios. ITU-R Recommendation BT.601-7 (2011)
31. Jin, Z., Yang, J.Y., Hu, Z.S., Lou, Z.: Face recognition based on the uncorrelated discriminant transformation. Patt. Recognit. **34**(7), 1405–1416 (2001)
32. Jolliffe, I.: Principal Component Analysis, 2nd edn. Springer, Berlin (2002)
33. Jones, G.A., Paragios, N., Regazzoni, C.S. (eds.): Video-Based Surveillance Systems: Computer Vision and Distributed Processing. Kluwer Academic, Dordrecht (2002)
34. Jošth, R., Antikainen, J., Havel, J., Herout, A., Zemčík, P., Hauta-Kasari, M.: Real-time PCA calculation for spectral imaging (using SIMD and GP-GPU). J. Real-Time Image Process. **7**(2), 95–103 (2012)

35. Kassim, A.A., Lee, W.S.: Embedded color image coding using SPIHT with partially linked spatial orientation trees. IEEE Transac. Circuit. Syst. Video Technol. **13**(2), 203–206 (2003)
36. Kim, H.M., Kim, W.S., Cho, D.S.: A new color transform for RGB coding. In: International Conference on Image Processing (ICIP), pp. 107–110 (2004)
37. Kouassi, R.K., Devaux, J.C., Gouton, P., Paindavoine, M.: Application of the Karhunen-Loeve transform for natural color images analysis. In: Asilomar Conference on Signals, Systems, and Computers, vol. 2, pp. 1740–1744 (1997)
38. Kountchev, R., Kountcheva, R.: New method for adaptive Karhunen-Loeve color transform. In: International Conference on Telecommunication in Modern Satellite, Cable, and Broadcasting Services (TELSIKS), pp. 209–216 (2009)
39. Kulis, B.: Metric learning: A survey. Found. Trends Mach. Learn. **5**(4), 287–364 (2013)
40. Limb, J.O., Rubinstein, C.B., Thompson, J.: Digital coding of color video signals — a review. IEEE Transac. Commun. **25**(11), 1349–1385 (1977)
41. Majchrowicz, M., Kapusta, P., Was, L., Wiak, S.: Application of general-purpose computing on graphics processing units for acceleration of basic linear algebra operations and principal components analysis method. In: Man-Machine Interactions 3, Advances in Intelligent Systems and Computing, vol. 242, pp. 519–527 (2014)
42. Malvar, H.S., Sullivan, G.J., Srinivasan, S.: Lifting-based reversible color transformations for image compression. In: Proceedings of SPIE, Applications of Digital Image Processing XXXI, vol. 7073 (2008)
43. Marpe, D., Kirchhoffer, H., George, V., Kauff, P., Wiegand, T.: Macroblock-adaptive residual color space transforms for 4:4:4 video coding. In: International Conference on Image Processing (ICIP), pp. 3157–3160 (2006)
44. Minervini, M., Tsaftaris, S.A.: Application-aware image compression for low cost and distributed plant phenotyping. In: International Conference on Digital Signal Processing (DSP), pp. 1–6 (2013)
45. Minervini, M., Abdelsamea, M.M., Tsaftaris, S.A.: Image-based plant phenotyping with incremental learning and active contours. Ecol. Inform. **23**, 35–48 (2014). (Special Issue on Multimedia in Ecology and Environment).
46. Minervini, M., Rusu, C., Tsaftaris, S.A.: Unsupervised and supervised approaches to color space transformation for image coding. In: International Conference on Image Processing (ICIP) (2014)
47. Palus, H.: Representations of colour images in different colour spaces. In: The Colour Image Processing Handbook, pp. 67–90. Springer, Berlin (1998)
48. Penna, B., Tillo, T., Magli, E., Olmo, G.: A new low complexity KLT for lossy hyperspectral data compression. In: International Geoscience and Remote Sensing Symposium (IGARSS), vol. 7, pp. 3525–3528 (2006)
49. Porikli, F., Tuzel, O.: Fast construction of covariance matrices for arbitrary size image windows. In: International Conference on Image Processing (ICIP), pp. 1581–1584 (2006)
50. Poynton, C.: A guided tour of color space. In: New Foundations for Video Technology, pp. 167–180 (1995)
51. Pratt, W.K.: Spatial transform coding of color images. IEEE Transac. Commun. Tech. **19**(6), 980–992 (1971)
52. Rubinstein, C.B., Limb, J.O.: Statistical dependence between components of a differentially quantized color signal. IEEE Transac. Commun. **20**(5), 890–899 (1972)
53. Scharr, H., Minervini, M., Fischbach, A., Tsaftaris, S.A.: Annotated image datasets of rosette plants. Tech. Rep. FZJ-2014-03837, Forschungszentrum Jülich GmbH, Jülich, Germany (2014)
54. Schönemann, P.H.: A generalized solution of the orthogonal Procrustes problem. Psychometrika **31**(1), 1–10 (1966)
55. Shi, Y.Q., Sun, H.: Image and Video Compression for Multimedia Engineering: Fundamentals, Algorithms, and Standards, 2nd edn. CRC, Boca Raton (2008)
56. Skodras, A., Christopoulos, C., Ebrahimi, T.: The JPEG 2000 still image compression standard. IEEE Signal Process. Mag. **18**(5), 36–58 (2001)

57. Soyak, E., Tsaftaris, S.A., Katsaggelos, A.K.: Low-complexity tracking-aware H.264 video compression for transportation surveillance. IEEE Transac. Circuits Syst. Video Tech. **21**(10), 1378–1389 (2011)
58. Staal, J., Abramoff, M., Niemeijer, M., Viergever, M., van Ginneken, B.: Ridge-based vessel segmentation in color images of the retina. IEEE Transac. Med. Imaging **23**(4), 501–509 (2004)
59. Theodoridis, S., Koutroumbas, K.: Pattern Recognition, 4th edn. Academic, New York (2009)
60. Vaquero, D., Turk, M., Pulli, K., Tico, M., Gelfand, N.: A survey of image retargeting techniques. In: SPIE Applications of Digital Image Processing XXXIII, vol. 7798, p. 779814 (2010)
61. Wang, Z., Bovik, A.C.: Mean squared error: Love it or leave it? A new look at signal fidelity measures. IEEE Signal Process. Mag. **26**(1), 98–117 (2009)
62. Wang, Z., Bovik, A.C., Sheikh, H.R., Simoncelli, E.P.: Image quality assessment: from error visibility to structural similarity. IEEE Transac. Image Process. **13**(4), 600–612 (2004)
63. Weinberger, K.Q., Saul, L.K.: Distance metric learning for large margin nearest neighbor classification. J. Mach. Learn. Res. **10**, 207–244 (2009)
64. Wolf, S.G., Ginosar, R., Zeevi, Y.Y.: Spatio-chromatic model for colour image processing. In: International Conference on Pattern Recognition (ICPR), vol. 1, pp. 599–601 (1994)
65. Yang, Y., Yuhua, P., Zhaoguang, L.: A fast algorithm for YCbCr to RGB conversion. IEEE Transac. Consumer Elec. **53**(4), 1490–1493 (2007)
66. Zabalza, J., Ren, J., Ren, J., Liu, Z., Marshall, S.: Structured covariance principal component analysis for real-time onsite feature extraction and dimensionality reduction in hyperspectral imaging. Appl. Opt. **53**(20), 4440–4449 (2014)
67. Zhang, L., Dong, W., Zhang, D., Shi, G.: Two-stage image denoising by principal component analysis with local pixel grouping. Patt. Recognit. **43**(4), 1531–1549 (2010)
68. Zheng, W., Zhao, L., Zou, C.: Foley–Sammon optimal discriminant vectors using kernel approach. IEEE Transac. Neural Netw. **16**(1), 1–9 (2005)

MIX
Papier aus verantwortungsvollen Quellen
Paper from responsible sources
FSC® C105338

If you have any concerns about our products, you can contact us on
ProductSafety@springernature.com

In case Publisher is established outside the EU, the EU authorized representative is:
Springer Nature Customer Service Center GmbH
Europaplatz 3, 69115 Heidelberg, Germany

Printed by Libri Plureos GmbH
in Hamburg, Germany